Lecture Notes in Mechanical Engineering

ABCM Series on Mechanical Sciences and Engineering

More information about this series at http://www.springer.com/series/14172

Agenor de T. Fleury · Domingos A. Rade
Paulo R. G. Kurka
Editors

Proceedings of DINAME 2017

Selected Papers of the XVII International Symposium on Dynamic Problems of Mechanics

Editors
Agenor de T. Fleury
Escola Politécnica
Universidade de São Paulo
São Paulo, São Paulo
Brazil

Paulo R. G. Kurka
Faculdade de Engenharia Mecânica
Universidade Estadual de Campinas
Campinas, São Paulo
Brazil

Domingos A. Rade
Divisão de Engenharia Mecânica
Instituto Tecnológico de Aeronáutica
São José dos Campos, São Paulo
Brazil

ISSN 2195-4356 ISSN 2195-4364 (electronic)
Lecture Notes in Mechanical Engineering
ISSN 2524-6011 ISSN 2524-602X (electronic)
ABCM Series on Mechanical Sciences and Engineering
ISBN 978-3-030-08200-0 ISBN 978-3-319-91217-2 (eBook)
https://doi.org/10.1007/978-3-319-91217-2

Printed on acid-free paper

This Springer imprint is published by the registered company Springer International Publishing AG part of Springer Nature
The registered company address is: Gewerbestrasse 11, 6330 Cham, Switzerland

Foreword

This book is the first volume of *ABCM Series on Mechanical Sciences and Engineering—Proceedings of International Symposium on Dynamic Problems of Mechanics—DINAME* and brings together the work of some Brazilian leaders on research concerning the broad area of Dynamics. This book presents a compendium of works presented at DINAME 2017, covering traditional subjects in the area such as dynamic systems, vibration, control, as well as new and non-less exciting subjects in contemporary Dynamics such as robotics and intelligent materials.

The *ABCM Series on Mechanical Sciences and Engineering* is a result of an agreement set by Springer and ABCM and its first volume was an initiative of the organizers of DINAME 2017 supported by the ABCM Direction Board—biennium 2015–2017.

The ABCM Direction Board expects that these proceedings become a disclosure vehicle for the best works presented in our events. We also hope that in a close future, the continuity of this series of books becomes reference material to graduate students, professors, and professionals developing research in the area of Dynamics.

The ABCM Direction Board and the authors and participants in DINAME 2017 gratefully acknowledge the support to this event received from Coordination for the Improvement of Higher Education Personnel—CAPES, National Council for Scientific and Technological Development—CNPq, and São Paulo Research Foundation—FAPESP, recognizing that without the success of this symposium, this book would not be possible.

The Board of Directors of ABCM is also grateful to the ABCM Committee of Dynamics and the professionals that have actively participated in the elaboration of this book as organizers authors, co-authors, and reviewers. In particular, we and the ABCM community thank our colleagues who took the responsibility of being the chairs of the DINAME 2017 and editors of ABCM Series: Agenor de T. Fleury from University of São Paulo (USP), Domingos A. Rade from Aeronautics Institute of Technology (ITA), and Paulo R. G. Kurka from University of Campinas (UNICAMP).

São Carlos, SP, Brazil

Gherhardt Ribatski
(President of ABCM)

Preface

The International Symposium on Dynamic Problems of Mechanics—DINAME—is a biannual symposium promoted since 1986 by the Brazilian Society of Mechanical Sciences and Engineering (ABCM), and organized by its Committee of Dynamics.

Along the years, the Symposium became a vivid forum for scientists, academics, and practitioners to present and discuss developments related to dynamic problems of mechanics.

The meetings are traditionally held in quiet and pleasant sites, away from overcrowded areas, in a regime of immersion, which enables cross-fertilization of ideas, strong scientific exchanging and socialization among participants. In order to maximize the exchange of scientific ideas, achievements, and trends on topics related to the broad area of Dynamics, a single-session format is adopted for the oral presentations of the papers. The acceptance for presentation and inclusion of the papers in the proceedings is based on a two-phase peer-reviewing process of abstracts and full-length (10 pages) manuscripts.

The 2017 edition of DINAME was held from March 5 to March 10, 2017, at Beach Hotel Sunset and Cambury, in São Sebastião, which is located on the seashore of the State of São Paulo, in Brazil. The technical program comprised 100 regular papers, 2 invited papers, 7 keynote lectures, and 2 short courses, the latter intended for the education of young researchers in topics encompassed by the scope of the symposium.

In the context of an existing agreement between Springer and ABCM, an initiative was launched by ABCM Direction Board and the organizers of DINAME 2017 aiming at publishing the symposium proceedings as the first volume of the *ABCM Series on Mechanical Sciences and Engineering—Proceedings of DINAME 2017: Selected Papers of the XVII International Symposium on Dynamic Problems of Mechanics*. For this purpose, after DINAME 2017, the authors were invited to submit improved versions of their papers, which were subjected to another peer-reviewing process. This process led to the selection of 39 papers that compose the present volume, which are believed to be a representative sample of the best research works presented in the symposium.

The papers are organized according to the sessions established in the symposium, namely: Rotordynamics, Vibrations and Structural Dynamics, Robotics and Mechatronic Systems, Control of Mechanical Systems, Nonlinear Dynamics, Vehicle Dynamics and Multibody Systems, Wave Propagation, Acoustics and Vibroacoustics, and Uncertainty Quantification and Stochastic Mechanics.

The organization of DINAME 2017 had the financial support of the following Brazilian research agencies, to which the organizers are very grateful:

- CAPES Foundation (Brazilian Ministry of Education)
- National Council for Scientific and Technological Development—CNPq (Brazilian Ministry of Science, Technology and Innovation)
- São Paulo State Research Foundation—FAPESP

We sincerely hope that these proceedings will be useful to Brazilian and international readers interested in dynamic problems of mechanics.

São Paulo, Brazil — Agenor de T. Fleury
São José dos Campos, Brazil — Domingos A. Rade
Campinas, Brazil — Paulo R. G. Kurka

ABCM—Brazilian Society of Mechanical Sciences and Engineering

Board of Directors (term 2015–2017)

ABCM Committee of Dynamics in 2017

DINAME 2017 Chairmen

Prof. Agenor de Toledo Fleury
Prof. Domingos Alves Rade
Prof. Paulo R. G. Kurka

Contents

Contributors

A. N. Albuquerque Department of Mechanical Engineering, Pontifical Catholic University, Gávea, Rio de Janeiro, Brazil

Abel Castro Pontifical Catholic University of Rio de Janeiro, Rio de Janeiro, Brazil

Adriana E. L. Pereira Universidade Federal do Rio Grande FURG, Rio Grande, Brazil

Agenor de Toledo Fleury Polytechnic School, University of São Paulo, São Paulo, Brazil

Airton Nabarrete Instituto Tecnológico de Aeronáutica, São José dos Campos, Brazil

Alberto Amancio Jr. FEI University, São Bernardo do Campo, Brazil

Albert Lau Faculty of Engineering, Department of Civil and Environmental Engineering, Norwegian University of Science and Technology, Trondheim, Norway

Alejandro de Miguel Department of Mechanical Engineering, Solid Mechanics, Technical University of Denmark, Copenhagen, Denmark

Aldemir Ap Cavalini Jr. LMEst – Structural Mechanics Laboratory, School of Mechanical Engineering, Federal University of Uberlândia, Uberlândia, MG, Brazil

Amarildo Tabone Paschoalini Department of Mechanical Engineering, UNESP, Ilha Solteira, SP, Brazil

Ana Paula Pagotti Universidade Federal do ABC, Santo André, Brazil

André de Souza Mendes Escola Politécnica da USP, Cidade Universitária, São Paulo, Brazil

André Teófilo Beck University of São Paulo, São Carlos, SP, Brazil

Arcanjo Lenzi CTC EMC, LVA, UFSC, Florianópolis, SC, Brazil

Cassio T. Faria Siemens Industry Software N.V., Louvain, Belgium

Celso Pupo Pesce Offshore Mechanics Laboratory, Escola Politécnica, University of São Paulo, São Paulo, Brazil

Christian Wagner Chair of Applied Mechanics, Technical University of Munich, Garching, Germany

D. A. Pereira Division of Mechanical Engineering, ITA – Aeronautics Institute of Technology, São José dos Campos, Brazil

D. A. Rade Division of Mechanical Engineering, ITA – Aeronautics Institute of Technology, São José dos Campos, Brazil

Daniel Rixen Department of Mechanical Engineering, Chair of Applied Mechanics, Technical University of Munich, Garching, Germany

Daniel A. Castello Mechanical Engineering Department, Universidade Federal do Rio de Janeiro, Rio de Janeiro, Brazil

David F. C. Zúñiga Instituto Tecnológico de Aeronáutica, São José dos Campos, Brazil

Décio Crisol Donha University of São Paulo, São Paulo, Brazil

Diogo B. P. de Oliveira Instituto Tecnológico de Aeronáutica, São José dos Campos, Brazil

Diego Colón Automation and Control Laboratory, Telecom and Control Department, University of São Paulo, CEP, São Paulo, Brazil

Ederval de Souza Lisboa Mechanical Engineering Department, Federal University of Rio Grande do Sul, Porto Alegre, Brazil

Edson Jansen Pedrosa de Miranda Jr. University of Campinas, UNICAMP-FEM-DMC, Campinas, SP, Brazil; Federal Institute of Maranhão, IFMA-NIB-DEP, São Luís, MA, Brazil

Elói Rotava Petróleo Brasileiro S. A., UO-BS/ATP-PSMG/EEIS, Santos, SP, Brazil; Departamento de Engenharia Mecânica, Escola Politécnica, Universidade de São Paulo, São Paulo, Brazil

Elisane B. Zanela Universidade Federal do Rio Grande FURG, Rio Grande, Brazil

Elvira Rafikova Center for Engineering, Modeling and Social Science, Federal University of ABC, Santo Andre, Brazil

Emanuel Moutinho Cesconeto Mechanical Engineering Department, Federal University of Rio Grande do Sul, Porto Alegre, Brazil

Éverton Lins de Oliveira University of São Paulo, São Paulo, SP, Brazil

Fabio Luis M. dos Santos Siemens Industry Software N.V., Louvain, Belgium

Fábio Kroll de Lima Department of Mechanical Engineering, UNESP, Ilha Solteira, SP, Brazil

Fabrizio Leonardi Centro Universitário FEI, São Bernardo do Campo, SP, Brazil

Fabrício Lobato César de Almeida Department of Mechanical Engineering, UNESP, Tupã, SP, Brazil

Fernando Malvezzi Department of Mechanical Engineering, Maua Institute of Technology, São Caetano do Sul, Brazil

Florencia Reguera Consejo Nacional de Investigaciones Científicas y Técnicas, Buenos Aires, Argentina; Departamento de Ingeniería, Universidad Nacional del Sur, Bahía Blanca, Argentina

Franco E. Dotti Facultad Regional Bahía Blanca, Grupo de Investigación en Multifísica Aplicada, Universidad Tecnológica Nacional, Bahía Blanca, Argentina; Consejo Nacional de Investigaciones Científicas y Técnicas, Buenos Aires, Argentina

Francisco Paulo Lépore Neto Laboratory of Mechanical Systems, Faculty of Mechanical Engineering, Federal University of Uberlândia, Uberlândia, Minas Gerais, Brazil

Flavio Celso Trigo Departamento de Engenharia Mecânica, Escola Politécnica, Universidade de São Paulo, São Paulo, Brazil

Georg Rill Regensburg University of Applied Sciences, Regensburg, Germany

Glauco A. P. Caurin São Carlos School of Engineering, University of São Paulo, São Carlos, Brazil

Gregory Bregion Daniel Laboratory of Rotating Machinery, School of Mechanical Engineering, University of Campinas, Campinas, SP, Brazil

Guilherme Rinaldo Center for Engineering, Modeling and Social Science, Federal University of ABC, Santo Andre, Brazil

Guilhem Michon Institut Clemant Ader, ISAE, Université de Toulouse, Toulouse, France

Guilherme Sampaio Laboratótio de Dinâmica e Vibrações, PUC-Rio, Rio de Janeiro, Brazil

Gustavo J. G. Lahr São Carlos School of Engineering, University of São Paulo, São Carlos, Brazil

Hans I. Weber Laboratótio de Dinâmica e Vibrações, Pontifical Catholic University of Rio de Janeiro, Rio de Janeiro, Brazil

Henrique B. Garcia São Carlos School of Engineering, University of São Paulo, São Carlos, Brazil

Heraldo N. Cambraia Paraná Federal University, DEMEC, Curitiba, Brazil

Herman Van der Auweraer Siemens Industry Software N.V., Louvain, Belgium; K.U. Leuven, Louvain, Belgium

Humberto Tronconi Coelho Laboratory of Mechanical Systems, Faculty of Mechanical Engineering, Federal University of Uberlândia, Uberlândia, Minas Gerais, Brazil

Ilmar Santos Department of Mechanical Engineering, Solid Mechanics, Technical University of Denmark, Copenhagen, Denmark

Isadora R. Henriques Mechanical Engineering Department, Universidade Federal do Rio de Janeiro, Rio de Janeiro, Brazil

João Baptista Dias Moreira Mechanical Engineering Department, Federal University of Rio Grande do Sul, Porto Alegre, Brazil

João Flávio P. Coelho Institut Clemant Ader, ISAE, Université de Toulouse, Toulouse, France

Jorge Luis Baliño Departamento de Engenharia Mecânica, Escola Politécnica, Universidade de São Paulo, São Paulo, Brazil

José Maria Campos dos Santos University of Campinas, UNICAMP-FEM-DMC, Campinas, SP, Brazil

José Roberto de França Arruda Faculty of Mechanical Engineering, University of Campinas, Campinas, SP, Brazil

José Roberto F. Arruda University of Campinas UNICAMP-FEM-DMC, Campinas, SP, Brazil

José M. Balthazar Instituto Tecnológico de Aeronáutica, São José dos Campos, Brazil

Katia Lucchesi Cavalca Laboratory of Rotating Machinery, School of Mechanical Engineering, University of Campinas, Campinas, SP, Brazil

Lavinia A. Borges Mechanical Engineering Department, Universidade Federal do Rio de Janeiro, Rio de Janeiro, Brazil

Leonardo M. L. Contini Paraná Federal University, DEMEC, Curitiba, Brazil

Leonardo Sanches Universidade Federal de Uberlândia, Uberlândia, MG, Brazil

Luiz C. S. Góes Instituto Tecnológico de Aeronáutica, São José dos Campos, Brazil

M. Speranza Neto Department of Mechanical Engineering, Pontifical Catholic University, Gávea, Rio de Janeiro, Brazil

Marco Antônio Meggiolaro Department of Mechanical Engineering, Pontifical Catholic University, Gávea, Rio de Janeiro, Brazil

Manuel Kiener Department of Mechanical Engineering, Chair of Applied Mechanics, Technical University of Munich, Garching, Germany

Marat Rafikov Center for Engineering, Modeling and Social Science, Federal University of ABC, Santo Andre, Brazil

Marcos Silveira School of Engineering, São Paulo State University (UNESP), Bauru, Brazil

Marcelo Braga dos Santos Laboratory of Mechanical Systems, Faculty of Mechanical Engineering, Federal University of Uberlândia, Uberlândia, Minas Gerais, Brazil

Marília Maurell Assad Pontifical Catholic University, Rio de Janeiro, Brazil

Marko Ackermann Centro Universitário FEI, São Bernardo do Campo, SP, Brazil

Massimo Ruzzene School of Aerospace Engineering, Georgia Institute of Technology, Atlanta, GA, USA

Matheus Inguaggiato Nora Rosa Faculty of Mechanical Engineering, University of Campinas, Campinas, SP, Brazil

Mauro Speranza Neto Pontifical Catholic University, Rio de Janeiro, Brazil

Michael John Brennan Department of Mechanical Engineering, UNESP, Ilha Solteira, SP, Brazil

Nicolò Bachschmid Department of Mechanical Engineering, Politecnico di Milano, Milan, Italy

Oliver Hofmann Department of Mechanical Engineering, Chair of Applied Mechanics, Technical University of Munich, Garching, Germany

Paulo José Paupitz Gonçalves School of Engineering, São Paulo State University (UNESP), Bauru, Brazil

Paulo Martins CTC EMC, LVA, UFSC, Florianópolis, SC, Brazil

Paulo R. G. Kurka Campinas State University, Campinas, SP, Brazil

Pedro Christian Ayala Castillo Department of Mechanical Engineering, UNESP, Ilha Solteira, SP, Brazil

Petter Krus Division of Fluid and Mechatronic Systems, Department of Management and Engineering, Linköping University, Linköping, Sweden

Rafael Salles Offshore Mechanics Laboratory, Escola Politécnica, University of São Paulo, São Paulo, Brazil

Rafael B. Chaves Pontifical Catholic University of Rio de Janeiro, Rio de Janeiro, Brazil; Regensburg University of Applied Sciences, Regensburg, Germany

Rafael Fernandes Pinheiro Automation and Control Laboratory, Telecom and Control Department, University of São Paulo, CEP, São Paulo, Brazil

Raphaela C. M. G. Barbosa Instituto Tecnológico de Aeronáutica, São José dos Campos, Brazil

Renato M. M. Orsino Department of Mechanical Engineering, Polytechnic School, University of São Paulo, São Paulo, Brazil

Ricardo Sampaio University of São Paulo, São Paulo, Brazil

Rúbia Mara Bosse University of São Paulo, São Carlos, SP, Brazil

Sebastião C. P. Gomes Universidade Federal do Rio Grande FURG, Rio Grande, Brazil

Sebastián P. Machado Facultad Regional Bahía Blanca, Grupo de Investigación en Multifísica Aplicada, Universidad Tecnológica Nacional, Bahía Blanca, Argentina; Consejo Nacional de Investigaciones Científicas y Técnicas, Buenos Aires, Argentina

Steven Dom Siemens Industry Software N.V., Louvain, Belgium

T. A. M. Guimarães School of Mechanical Engineering, UFU - Federal University of Uberlândia, Uberlândia, Brazil

Tarcisio Antonio Hess Coelho Department of Mechatronics and Mechanical Systems Engineering, Polytechnic School, University of São Paulo, São Paulo, Brazil

Tobias Berninger Chair of Applied Mechanics, Technical University of Munich, Garching, Germany

Thales Freitas Peixoto Laboratory of Rotating Machinery, School of Mechanical Engineering, University of Campinas, Campinas, SP, Brazil

Theo Geluk Siemens Industry Software N.V., Louvain, Belgium

Thiago H. S. Silva São Carlos School of Engineering, University of São Paulo, São Carlos, SP, Brazil

Thomas Thümmel Chair of Applied Mechanics, Technical University of Munich, Garching, Germany

Tobias Souza Morais LMEst – Structural Mechanics Laboratory, School of Mechanical Engineering, Federal University of Uberlândia, Uberlândia, MG, Brazil

Valder Steffen Jr. LMEst – Structural Mechanics Laboratory, School of Mechanical Engineering, Federal University of Uberlândia, Uberlândia, MG, Brazil

Vinícius Dias de Lima University of Campinas UNICAMP-FEM-DMC, Campinas, SP, Brazil

Walter Jesus Paucar Casas Mechanical Engineering Department, Federal University of Rio Grande do Sul, Porto Alegre, Brazil

Wataru Tsunoda Department of Mechanical Engineering, Tokyo Institute of Technology, Yokohama, Japan

William S. Cardozo PUC-Rio, Rio de Janeiro, Brazil

Willian Minnemann Kuhnert School of Engineering, São Paulo State University (UNESP), Bauru, Brazil

Part I
Rotordynamics

Estimation of Rotordynamic Seal Coefficients Using Active Magnetic Bearing Excitation and Force Measurement

Christian Wagner, Wataru Tsunoda, Tobias Berninger, Thomas Thümmel and Daniel Rixen

Abstract In high-speed rotational machinery such as pumps or compressors, contactless seals are commonly used to separate different fluids or gases and pressure levels. However, the presence of a leakage flow through the seal gap exerts forces on a rotor. These can culminate in stiffening, restoring, and damping effects as well as in unstable, self-excited vibrational behavior. The JEFFCOTT rotor model and rotordynamic seal coefficients are put under investigation to prevent instability in the rotating machinery and to determine the rotor-seal systems dynamic behavior. This paper focuses on an experimental methodology, examined on a flexible rotor-seal test rig using an active magnetic bearing for excitation. Coefficient identification problems due to unknown random force (noise) in the experiment are shown and a solution is described in detail and validated on the test rig. The presented methodology leads to a calculation of rotordynamic seal coefficients during safe operating conditions. They are ultimately used to describe the system's behavior and to predict the onset speed of instability.

Keywords Rotordynamic · Seal · Instability · Self-excited vibration
Active magnetic bearing · Coefficient estimation

1 Introduction

Seals in turbopumps are mostly used to minimize leakage flow from high pressure areas to low pressure parts. Because of the high rotational speeds of common cen-

C. Wagner (✉) · T. Berninger · T. Thümmel · D. Rixen
Chair of Applied Mechanics, Technical University of Munich, Boltzmannstraße 15, 85748 Garching, Germany
e-mail: c.wagner@tum.de
URL: http://www.amm.mw.tum.de

W. Tsunoda
Department of Mechanical Engineering, Tokyo Institute of Technology, 4259 Nagatsuta-cho, Midori-ku, Yokohama 226-8503, Japan
e-mail: tsunoda.w.aa@m.titech.ac.jp

A. de T. Fleury et al. (eds.), *Proceedings of DINAME 2017*, Lecture Notes in Mechanical Engineering, https://doi.org/10.1007/978-3-319-91217-2_1

trifugal pumps or compressors, contactless seals, such as floating ring, labyrinth or small gaps are inserted between the rotating and the stationary parts. The ever-present clearance around these contactless seals permits fluid flow through the gap. For an eccentric rotor position, the fluid-velocity distribution inside the seal becomes unsymmetrical, which entails forces on the rotor. These can induce effects such as stiffening, damping, and added mass, but they can also end up in a rotor instability, a self-excited vibration which can destroy the machinery. Seal effects can engender the critical speeds that should be avoided in stationary operation. Thus, dry predictions of the rotor-system's behavior, such as those for critical speeds, damping or stability limits, are unusable under real operating conditions.

The seal forces within a rotor system are mostly modeled as rotordynamic coefficients:

$$-\mathbf{h}_s = \begin{bmatrix} m_{xx} & 0 \\ 0 & m_{yy} \end{bmatrix} \ddot{\mathbf{q}} + \begin{bmatrix} c_{xx} & c_{xy} \\ c_{yx} & c_{yy} \end{bmatrix} \dot{\mathbf{q}} + \begin{bmatrix} k_{xx} & k_{xy} \\ k_{yx} & k_{xx} \end{bmatrix} \mathbf{q} \tag{1}$$

with the motion of the rotor, $\mathbf{q}$, the seal reaction force, $\mathbf{h}_s$, and the rotordynamic seal coefficients m, c and k for added mass, the fluid's inertia, damping, and stiffness with cross-coupling parts. The coupling inertia terms in the mass matrix are neglected. The different types of seals in a pump can usually be simplified to a cylindrical annular seal for a rotordynamic analysis.

To ensure safe rotor-seal system operation, validated models and methods for characterizing the seals' behavior, the rotordynamic seal coefficients, in simulation and experiment are needed.

2 Literature Overview

The prediction of seal forces and effects mostly leads to a calculation of their rotordynamic coefficients. Several efforts have been made for theoretical and experimental prediction. Simple and fast models are based on bulk-flow theory, which is a simplification of the NAVIER-STOKES equations assuming a constant fluid velocity along the seals clearance.

One of the first efforts to determine restoring seal forces was made by [2]. They used the bulk-flow theory and incompressible fluid flow through a short annular seal. They express equilibrium through the axial momentum equation using turbulent wall friction models and the given pressure gradient over the seal as a boundary condition. For a centered rotor position, a perturbation analysis with small dynamic motion results in differential equations of the fluids motion. Moreover, a linearization of the fluid forces in reaction to the perturbation leads to the rotordynamic seal coefficients for the centered shaft position. The circumferential flow is supposed to be a fully developed, turbulent COUETTE flow. The assumption of constant fluid velocity in the axial and circumferential directions leads to a constant wall-friction factor, λ, for the whole seal [1].

Also based on the bulk-flow theory [3] introduced a closed-form analytical solution for the rotordynamic coefficients of a short, plain annular seal. He also used a perturbation analysis to solve the differential equations. The model's improvement is the consideration of fluid inertia terms and the inlet swirl, the circumferential fluid velocity at the seal's entrance [1]. The simulation agrees well overall with measurements [15].

Padavala and Palazzolo [10] developed a more detailed model, but with higher computational costs. Based on bulk-flow theory, the model discretized the annulus into finite parts to consider a variation of wall-friction factors in the circumferential and axial directions. In contrast to the finite difference methods used in [5] or [12], Padavala's model uses continuous functions, created by cubic splines to fit the distribution of the variables, pressure, velocity, and so forth. With this technique, it is possible to solve the bulk-flow equations for every finite part to get the pressure and velocity distribution within the seal. Although this model gives good to excellent agreement with measurements, its computational costs are high [15].

Further methods, based on finite volume CFD calculations like those in [17], or finite difference methods, or using the REYNOLDS equation known for journal bearings with turbulence correction factors and solved it with finite element methods, see [13], gives high quality results.

The consequence of seal forces acting on the rotordynamics of the whole system are well described in [4, 6]. The effect of self-excited vibrations and rotor instability like the "oil-whip" phenomenon are illustrated in [9] for a system with two degrees of freedom. Parametrization and variable description of the JEFFCOTT rotor model used are attributable to [11, 14].

Others focus on coefficient measurements using an AMB rotor system to measure transfer functions. Zutavern [18] for example gets good measurement results for frequency domain identification methods. The use of a rotor seal system in journal bearings with AMB excitation, in [7], with the variation of stiffness and damping of the AMB controller leads to the rotordynamic seal coefficients for steam turbine seals.

3 Modeling: Jeffcott Rotor Model

The simplified JEFFCOTT rotor model (see Fig. 1) with liquid annular seals is used for theoretical explanation, simulation and as far as possible, experiments on the test rig.

The JEFFCOTT rotor models a flexible, massless shaft with a mass disk symmetrically arranged between rigid bearings; see [6]. Here, the center of mass, S, has the distance ϵ from the disk's geometric center, M. Hence, $\mathbf{r}_M$ gives the position for M . $\mathbf{r}_S = \mathbf{r}_M + \epsilon$ is the position of S. Lumping the shaft stiffness, k_r, onto the rotor's center, M, and taking as degrees of freedom the two translations $\mathbf{q} = \mathbf{r}_M$ leads to dynamic equilibrium for the rotor with mass m_r and rotational speed Ω:

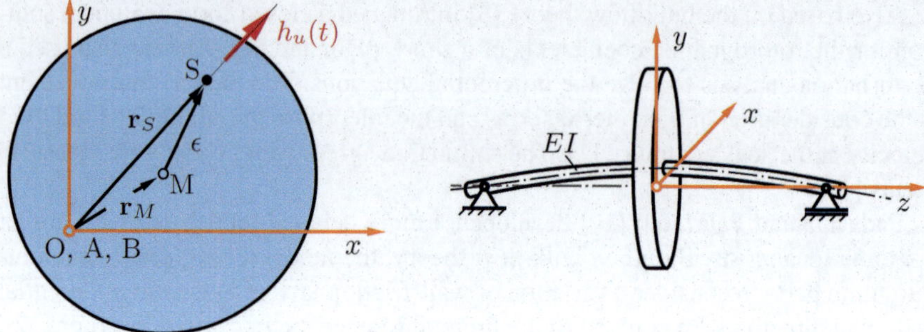

Fig. 1 JEFFCOTT rotor model according to [11, 14]

$$\begin{bmatrix} m_r & 0 \\ 0 & m_r \end{bmatrix} \ddot{\mathbf{q}} + \begin{bmatrix} k_r & 0 \\ 0 & k_r \end{bmatrix} \mathbf{q} = \mathbf{h} \tag{2}$$

$$\mathbf{h}_u = m_r \epsilon \Omega^2 [cos(\Omega t)\ sin(\Omega t)]^T \tag{3}$$

with the equivalent forces $\mathbf{h} = \mathbf{h}_u + \mathbf{h}_e + \mathbf{h}_s ...$ (unbalance, external forces, seal forces, and so forth). The rotor's natural frequency is $\omega_{crit} = \sqrt{\frac{k_r}{m_r}}$, its critical speed.

3.1 Contactless Seal: Minimal Model and Coupling to Rotor System

Defining the seal as system with spring, mass, damper and coupling to the rotor using force $\mathbf{h}_s$ leads to the dynamic equilibrium for the whole rotor seal system:

$$\begin{bmatrix} m_r + m_{xx} & 0 \\ 0 & m_r + m_{yy} \end{bmatrix} \ddot{\mathbf{q}} + \begin{bmatrix} c_{xx} & c_{xy} \\ c_{yx} & c_{yy} \end{bmatrix} \dot{\mathbf{q}} + \begin{bmatrix} k_r + k_{xx} & k_{xy} \\ k_{yx} & k_r + k_{yy} \end{bmatrix} \mathbf{q} = 0 \tag{4}$$

Assuming that

$$\mathbf{q} = \hat{\mathbf{q}} e^{\lambda t} \tag{5}$$

yields the eigenvalue problem with eigenvalues $\lambda = \delta \pm j\omega$. For self-excited vibration, i.e. rotor instability, the positive real parts, δ, must be observed. The seal coefficients' speed dependency, mainly the increasing of cross-coupled parts of the stiffness, k_{xy} and k_{yx}, sets a speed limit for safe operation: the onset speed. A sub-synchronous, self-excited vibration at the rotors natural frequency arises when the onset speed is reached.

Calculating the seal coefficients is essential in this case to avoid rotor instability for safe operation.

3.2 Bulk-Flow Modeling and Seal Simulation

The bulk-flow theory is derived from the NAVIER-STOKES equations by neglecting all changes to the fluid flow parameters in radial direction and setting them to constant values or zero.

These assumptions lead to a pressure- and shear-driven fluid flow and a perturbation analysis about a steady state position leads to the fluid forces as a function of the rotor's movement. Simplifications made by Black and Jenssen [2], Childs [3] and Padavala and Palazzolo [10] are used to solve the fluid momentum equations to get the rotordynamic seal coefficients. The detailed description of the used equations and the solving process is well explained in the cited literature. The three models are implemented in MATLAB and called now as Black, Childs and Padavala model. The simulation results will be discussed in later chapters.

4 Experiments: Test Rig Setup

The test rig design is shown in Figs. 2 and 3. It is based on a flexible shaft 1 and a mass disc 2 rotating within a pressurized chamber with length l_c and clearance c_c. The shaft support is realized with stiff ball bearings 3 and the rig is driven by a servo motor 4. The fluid is injected into the chamber and flows through two symmetric

Fig. 2 Seals test rig, photograph

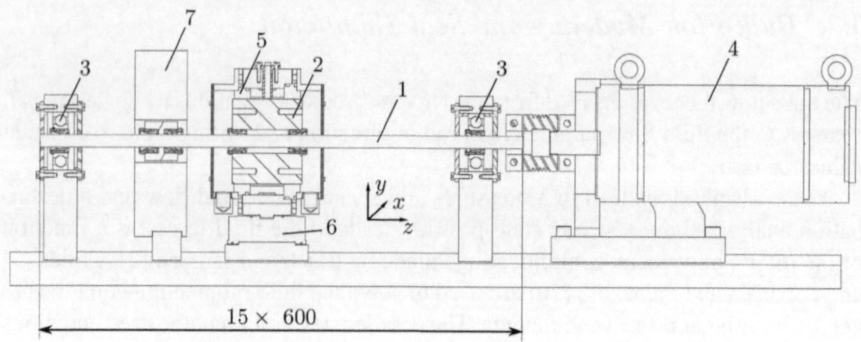

Fig. 3 Seals test rig, see [16]

Table 1 Test rig and seal parameters

Name	Value	Name	Value
Seal clearance	0.17 mm	Rotational speed Ω	0–100 rps
Chamber clearance c_c	2 mm	Rotor mass m_r	5 kg
Seal length l_c	20 mm	Shaft stiffness k_r	2.93×10^5 N/m
Chamber length	40 mm	Density at 40 °C	880 kg/m^3
Pressure difference at the seals Δp	2×10^5 Pa	Viscosity at 40 °C	0.04048 Pa s
Sealless natural frequency ω_0	38.6 Hz	Seal diameter	0.1 m

annular seals 5 to the environment. Two eddy current sensors measure the rotor's motion. A dynamometer 6 under the stator seal is used to get the seal's reaction forces. Further, the fluid inlet pressure, temperature, rotational speed, leakage flow and torque are measured. An active magnetic bearing 7 is used as an actuator for dynamic system excitation.

Table 1 list the test rig, the fluid and seal parameters used for simulation and measurements. The rotor's "dry" first natural frequency ω_0 is at 38.6 Hz. It is decreased by the seal influence to about half the rotational speed Ω, see Fig. 7.

4.1 Measurement Methods for Seal Coefficients

For a symmetrical rotor seal system, the coefficients in Eq. (1) can be written, according to [6]: $M_s = m_{xx} = m_{yy}$, $C_s = c_{xx} = c_{yy}$, $c_s = c_{xy} = -c_{yx}$, $K_s = k_{xx} = k_{yy}$ and $k_s = k_{xy} = -k_{yx}$. To determine the rotordynamic seal coefficients, the seal reaction forces in Eq. (1) are FOURIER transformed into the frequency, see [8]:

$$\underbrace{\begin{bmatrix} K_s + j\omega C_s - \omega^2 M_s & k_s + j\omega c_s \\ -k_s - j\omega c_s & K_s + j\omega C_s - \omega^2 M_s \end{bmatrix}}_{\tilde{\mathbf{A}}} \underbrace{\begin{bmatrix} \tilde{x} \\ \tilde{y} \end{bmatrix}}_{\tilde{\mathbf{q}}} = \underbrace{\begin{bmatrix} \tilde{f}_x \\ \tilde{f}_y \end{bmatrix}}_{\tilde{\mathbf{h}}_s} \tag{6}$$

$\tilde{\mathbf{A}}$ is the system's dynamic stiffness matrix for every frequency, ω, $\tilde{\mathbf{q}}$ the rotor complex displacement amplitudes, and $\tilde{\mathbf{h}}_s$ the seal complex reaction force amplitudes. Using the notation

$$\begin{aligned} a(\omega) &= K_s + j\omega C_s - \omega^2 M_s \\ b(\omega) &= k_s + j\omega c_s \end{aligned} \tag{7}$$

Equation (6) becomes:

$$\begin{bmatrix} a & b \\ -b & a \end{bmatrix} \begin{bmatrix} \tilde{x} \\ \tilde{y} \end{bmatrix} = \begin{bmatrix} \tilde{f}_x \\ \tilde{f}_y \end{bmatrix} \Leftrightarrow \begin{bmatrix} \tilde{x} & \tilde{y} \\ \tilde{y} & -\tilde{x} \end{bmatrix} \begin{bmatrix} a \\ b \end{bmatrix} = \begin{bmatrix} \tilde{f}_x \\ \tilde{f}_y \end{bmatrix} \Leftrightarrow \begin{bmatrix} a \\ b \end{bmatrix} = \frac{1}{\tilde{x}^2 + \tilde{y}^2} \begin{bmatrix} \tilde{x}\tilde{f}_x + \tilde{y}\tilde{f}_y \\ \tilde{y}\tilde{f}_x - \tilde{x}\tilde{f}_y \end{bmatrix} \tag{8}$$

Equation (8) can be solved for a and b at several excitation frequencies and for multiple measurements using a least squares method. Separation into the real (*Re*) and imaginary (*Im*) part let us use linear and quadratic fit curves to calculate the rotordynamic coefficients:

$$\begin{aligned} Re\{a(\omega)\} &= K_s - \omega^2 M_s \\ Im\{a(\omega)\} &= \omega C_s \\ Re\{b(\omega)\} &= k_s \\ Im\{b(\omega)\} &= \omega c_s \end{aligned} \tag{9}$$

Measuring seal forces, $\tilde{\mathbf{h}}_s$, and displacements, $\tilde{\mathbf{q}}$, with one directional excitation in the y direction at several frequencies, ω, using the AMB yields the values for a and b shown in Fig. 4 for the test-rig parameters in Table 1.

The blue line in Fig. 4, which represents $Re\{a(\omega)\} = K_s - \omega^2 M_s$, should be a quadratic curve. At about 42 (rps) $\approx \Omega/2$, a discontinuous point—a "jump-effect"—disturbs the curve fitting and coefficient identification. Even when the rotational speed changes, the same noise effect occurs in every real and imaginary part of a and b at about half rotational speed; see Fig. 4.

Instead of using the x-y coordinate system, Eq. (6) can be transformed to complex coordinate in the forward whirl direction $z = x + jy$ by summing up the first equation of (6) and the second equation multiplied by j, one obtains after rearranging:

$$[-\omega^2 M_s + \omega c_s + K_s + j(C_s\omega - k_s)] \cdot \tilde{z} = \tilde{f}_z \tag{10}$$

$$\frac{\tilde{f}_z}{\tilde{z}} = -\omega^2 M_s + \omega c_s + K_s + j(C_s\omega - k_s) \tag{11}$$

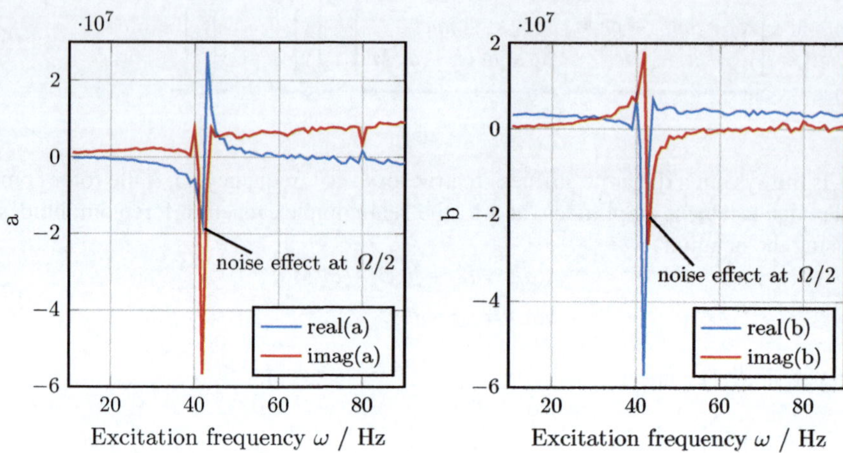

Fig. 4 Noise effect: *a* and *b* measured at $\Omega = 80$ rps

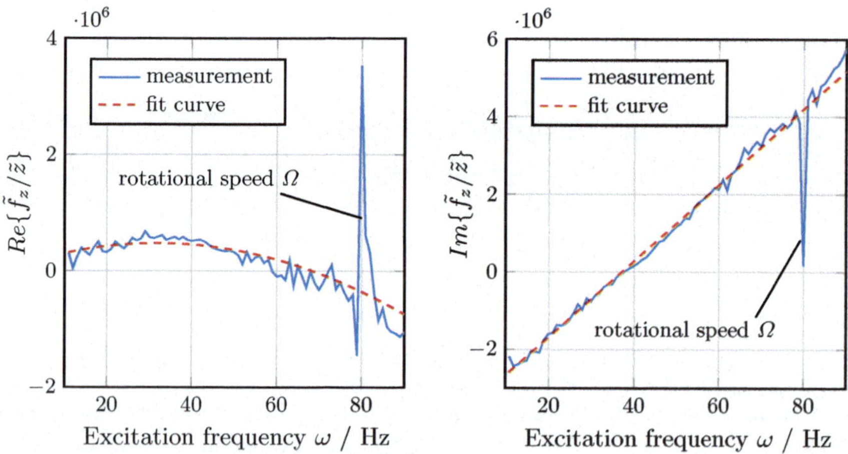

Fig. 5 Real and imaginary part of $\frac{\tilde{f}_z}{\tilde{z}}$ measured at $\Omega = 80$ rps

Separated in real and imaginary parts:

$$\begin{aligned} Re\left(\frac{\tilde{f}_z}{\tilde{z}}\right) &= -\omega^2 M_s + \omega c_s + K_s \\ Im\left(\frac{\tilde{f}_z}{\tilde{z}}\right) &= C_s\omega - k_s \end{aligned} \tag{12}$$

Figure 5 shows that fitting the rotordynamic coefficients with linear and quadratic curves is possible using the coordinate transformation. No jump effect at half rotational speed occurs in the complex coordinate description and the blue lines agree well with the dotted red quadratic and linear fitted functions. The noise at 80 Hz occurs due to the dominant unbalance force response polluting the measurement at rotational speed $\Omega = 4800$ rpm.

4.2 Discussion of Noise Effect

The experiments show a noise effect at half rotational speed in the x-y coordinate measurement. A transformation to complex coordinate solves this problem. The explanation can be found analyzing the block diagram for Eq. (8), shown in Fig. 6, with the unknown noise force, $\Delta\tilde{f}(t)$, acting on the rotor. The relation for the displacement errors $\tilde{\Delta x}$ and $\tilde{\Delta y}$, is given by:

$$\tilde{\Delta x} = -\frac{b}{a}\tilde{\Delta y} \qquad \text{and} \qquad \tilde{\Delta y} = \Delta\tilde{f}(t)/a \tag{13}$$

Using for the rotor displacement:

$$\tilde{x} = \tilde{x_0} + \tilde{\Delta x} \qquad \text{and} \qquad \tilde{y} = \tilde{y_0} + \tilde{\Delta y} \tag{14}$$

For the displacement and forces these substitutions can be done:

$$\begin{aligned} q &= \frac{\tilde{y}}{\tilde{f}_y} = \frac{\tilde{y_0} + \tilde{\Delta y}}{\tilde{f}_y} = q_0 + \Delta q \\ p &= \frac{\tilde{x}}{\tilde{f}_y} = \frac{\tilde{x_0} + \tilde{\Delta x}}{\tilde{f}_y} = p_0 + \frac{\tilde{\Delta x}}{\tilde{f}_y} = p_0 - \frac{b}{a}\frac{\Delta\tilde{y}}{\tilde{f}_y} = p_0 - \frac{b}{a}\Delta q \\ r &= \frac{\tilde{f}_x}{\tilde{f}_y} \end{aligned} \tag{15}$$

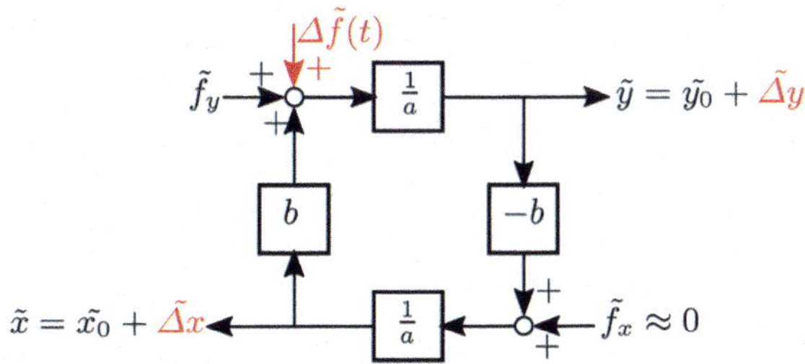

Fig. 6 Block diagram of random noise influence

Set into Eq. (8) leads to a new description for a and b:

$$\begin{bmatrix} a \\ b \end{bmatrix} = \frac{1}{p^2 + q^2} \begin{bmatrix} p \cdot r + q \\ q \cdot r - p \end{bmatrix} \tag{16}$$

Making the assumption $f_y \gg f_x$, $(4 \times 10^1 N \gg 0)$ and AMB excitation in y direction, so $r \approx 0$. Substitution into Eq. (11):

$$\frac{\tilde{f}_z}{\tilde{z}} = a - jb = \frac{1}{p^2 + q^2}[q + jp] = \frac{1}{p^2 + q^2}[q_0 + \Delta q + jp_0 - j\frac{b}{a}\Delta q] \tag{17}$$

Looking back at the block diagram in Fig. 6 and Eq. (13), it follows that

$$\tilde{x}_0 = -\frac{b}{a}\tilde{y}_0 \qquad \text{and} \qquad p_0 = -\frac{b}{a}q_0 \tag{18}$$

For the part,

$$\frac{b}{a} = \frac{jc_s\omega + k_s}{-M_s\omega^2 + K_s + jC_s\omega} \tag{19}$$

At around the frequency $\omega = \Omega/2$, where the former "jump effect" occurs, the assumptions $k_s \gg c_s\omega$, $(4 \times 10^6\,\text{N/m} \gg 6 \times 10^5\,\text{N/m})$ and $C_s\omega \gg (-M_s\omega^2 + K_s)$, $(4 \times 10^6\,\text{N/m} \gg 5 \times 10^5\,\text{N/m})$ can be made, so:

$$\frac{b}{a} \approx \frac{k_s}{jC_s\omega} = -j\frac{k_s}{C_s\omega} \tag{20}$$

using the rotational speed dependency of the coupled stiffness [9], $k_s \approx C_s \cdot \Omega/2$, $(3.7 \times 10^6\,\text{N/m} \approx 3.9 \times 10^6\,\text{N/m})$, $\frac{b}{a} \approx -j$ at about $\omega = \Omega/2$. So Eq. (17) can be rewritten and the displacement error gets subtracted out:

$$\frac{\tilde{f}_z}{\tilde{z}} = \frac{1}{p^2 + q^2}[q_0 + \Delta q + jp_0 - j(-j\Delta q)] = \frac{1}{p^2 + q^2}[q_0 + jp_0] \tag{21}$$

For the fraction $\frac{1}{p^2+q^2}$ of Eq. (21), the displacement error is subtracted out, too:

$$p^2 + q^2 = p_0^2 + q_0^2 + 2\Delta q(q_0 - \frac{b}{a}p_0) = p_0^2 + q_0^2 + 2\Delta q(q_0 + \left(\frac{b}{a}\right)^2 q_0) = p_0^2 + q_0^2 \tag{22}$$

By neglecting terms of second order, $(\Delta q)^2$. The displacement error Δx and Δy due to the unknown noise force $\Delta f(t)$ can be eliminated using the complex coordinate, as shown. However, it is possible to apply this method to determine rotordynamic seal coefficients using one-directional, active magnetic bearing excitation in a flexible rotor system with additional unknown forces. The determined coefficients are needed to validate simulation models and to describe the rotor-seal systems behavior including stability prediction.

5 Instability Prediction

5.1 Simulation and Eigenvalue Analysis

The rotor-seal system eigenvalues, $\lambda = \delta \pm i\omega$, are calculated using the simulated seal coefficients. The zero crossing of the real part gives the rotors onset speeds of instability at about $\Omega = [160, 124$ and $136]$ rps for the three simulation models: Black, Childs and Padavala. The systems simulated natural frequency agrees with the test rig behavior at higher rotational speeds, where the natural frequency is about half the rotational speed, $\Omega/2$.

5.2 Coefficient Measurement and Experimental Predictive Eigenvalue Analysis

The system's eigenvalues can be calculated using the measured seal coefficients and the dry test rig parameters; see Fig. 7. Using "least squares" to fit a linear function to the real eigenvalues enables you to extrapolate the zero crossing and the onset speed of instability to $\Omega = 171$ rps.

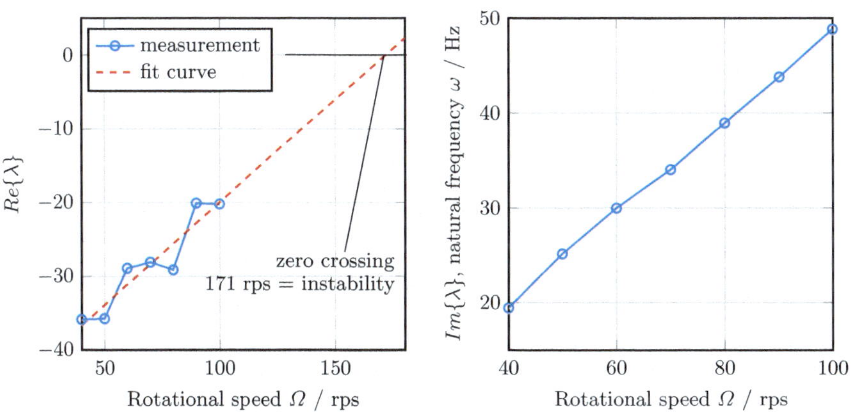

Fig. 7 Measured system: calculated real and imaginary parts of the eigenvalues

Comparing the estimated onset speed shows that the experimental method predicts 38% higher rotational speed than Childs', 26% higher than Padavala's, and 7% higher than Blacks simulation model. The systems natural frequency is always about $\Omega/2$ –half the rotational speed, see Fig. 7, which agrees well with the real test rig behavior.

6 Conclusion

The influence of seals on rotordynamics is shown in this paper. Models from the well-known literature for calculating rotordynamic coefficients are examined and used to determine the rotor system's vibrational behavior. A test-rig design, the measurement concept, details for solving noise problems to get a robust method, and calculated eigenvalues using the determined coefficients are shown. The model coupling of seal and rotor leads to a prediction of the stability limit and rotordynamic behavior. Because of the early state, these are preliminary results for the test-rig and the implemented simulation models. The differences between simulated and measured results are mostly caused by anisotropic and misalignment effects at the test rig. This is confirmed by the movement of the unloaded rotor from the complete rotor-stator contact at $\Omega = 0$ rps nearly to the seal center at $\Omega = 100$ rps. Additional investigations to improve the test rig design and to avoid disturbances such as increasing fluid temperature, misalignment, and so forth are planned. Future works also involves increasing the rotational speed to get the real onset speed of instability. On the simulation side, improved models will be implemented for the fluid motion and the resulting forces in future.

Acknowledgements This project is supported by Ludwig Bolkow Campus and funded by the Bavarian government. Good cooperation with the consortium partners is acknowledged.

References

1. Barrett, L.: Turbulent flow annular pump seals: a literature review. Shock Vib. Dig. **16**(2), 313 (1984)
2. Black, H., Jenssen, D.: Dynamic hybrid bearing characteristics of annular controlled leakage seals. Proc. Inst. Mech. Eng. Paper 9, **184**(14), 92–100 (1969)
3. Childs, D.W.: Dynamic analysis of turbulent annular seals based on Hirs lubrication equation. J. Lubr. Technol. **105**(3), 429–436 (1983)
4. Childs, D.W.: Turbomachinery Rotordynamics: phenomena, modeling, and analysis. Wiley (1993)
5. Dietzen, F., Nordmann, R.: Calculating rotordynamic coefficients of seals by finite-difference techniques. J. Tribol. **109**(3), 388–394 (1987)
6. Gasch, R., Nordmann, R., Pfützner, H.: Rotordynamik. Springer (2006)
7. Gaszner, M.: Rotordynamische Charakterisierung von Dichtungssystemen zur Anwendung in Kraftwerksdampfturbinen. Ph.D. thesis, Dr. Hut (2014)

8. Massmann, H.: Ermittlung der dynamischen Parameter turbulent durchströmter Ringspalte bei inkompressiblen Medien. Ph.D. thesis (1986)
9. Muszynska, A.: Whirl and whiprotor/bearing stability problems. J. Sound Vib. **110**(3), 443–462(1986)
10. Padavala, S., Palazzolo, A.: Enhanced simulation software for rocket turbopump, turbulent, annular liquid seals. NASA Report (1994)
11. Roßner, M.: Modellbasiertes monitoring von rotoren mit mehreren gleichzeitigen Fehlern. Ph.D, thesis, Technical University of Munich (2015)
12. San Andres, L., Adolfo D.: Analysis of variable fluid properties, turbulent annular seals. J. Tribol. **113**(4), 694–702 (1991)
13. San Andres, L., Adolfo D.: A Novel bulk-flow model for improved predictions of force coefficients in grooved oil seals operating eccentrically. J. Eng. Gas Turbines Power **134**(5) 052509 (2012)
14. Thümmel, T., Roßner, M., Ulbrich, H., Rixen, D.: Unterscheidung verschiedener Fehlerarten beim modellbasierten Monitoring. In: Tagungsband SIRM 2015 in Magdeburg (2015)
15. Tiwari, R., Manikandan, S., Dwivedy, S.K.: A review of the experimental estimation of the rotor dynamic parameters of seals. Shock Vib. dig. **37**(4), 261–284 (2005)
16. Wagner, C., T. Berninger, T., Thümmel, T. Rixen, D.: Rotordynamic effects in turbopumps. In: Space Propulsion Conference Rome (2016)
17. Yan, X., He, K., Li, J., Feng Z.: A generalized prediction method for rotordynamic coefficients of annular gas seals. In: J. Eng. Gas Turbines Power **137**(9), 092–506 (2015)
18. Zutavern, Z.S.: Identification of rotordynamic forces in a flexible rotor system using magnetic bearings. Ph.D. thesis, Texas A&M University (2006)

Experimental Estimation of Equivalent Damping Coefficient of Thrust Bearings

Thales Freitas Peixoto, Gregory Bregion Daniel and Katia Lucchesi Cavalca

Abstract A specific class of rotary machines is the high rotation turbochargers, to automotive application, wherein the shaft is continually subjected to axial forces of different magnitudes due to gas flows in the turbine and the compressor. These forces are supported by axial lubricated thrust bearings. The thrust bearings are modeled through equivalent stiffness and damping coefficients and the objective of the work is to get good estimates of these coefficients, comparing simulated results with experimental results. The stiffness coefficient is first obtained by small perturbation around the equilibrium position and used in a finite element model of the system at specific rotational speeds, and this value is compared to experimental results. Then, the damping coefficient is estimated, by running an optimization problem on this parameter, to approximate the simulated dynamic response of the system to experimental results of the turbocharger excited by an impact hammer, where both the displacement and force were measured.

Keywords Hydrodynamic thrust bearing · Stiffness coefficient
Damping coefficient

1 Introduction

A shaft is a rotating member of circular cross section used to transmit power and motion. A rotary machine is the assembly of a rotating shaft supported by bearings, with one or more rotors. A very specific class of rotary machines is the high rotation turbochargers, to automotive application. The turbocharger is an equipment added to internal combustion engines to raise its power or efficiency, using the exhausting gases of the engine to move the turbine attached to a compressor. The compressor

T. F. Peixoto (✉) · G. B. Daniel · K. L. Cavalca
Laboratory of Rotating Machinery, School of Mechanical Engineering,
University of Campinas, Campinas, SP, Brazil
e-mail: thalesfp@fem.unicamp.br

A. de T. Fleury et al. (eds.), *Proceedings of DINAME 2017*, Lecture Notes in Mechanical Engineering, https://doi.org/10.1007/978-3-319-91217-2_2

raises the air pressure entering the combustion chamber of the engine, allowing a larger mass flow rate with respect to naturally aspirated engines [1].

The turbochargers have, essentially, four elements: the rotating shaft, the turbine, the compressor, and the bearings. The radial inflow turbine is the element that drives the whole system. The centrifugal compressor is the element that improves the combustion process and the shaft is responsible for transferring the energy produced by the turbine to the compressor.

The bearings are responsible for supporting the loadings in the shaft. Because of unbalanced mass, inherently to every rotating system, there is radial vibration in the shaft, which are supported by journal bearings. Moreover, the gas flows in the turbine and the compressor are not constant and these flows produce radial and axial forces of different amplitudes. The axial forces produce axial displacements and the use of thrust bearings is necessary, to absorb the shaft axial loadings and displacements. The thrust bearings must be designed to let the oil film between the bearing and the collar attached to the shaft sustain the shaft axially, avoiding contact between the surfaces in relative motion, to mitigate friction and premature wear of these surfaces [2].

The axial vibration of the entire turbocharger is the object of analysis. It is assumed that this vibration can be modeled by concentrated parameters and the thrust bearings can be approximated by its dynamic characteristics, i.e., its equivalent damping and stiffness coefficients. Lund [3] introduced the concept to approach the dynamic characteristics of bearings through its equivalent coefficients, which basically consists in solve the governing equation for the pressure distribution to find the equilibrium position of the system and recalculate the pressure distribution applying small perturbations around the previously found equilibrium position.

The governing equation for the pressure distribution in lubricated bearings is the Reynolds' Equation [4], a second order partial differential equation governing the pressure distribution of thin viscous fluid films, which can be derived from the continuity and Navier-Stokes equations, neglecting the higher order terms that consider the (small) thickness of the oil film. Specifically to thrust bearings, Hamrock et al. [5] obtained an analytical solution to bearings whose pads radial dimensions were much higher than its circumferential dimensions. With this assumption, the radial flows crossing the interface between the bearing and the collar (shaft) can be neglected. This solution, however, is not always adequate, since the oil flow in the radial direction can be considerable, depending on the radial and circumferential dimensions of the thrust bearings. Pinkus and Lynn [6] was the first to obtain a numerical solution to the Reynolds' Equation, by the Finite Difference Method, to thrust bearings, whose oil film thickness linearly varies with the circumferential length.

Pinkus idea to apply numerical methods to solve the Reynolds' Equation was not restricted to hydrodynamic analysis (HD analysis), considering other effects in the bearings, besides the pressure distribution and the hydrodynamic forces. Numerical methods began to consider thermal effects on the bearings creating the thermo-hydrodynamic analysis (THD analysis). Dowson [7] published one of the main works in the THD analysis, deriving the generalized Reynolds' Equation, which takes into account the variation of fluid properties, such as the viscosity and density and his work contributed to future works that covered the THD lubrication. Recent works in solving the Reynolds' Equation in thrust bearings can be cited, such as Arghir et al. [8], that utilized the Finite Volume Method to obtain the pressure distribution in thrust bearings with discontinuities in the oil film and Alqmvist et al. [9], that compared simulated THD results with experimental results. Dadouche et al. [10] empirically observed the influence of various parameters, such as axial load applied, rotation speed and the replacement oil temperature, over the pressure and temperature field in the bearings and Dadouche et al. [11] compared experimental results obtained in the previous experimental bench with simulated results. Finally, Ahmed et al. [12] based their work in Dowson equation to obtain a solution to the THD problem using the generalized Reynolds' Equation to thrust bearings.

In this paper, the coefficients are estimated using experimental results. Daniel et al. [13] estimated the stiffness coefficient, solving the governing equations for the journal bearings to estimate the oil inlet temperature in the thrust bearing, necessary to solve the Reynolds' Equation taking into account thermal effects. The objective of this paper is to estimate the equivalent damping coefficient of the system, by running an optimization algorithm so that a simulated transient response of the shaft displacement matches the measured displacement of the shaft.

2 Materials and Methods

A scheme of the turbocharger is shown in Fig. 1a. The turbocharger in Fig. 1a consists of a turbine and a compressor attached to the shaft, supported by two thrust bearings. This turbocharger can be modeled by the equivalent system in Fig. 1b, considering that there is only relative motion between the collar and the thrust bearings, increasing or decreasing the oil film thickness. It is assumed that the thrust bearings are clamped and the collar moves only in the axial direction (defined as the x coordinate). The thrust bearings support the external forces ΔF to maintain the system working properly. The springs and viscous dampers are the equivalent

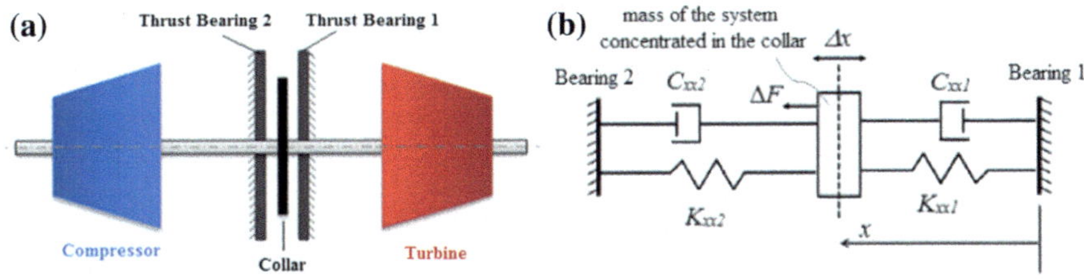

Fig. 1 **a** Scheme of turbocharger **b** equivalent system of the thrust bearing

Table 1 Geometric ratios of the bearings [2]

Variable	Bearing 1	Bearing 2
r_o/r_i	1.5	1.7
r_i/h_0	350	340
h_{max}/h_0	2.4	2.0
$\theta_{ramp}/(\theta_0 - \theta_{ramp})$	5.0	4.0
n_{pad}	3	3

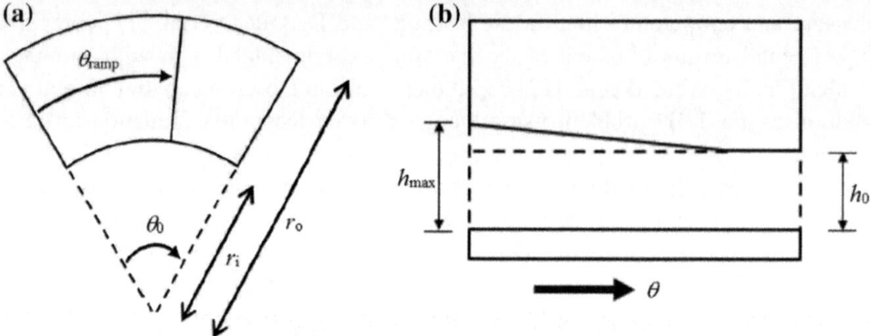

Fig. 2 **a** Bearing pad and **b** fluid thickness profile (adapted from [2])

coefficients of the oil film between the bearings and the collar. The geometric dimensions of the thrust bearings are shown in Table 1, through the ratios of the main design parameters. The main variables of the thrust bearings, also shown schematically in Fig. 2, are the inner (r_i) and outer (r_o) radius, the angular length of the pad ramp (θ_{ramp}), the angular length of the pad (θ_0), the minimum (h_0) and maximum (h_{max}) oil thickness and the number of pads (n_{pad}) of the bearing.

The turbocharger axial vibration is modeled as a one degree of freedom (DOF) system [2], admitting that its entire mass M is concentrated in the collar of the shaft. The springs and dampers of Fig. 1b are in parallel to each other, i.e., the equivalent stiffness and damping coefficients are the sum of those coefficients, $K_{xx} = K_{xx1} + K_{xx2}$ and $C_{xx} = C_{xx1} + C_{xx2}$. Therefore, the equation of motion of the system is

$$M\ddot{x}(t) + C_{xx}\dot{x}(t) + K_{xx}x(t) = F(t) \tag{1}$$

which can be numerically evaluated, by integrating Eq. (1) using the state space model defined as [14]:

$$\begin{Bmatrix} \dot{x}(t) \\ \ddot{x}(t) \end{Bmatrix} = \begin{bmatrix} 0 & 1 \\ -\frac{K_{xx}}{M} & -\frac{C_{xx}}{M} \end{bmatrix} \begin{Bmatrix} x(t) \\ \dot{x}(t) \end{Bmatrix} + \begin{Bmatrix} 0 \\ \frac{F(t)}{M} \end{Bmatrix} \tag{2}$$

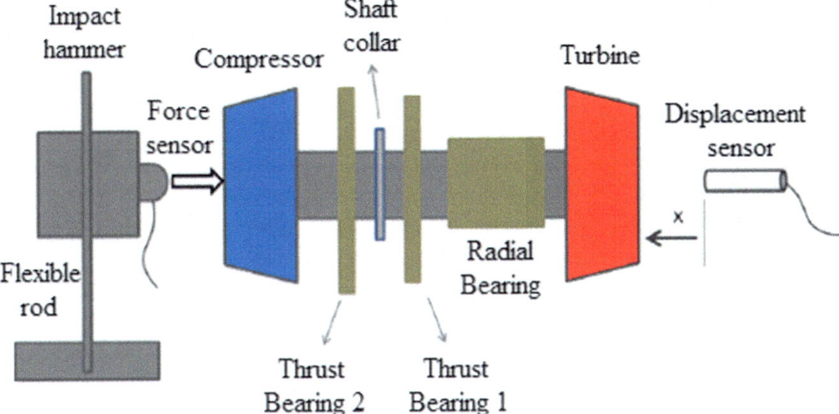

Fig. 3 Scheme of experimental setup of the turbocharger (adapted from [2])

In the test condition, the turbine is driven by compressed air, while the compressor is open to the atmosphere. An impact hammer, in the shaft end, axially excites the system. The axial displacement of the system is measured, as well as the force applied by the hammer, in time domain, as illustrated schematically in Fig. 3.

The aim of this work is to estimate the equivalent coefficients of the bearings, so the simulated response is compatible with the experimental results obtained, due to an impulsive excitation force. The stiffness coefficient can be estimated as described by [13], which consists in solving the Reynolds' Equation utilizing a THD model, by the Finite Volume Method (FVM), to obtain the pressure distribution of the oil film circulating in the thrust bearings, accounting for parameters like the geometry of the bearings and the temperature distribution in the oil film.

The Reynolds' Equation is the governing equation for pressure distribution in the oil film. To account for the temperature variation of the oil film, which changes the oil viscosity along the bearing, the generalized Reynolds' Equation must be utilized. This equation was introduced by [7] and is written in cylindrical coordinates as

$$\frac{1}{r}\frac{\partial}{\partial\theta}\left(F_2\frac{\partial p}{\partial\theta}\right)+\frac{\partial}{\partial r}\left(rF_2\frac{\partial p}{\partial r}\right)=\Omega r\frac{\partial}{\partial\theta}\left(\frac{F_1}{F_0}\right)+r\frac{\partial h}{\partial t} \tag{3}$$

in which Ω is the rotational speed of the shaft and the functions F_0, F_1 and F_2 are introduced to account for the temperature distribution in the oil film, which changes the oil viscosity (μ) along the mesh. These functions are defined as

$$F_0=\int_0^h\frac{1}{\mu}dx,\quad F_1=\int_0^h\frac{x}{\mu}dx,\quad F_2=\int_0^h\frac{x}{\mu}\left(x-\frac{F_1}{F_0}\right)dx \tag{4}$$

These integrals are responsible for the coupling between the viscosity variation along the oil film and the pressure to be calculated by the Reynolds' Equation (Eq. 3).

Daniel et al. [13] approached the problem to obtain the equivalent stiffness coefficient of the thrust bearing as suggested by Lund [3], by solving the Reynolds' Equation applying small perturbations around the previously calculated equilibrium position, disregarding the fluid film thickness variation with the time ($\partial h/\partial t = 0$). However, to estimate this coefficient, it is necessary to observe that the circulating oil in the thrust bearing enters the turbocharger through journal bearings, which causes an increase in its temperature. A THD model of the journal bearing is first solved to estimate the temperature of the oil leaving the journal bearing, assumed as the temperature of the oil inlet in the thrust bearing.

The results obtained by [13] are checked using the one DOF equation (Eq. 1) and a finite element method (FEM) to discretize the turbocharger, following the method suggested by [15, 16]. Peixoto et al. [15] compared the natural frequency of the system obtained by the one DOF system, by a FEM solution and the equation of longitudinal vibration of uniform bars with discontinuities of concentrated masses and springs. Peixoto [16] calculated the oil thickness of the fluid in the thrust bearing, using a step force and the equivalent stiffness coefficient calculated in the simulations of [13], comparing the values with measured, empirical results. The stiffness coefficient is estimated from the experimental results by dividing the measured force by the measured axial displacement of the system.

The oil thickness was calculated according to Fig. 4. The oil thickness of each thrust bearing is originally h_{01} and h_{02}, for thrust bearings 1 and 2, respectively. The collar attached to the shaft changes its equilibrium position by Δx after an external excitation, modeled as a step excitation, acts on the system. The collar changes its static equilibrium position due to the step excitation, which causes the oil thickness to change an amount equal to $h_0 \pm \Delta x$. The minimum estimated oil thickness of the bearing is $h_0 - |\Delta x|$ and this value is compared to the experimental measurements.

To check the oil thickness change of the model means to verify if the estimated stiffness coefficient is in agreement with experimental results. However, the stiffness coefficient changes only the static equilibrium position of the system, but gives little information on its dynamic response. To fully add the flexibility of the oil film

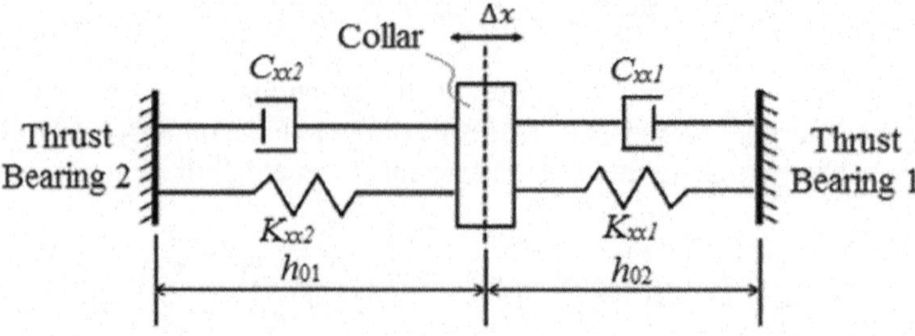

Fig. 4 Simplified scheme of the assembly bearings and collar (adapted from [16])

in the dynamic response, it is also necessary to obtain the equivalent damping coefficient. This coefficient is estimated by optimizing this parameter, setting as the objective function the difference between the simulated and the experimental response, during the transient response of the system.

The optimization problem is constructed admitting that the objective function to minimize is the maximum absolute difference between the experimental response $\{x_{exp}\}$ and the simulated response $\{x_{sim}\}$:

$$\min_{c_{\min} \leq C_{xx} \leq c_{\max}} \left(\max\left(\left|\{x_{sim}\} - \{x_{exp}\}\right|\right)\right) \tag{5}$$

The vector $\{x_{exp}\}$ is obtained from experimental measured results [2], along with the time vector $\{t\}$, while the simulated response $\{x_{sim}\}$ is obtained from the numerical integration of the system (Eq. 2), which is a function of the damping coefficient C_{xx}. The input force for the simulation is the force measured in the impact hammer, during the experiment. It is important to notice that the `max` function returns the largest element of the array, $|\cdot|$ is the element-wise absolute value function and the $\min(\cdot)$ function is the `min` operator, which returns the minimum value of the function.

The system is numerically integrated, with initial conditions $x_0 = 0$, i.e., the system starts from rest, and $\dot{x}_0 = v_0$, i.e., with an initial velocity given by

$$v_0 = \frac{dx}{dt} \cong \frac{x[2] - x[1]}{t[2] - t[1]} \tag{6}$$

so, the initial velocity of the system is estimated by the forward difference approximation, utilizing the measured values of displacement and time. The optimization problem is solved using the `fmincon` function from MATLAB®, using the interior point algorithm [17]. The range admitted for the damping coefficient, in the case studied here, is $0 < C_{xx} < 10^6$ N s/m.

3 Results

The turbocharger analyzed here has a total mass of 165 g. For a rotational speed of 14,100 rpm, the results obtained in the experiment schemed in Fig. 3 are shown in Fig. 5. Figure 5a shows the applied force by the impact hammer and Fig. 5b shows the axial displacement measured, both as a function of time. The collar bearing has an outer diameter of 30 mm, a thickness of 1.5 mm and a clearance of 49.9 μm (equal for both bearings). For this operation condition, the measured minimum oil thickness is 23.8 μm, the replacement oil temperature entering the turbocharger was measured in 37 °C and the temperature of the inlet oil in the thrust bearing was estimated in 60.6 °C, given by [13]. The equivalent stiffness coefficient and the

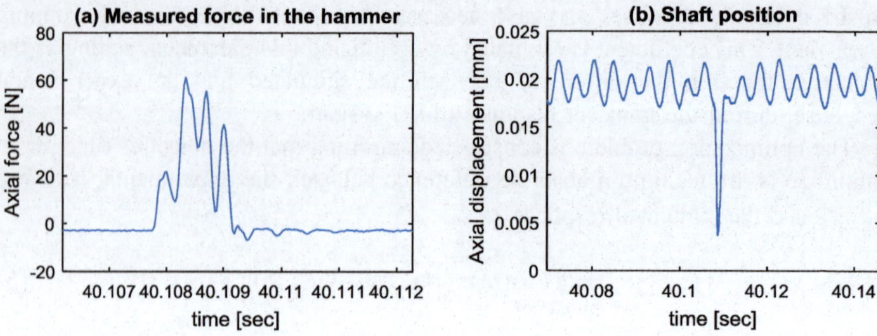

Fig. 5 Experimental results: **a** measured force of hammer **b** axial displacement of shaft

Table 2 Estimated supported forces and equivalent stiffness coefficients of the thrust bearings, for a rotation speed of 14,100 rpm (adapted from [13])

Thrust Bearing	Supported force (N)	Equivalent stiffness coefficient (10^6 N/m)
1	32.75	2.264
2	−38.03	4.460

force supported by each thrust bearing in Fig. 4 were estimated and are shown in Table 2.

The equivalent stiffness coefficient is, therefore, 6.724 MN/m (the sum of the values shown in Table 2). With this value of the equivalent stiffness coefficient of the thrust bearings, the natural frequency of the system was estimated for a lumped parameters model (the one DOF equation, Eq. 1), by the FEM (using linear and quadratic elements) and by the equation of longitudinal vibration of bars (known as the distributed parameters models, following the work of [15]). These values are shown in Table 3. Since the distributed parameters is the actual governing equation for axial vibration of the turbocharger, this value is considered the reference in calculating the percentage difference between the approximated models.

It can be argued from Table 3 that the lumped parameters model is good enough to observe the turbocharger axial vibration, since the natural frequency is almost the same for every model. The linear differential equation to model the one DOF system requires less computational effort of the solver in order to find the optimum

Table 3 Estimated first natural frequency of the turbocharger axial vibration ($\Omega = 14{,}100$ rpm)

Model	ω_n (rad/s)	Percentage difference
Distributed parameters	6,375	–
Lumped parameters	6,383	0.133%
Linear finite element	6,684	4.85%
Quadratic finite element	6,707	5.22%

value of the damping coefficient and is utilized instead of the more onerous equations.

The value of the stiffness coefficient can also be checked calculating the new static equilibrium position under a step excitation with the same magnitude of the resultant of the forces shown in Table 1 acting on the system, as in [16]. The estimated oil film thickness was 24.2 μm, while the measured minimum oil thickness is 23.8 μm (equal to the measured shaft displacement), which confirms the estimation of the equivalent stiffness coefficient. This value is adequate while running the optimization problem.

It is important to notice that the estimated stiffness coefficient is constant for the rotation speed of 14,100 rpm. This coefficient is estimated based on the system static equilibrium position, for each rotational speed, after it reaches steady state. The minor fluctuations in the rotational speed, typical of a turbocharger, does not alter the estimated stiffness coefficient. Vieira [2] inspect this coefficient variation with rotational speed and the minor fluctuations in the speed are not big enough to observe a difference on the estimated stiffness coefficient. Furthermore, since this coefficient is estimated from the static equilibrium position, this parameter can be estimated disregarding the dynamic characteristics of the system, which is also the reason to why it is not taken into account on the optimization process. The only optimization variable is the damping coefficient, which is fundamentally a dynamic quantity, and the optimization problem takes into account the dynamic response of the system and the already estimated stiffness coefficient to estimate the damping coefficient.

The first estimated damping coefficient is obtained (by [13]) solving the Reynolds' Equation, considering the $\partial h/\partial t$ term, applying a perturbation of velocity $\dot{h}$, giving a value of 6.165 kN s/m. Due to the lack of information on the variables that influence the damping coefficient, the estimated value for this coefficient is inadequate. This can be observed integrating the equation of motion of the system, with the values obtained for the mass and the stiffness and damping coefficients (Fig. 6). Consequently, it is necessary to find the correct damping coefficient, which is achieved optimizing this parameter, so the simulated response gets closer to the experimental results.

The optimization problem (Eq. 5) estimates the damping coefficient, setting Eq. 2 as the equation of motion of the system, using the values of the turbocharger mass and the previously estimated stiffness coefficient. The damping coefficient is identified during the transient response of the system, which limits the time duration of the impulse applied by the hammer (Fig. 7a) and the measured axial displacement (Fig. 7b). Also, due to fluctuations in the measured force, this coefficient identification was carried on under two conditions of external forces, as shown in Fig. 7a: considering the force measurements of the impact hammer (dotted blue line) and considering an adjusted force, disregarding those fluctuations (solid red line). Figure 7b shows the displacement of the shaft considered in the optimization, covering the same amount of time of the force, and is only the transient response of the shaft.

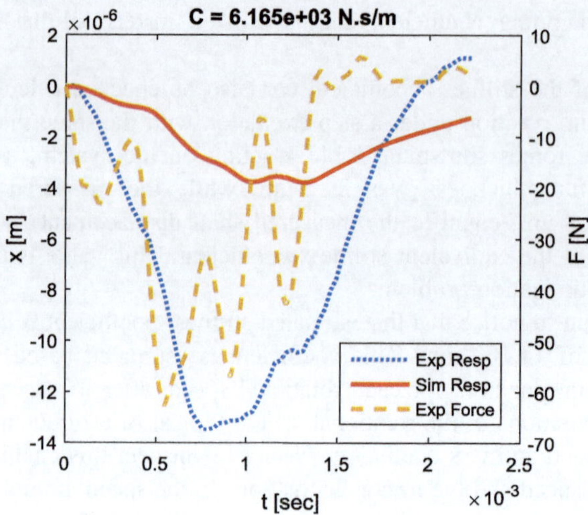

Fig. 6 Simulated and experimental results (first estimation of equivalent damping coefficient)

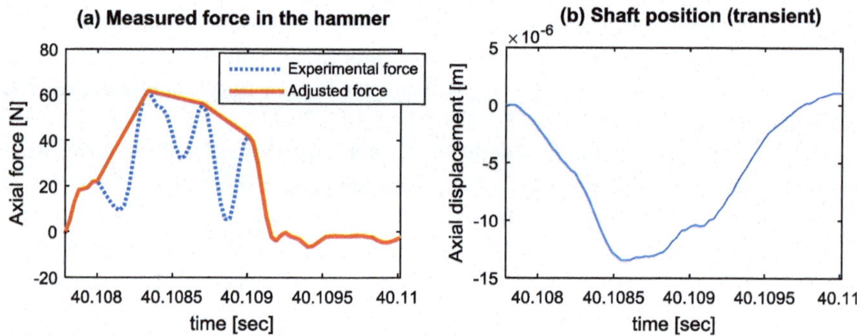

Fig. 7 Parameters considered during the optimization problem: **a** adjusted experimental force and **b** transient shaft displacement during the force application

The first optimization of the damping coefficient, utilizing the experimental force (dotted line, Fig. 7a), returns a value of 1.090 kN s/m. The dynamic (simulated) response of the system with this value, to the external excitation, is shown in Fig. 8a. The optimization was run with a small tolerance for the objective function, to ensure the value obtained is the global minimum, not a local minimum. The initial value used for the algorithm was the previously calculated damping coefficient of 6.165 kN s/m.

It can be seen from Fig. 8a that the general behavior tendency of the shaft displacement given by the simulations is the same as the experimental results. However, the displacement amplitude is not in good accordance between both curves. A second optimization was carried on, using the adjusted force shown in

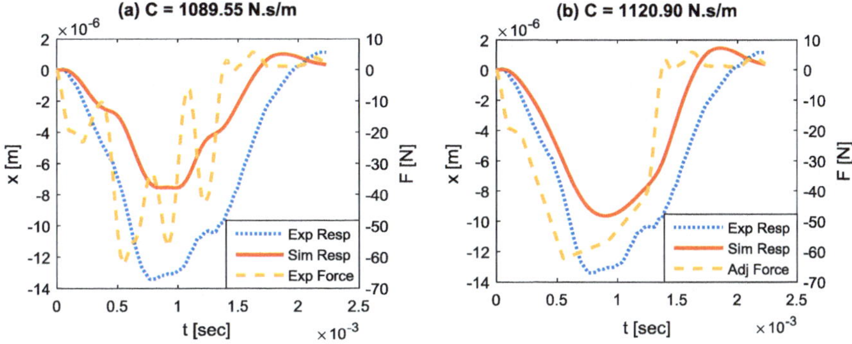

Fig. 8 Comparison of simulated (Sim Resp) and experimental (Exp Resp) results: optimization run with **a** experimental force (Exp Force) and **b** adjusted force (Adj Force)

Fig. 7a, disregarding the minor fluctuations in the experimental measured forces. This optimization returns a value for the damping coefficient of 1.121 kN s/m and the dynamic response for this system is shown in Fig. 8b, bringing a better accordance between the experimental and simulated displacement of the collar clamped to the shaft.

Using the value of the initial estimated equivalent damping coefficient, the damping factor ξ of the system is estimated in 2.92, whereas the damping factor utilizing the value for the damping coefficient obtained from the second optimization (Fig. 8b) is 0.532. From the agreement between the curves of Fig. 8b and the more plausible value for the damping factor, it can be said that the original value estimated for the damping coefficient was not adequate. One can assume that, for this particular system, the perturbation model utilized to approach the damping factor may be not well suited. Moreover, in turbochargers, one aim is to reduce vibrations to a minimum. For a desirable transient response, the damping factor must be between 0.4 and 0.8, which was satisfactorily obtained, because small values of $\xi(\xi < 0.4)$ yield excessive overshoot in the transient response, and a system with a large value of $\xi(\xi > 0.8)$ responds sluggishly [18]. This is achieved by designing a thrust bearing with a relative high damping factor. However, overdamped systems, although do not present vibrations, respond slower, i.e., the settling time is usually higher than the settling time of systems with a damping ratio close to unity. Therefore, the value of 0.532 for this system is more acceptable and suitable for the thrust bearing performance.

Finally, it is also important to notice that the adjusted experimental force utilized on the simulations seems more appropriate. The peaks observed on the measured experimental force (Fig. 7a) are probably measurement errors, since the experimental response of the system does not have peaks accompanying the input force. Moreover, the simulations disregarding the peaks on the input force provide a better response, which is another indication that the correct force applied by the hammer is just one single impact on the turbocharger.

4 Conclusions

With the damping coefficient obtained from the second optimization, the damping factor is 0.532, which indicates that this system is underdamped and its impulse response decays fast enough. The amplitude difference of the measured displacement and the simulated response with the adjusted force is about 4 μm. Besides, the oscillation frequency of the simulated response approaches the experimental results frequency, as observed in Figs. 7b and 8b, because both curves have the same shape.

It must also be noted that the numerical integration must be made utilizing the actual experimental force as an input. Although the force applied by the impact hammer approaches an impulsive excitation, which could be modeled using the Dirac delta function, the fundamental period of the system (the inverse of the natural frequency) is about 1×10^{-3} s, so the input of the system cannot be approximated by an impulse. This approximation is only valid if the time duration of the force is much smaller than the fundamental period of the system, which is not true for the case in analysis.

Furthermore, two other important points must be noted. First, because of the high rotational speed of the system, the estimation of the stiffness coefficient must be made utilizing a THD model, because heat is generated by viscous dissipation, due to fluid shear. This must be taken into account, even for steady state operation, which gives a good estimate for the stiffness coefficient. To do a THD analysis on the thrust bearing, the inlet temperature of the oil must be known. This temperature was estimated utilizing a THD model to obtain the outlet temperature of the oil leaving the journal bearings of the turbocharger. Once the stiffness coefficient is obtained, the damping coefficient can be estimated.

The second important point here is that the damping coefficient is estimated by an optimization problem, requiring that the dynamic response of the system must be close enough to experimental measured results. These estimated coefficients are nearly constant for the rotational speed of the system, since minor variations on the rotational speed provokes little variation on these coefficients. The order of magnitude of these variations are much smaller than the order of magnitude of these parameters, so the estimates are good enough to simulate dynamic response of systems in steady state conditions, for constant rotational speeds.

Acknowledgements The authors would like to thank the BorgWarner, Inc company and the Student Support Service at Unicamp (SAE) for the financial support of this project and the Technische Universität Darmstadt, where the experimental tests were performed.

References

1. Korpela, S.A.: Principles of Turbomachinery. Wiley (2011)
2. Vieira, L.C.: Analysis of a Thermohydrodynamic Model for Thrust Bearings [s.n.] (2014)

3. Lund, J.W.: Review of the concept of dynamic coefficients for fluid film journal bearings. J. Tribol. **109**(1), 37 (1987)
4. Reynolds, O.: On the theory of lubrication and its application to Mr. Beauchamp Tower's experiments, including an experimental determination of the viscosity of olive oil. Philos. Trans. R. Soc. Lond. **177**, 157–234 (1886)
5. Hamrock, B.J., Schmid, S.R., Jacobson, B.O.: Fundamentals of Fluid Film Lubrication. Marcel Dekker (2004)
6. Pinkus, O., Lynn, W.: Solution of the tapered-land sector thrust bearing. Trans. ASME **80**, 1510–1516 (1958)
7. Dowson, D.: A generalized Reynolds equation for fluid-film lubrication. Int. J. Mech. Sci. **4** (2), 159–170 (1962)
8. Arghir, M., Alsayed, A., Nicolas, D.: The finite volume solution of the Reynolds equation of lubrication with film discontinuities. Int. J. Mech. Sci. **44**(10), 2119–2132 (2002)
9. Almqvist, T., Glavatskih, S.B., Larsson, R.: THD analysis of tilting pad thrust bearings—comparison between theory and experiments. J. Tribol. **122**(2), 412–417 (2000)
10. Dadouche, A., Fillon, M., Bligoud, J.: Experiments on thermal effects in a hydrodynamic thrust bearing. Tribol. Int. **33**(3–4), 167–174 (2000)
11. Dadouche, A., Fillon, M., Dmochowski, W.: Performance of a hydrodynamic fixed geometry thrust bearing: comparison between experimental data and numerical results. Tribol. Trans. **49** (3), 419–426 (2006)
12. Ahmed, S.A., Fillon, M., Maspeyrot, P.: Influence of pad and runner mechanical deformations on the performance of a hydrodynamic fixed geometry thrust bearing. Proc. Inst. Mech. Eng. Part J J. Eng. Tribol. **224**(4), 305–315 (2009)
13. Daniel, G.B., Vieira, L.C., Cavalca, K.L.: Sensitivity analysis of the dynamic characteristics of thrust bearings. In: Procedings of the 3rd International Symposium on Uncertainty Quantification and Stochastic Modeling, pp. 1–10 (2016)
14. Müller, P.C., Schiehlen, W.O.: Linear Vibrations: A Theoretical Treatment of Multi-degree-of-Freedom Vibrating Systems. Springer, Netherlands (1985)
15. Peixoto, T.F., Vieira, L.C., Cavalca, K.L.: Analysis of rotors supported by thrust bearings. In: 23rd ABCM International Congress of Mechanical Engineering, pp. 1–8 (2015)
16. Peixoto, T.F.: Analysis of Rotors Supported by Thrust Bearings [s.n.] (2016)
17. Byrd, R.H., Hribar, M.E., Nocedal, J.: An interior point algorithm for large-scale nonlinear programming. SIAM J. Optim. **9**(4), 877–900 (1999)
18. Ogata, K.: Modern Control Engineering, 5th ed. Prentice-Hall (2010)

Analysis of the Dynamic Behavior of a Cracked Rotating Shaft by Using the Harmonic Balance Approach

Aldemir Ap Cavalini Jr., Tobias Souza Morais, Nicolò Bachschmid and Valder Steffen Jr.

Abstract There are several SHM techniques proposed in the literature for crack detection in rotating machines. Among them, the ones based on vibration measurements are recognized as useful tools in the industrial context. Although widely used, when applied under non-ideal conditions, such techniques can only detect cracks that eventually have already spread significantly along the cross section of the shaft. Therefore, currently, the researchers' attention is turning to more sophisticated methods capable of identifying incipient cracks, which represent a type of damage that are hardly observable in classical vibration analysis. In a previous contribution, a crack identification methodology based on a nonlinear approach was proposed. The technique uses external applied diagnostic forces at certain frequencies attaining combinational resonances, together with a pseudo-random optimization code, known as Differential Evolution, in order to characterize the signatures of the crack in the spectral responses of flexible rotor. In the present paper, the favorable conditions to apply the proposed methodology are investigated. The analysis procedure is confined to the operating parameters of the system, being characterized by the rotation speed of the rotor and the amplitude and frequency of the diagnostic forces. The harmonic balance approach is used to determine the vibration responses of the cracked rotor system and the open crack behavior is simulated according to the FLEX model. For illustration purposes, a rotor composed by a horizontal flexible shaft, two rigid discs, and two self-aligning ball bearings is used to compose a FE model of the system.

Keywords Rotordynamics · Crack detection and identification
Harmonic balance approach

A. A. Cavalini Jr. (✉) · T. S. Morais · V. Steffen Jr.
LMEst – Structural Mechanics Laboratory, School of Mechanical Engineering, Federal University of Uberlândia, Av. João Naves de Ávila, 2121, Uberlândia, MG 38408-196, Brazil
e-mail: aacjunior@ufu.br

N. Bachschmid
Department of Mechanical Engineering, Politecnico di Milano, Via La Masa 1, Milan 20156, Italy

A. de T. Fleury et al. (eds.), *Proceedings of DINAME 2017*, Lecture Notes in Mechanical Engineering, https://doi.org/10.1007/978-3-319-91217-2_3

1 Introduction

Shaft crack detection is an important issue in rotor dynamics and machines that are suspect of having a crack must be treated with the upmost concern [1]. The importance attributed to this problem is addressed to the serious consequences when cracks are not early identified in rotating systems. Thus, manufacturers have adopted design concepts, as well as special procedures for start-up, operation, monitoring, and maintenance, in order to minimize the appearance and growth of cracks in different rotors, such as steam turbines, centrifugal compressors, and generator units found in power plants. Various structural health monitoring (SHM) techniques devoted to crack detection in rotating machines have been proposed in the last decade. Therefore, the methodologies that use harmonic excitations as diagnostic forces has attracted the attention of several researchers and two interesting results are here recalled.

Mani et al. [2] presented a theoretical analysis considering a simple rotor model with 2 degrees of freedom containing a breathing crack. The method of multiple scales was used to solve the equations of motion of the system, in which the stiffness of the shaft was affected by the nonlinearity (i.e., the breathing crack). The so-called combination vibrations were defined in the context of rotating cracked shafts. It was shown that the vibration amplitudes associated with the combination vibrations are directly proportional to the time dependent stiffness; in other words, to the crack depth. Ishida and Inoue [3] made accurate numerical and analytical analyzes on a cracked Jeffcott rotor. The stiffness of the cracked shaft has been modeled by using two different approaches, namely, (i) a piecewise linear stiffness, and (ii) by using power series. The effects of the excitation intensity (diagnostic forces) on the forward and backward whirl vibration responses of the rotor system at the combination frequencies were evaluated according to the crack severity. An experimental validation of the proposed method was presented, in which the combination vibrations were demonstrated on the vibration responses of the considered rotating machine.

More recently, Cavalini Jr. et al. [4] have analyzed the possibility of identifying the severity of transverse cracks (i.e., position and depth) in rotating shafts by using the so-called diagnostic forces and the combination vibrations. The frequencies of the diagnostic forces were determined by using the method of multiple scales. This model based approach (i.e., considering the finite element model of the system—FE model) was applied in a rotor test rig composed by a horizontal shaft, two rigid discs, two self-alignment ball bearings, and one electromagnetic actuator used to apply the harmonic excitations. The horizontal vibration responses of the rotating machine were measured by using displacement sensors located near to the discs. The dynamic behavior of the system was investigated considering the breathing and open crack models. The crack models were formulated from the Mayes model (breathing crack) allied to the linear fracture mechanics approach (breathing and open cracks). Vibration responses in the time domain have been determined for different crack positions and depths. In a given test case, the proposed methodology was able to identify, with good accuracy, the severity of the crack by using the Differential Evolution optimization method [5]. In that contribution, constant rotation speed and various diagnostic excitations at frequencies suitable for exciting two combination vibrations were considered.

The purpose of applying the diagnostic forces in a cracked shaft at frequencies Ω d different from the rotation speed Ω is to exciting combination vibrations. The natural frequency (Ωn; i.e., forward and backward natural frequencies of the shaft operating at Ω) and rotation speed of the rotor are used to determine the conditions in which the combination vibrations appear, which results in the frequencies of the diagnostic forces (i.e., $\Omega d = 2\Omega - \Omega n$, $\Omega d = -2\Omega + \Omega n$, $\Omega d = 4\Omega - \Omega n$, $\Omega d = -4\Omega + \Omega n$, etc.—in the case of open crack). Significant vibration amplitudes can be observed at these frequencies, allowing to distinguish the resonance peak from other vibration components and noise. The natural frequencies of the shaft must be known, so that the frequencies of the diagnostic forces can be selected in advance. However, the problem consists in determining the amplitude and frequency of the diagnostic forces to generate measurable peaks on the vibration spectrum at the combination vibrations. This is an important issue, mainly when the proposed technique is applied in industrial machinery due to the limitations regarding the applicable force amplitude and position.

Therefore, the vibration response of the system at the combination vibrations depends on the damping, the location of the crack in the shaft, the locations where the diagnostic forces are applied, and the amplitude of the diagnostic forces. In this paper, the dynamic behavior of a cracked rotating shaft is analyzed to determine the most favorable conditions to apply the mentioned SHM technique by using the harmonic balance approach. This quasi-linear methodology is able to determine the vibration amplitudes at the combination vibrations generated by the presence of the crack when the external diagnostic forces are applied in the rotor system. Additionally, the obtained results are compared to the ones determined from the trapezoidal rule integration scheme, which was coupled with the Newton-Raphson iterative method for nonlinear analysis [6].

It is worth mentioning that cracks may be always open or they can breathe depending on the rotating machine and operating conditions. Shafts affected by open cracks behave according to linear systems with parametric excitation. Differently, shafts with breathing cracks becomes really non-linear when dominated by vibrations. It may occur in vertical shafts or in horizontal light and weakly damped shafts. However, shafts with breathing cracks may also be considered linear systems when the dynamic behavior is weight dominated, as occurs in horizontal rotating heavy shafts. In this contribution, the study is restricted to linear systems with parametric excitation. Thus, the so-called FLEX model for open cracks is used [7].

2 Rotor Test Rig

Figure 1a shows the rotor test rig used to represent the analyzed rotor system, leading to the numerical simulations shown in this work. Thus, a model with 33 finite elements (Timoshenko's beam elements with 4 degrees of freedom per node; Fig. 1b) was used to mathematically characterize the system. It is composed of a flexible steel shaft with 860 mm length and 17 mm diameter ($E = 205$ GPa, $\rho = 7850$ kg/m^3, $\upsilon = 0.29$), two rigid discs D_1 (node #13; 2.637 kg; according to the FE model) and D_2 (node #23; 2.649 kg), both of steel and with 150 mm diameter and 20 mm thickness ($\rho = 7850$ kg/m^3), and two roller bearings (B_1 and B_2, located at nodes #4 and #31, respectively). Displacement sensors are

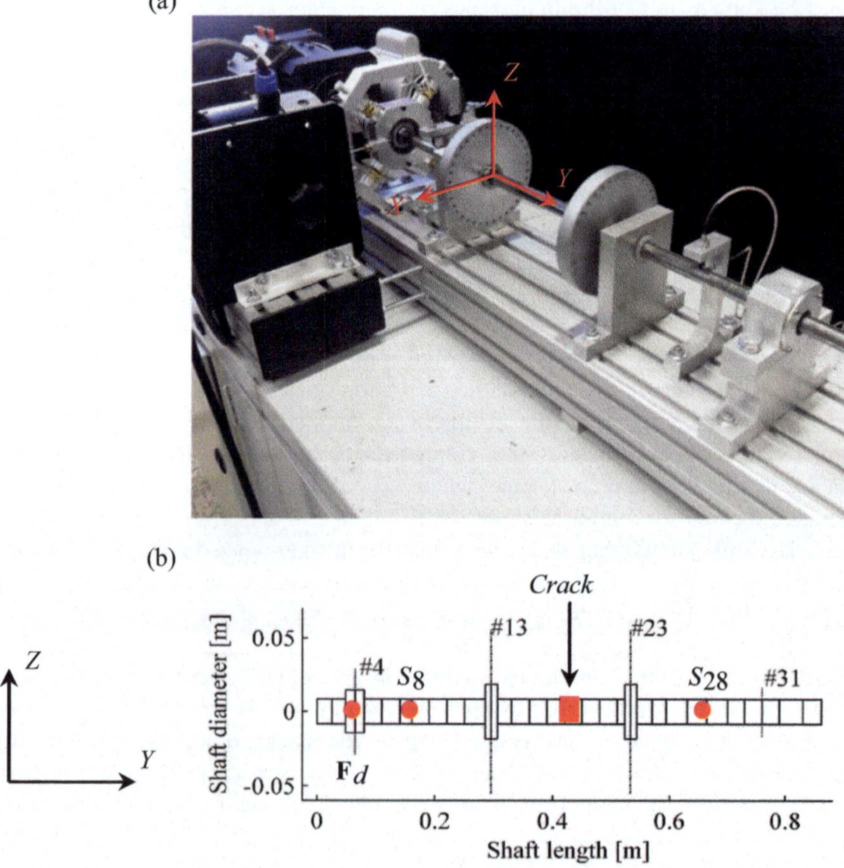

Fig. 1 Rotor test rig used in the numerical simulations of the SHM technique: **a** Test rig; **b** FE model

orthogonally mounted at nodes #8 (S_{8X} and S_{8Z}) and #28 (S_{28X} and S_{28Z}) to measure the shaft vibration. The system is driven by an electric DC motor.

Equation (1) governs the dynamic behavior of the cracked flexible rotor supported by roller bearings [8].

$$\mathbf{M}\ddot{\mathbf{q}} + \left[\mathbf{D} + \Omega\,\mathbf{D}_g\right]\dot{\mathbf{q}} + \mathbf{K}(\Omega t)\mathbf{q} = \mathbf{W} + \mathbf{F}_u + \mathbf{F}_d \tag{1}$$

where $\mathbf{M}$ is the mass matrix, $\mathbf{D}$ is the damping matrix, $\mathbf{D}_g$ is the gyroscopic matrix, and $\mathbf{K}(\Omega t)$ is the periodic stiffness matrix with variable values due to the crack (i.e., Ωt is the angular position of the shaft). $\mathbf{W}$ stands for the weight of the rotating parts, $\mathbf{F}_u$ represents the rotating unbalance forces, and $\mathbf{F}_{diag}$ represents the diagnostic force applied in the rotor (force fixed in space), and $\mathbf{q}$ is the generalized displacement vector.

A model updating procedure was used in order to obtain a representative FE model, considering the rotor system in a pristine condition (Fig. 1). In this sense, a heuristic optimization technique (Differential Evolution) was used to determine the unknown parameters of the model, namely the stiffness and damping coefficients of the bearings, the proportional damping added to $\mathbf{D}$ (coefficients γ and β; $\mathbf{D}_p = \gamma\,\mathbf{M} + \beta\,\mathbf{K}$), and the angular stiffness k_{ROT} due to the coupling between the electric motor and the shaft (added around the orthogonal directions X and Z of the node #1).

The proposed identification process (i.e., the comparison between simulated and experimental frequency response functions, FRF) was performed 10 times, considering 100 individuals in the initial population of the optimizer. However, in this case only the regions close to the peaks associated with the natural frequencies were taken into account. Table 1 summarizes the parameters determined at the end of the minimization process associated with the smaller fitness value (i.e. objective function value). Figure 2 presents the Campbell diagram of the rotating machine, in which the first two forward critical speeds were determined at, approximately, 1714 rev/min and 5912 rev/min. More details about the model updating procedure adopted in this work can be found in Cavalini Jr. et al. [4].

Table 1 Parameters determined by the model updating procedure

Parameters	Values	Parameters	Values	Parameters	Values
k_X/B_1	8.551×10^5	k_X/B_2	5.202×10^7	γ	2.730
k_Z/B_1	1.198×10^6	k_Z/B_2	7.023×10^8	β	4.85×10^{-6}
d_X/B_1	7.452	d_X/B_2	25.587	k_{ROT}	770.442
d_Z/B_1	33.679	d_Z/B_2	91.033		

k stiffness (N/m)
d damping (Ns/m)

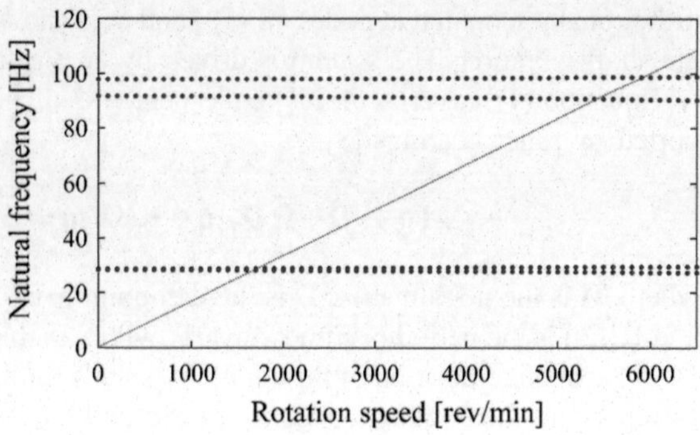

Fig. 2 Campbell diagram of the rotating machine

3 Open Crack Model

Three models are the most currently used to represent the breathing behavior. The models proposed by Gasch [9] and Mayes and Davies [10] are weight dominated. In both, the mechanism for opening and closing the crack is described by simple mathematical functions. The Gasch's model considers the crack as opening and closing abruptly, while the Mayes' model allows a smooth transition between the fully opened and fully closed crack. Finally, there is the more sophisticate model, known as FLEX model, as proposed by Bachschmid et al. [7].

The FLEX model was originally formulated for breathing cracks, in which the dynamic phenomena is characterized by the stress distribution on the crack cross-section [σ_{FLEX}; see Eq.(2)]. For a given angular position Ωt of the shaft, the locations of the crack cross-section presenting tensile stresses are considered open. Differently, compressive stresses represent crack closed regions.

$$\sigma_{FLEX} = \frac{M_Z I_{XX} + M_X I_{XZ}}{I_{XX} I_{ZZ} - I_{XZ}^2} X_{cr} - \frac{M_X I_{ZZ} + M_Z I_{XZ}}{I_{XX} I_{ZZ} - I_{XZ}^2} Z_{cr} \tag{2}$$

where I_{XX}, I_{ZZ}, and I_{XZ}, are the area inertia moments related to the geometric center (GC) of the shaft cross-section with crack. X_{cr} and Z_{cr} are the distances obtained along the same cross section along the directions X and Z, respectively, from GC to the position where tension is calculated. M_X and M_Z are the dynamic moments around the X and Z directions, respectively. These moments are given by the strength of materials theory and dependent of the external loads applied to the shaft.

The FLEX model results in a symmetrical 8×8 stiffness matrix named $\mathbf{K}_{FLEX}$ [see Eq. (3)], which is included in the FE model presented by Fig. 1b and Eq. (1) to evaluate the dynamic behavior of the cracked rotor system.

$$\mathbf{K}_{FLEX} = \begin{bmatrix} b_F & p_F & -q_F & -d_F & -b_F & -p_F & -q_F & -d_F \\ & a_F & c_F & q_F & -p_F & -a_F & c_F & q_F \\ & & e_F & r_F & q_F & -c_F & f_F & s_F \\ & & & h_F & d_F & -q_F & s_F & g_F \\ & & & & b_F & p_F & q_F & d_F \\ & & & & & a_F & -c_F & -q_F \\ & & & & & & e_F & r_F \\ SYM & & & & & & & h_F \end{bmatrix} \quad (3)$$

in which,

$$\begin{aligned} a_F &= \frac{12EI_{ZZ}}{L_{FLEX}^3(1+\vartheta_{FZZ})} & d_F &= \frac{6EI_{XX}}{L_{FLEX}^2(1+\vartheta_{FXX})} & g_F &= \frac{EI_{XX}(2-\vartheta_{FXX})}{L_{FLEX}(1+\vartheta_{FXX})} \\ b_F &= \frac{12EI_{XX}}{L_{FLEX}^3(1+\vartheta_{FXX})} & e_F &= \frac{EI_{ZZ}(4+\vartheta_{FZZ})}{L_{FLEX}(1+\vartheta_{FZZ})} & h_F &= \frac{EI_{XX}(4+\vartheta_{FXX})}{L_{FLEX}(1+\vartheta_{FXX})} \\ c_F &= \frac{6EI_{ZZ}}{L_{FLEX}^2(1+\vartheta_{FZZ})} & f_F &= \frac{EI_{ZZ}(2-\vartheta_{FZZ})}{L_{FLEX}(1+\vartheta_{FZZ})} & p_F &= \frac{12EI_{XZ}}{L_{FLEX}^3(1+\vartheta_{FXZ})} \end{aligned} \quad (4a)$$

$$\begin{aligned} q_F &= \frac{6EI_{XZ}}{L_{FLEX}^2(1+\vartheta_{FXZ})} & \vartheta_{FXX} &= \frac{12EI_{XX}}{GSL_{FLEX}^2} \\ r_F &= \frac{EI_{XZ}(4+\vartheta_{FXZ})}{L_{FLEX}(1+\vartheta_{FXZ})} & \vartheta_{FZZ} &= \frac{12EI_{ZZ}}{GSL_{FLEX}^2} \\ s_F &= \frac{EI_{XZ}(2-\vartheta_{FXZ})}{L_{FLEX}(1+\vartheta_{FXZ})} & \vartheta_{FXZ} &= \frac{12EI_{XZ}}{GSL_{FLEX}^2} \end{aligned} \quad (4b)$$

with G being the shear modulus, S is the cross-section area of the shaft, and L_{FLEX} is the length of the element as a function of the crack depth [7].

For the breathing crack model, the stiffness matrix $\mathbf{K}_{FLEX}$ is updated according to the applied external loads for each angular position of the shaft. The iterative process is based on the identification of the remaining cracked area cross section, which is given by the stress distribution σ_{FLEX} determined by Eq. (2). Thus, the area moments of inertia presented in Eq. (4a) and (4b) are determined allowing the stiffness matrix calculation of the cracked element. In the open crack model, the stiffness matrix $\mathbf{K}_{FLEX}$ is only dependent of the angular position Ωt and the area moments of inertia can be written as follows:

$$\begin{aligned} I_{XX}(\Omega t) &= \frac{I_{xx}+I_{zz}}{2} + \frac{I_{xx}-I_{zz}}{2}\cos(2\Omega t) \\ I_{ZZ}(\Omega t) &= \frac{I_{xx}+I_{zz}}{2} - \frac{I_{xx}-I_{zz}}{2}\cos(2\Omega t) \\ I_{XZ}(\Omega t) &= -\frac{I_{xx}-I_{zz}}{2}\sin(2\Omega t) \end{aligned} \quad (5)$$

where I_{xx} and I_{zz} are the area moments of inertia of the shaft cross-section with crack about rotating x- and z-axes as defined by Al-Shudeifat [11].

As mentioned, the open crack behavior is a function of the shaft angular position Ωt only. Therefore, the periodic stiffness matrix of the shaft $\mathbf{K}(\Omega t)$ [see Eq.(1)] can be written according to Eq. (6).

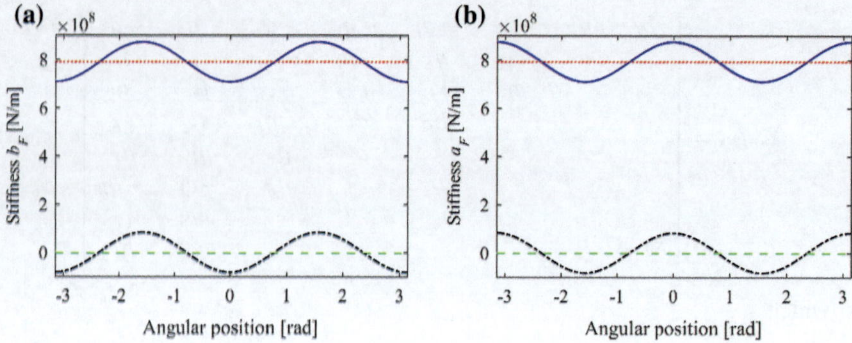

Fig. 3 Campbell diagram of the rotating machine. Stiffness behavior in fixed coordinates (— K (Ωt); - - - Km; -- -- ΔK1; -- - -- ΔK2; -- -- ΔK3): **a** stiffness coefficient b_F; **b** stiffness coefficient a_F [see Eq. (3)]

$$\mathbf{K}(\Omega t) = \mathbf{K}_m + \sum_{j=1}^{3} \frac{1}{2} \left(\Delta\mathbf{K}_j e^{ij\Omega t} + \Delta\mathbf{K}_j^* e^{-ij\Omega t} \right) \tag{6}$$

where $\mathbf{K}_m$ is the mean stiffness matrix of the shaft with the crack and $\Delta\mathbf{K}_j$ (j = 1, 2, and 3) are the stiffness variation related to the 1X, 2X, and 3X vibration components of the rotor speed. $\Delta\mathbf{K}_j^*$ is the complex conjugate of $\Delta\mathbf{K}_j$. The Fourier expansion of the periodic stiffness is truncated at the third harmonic component. Figure 3 shows the behavior of the stiffness terms described in Eq. (6) considering a 50% depth open crack located at the element #18 of the FE model (see Fig. 1b). Note that the open crack behavior is represented only by the mean stiffness matrix and the component $\Delta\mathbf{K}_2$, which is in accordance with the area inertia moments of Eq. (5).

4 Harmonic Balance Approach

As a result of the described open crack model, the equation of motion that governs the dynamic behavior of the flexible rotor can be rewritten as follows:

$$\mathbf{M}\ddot{\mathbf{q}} + \left[\mathbf{D} + \Omega\mathbf{D}_g\right]\dot{\mathbf{q}} + \mathbf{K}_m\mathbf{q} = \mathbf{W} + \mathbf{F}_u + \mathbf{F}_d - \frac{1}{2}\left(\Delta\mathbf{K}_2 e^{i2\Omega t} + \Delta\mathbf{K}_2^* e^{-i2\Omega t}\right)\mathbf{q} \tag{7}$$

in which, the vector of degrees of freedom $\mathbf{q}$ can be expressed as a Fourier series, as shows Eq. (8).

$$\mathbf{q} = \mathbf{q}_{st} + \frac{1}{2}\left(\mathbf{q}_d e^{i\Omega_d t} + \mathbf{q}_d^* e^{-i\Omega_d t}\right) + \frac{1}{2}\left(\mathbf{q}_1 e^{i\Omega t} + \mathbf{q}_1^* e^{-i\Omega t}\right) \tag{8}$$

where $\mathbf{q}_{st}$ is the static displacement, $\mathbf{q}_j$ ($j = d$ and 1) are the dynamic displacements related to the diagnostic force and 1X vibration components of the rotor speed due to the unbalance ($\mathbf{q}_j^*$ is the complex conjugate of $\mathbf{q}_j$).

Equation (8) is included in Eq. (7), resulting in new vibration components [i.e., from the last RHS term in Eq. (7)]. Therefore, the vector of degrees of freedom must be updated as presents Eq. (9).

The two last terms of Eq. (9) are the so-called combination vibrations. The resonant condition is induced in the cracked rotating shaft when one of the combination frequencies equals one of the critical speeds (i.e., $\Omega_c = 2\Omega + \Omega_d$ and $\Omega_c = 2\Omega - \Omega_d$). The substitution process continues including Eq. (9) in Eq. (7). Different vibration components and combination vibrations are determined.

$$\begin{aligned}\mathbf{q} = \mathbf{q}_{st} &+ \frac{1}{2}\left(\mathbf{q}_d e^{i\Omega_d t} + \mathbf{q}_d^* e^{-i\Omega_d t}\right) + \frac{1}{2}\left(\mathbf{q}_1 e^{i\Omega t} + \mathbf{q}_1^* e^{-i\Omega t}\right) + \frac{1}{2}\left(\mathbf{q}_2 e^{i2\Omega t} + \mathbf{q}_2^* e^{-i2\Omega t}\right)\\ &+ \frac{1}{2}\left(\mathbf{q}_3 e^{i3\Omega t} + \mathbf{q}_3^* e^{-i3\Omega t}\right) + \frac{1}{2}\left(\mathbf{q}_{2\Omega+\Omega_d} e^{i(2\Omega+\Omega_d)t} + \mathbf{q}_{(2\Omega+\Omega_d)}^* e^{-i(2\Omega+\Omega_d)t}\right)\\ &+ \frac{1}{2}\left(\mathbf{q}_{2\Omega-\Omega_d} e^{i(2\Omega-\Omega_d)t} + \mathbf{q}_{(2\Omega-\Omega_d)}^* e^{-i(2\Omega-\Omega_d)t}\right)\end{aligned} \tag{9}$$

Equation (10) presents the amplitudes of the equivalent forces $\mathbf{F}_j$ associated with the vibration components 0, Ω_d, Ω, 2Ω, 3Ω, and 4Ω, and the combination vibrations $2\Omega + \Omega_d$, $2\Omega - \Omega_d$, $4\Omega + \Omega_d$, $4\Omega - \Omega_d$, $6\Omega + \Omega_d$, and $6\Omega - \Omega_d$ ($j = st$, Ω_d, 1, 2, …, $6\Omega + \Omega_d$, and $6\Omega - \Omega_d$). The forces were obtained from subsequent substitutions of the vector of degrees of freedom in the last RHS term of Eq. (7). Note that the forces at Ω and 3Ω ($\mathbf{F}_1$ and $\mathbf{F}_3$, respectively) are obtained from $\mathbf{q}_1$, $\mathbf{q}_3$, $\mathbf{q}_1^*$ $\mathbf{q}_1^*$, and $\mathbf{q}_5$; i.e., odd vibration components. Differently, $\mathbf{F}_{st}$, $\mathbf{F}_2$, and $\mathbf{F}_4$ are proportional to even vibration components and $\mathbf{q}_{st}$. Additionally, the equivalent forces at the combinations resonances depend on the frequency of the diagnostic force Ω_d and the even vibration components 2Ω, 4Ω, 6Ω, and 8Ω. Similar behavior is observed considering the amplitudes at $-\Omega_d$, $-\Omega$, -2Ω, -3Ω, -4Ω, $-(2\Omega + \Omega_d)$, $-(2\Omega - \Omega_d)$, $-(4\Omega + \Omega_d)$, $-(4\Omega - \Omega_d)$, $-(6\Omega + \Omega_d)$, and $-(6\Omega - \Omega_d)$. Therefore, the odd vibration components, as well as the unbalance excitation [$\mathbf{F}_u$ in Eq. (7)], can be disregarded in the analysis of combination vibrations induced in rotating shafts affected by open cracks.

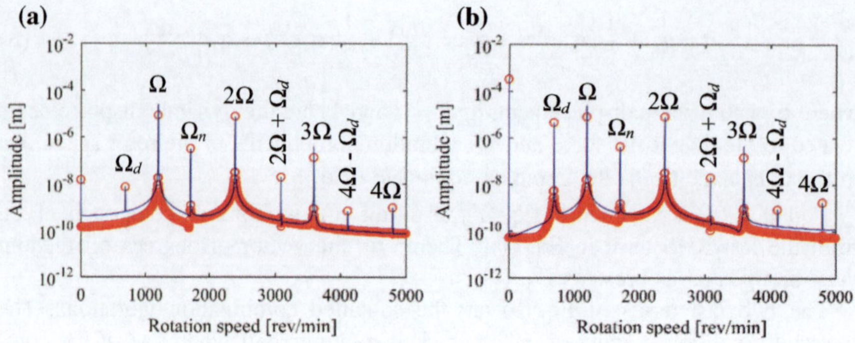

Fig. 4 Vibration responses of the rotating shaft operating at 1200 rev/min (O 100 g.mm/0°; — 300 g.mm/0°): **a** S_{28X}; **b** S_{28Z}

$$
\begin{aligned}
&\mathbf{F}_{st} = \tfrac{1}{4}\Delta\mathbf{K}_2\mathbf{q}_2^* + \tfrac{1}{4}\Delta\mathbf{K}_2^*\mathbf{q}_2 && \mathbf{F}_{2\Omega+\Omega_d} = \left(\tfrac{1}{4}\Delta\mathbf{K}_2\mathbf{q}_d + \tfrac{1}{4}\Delta\mathbf{K}_2^*\mathbf{q}_{4\Omega+\Omega_d}\right)\mathrm{e}^{i(2\Omega+\Omega_d)t} \\
&\mathbf{F}_d = \left(\tfrac{1}{4}\Delta\mathbf{K}_2\mathbf{q}_{2\Omega-\Omega_d}^* + \tfrac{1}{4}\Delta\mathbf{K}_2^*\mathbf{q}_{2\Omega+\Omega_d}\right)\mathrm{e}^{i\Omega_d t} && \mathbf{F}_{2\Omega-\Omega_d} = \left(\tfrac{1}{4}\Delta\mathbf{K}_2\mathbf{q}_d^* + \tfrac{1}{4}\Delta\mathbf{K}_2^*\mathbf{q}_{4\Omega-\Omega_d}\right)\mathrm{e}^{i(2\Omega-\Omega_d)t} \\
&\mathbf{F}_1 = \left(\tfrac{1}{4}\Delta\mathbf{K}_2\mathbf{q}_1^* + \tfrac{1}{4}\Delta\mathbf{K}_2^*\mathbf{q}_3\right)\mathrm{e}^{i\Omega t} && \mathbf{F}_{4\Omega+\Omega_d} = \left(\tfrac{1}{4}\Delta\mathbf{K}_2\mathbf{q}_{2\Omega+\Omega_d} + \tfrac{1}{4}\Delta\mathbf{K}_2^*\mathbf{q}_{6\Omega+\Omega_d}\right)\mathrm{e}^{i(4\Omega+\Omega_d)t} \\
&\mathbf{F}_2 = \left(\tfrac{1}{2}\Delta\mathbf{K}_2\mathbf{q}_{st} + \tfrac{1}{4}\Delta\mathbf{K}_2^*\mathbf{q}_4\right)\mathrm{e}^{i2\Omega t} && \mathbf{F}_{4\Omega-\Omega_d} = \left(\tfrac{1}{4}\Delta\mathbf{K}_2\mathbf{q}_{2\Omega-\Omega_d} + \tfrac{1}{4}\Delta\mathbf{K}_2^*\mathbf{q}_{6\Omega-\Omega_d}\right)\mathrm{e}^{i(4\Omega-\Omega_d)t} \\
&\mathbf{F}_3 = \left(\tfrac{1}{4}\Delta\mathbf{K}_2\mathbf{q}_1 + \tfrac{1}{4}\Delta\mathbf{K}_2^*\mathbf{q}_5\right)\mathrm{e}^{i3\Omega t} && \mathbf{F}_{6\Omega+\Omega_d} = \left(\tfrac{1}{4}\Delta\mathbf{K}_2\mathbf{q}_{4\Omega+\Omega_d} + \tfrac{1}{4}\Delta\mathbf{K}_2^*\mathbf{q}_{8\Omega+\Omega_d}\right)\mathrm{e}^{i(6\Omega+\Omega_d)t} \\
&\mathbf{F}_4 = \left(\tfrac{1}{4}\Delta\mathbf{K}_2\mathbf{q}_2 + \tfrac{1}{4}\Delta\mathbf{K}_2^*\mathbf{q}_6\right)\mathrm{e}^{i4\Omega t} && \mathbf{F}_{6\Omega-\Omega_d} = \left(\tfrac{1}{4}\Delta\mathbf{K}_2\mathbf{q}_{4\Omega-\Omega_d} + \tfrac{1}{4}\Delta\mathbf{K}_2^*\mathbf{q}_{8\Omega-\Omega_d}\right)\mathrm{e}^{i(6\Omega-\Omega_d)t}
\end{aligned}
\tag{10}
$$

Figure 4 presents the vibration responses of the cracked rotating shaft (see Fig. 1b; 50% depth crack located at the element #18) determined by using the trapezoidal rule integration scheme. In this case, two unbalance forces are applied separately to the disc D_1. The diagnostic force was applied along the X direction at the node #4 of the FE model ($\Omega_d = 2\Omega - \Omega_n = 40 - 28.5 = 11.5$ Hz; amplitude of 25 N). Note that only the vibration amplitudes at Ω and 3Ω changed according to the unbalance levels, as previously announced.

The equation of motion [see Eq.(7)] can be rewritten in a matrix form, according to the considered vibration components and combination vibrations. Equation (11) presents the problem formulated to determine the vibration responses of the cracked rotating shaft, in which $\mathbf{H}_j$ ($j = d$, 2Ω, 4Ω, 6Ω, …, $2\Omega + \Omega_d$, $2\Omega - \Omega_d$, $4\Omega + \Omega_d$, $4\Omega - \Omega_d$, etc.) is given by Eq. (12) ($\mathbf{H}_j^*$ is the complex conjugate of $\mathbf{H}_j$).

$$\begin{bmatrix} \mathbf{K}_m & \mathbf{0} & \mathbf{0} & \frac{1}{4}\Delta\mathbf{K}_2^* & \frac{1}{4}\Delta\mathbf{K}_2 & \cdots & \mathbf{0} & \mathbf{0} & \cdots \\ \mathbf{0} & \mathbf{H}_d & \mathbf{0} & \mathbf{0} & \mathbf{0} & \cdots & \frac{1}{4}\Delta\mathbf{K}_2^* & \mathbf{0} & \cdots \\ \mathbf{0} & \mathbf{0} & \mathbf{H}_d^* & \mathbf{0} & \mathbf{0} & \cdots & \mathbf{0} & \frac{1}{4}\Delta\mathbf{K}_2 & \cdots \\ \frac{1}{2}\Delta\mathbf{K}_2 & \mathbf{0} & \mathbf{0} & \mathbf{H}_2 & \mathbf{0} & \cdots & \mathbf{0} & \mathbf{0} & \cdots \\ \frac{1}{2}\Delta\mathbf{K}_2^* & \mathbf{0} & \mathbf{0} & \mathbf{0} & \mathbf{H}_2^* & \cdots & \mathbf{0} & \mathbf{0} & \cdots \\ \vdots & \vdots & \vdots & \vdots & \vdots & \ddots & \vdots & \vdots & \cdots \\ \mathbf{0} & \frac{1}{4}\Delta\mathbf{K}_2 & \mathbf{0} & \mathbf{0} & \mathbf{0} & \cdots & \mathbf{H}_{2\Omega+\Omega_d} & \mathbf{0} & \cdots \\ \mathbf{0} & \mathbf{0} & \frac{1}{4}\Delta\mathbf{K}_2^* & \mathbf{0} & \mathbf{0} & \cdots & \mathbf{0} & \mathbf{H}_{2\Omega+\Omega_d}^* & \cdots \\ \vdots & \vdots & \vdots & \vdots & \vdots & \vdots & \vdots & \vdots & \ddots \end{bmatrix} \begin{Bmatrix} \mathbf{q}_{st} \\ \mathbf{q}_d \\ \mathbf{q}_d^* \\ \mathbf{q}_2 \\ \mathbf{q}_2^* \\ \vdots \\ \mathbf{q}_{2\Omega+\Omega_d} \\ \mathbf{q}_{2\Omega+\Omega_d}^* \\ \vdots \end{Bmatrix} = \begin{Bmatrix} \mathbf{W} \\ \frac{1}{2}\mathbf{F}_{1d} \\ \frac{1}{2}\mathbf{F}_{1d} \\ 0 \\ 0 \\ \vdots \\ \mathbf{0} \\ \mathbf{0} \\ \vdots \end{Bmatrix} \tag{11}$$

$$\mathbf{H}_j = -\frac{j^2}{2}\mathbf{M} + i\frac{j}{2}\left[\mathbf{D} + \Omega\,\mathbf{D}_g\right] + \frac{1}{2}\mathbf{K}_m \tag{12}$$

The diagnostic force $\mathbf{F}_d$ is also expressed as a Fourier series. Thus,

$$\mathbf{F}_d = \mathbf{F}_{1d}\cos(\Omega_d t) = \frac{1}{2}\mathbf{F}_{1d}\left(\mathrm{e}^{i\Omega_d t} + \mathrm{e}^{-i\Omega_d t}\right) \tag{13}$$

where $\mathbf{F}_{1d}$ is the amplitude of the diagnostic force.

5 Numerical Results

Figure 5 compares the vibration responses of the rotating machine (measuring plane S_{28}) determined by using the harmonic balance approach and the trapezoidal rule integration scheme.

This analysis was performed for the rotor under two different structural conditions. The first one comprises the shaft with a crack located at the element #18 with 25% depth. The second test was performed for the shaft with a crack located at the same element with 50% depth. The operational rotation speed of the rotor Ω was fixed to 1200 rev/min and the unbalance forces were disregarded in these results. The diagnostic force was applied along the X direction in the node #4 of the FE model ($\Omega_d = 2\Omega - \Omega_n = (40 - 28.5)$ Hz $= 11.5$ Hz $= 690$ rev/min, and amplitude of 25 N). Note that the obtained vibration responses are very close, thus validating the formulation based on the harmonic balance approach. The responses determined along the plane S_8 are similar to the previous ones. It is worth mentioning that vector of degrees of freedom used in this work encompasses the following vibration components: $\mathbf{q}_{st}$, $\mathbf{q}_{\Omega d}$, $\mathbf{q}_{2\Omega}$, $\mathbf{q}_{4\Omega}$, …, $\mathbf{q}_{10\Omega}$, $\mathbf{q}_{2\Omega+\Omega d}$, $\mathbf{q}_{2\Omega-\Omega d}$, $\mathbf{q}_{4\Omega+\Omega d}$, $\mathbf{q}_{4\Omega-\Omega d}$, …, $\mathbf{q}_{10\Omega+\Omega d}$, $\mathbf{q}_{10\Omega-\Omega d}$ [see Eq. (11)].

Figures 4 and 5 showed that the amplitudes at the combination vibrations are too small (<1.0 μm), affecting the applicability of the considered dynamic phenomenon in crack detection or identification techniques [4, 12]. As mentioned, the problem consists in determining the amplitude and frequency of the diagnostic forces to

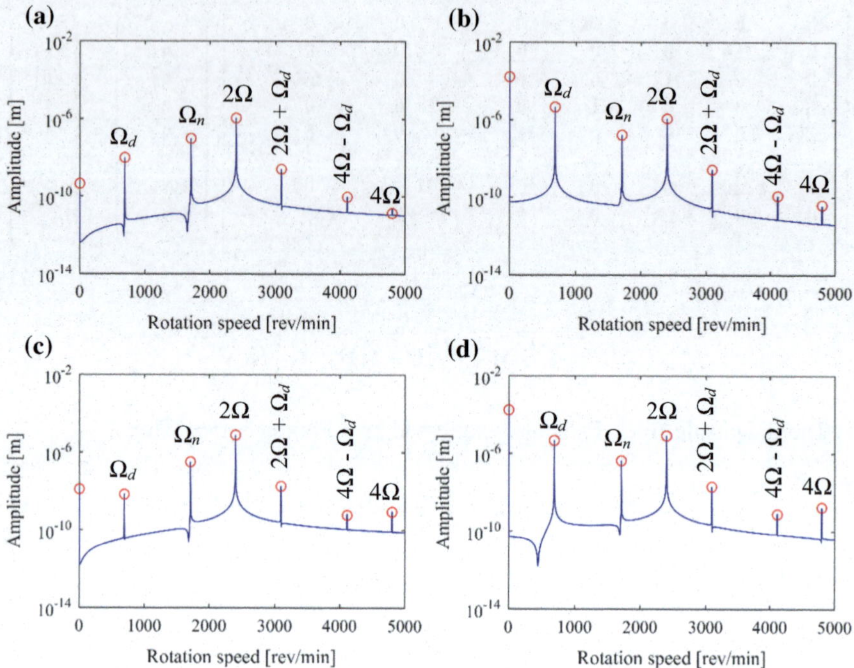

Fig. 5 Vibration responses of the cracked rotating shaft (— time integration; O harmonic balance): **a** 25% crack depth/S_{28X}; **b** 25% crack depth/S_{28Z}; **c** 50% crack depth/S_{28x}; **d** 50% crack depth/S_{28Z}

generate measurable peaks along the vibration spectrum at the combination vibrations.

Figure 6 presents the vibration responses of the rotor obtained by the sensor S_{28Z} at the frequencies $2\Omega + \Omega_d$, $2\Omega - \Omega_d$, $4\Omega + \Omega_d$, $4\Omega - \Omega_d$, $6\Omega + \Omega_d$, and $6\Omega - \Omega_d$, varying Ω_d from 0 to 85 Hz in steps of 0.1 Hz. The diagnostic forces were applied along the X direction in the node #4 of the FE model with 25 N, 50 N, and 100 N of amplitude, separately. These tests were performed for the shaft with a crack located at the element #18 with 50% depth. The operational rotation speed of the rotor Ω was fixed to 1200 rev/min. No unbalance forces are being considered. Note that measurable peaks (>1.0 μm) are obtained only at the combinations $2\Omega + \Omega_d$ and $2\Omega - \Omega_d$ (Fig. 6a and Fig. 6b, respectively). Small vibration amplitudes were obtained for the higher combination frequencies (i.e., $8\Omega + \Omega_d$, $8\Omega - \Omega_d$, $10\Omega + \Omega_d$, $10\Omega - \Omega_d$, etc.). Similar results were observed for the remaining sensors.

Figure 6a shows highest vibration amplitudes at 27.5 and 57.6 Hz. In Fig. 6b, significant vibration responses are observed at 13.0, 27.0, and 67.5 Hz. It is worth mentioning that the first five natural frequencies of the cracked rotor operating at 1200 rev/min are given by (frequencies in Hz): $26.46 \leq \Omega_{1n} \leq 26.71$,

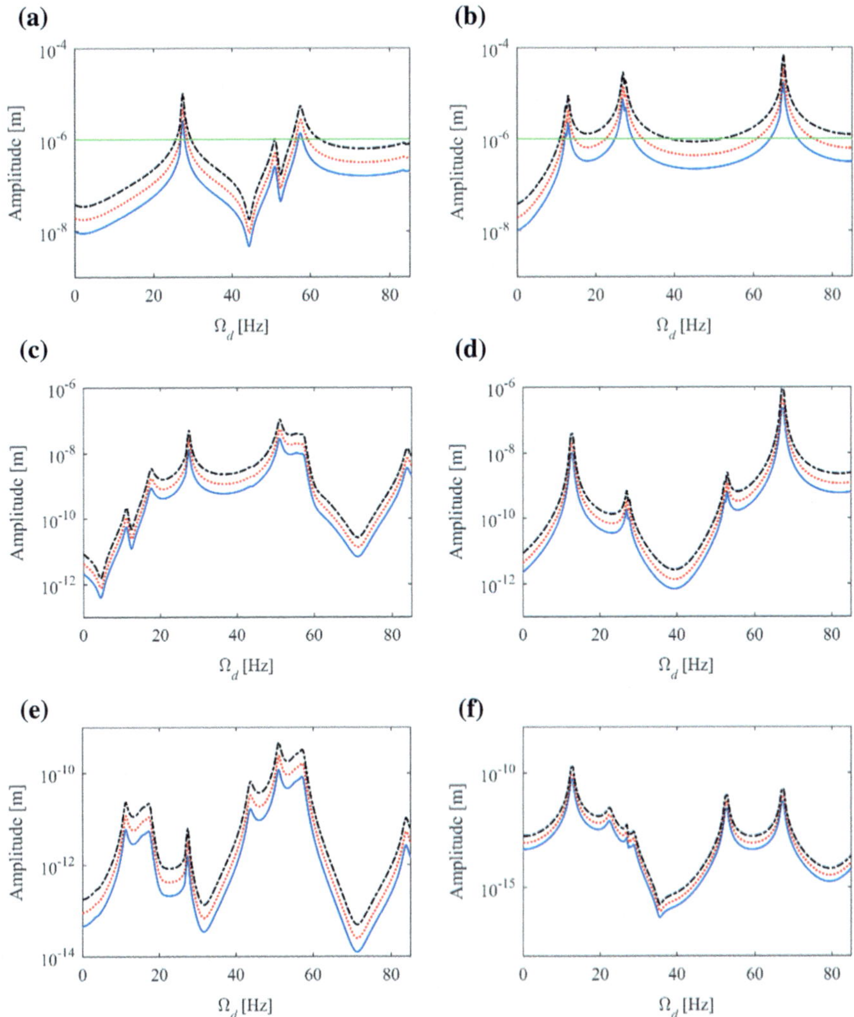

Fig. 6 Vibration responses obtained by the sensor S_{28Z} at the combination vibrations according to Ω_d (—— F_{1d} = 25 N; - - - F_{1d} = 50 N; -- - -- F_{1d} = 100 N): **a**a$2\Omega + \Omega_d$; **b** $2\Omega - \Omega_d$; **c** $4\Omega + \Omega_d$; **d** $4\Omega - \Omega_d$; **e** $6\Omega + \Omega_d$; **f** $6\Omega - \Omega_d$

$27.85 \leq \Omega_{2n} \leq 28.14$, $91.03 \leq \Omega_{3n} \leq 91.34$, $97.43 \leq \Omega_{4n} \leq 97.62$, and $123.76 \leq \Omega_{5n} \leq 123.84$ (variation according to the angular position of the shaft). A resonance condition is obtained for $\Omega_d = 27.5$ Hz (frequency close to Ω_{1n} and Ω_{2n}), as well as for $\Omega_d = 27.0$ Hz. Differently, combination vibrations are observed at $\Omega_d = 57.6$ Hz $\sim 2\Omega + \Omega_{2n}$, $\Omega_d = 13.0$ Hz $\sim 2\Omega - \Omega_{1n}$, and $\Omega_d = 67.5$ Hz $\sim 2\Omega + \Omega_{2n}$. Figure 7 presents the vibration responses of the rotor obtained by the sensor S_{28X}. The highest vibration amplitudes are obtained at 27.5, 51.2, 83.7 Hz

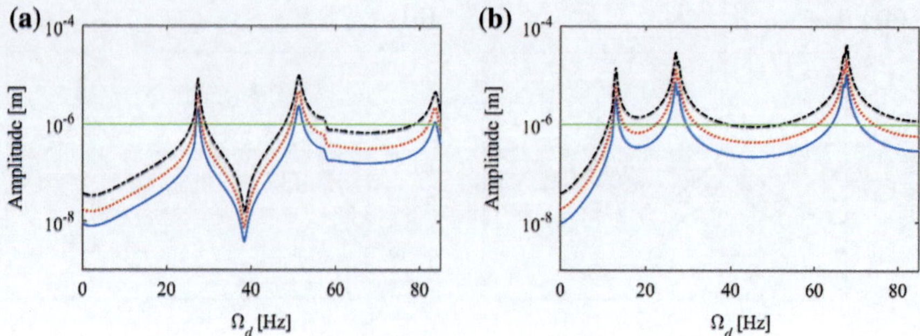

Fig. 7 Vibration responses obtained by the sensor S_{28X} at the combination vibrations according to Ω_d (— $F_{1d} = 25$ N; - - - $F_{1d} = 50$ N; -- -- -- $F_{1d} = 100$ N): **a** $2\Omega + \Omega_d$; **b** $2\Omega - \Omega_d$

(see Fig. 7a), 13.0, 27.0, and 67.6 Hz (see Fig. 7b). Note that along the *X* direction, new combination vibrations are observed at $\Omega_d = 51.2$ Hz $\sim \Omega_{3n} - 2\Omega$ and $\Omega_d = 83.7$ Hz $\sim \Omega_{5n} - 4\Omega$.

6 Conclusions

The presented results validate a promising methodology of fault detection, based on the application of a diagnostic force to detect cracks in rotating machines, coupled with a quasi-linear harmonic balancing methodology. This approach exhibits considerable computational time saving as compared with a pure integration in time; besides, it promotes better understanding about the components of the forces acting on the rotating system. Considering the case of a shaft with an open crack, it was demonstrated that the observed effect on the combination vibrations is just due to the even vibration components. Consequently, in the present case, the odd vibration components, as well as the unbalance influence have been neglected in the analysis. When it was produced a sweep of the diagnostic force frequency, it was observed the combinations $2\Omega + \Omega_d$ and $2\Omega - \Omega_d$ produced the highest measurable peaks exciting the critical speeds below 85 Hz. Finally, the evaluation of the diagnostic force level showed important contribution on the combination vibration emerging. Although, when applied to structure under operating conditions, the amplitude of the diagnostic forces must be regulated to keep the system functioning safely on an acceptable vibration level. The potential of the conveyed techniques will be explored in future research work using combination vibrations.

Acknowledgements The authors are thankful to the Brazilian Research Agencies FAPEMIG and CNPq (INCT-EIE) for the financial support provided for this research effort. The authors are also thankful to the companies CERAN, BAESA, ENERCAN, and Foz do Chapecó for the financial support through the R&D project *Robust Modeling for the Diagnosis of Defects in Generating Units* (02476-3108/2016). The first author is also grateful to his post-doc grant from CNPq (Proc. Nb. 152558/2016-0).

References

1. Bently, D.E., Hatch, C.T.: Fundamentals of Rotating Machinery Diagnostics, 1st edn. Bently Pressurized Bearing Company, USA (2002)
2. Mani, G., Quinn, D.D., Kasarda, M.: Active health monitoring in a rotating cracked shaft using active magnetic bearings as force actuators. J. Sound Vib. **1**(294), 454–465 (2006)
3. Ishida, Y., Inoue, T.: Detection of a rotor crack using a harmonic excitation and nonlinear vibration analysis. J. Vib. Acoust. **128**(1), 741–749 (2006)
4. Cavalini Jr., A.A., Sanches, L., Bachschmid, N., Steffen Jr., V.: Crack identification for rotating machines based on a nonlinear approach. Mech. Syst. Signal Process. **1**(1), 1–14 (2016)
5. Storn, R., Price, K.: Differential evolution—a simple and efficient heuristic for global optimization over continuous spaces. J. Global Optim. Kluwer Acad. Publ. **11**(1), 341–359 (1995)
6. Cavalini Jr., A.A., Lara-Molina, F.A., Sales, T.P., Koroishi, E.H., Steffen Jr., V.: Uncertainty analysis of a flexible rotor supported by fluid film bearings. Lat. Am. J. Solids Struct. **12**(8), 1487–1504 (2015)
7. Bachschmid, N., Pennacchi, P., Tanzi, E.: Cracked Rotors: A Survey on Static and Dynamic Behaviour Including Modelling and Diagnosis. Springer, Berlin, Germany (2010)
8. Lalanne, M., Ferraris, G.: Rotordynamics Prediction in Engineering, 2nd edn. Wiley, England (1998)
9. Gasch, R.: Dynamic behaviour of a simple rotor with a cross sectional crack. In: International Conference on Vibrations in Rotating Machinery, IMECHE 1976, vol.1, pp. 123–128 (1976)
10. Mayes, I.W., Davies, W.G.R.: Analysis of the response of a multi-rotor-bearing system containing a transverse crack in a rotor. J. Vib. Acoust. Stress Reliab. Des. **106**(1), 139–145 (1984)
11. Al-Shudeifat, M.A.: On the finite element modeling of the asymmetric cracked rotor. J. Sound Vib. **332**(11), 2795–2807 (2013)
12. Sawicki, J.T., Storozhev, D.L., Lekki, J.D.: Exploration of NDE properties of AMB supported rotors for structural damage detection. J. Eng. Gas Turbines Power **133**(1), 1–9 (2011)

Part II
Vibrations and Structural Dynamics

Tunable Auxiliary Mass Damper with Friction Joint: Numerical Assessment

Humberto Tronconi Coelho, Francisco Paulo Lépore Neto and Marcelo Braga dos Santos

Abstract Auxiliary Mass Dampers (AMD) are often used to reduce excessive vibration amplitude in mechanical systems. It is also known that their performances are susceptible to changes in the frequency or in the nature of the excitation force. Therefore, to improve the robustness of the AMD it is necessary to design new systems which are adaptable to the excitation, i.e., tunable devices that could be used over large frequency range. In this work a friction damper, which is the association of an elastic element and a scratcher, is used to tune the AMD by changing the normal force in the scratcher at the same time that it dissipates the mechanical energy of the principal mass. This AMD is named Tunable Auxiliary Mass Damper (TAMD). Three normal force control strategies, and two combinations of them, are studied: (i) The normal force is assumed constant; (ii) The normal force is obtained from the solution of the equation of motion assuming null displacement for the principal mass; (iii) The normal force is obtained based on the vibratory system's movement, warranting that the direction of the friction force promotes the movement of the principal mass toward its static equilibrium position. The effectiveness of the proposed TAMD is numerically evaluated based on mass and frequency ratios variations for each strategy. Therefore, a multi-degree-of-freedom (MDOF) system analysis is made in order to verify the TAMD's robustness and efficiency.

Keywords Tunable auxiliary mass damper · Variable damped absorber
Semi-active device control strategies · Semi-active friction damper

H. T. Coelho (✉) · F. P. Lépore Neto · M. B. dos Santos
Laboratory of Mechanical Systems, Faculty of Mechanical Engineering, Federal University of Uberlândia, Av. João Naves de Ávila 2121, Campus Santa Mônica, Bloco 1M, Uberlândia, Minas Gerais 38400-902, Brazil
e-mail: humbertohtc@gmail.com

A. de T. Fleury et al. (eds.), *Proceedings of DINAME 2017*, Lecture Notes in Mechanical Engineering, https://doi.org/10.1007/978-3-319-91217-2_4

1 Introduction

Auxiliary masses are frequently attached to vibrating systems by means of elastic elements and damping devices aiming to reduce the excessive vibration amplitude. Depending on the application, these auxiliary mass systems fall into one of two distinct classes. A Dynamic Vibration Absorber (DVA) that is an auxiliary mass on a compliant suspension, which has a damping factor as lower as possible, and once it is tuned to the frequency of the excitation force a system's antiresonance is introduced, reducing the primary system's vibration amplitudes. If it is necessary to provide damping, an auxiliary mass with viscous damping is attached to the structure, so that, the auxiliary mass system works as a particular type of damper. It is called Damped Absorber or Auxiliary Mass Damper (AMD) being an extension of the DVA concept [1].

AMD enables to reduce vibration amplitudes without energy consumption, within a narrow frequency band for which the AMD has been tuned. Unfortunately, when changes in the excitation's nature or in the system's parameters occur, its performance drops drastically. To improve the AMD's performance in the vibration attenuation some researchers use an active device on its suspension. These devices are active springs, made with memory shape alloys, piezo stacks and other actuators able to tune the AMD in the desired frequency [2]. Other solutions are the semi-active systems, which use friction dampers [3, 4] or magnetorheological dampers [5] to tune the frequency of the AMD system and simultaneously dissipate the mechanical energy.

Such a device has its physical parameters, as consequence also its impedance, adjustable. Associated to a suitable control law it is possible to adapt the system to a variety of excitations reducing the vibration amplitude. In this way, this system becomes a Tunable Auxiliary Mass Damper (TAMD).

In this work, a semi-active mechanism called semi-active friction damper, which is an association in series of a friction joint and an elastic element, is used in the TAMD's suspension. This semi-active mechanism has been deeply studied by Santos et al. [6] due its capacity to dissipate the mechanical energy of a vibratory system in the friction joint.

This approach is interesting since the energy necessary to tune the TAMD is much less than the energy necessary to achieve the same attenuation using active actuators, once for the active systems the energy is expended to work against the excitation force.

This work aims to evaluate the proposed TAMD's adaptability that is obtained by controlling the normal force of a smart friction damper. The numerical study demonstrates the effectiveness of the developed strategies and indicates the mass and frequency ratios to be used. These chosen ratios are used in a multi-degree-of-freedom (MDOF) system analysis in order to verify the TAMD's robustness and its vibration attenuation capacities.

2 Theoretical Approach

The TAMD has been initially applied to a one-degree-of-freedom (DOF) vibratory system as shown on Coelho et al. [7] and Guerineau et al. [8], the schema of the studied system is shown in Fig. 1a. It is a two-DOF system that can be modeled as a one-DOF linear vibratory system (m_1, in black) with an AMD coupled to it using a semi-active friction damper in its suspension (m_a, in red).

The adaptability is achieved using a semi-active friction damper (Fig. 1b). This damper is an association in series of an elastic element and a scratcher which will tune the TAMD. This system enables to dissipate the mechanical energy of the principal mass at same time that, by changing the normal force on the scratcher, the apparent damping coefficient and stiffness of the TAMD's suspension are modulated [6]. The force between nodes (1) and (3), indicated in Fig. 1b, is F_{13} and can be written as presented in Eq. (1) [9].

$$F_f = F_{13} = \begin{cases} k_T(x_3 - x_1) & if\ |k_T(x_3 - x_1)| \leq \mu N \\ \mu N & otherwise \end{cases} \tag{1}$$

Points (1) and (3) from friction damper, as shown in Fig. 1b, are attached to mass m_1 and m_a of the vibratory system, respectively. The TAMD's suspension is composed by the stiffness k_a and the damping c_a, as linear elements, and the nonlinear component characterized by the tangential stiffness k_T and the scratcher which has its force as defined in Eq. (1). The suspension between m_1 and the inertial frame is composed by the stiffness k_1 and the damping c_1. The equation of motion for the entire system becomes:

$$\begin{bmatrix} m_1 & 0 \\ 0 & m_a \end{bmatrix} \begin{Bmatrix} \ddot{x}_1 \\ \ddot{x}_a \end{Bmatrix} + \begin{bmatrix} c_1 + c_a & -c_a \\ -c_a & c_a \end{bmatrix} \begin{Bmatrix} \dot{x}_1 \\ \dot{x}_a \end{Bmatrix} + \begin{bmatrix} k_1 + k_a & -k_a \\ -k_a & k_a \end{bmatrix} \begin{Bmatrix} x_1 \\ x_a \end{Bmatrix} = \begin{bmatrix} 1 \\ -1 \end{bmatrix} F_f + \begin{bmatrix} 1 \\ 0 \end{bmatrix} F_{exc} \tag{2}$$

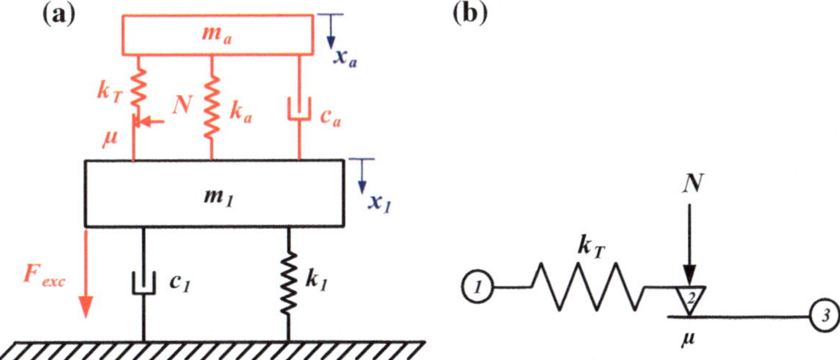

Fig. 1 **a** Schema of the studied system and **b** semi-active friction damper model

The equation of motion has been integrated using the methodology proposed by Lu et al. [3] which uses the state space formulation for the linear system and dispose the nonlinear force from friction damper as part of the excitation forces.

Knowing that mass (m_a/m_1) and frequency (ω_a/ω_1) ratios affect DVA's and AMD's behavior, the numerical assessment aimed to determine the best ones to be used for the future experimental workbench, under design. As mentioned before TAMD works associated normal force control law. Five control laws are used in an attempt to minimize the vibration amplitude of the principal mass by directing the energy to the TAMD and use the relative displacement promoted in the friction damper to dissipate the mechanical energy. All control laws aim to produce an efficient energy sink.

There are three main normal force control strategies, and two combinations of them: (i) The normal force is assumed constant [S1]; (ii) The normal force is obtained, from Eq. (3), which is the solution of the equation of motion assuming null displacement for the principal mass, i.e. $x_1 = 0$ [S2]; (iii) When F_f induces a movement of m_1 towards its static equilibrium position the normal force is stemming from Eq. (4), otherwise it is null [S3]. The strategy [S4] uses the same logic described for [S3], but now the normal force is calculated using Eq. (3). And [S5] also uses the logic developed for [S3], however, as for [S1], it uses a constant value for the normal force N_1.

$$N_1 = \frac{|m_a\ddot{x}_a + c_a\dot{x}_a + k_a x_a|}{\mu} \tag{3}$$

$$N_1 = \frac{k_T(x_a - x_1)}{\mu} \tag{4}$$

Table 1 summarizes the control strategies operation.

To compare all control methodologies, it is necessary to establish a criterion that enables to show the reduction of the resonance amplitude peak simultaneously to the reduction of the amplitude of the receptance over the entire interested frequency

Table 1 Control strategies operation

Control strategy	Condition	Normal force
S1	Applied all the time	$N = N_1$ (a constant value)
S2		$N = N_1$ from Eq. (3)
S3	Conditional applicability (applied when F_f induces a movement of m_1 towards its static equilibrium position)	$N = \begin{cases} N_1 \text{ from Eq. (4)} \\ 0 \end{cases}$
S4		$N = \begin{cases} N_1 \text{ from Eq. (3)} \\ 0 \end{cases}$
S5		$N = \begin{cases} N_1 \text{(a constant value)} \\ 0 \end{cases}$

band. It is clear from the literature that DVA split the original resonant peak in two new resonant peaks, which, can be disastrous to the vibratory system if the excitation force contains harmonics with these new resonant frequencies. Normally, two parameters are used simultaneously to describe the performance of the control systems: Maximum Value (L_∞ *norm*) and the L_2 *norm*. The former indicates the maximum amplitude expected for the system response and the second the overall mean value of the response. A similar statement has been used in the control technique H_∞/H_2 [10]. Therefore, in this work the performance parameter $\left(P_p\right)$ is defined as the ratio $(L_\infty\,norm)/(L_2\,norm)$ where the best control law performance will provide the lowest values for both norms. In this work the performance parameter $\left(P_p\right)$ is defined as:

$$P_p = \frac{maximum\ receptance\ amplitude}{receptance\ amplitude\ norm} \tag{5}$$

Assuming a column vector $A = [\varepsilon\,\varepsilon\,\varepsilon \cdots A_1 \cdots \varepsilon\,\varepsilon\,\varepsilon]_{1\times N}$, where ε is a real constant closest to zero and A_1 a real positive constant, which represents the receptance with one peak only on the frequency spectrum $A = A_1\delta(f_1)$, where $\delta(f_1)$ is the Dirac function on the frequency f_1, the parameter P_p can be written as:

$$P_p = \lim_{\varepsilon\to 0} P_p = \lim_{\varepsilon\to 0} \frac{A_1}{\sqrt{(N-1)\varepsilon^2 + A_1^2}} = \frac{A_1}{A_1} = 1 \tag{6}$$

The other extreme is a constant amplitude receptance, which are represented by the column vector $A = [\varepsilon\,\varepsilon\,\varepsilon \cdots \varepsilon\,\varepsilon\,\varepsilon]_{1\times N}$ for which the parameter P_p is written as:

$$P_p = \lim_{\varepsilon\to 0} P_p = \lim_{\varepsilon\to 0} \frac{\varepsilon}{\sqrt{N\varepsilon^2}} = \frac{1}{\sqrt{N}} \tag{7}$$

Equations (6) and (7) give the maximum and minimum values of the performance parameter as defined in Eq. (5).

3 Numerical Results for a Single-Degree-of-Freedom System

To obtain a reasonable comparison among the obtained results and those of the literature, mass and frequency ratios are the same described by Harris and Piersol [1]. The effectiveness of the proposed TAMD, where the adaptability is obtained by controlling the normal force on the smart friction damper, is evaluated based on mass and frequency ratios variations for each strategy. In this work, the mass ratios studied are $m_a/m_1 = [0.1, 0.2, 0.3, 0.4, 0.5]$ and for the frequency are $\omega_a/\omega_1 = [0.1, 0.5, 1]$. These ratios with the five control strategies are compared to the correspondent optimum viscous damping AMD, to a well-tuned DVA and to the 1 DOF uncontrolled vibration response.

The numerical results presented in this section have been obtained using the following physical properties values, which represent the parameters from a designed modification of the experimental workbench used on previously works [7, 8], $m_1 = 4.14\,\mathrm{kg}, c_1 = 3.93\,\mathrm{N\,s/m}$ and $k_1 = 70.3\,\mathrm{kN/m}$. The physical properties for the secondary system ($m_\mathrm{a}, c_\mathrm{a}$ and k_a) are deduced from the mass and frequency ratios aforementioned. The contact parameters are the tangential stiffness $k_T = 1.16\,\mathrm{MN/m}$ and the friction coefficient $\mu = 0.33$. The auxiliary damping c_a used in the simulations is fixed and equal to $1.0\,\mathrm{N\,s/m}$, a small damping to ensure that most of the damping promoted by the TAMD becomes from the friction damper. And the optimum viscous damping for the AMD is $c_{opt} = 18.1\,\mathrm{N\,s/m}$ obtained following Den Hartog's procedure as Harris and Piersol [1].

All receptances have been obtained using a harmonic force excitation with an amplitude of 10 N. The excitation force frequency has been swept from 5 Hz up to 100 Hz, in steps of 0.1 Hz. For strategies which use constant normal force value, $N_1 = 20\,\mathrm{N}$ has been applied. Again, these values come from previous tests, which also had determined that the ratio $N_1/F_{exc} = 2$ is the best to be used for constant normal force [7, 8].

The responses for all combinations of control strategies, mass and frequency ratios are placed in Fig. 2. There are 15 combinations for each strategy, only the best of each is indicated as a filled circle. Additionally, for comparison reasons, the performance parameter for 1 DOF uncontrolled vibration system, the vibratory system coupled to AMD with optimum viscous damping and the system coupled to a well-tuned DVA are also shown in Fig. 2 as a filled circle.

In Fig. 2 the color bar indicates the value of P_p and the arrows indicate the location of the best result for each control strategy, also for the location of the DVA,

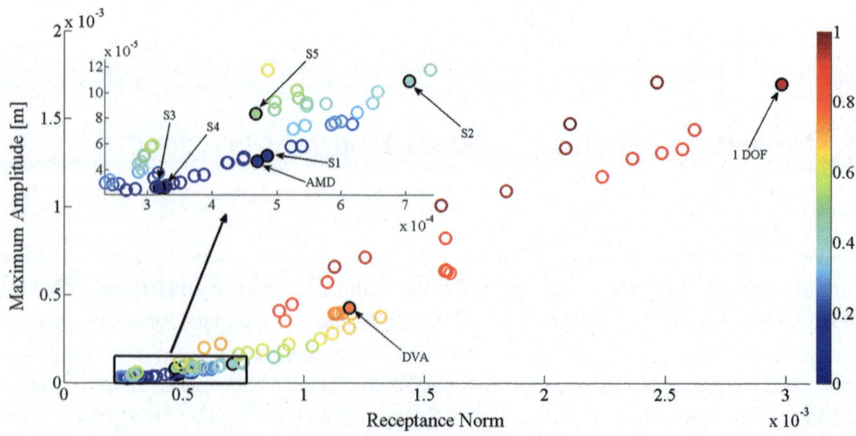

Fig. 2 Maximum amplitude and receptance norm chart

optimum viscous damping AMD and the 1DOF uncontrolled vibration. The last one refers to the system composed only by m_1, k_1 and c_1.

It is possible to observe that the performance parameter is efficient in locate the responses with the lowest norms and smaller maximum amplitudes, i.e., that are located in the lower left corner in the figure. At the enlargement is noted that the best responses from strategies S3 and S4 are better than the other approaches and much better than the DVA and the 1 DOF uncontrolled vibration response. It is also possible to see that the optimum viscous damping AMD is nowadays the most efficient passive approach.

In Fig. 3 the responses for the best of each control strategy are presented and compared with the best optimum viscous damping AMD result, the well-tuned DVA and the 1 DOF uncontrolled vibration response.

The best receptances are for [S3] and [S4], they present, at the same time, the lowest peaks and the lowest norm. Based on these receptance results, it can be concluded that the proposed semi-active suspension almost entirely suppressed the resonance peaks. Should be also observed that the receptances for [S3] and [S4] had almost their entirely values at the same levels or lower than the static response, which is a great advantage for the proposed TAMD.

These results demonstrate that the TAMD can be effective in a wide frequency range, since all strategies promote an improvement in the attenuation of the resonant peaks as well as in the $L_2\,norm$ value. They are also promoting a better response than that obtained by optimal viscous damped AMD. Strategies [S1], [S2] and [S5] also present good receptances, almost as good as the optimum damping AMD, their highest values for P_p are due to the maximum amplitude of the receptance, which are a little higher than [S3] and [S4] maximum amplitude. It should be noticed that all strategies give better results than the well-tuned DVA.

Table 2 summarizes the mass and frequency ratio combinations on which the lowest values for P_p for each strategy were obtained. This table also presents the

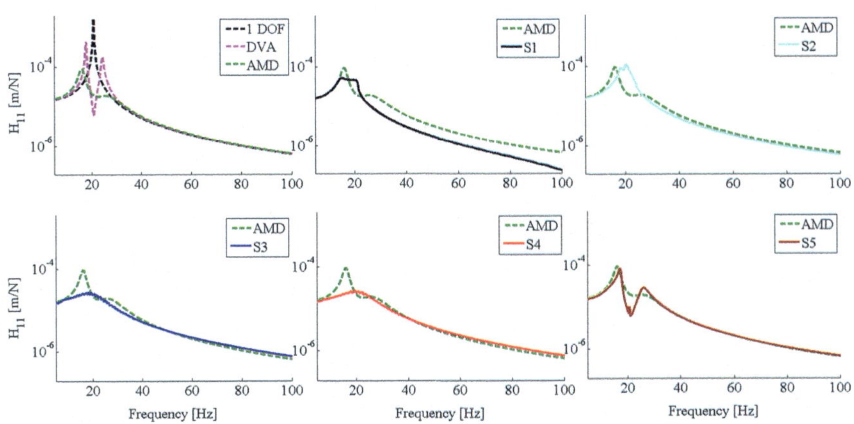

Fig. 3 Receptance results for the best of each strategy

Table 2 Ratios combination and P_p value for the bests of each strategy

Control strategy	Ratios combination		P_p
	ω_a/ω_1	m_a/m_1	
1 DOF	(–)	(–)	0.57
DVA	1.0	0.1	0.36
AMD	0.5	0.5	0.09
Strategy S1	0.5	0.5	0.11
Strategy S2	1.0	0.1	0.15
Strategy S3	0.1	0.5	0.08
Strategy S4	0.1	0.5	0.08
Strategy S5	1.0	0.1	0.18

ratios used to tune the DVA, those that presented the best optimum viscous damping AMD response and their respective performance parameter value. The 1 DOF uncontrolled vibration also had its receptance quantified by using the performance parameter P_p.

As observed, strategies [S3] and [S4] presented excellent results, however with mass ratio $m_a/m_1 = 0.5$, which is too much mass to be added. Strategies [S2] and [S5] have their best performance for mass ratio $m_a/m_1 = 0.1$ and frequency ratio $\omega_a/\omega_1 = 1$. So, for this ratio combination the P_p values for the optimum viscous damping AMD, [S1], [S3] and [S4] were $0.15, 0.17, 0.11$ and 0.12 respectively, which are excellent values when compared to the mass reduction that was possible to achieved. Note that strategies [S3] and [S4] performance parameter values remain smaller than for [S2] and [S5] presented on Table 2.

Here it becomes clear that there is a compromise solution between the mass ratio and the performance parameter. Changing the mass ratio for 10% and recalculating again the receptances it is possible to verify that strategies [S3] and [S4] remain the better ones, as shown in Fig. 4.

The symbol (*) in the legend indicates the optimum viscous damping AMD receptance previously obtained with a mass ratio $m_a/m_1 = 0.5$. Besides the good aspect of the receptances, presented in the Fig. 4, they are little worse than those presented in Fig. 3. Concerning the compromise solution between the mass ratio and the performance of the proposed TAMD, the worsening of the receptances is justified by the great reduction in the mass to be added in the system. Besides, strategies [S3] and [S4] receptance remains better than optimum viscous damping AMD in almost one order of magnitude attenuation for the receptance maximum value on the chosen ratios.

The natural frequency of the TAMD is 20.73 Hz, which is close to the resonant frequency of the original vibratory system. The TAMD acts similarly to a well-tuned DVA; however, introducing two new resonant peaks less significative than as is expected in the application of DVAs. The discontinuity in the receptance shows that the selected physical properties for the auxiliary system makes more difficult for the TAMD to work against the resonance frequency and maintain lower amplitude levels.

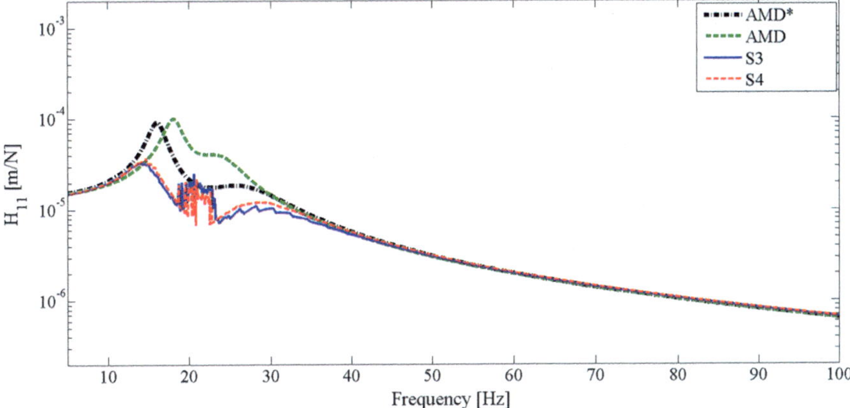

Fig. 4 Receptance results for $\boldsymbol{m_a/m_1 = 0.1}$ and $\boldsymbol{\omega_a/\omega_1 = 1}$. Optimum viscous damping AMD response with a mass ratio $\boldsymbol{m_a/m_1 = 0.5}$ (AMD*), optimum viscous damping AMD response (AMD), TAMD response using strategy S3 (S3) and TAMD response using strategy S4 (S4)

To evaluate the robustness of the TAMD with the chosen mass and frequency ratios, a numerical study with other types of excitation force for strategies [S3] and [S4] were also performed. These are clearly better than strategies [S1], [S2] and [S5]. Figure 5 presents the system's time response of m_1 displacement to a 10 N impact excitation applied at 0.5 s to the 2 DOF presented in Fig. 1a.

It can be noted that strategies [S3] and [S4] presents the lowest settling time showing an excellent damping capability. They show better responses than

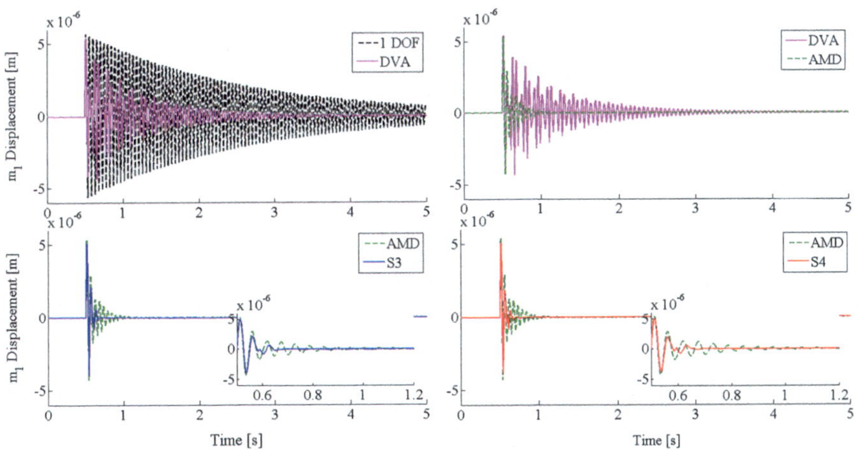

Fig. 5 Time response to an impact excitation for $\boldsymbol{m_a/m_1 = 0.1}$ and $\boldsymbol{\omega_a/\omega_1 = 1}$. Uncontrolled vibration response of without auxiliary mass (1 DOF), well-tuned DVA response (DVA), optimum viscous damping AMD response (AMD), TAMD response using strategy S3 (S3) and TAMD response using strategy S4 (S4)

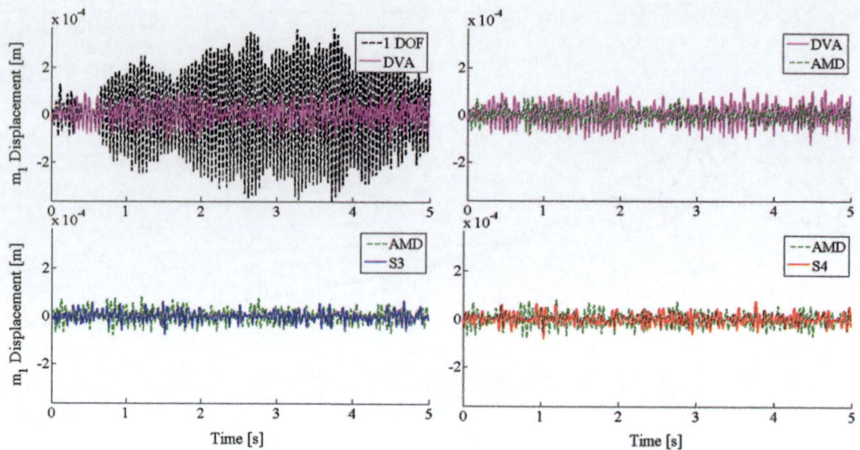

Fig. 6 Time response to a random excitation for $\boldsymbol{m_a/m_1 = 0.1}$ and $\boldsymbol{\omega_a/\omega_1 = 1}$. Uncontrolled vibration response of without auxiliary mass (1 DOF), well-tuned DVA response (DVA), optimum viscous damping AMD response (AMD), TAMD response using strategy S3 (S3) and TAMD response using strategy S4 (S4)

optimum viscous damping AMD, which in turn is better than the DVA's response. Figure 6 presents the time response to a random excitation varying up to 10 N.

Normally it is too difficult to deal with random excitations due its nature, especially for semi-active systems, which try to tune the vibratory system to the excitation force, maximizing the performance of the TAMD. It can be noted that the DVA presents a better response than 1 DOF system's uncontrolled vibration, with RMS amplitude of 45.1 μm and 162.2 μm, respectively. A good improvement is obtained with optimal viscous damped AMD response with RMS amplitude of 27.2 μm and even better responses can be observed for the results obtained with [S3] and [S4]. The RMS values for these responses are 19.8 μm and 21.2 μm, respectively for [S3] and [S4], which are less than half of the DVA's RMS and eight times smaller than the 1 DOF system's uncontrolled vibration RMS.

The time response of each strategy and DVA for chirp excitation with its frequency sweeping from 5 Hz up to 100 Hz, changing in a ratio of 19 Hz/s, are presented in Fig. 7. The 1 DOF response presents a maximum amplitude of 2.28 mm near to 1.0 s, which is bigger than DVA's response with 1.48 mm around 1.5 s. The optimal viscous damped AMD maximum amplitude is 0.82 mm. The strategies [S3] and [S4] presents better results than AMD's response with the lowest amplitudes, in which the passage through resonance is almost imperceptible, with maximum amplitudes of 0.37 mm and 0.39 mm, respectively. This last fact could be useful for applications in rotating machines, permitting a smoother passage through critical speeds.

Therefore, in this numerical study of the TAMD it was also verified that the proposed control strategies' efficiencies are independent of the excitation force nature. It was also verified that strategies [S3] and [S4] present better results than

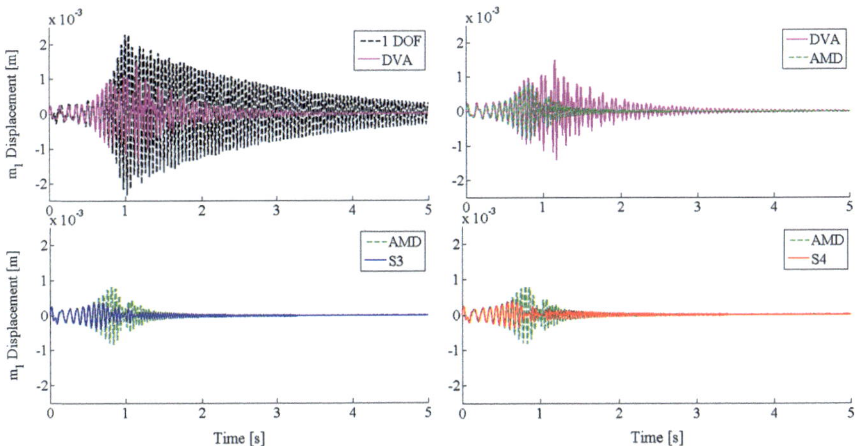

Fig. 7 Time response to a chirp excitation for $\boldsymbol{m_a/m_1 = 0.1}$ and $\boldsymbol{\omega_a/\omega_1 = 1}$. Uncontrolled vibration response of without auxiliary mass (1 DOF), well-tuned DVA response (DVA), optimum viscous damping AMD response (AMD), TAMD response using strategy S3 (S3) and TAMD response using strategy S4 (S4)

the optimal viscous damped AMD, which is nowadays some of the best design alternatives for this vibration attenuation approach.

Since they make use of friction dampers, strategies [S3] and [S4] can be employed in heavy and big structures applications, even under low velocities of vibration, were viscous dampers are inefficient or unfeasible due to their size.

4 Numerical Results for a Multiple-Degree-of-Freedom System

In their work Santos et al. [6] have used a semi-active friction damper between the inertial frame and the first DOF of a three-DOF vibratory system. In this section, the same vibratory system is used to analyze the system's behavior when the proposed TAMD is placed above the first DOF as presented in Fig. 8. It is a four-DOF system that can be modeled as a three-DOF linear vibratory system (composed by m_1, m_2 and m_3 in black) with the TAMD (m_a, in red) coupled to m_1.

This assembly aims to dissipate the energy introduced by the work done by F_{exc} on mass m_1, preventing the DOFs to receive too much energy; therefore, reducing the vibration amplitude levels. In this way, the equation of motion for the entire four-DOF system becomes as Eq. (8).

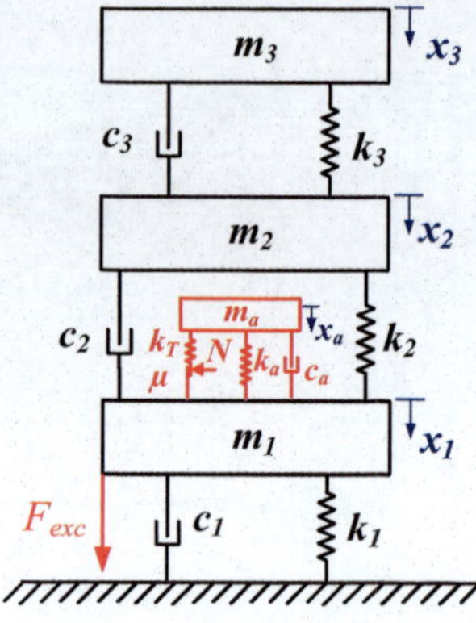

Fig. 8 Schema of the studied three-DOF vibratory system with the TAMD

$$
\begin{bmatrix} m_1 & 0 & 0 & 0 \\ 0 & m_a & 0 & 0 \\ 0 & 0 & m_2 & 0 \\ 0 & 0 & 0 & m_3 \end{bmatrix} \begin{Bmatrix} \ddot{x}_1 \\ \ddot{x}_a \\ \ddot{x}_2 \\ \ddot{x}_3 \end{Bmatrix} \begin{bmatrix} c_1 + c_a + c_2 & -c_a & -c_2 & 0 \\ -c_a & c_a & 0 & 0 \\ -c_2 & 0 & c_2 + c_3 & -c_3 \\ 0 & 0 & -c_3 & c_3 \end{bmatrix} \begin{Bmatrix} \dot{x}_1 \\ \dot{x}_a \\ \dot{x}_2 \\ \dot{x}_3 \end{Bmatrix}
$$
$$
+ \begin{bmatrix} k_1 + k_a + k_2 & -k_a & -k_2 & 0 \\ -k_a & k_a & 0 & 0 \\ -k_2 & 0 & k_2 + k_3 & -k_3 \\ 0 & 0 & -k_3 & k_3 \end{bmatrix} \begin{Bmatrix} x_1 \\ x_a \\ x_2 \\ x_3 \end{Bmatrix} = \begin{bmatrix} 1 \\ -1 \\ 0 \\ 0 \end{bmatrix} F_f + \begin{bmatrix} 1 \\ 0 \\ 0 \\ 0 \end{bmatrix} F_{exc} \tag{8}
$$

The complementary physical properties values are presented in Table 3. The physical properties for the auxiliary system (m_a and k_a) are deduced from the aforementioned mass and frequency ratios, i.e., $m_a = 0.41$ kg, $k_a = 7.03$ kN/m and $c_a = 1$ N s/m. The contact parameters remain the same, the tangential stiffness is $k_T = 1.16$ MN/m and the friction coefficient is $\mu = 0.33$.

Figure 9 show the receptances obtained as in previous section, however now with excitation force frequency sweeping from 5 Hz up to 50 Hz, also in steps of 0.1 Hz. For the multi degree of freedom case, only results obtained with the strategies S3 and S4 are presented. Once again, these were the best strategies to modulate the normal force in a TAMD.

As can be observed in Fig. 9, the use of the proposed TAMD and strategies promote excellent results suppressing the second resonant peak of the MDOF vibratory system. This happens due to the natural frequency of the TAMD, which is 20.73 Hz, is close to one of the resonant frequencies of the original MDOF

Table 3 Complementary physical properties values of the three-DOF system

Physical properties	1st DOF	2nd DOF	3rd DOF
Mass (kg)	$m_1 = 4.14$	$m_2 = 1.97$	$m_3 = 0.93$
Stiffness (kN/m)	$k_1 = 70.3$	$k_2 = 9.67$	$k_3 = 24.44$
Damping (N s/m)	$c_1 = 3.93$	$c_2 = 2.35$	$c_3 = 0.99$

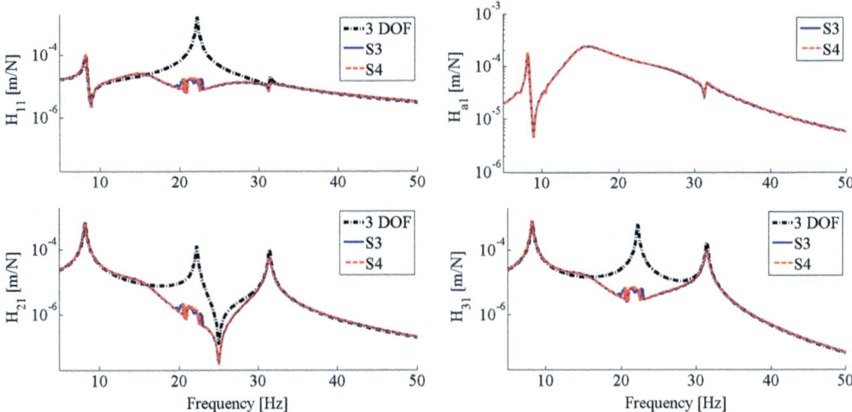

Fig. 9 Receptance results for the three-DOF system with $\boldsymbol{m_a/m_1 = 0.1}$ and $\boldsymbol{\omega_a/\omega_1 = 1}$. Uncontrolled vibration response (3 DOF), TAMD response using strategy S3 (S3) and TAMD response using strategy S4 (S4)

vibratory system. Acting similarly to a well-tuned DVA, however, without introducing two new resonant peaks as is expected in the application of DVAs. Over again the discontinuity in the receptance shows the TAMD working against the resonance frequency to maintain lower amplitude levels.

The semi-active friction damper is an energy sink in the proposed TAMD, dissipating the energy in the frequency range for which the TAMD has been previously adjusted.

The proposed TAMD do not present a significant attenuation of the resonant peaks around 8 and 32 Hz. This occurs due to the TAMD's natural frequency, as mentioned before. Someone can design the parameters of the TAMD using an optimization procedure aiming to widen the frequency band in which the TAMD is efficient.

To evaluate the robustness of the TAMD in a MDOF vibratory system, when it is positioned at m_1, the system was subjected to other types of excitation force. Figure 10 presents the envelope of the absolute time response to a 10 N impact force applied at 0.5 s on m_1.

It can be noted in Fig. 10 that the TAMD placed on m_1 can be extremely effective. The amplitude of the mass m_1 has been reduced from 11 to 9 μm, the vibration damping was impressive, but it was reduced to insignificant levels in fractions of second. The RMS amplitude levels were reduced from 3.4 to 0.6 μm. The proposed TAMD can also promote a good attenuation in the displacement of m_3 with the maximum amplitude coming from 10 to 9 μm and RMS amplitude levels from 2.2 to 1.7 μm. Also, a little vibration suppression was observed for m_2 with the maximum amplitude value coming from 6.1 to 5.9 μm and RMS levels from 1.58 to 1.53 μm.

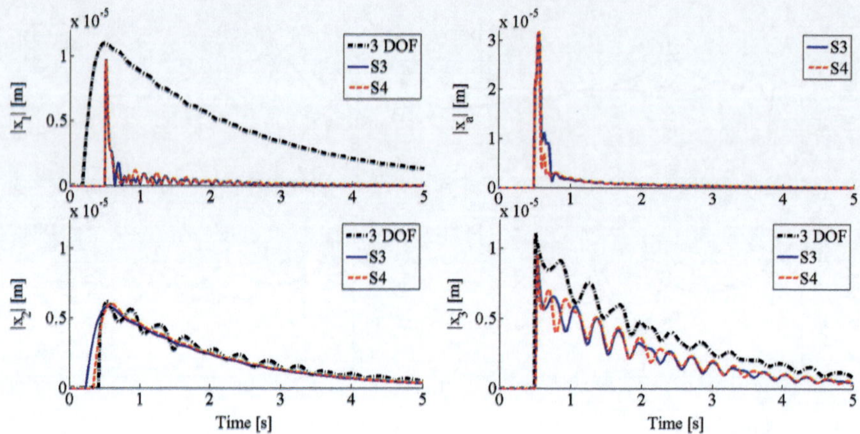

Fig. 10 Envelope of the absolute time response to an impact excitation for the three-DOF system with $m_a/m_1 = 0.1$ and $\omega_a/\omega_1 = 1$. Uncontrolled vibration response (3 DOF), TAMD response using strategy S3 (S3) and TAMD response using strategy S4 (S4)

Figure 11 presents the envelope of the absolute time response to a random excitation varying up to 10 N. It can be observed that the proposed TAMD achieves again an excellent performance for m_1, with the RMS value coming from 0.12 to 0.02 mm. A reasonably attenuation of the displacement of m_3 is also observed, with RMS level reducing from 0.06 to 0.05 mm and maintaining the RMS levels of m_2.

The envelope of the absolute time responses of each strategy to a chirp excitation with its frequency sweeping from 5 Hz up to 50 Hz, changing in a ratio of 9 Hz/s,

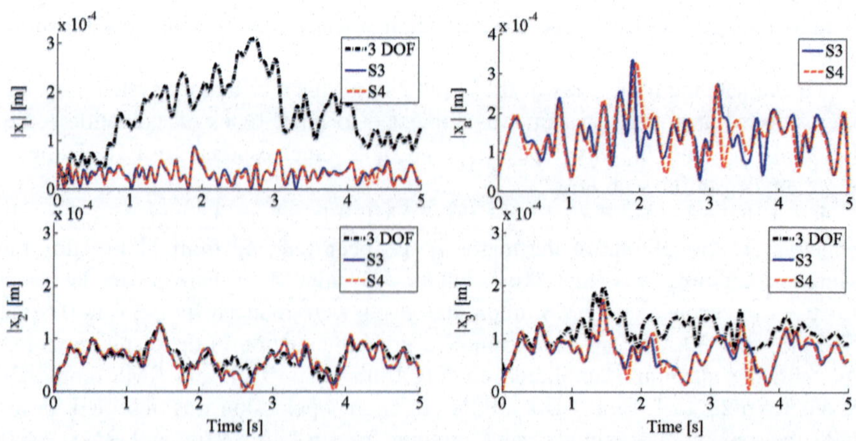

Fig. 11 Envelope of the absolute time response to a random excitation for the three-DOF system with $m_a/m_1 = 0.1$ and $\omega_a/\omega_1 = 1$. Uncontrolled vibration response (3 DOF), TAMD response using strategy S3 (S3) and TAMD response using strategy S4 (S4)

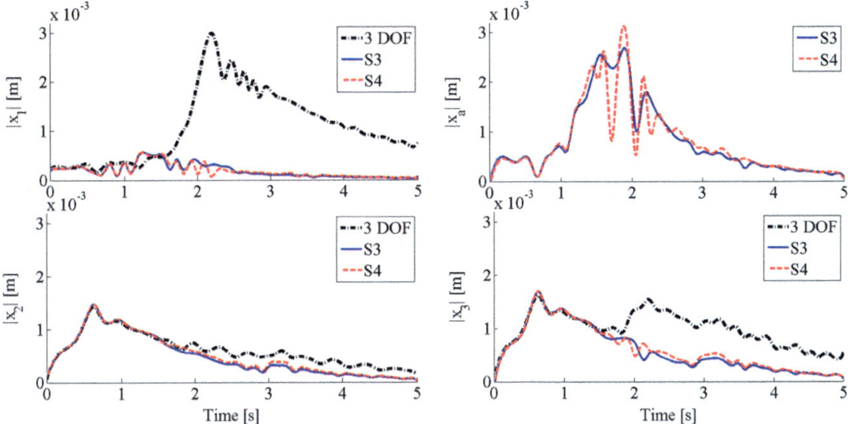

Fig. 12 Envelope of the absolute time response to a chirp excitation for the three-DOF system with $\boldsymbol{m_a/m_1 = 0.1}$ and $\boldsymbol{\omega_a/\omega_1 = 1}$. Uncontrolled vibration response (3 DOF), TAMD response using strategy S3 (S3) and TAMD response using strategy S4 (S4)

are presented in Fig. 12. Over again the efficiency over x_1 is excellent, m_1 is almost insensitive to the passage through the resonances, the maximum displacement value is reduced from 3.0 to 0.5 mm, and the RMS level is reduced from 0.89 to 0.13 mm, using the proposed TAMD. Despite the maximum displacement value was not significantly reduced for m_3, the TAMD is effective, especially after 2.0 s, as it reduces the RMS value from 0.60 to 0.48 mm. And for m_2, although the maximum displacement value was maintained the same, a small attenuation of the RMS levels, from 0.44 to 0.42 mm, is achieved.

Although the proposed TAMD could not prevent the energy to achieve the others DOF, it was extremely effective in attenuating the vibration amplitude from the DOF in which it was attached. The TAMD removes a significative amount of energy from m_1, but not enough to present the same efficient in the others DOF.

5 Conclusions

A performance parameter has been defined and was shown to be effective as a metric for optimization procedures, identifying the receptance with the lowest peak and the lowest L_2 *norm* over the analyzed frequency band. Through the presented numerical simulations, it is possible to state that the proposed TAMD model presents better results than the traditional DVA.

TAMD's selected strategies are capable of achieving excellent performance, with low mass ratios, for different kinds of excitation. For the 1 DOF vibratory systems, the proposed TAMD presented results even better than the optimal viscous damped AMD, which is nowadays one of the best design passive alternatives for vibration attenuation. Results presented herein suggest that the TAMD is able to be used in a wide range of applications, and the fact that it requires only a 10% mass addition means they can be used on embedded systems with weight restriction.

Great results were also obtained for a MDOF vibratory system where the vibration suppression achieved in the receptances clarifies the device's and strategies' potential to be effective in a wide frequency range and demonstrate its capacity to work in cases of different types of excitations, especially for the DOF in which the TAMD is attached.

Future works will investigate positioning of the TAMD in a MDOF vibratory system, intending to make the TAMD effective for all DOFs and vibration modes.

Acknowledgements The authors are grateful to the agencies and bureaus which have supported this research project: Capes, Fapemig, CNPq and the Federal University of Uberlândia, by means of its Mechanical Engineering Postgraduate Program.

References

1. Harris, C.M., Piersol, A.G.: Harris' Shock and Vibration Handbook, 5th ed. McGraw-Hill Handbooks (2002) (Chapter 6)
2. Chatterjee, S.: Optimal active absorber with internal state feedback for controlling resonant and transient vibration. J. Sound Vib. **329**, 5397–5414 (2010). https://doi.org/10.1016/j.jsv.2010.07.017
3. Lu, L.-Y., Chung, L.-L., Wu, L.Y. and Lin, G.-L.: Dynamic analysis of structures with friction devices using discrete-time state-space formulation. Comput. Struct. (Taiwan) **84**, 1049–1071 (2006). https://doi.org/10.1016/j.compstruc.2005.12.005
4. Lin, C.-C., Lin, G.-L., Wang, J.-F.: Protection of seismic structures using semi-active friction TMD. Earthq. Eng. Struct. Dyn. **39**, 635–659 (2010). https://doi.org/10.1002/eqe.961
5. Weber, F.: Semi-active vibration absorber based on real-time controlled MR damper. Mech. Syst. Signal Process. **46**, 272–288 (2014). https://doi.org/10.1016/j.ymssp.2014.01.017
6. Santos, M.B., Coelho, H.T., Lepore Neto, F.P., Mahfoud, J.: Assessment of semi-active friction dampers. Mech. Syst. Signal Process. **94**, 33–56 (2017). https://doi.org/10.1016/j.ymssp.2017.02.034
7. Coelho, H.T., Lepore Neto, F.P., Santos, M.B.: Numerical Assessment of a Tunable Auxiliary Mass Damper Using a Friction Damper. Society for Experimental Mechanics—IMAC-XXXIV—Dynamics of Multiphysical Systems: From Active Materials to Vibroacoustics (2016)
8. Guerineau, E.L.C., Coelho, H.T., Lepore Neto, F.P., Santos, M.B. and Mahfoud, J.: On the assessment of a tunable auxiliary mass damper with a friction damper in its suspension: numerical study. In: Third International Conference on Structural Nonlinear Dynamics and Diagnosis—CSNDD (2016). https://doi.org/10.1051/matecconf/20168301005
9. Gola, M.M., Liu, T., Dos Santos, M.B.: Investigation of under-platform damper kinematics and its interaction with contact parameters (nominal friction coefficient). In: 5th World Tribology Congress, WTC 2013. Politecnico di Torino (DIMEAS) (2013)
10. Rotea, M.A., Khargonekar, P.P.: H^2-optimal control with an H^∞-constraint: the state feedback case. Automatica **27**(2), 307–316 (1991). https://doi.org/10.1016/0005-1098(91)90079-H

The Mechanical Behavior of Viscoelastic Materials in the Frequency Domain

Isadora R. Henriques, Lavinia A. Borges and Daniel A. Castello

Abstract In the last few decades, a growing need for new materials for several applications led to the development and increase of studies in new theories such as viscoelasticity. Many efforts have been done to understand and characterize the mechanical behavior of these materials. The purpose of this work is to determine the viscoelastic Poisson's ratio in the frequency domain, $\nu^*(\omega)$, for a viscoelastic material in order to characterize its three-dimensional behavior. To do so, the work is based on the elastic-viscoelastic correspondence principle (EVCP) and the time-temperature superposition principle (TTSP). Measurements of the complex shear modulus, $G^*(\omega)$, and the complex modulus, $E^*(\omega)$, were performed using a dynamic mechanical analyzer (DMA). To consider eventual uncertainties, each specific mechanical test was carried out using three test-specimens.

Keywords Viscoelastic poisson's ratio · Mechanical characterization
Dynamical mechanical analysis

1 Introduction

Viscoelastic materials are used in many applications such as structural components to control vibrations [5, 15, 28], tooth reconstruction [9], adhesives [12], among others. As their fields of applications have been continuously enlarging, several studies attempt to characterize and predict the mechanical behavior of these materials focusing mainly on properties such as the complex moduli [13]. Nevertheless, few studies investigated the Poison's ratio (PR), or a property analogous to it, on this type of materials. In fact, to the authors's best knowledge, its own definition seems to

I. R. Henriques · L. A. Borges · D. A. Castello (✉)
Mechanical Engineering Department, Universidade Federal do Rio de Janeiro,
Rio de Janeiro, RJ 21941-972, Brazil
e-mail: dnl.castello@gmail.com

I. R. Henriques
e-mail: ir.henriques@mecanica.coppe.ufrj.br

A. de T. Fleury et al. (eds.), *Proceedings of DINAME 2017*, Lecture Notes
in Mechanical Engineering, https://doi.org/10.1007/978-3-319-91217-2_5

remain ambiguous; moreover, one would say that investigations on the experimental procedures to its determination are not closed yet.

The viscoelastic Poisson's ratio (PR) does not present a well established definition. In most cases, this ratio is considered to be constant but its thermo-temporal dependency has been supported by experimental evidence as early as 1987. According to Lakes and Wineman [20], its time (or frequency) dependence varies even with the test modality chosen to evaluate it. Furthermore, it also depends on the stress and strain states, and the environmental conditions [16].

Due to its frequency dependence, it can also be seen as a complex number known as complex Poisson's ratio with its real part known as dynamic Poisson's ratio and its imaginary part known as loss component. The later is related to a phase lag between the lateral and axial strains. Both parts of this ratio are frequency and temperature dependent, presenting a much more complex behavior that leads to additional difficulty on its experimental characterization [25]. Further, it can be determined once one obtains two other complex properties such as the complex modulus and the complex shear modulus [31].

Many techniques to measure viscoelastic properties as a function of temperature and frequency are available in the literature and may be found in the review article by Lakes [21]. Among them, one of the most common methods is the one based on Dynamic Mechanical Analysis (DMA). This technique consists of the application of a force to a test-specimen in a cyclic manner in order to obtain the material's response. It can be performed at different frequencies and temperatures, allowing us to obtain information about the thermo-mechanical behavior of some specific material. Additionally, one can perform different tests at the same conditions and also, study the composition and its physical properties [19]. It should be emphasized that although DMA may provide both the complex modulus and the complex shear modulus, they do not provide any information about the PR.

As for the viscoelastic PR characterization, some recent works should be mentioned here. Bonfiglio and Pompoli [4] proposed an experimental procedure for determining the viscoelastic PR over the frequency domain. In their methodology, the material was excited by an electromagnetic shaker and then, the axial and the radial velocities were measured over the frequency using laser vibrometers. Yu et al. [35] modeled the viscoelastic PR after measuring the lateral expansion as a function of the axial compression in a flat punch test.

The purpose and contribution of this work is to assess the estimates of the complex Poisson's ratio (PR) $\nu^*(\omega)$ that are obtained by the manipulations of the mechanical properties provided by the Dynamic Mechanical Analyzer (DMA). More especifically, the complex modulus, $E^*(\omega)$, and the complex shear modulus, $G^*(\omega)$, were measured using two different operational modes: single cantilever bending and simple shear modes, respectively. The elastic-viscoelastic correspondence principle (EVCP) [22] was employed to obtain the viscoelastic relation between these complex properties. Additionally, the thermorheologically simple behavior was verified by both the Cole-Cole Diagram and Black Space and consequently, the time-temperature superposition principle (TTSP) [10] was applied to build the master curves for the viscoelastic material properties.

The paper is organized as follows. Section 2 gives the principal aspects related to the complex Poissons ratio, and the experimental set-up is explained in Sect. 3. The experimental results and the master curves are shown in Sect. 4, followed by conclusions.

2 Fundamentals

According to the theory of viscoelasticity, the Poisson's ratio does not have unique definition. It is rather defined in several ways ranging from seven different categories sorted by their physical meaning as stated by Hilton in [17]. However, all the definitions present an important similarity: the viscoelastic Poisson's ratio is not a constant value; it is a time- or frequency-dependent function as well as other viscoelastic material properties.

Its time-domain dependence can even vary with the modality of the mechanical test that is chosen to evaluate it [1, 6, 20, 31]. The viscoelastic PR in creep and in relaxation present different time dependence. However, it can be neglected for small to moderate relaxation levels. In this case, the differences are minor as stated by Lakes and Wineman [20].

Unlike Tschoegl et al. [31], Lakes and Wineman [20] established that the viscoelastic Poisson's ratio need not increase with time and it need not be monotonic. Later on, Charpin and Sanahuja [6] showed both theoretical and experimentally that it is possible to have this property presenting any evolution over time: increasing, decreasing and even non-monotonic.

Therefore, the assumption of a constant PR is inconsistent with the theory developed so far and with the experimental data as well. Even if it sometimes simplifies some mathematical formulations, this assumption may cause errors in engineering design.

2.1 Complex Poisson's Ratio, $\nu^*(\omega)$

According to Tschoegl et al. [31], the complex Poisson's ratio corresponds to the lateral contraction ratio measured in an infinitesimally small uniaxial deformation of a viscoelastic material in response to a steady-state sinusoidally oscillating axial strain. Mathematically, this statement may be expressed as

$$\nu_{j1}^*(\omega) = -\frac{\varepsilon_{jj}^*(\omega)}{\varepsilon_{11}^0}, j \neq 1 \tag{1}$$

where ε_{11}^{0} is the amplitude of the sinusoidal steady-state axial strain, $\varepsilon_{jj}^{*}(\omega)$ is the complex sinusoidal steady-state lateral contraction and ω, the angular frequency in radians per second.

Being $\nu^*(\omega)$ a complex property, it can be decomposed in its real and imaginary parts. The former is associated with the elastic response (stored energy), whereas the latter one is associated with the viscous response (energy dissipation). Therefore, the PR may be recast as

$$\nu^*(\omega) = \nu'(\omega) - j\nu''(\omega) = \nu'(\omega)[1 - j\eta_\nu(\omega)], \tag{2}$$

where $\nu'(\omega)$ is the dynamic Poisson's ratio and $\nu''(\omega)$ is the loss component and $\eta_\nu(\omega)$ is the Poisson's loss factor defined as [25]

$$\eta_\nu(\omega) = \frac{\nu''(\omega)}{\nu'(\omega)}. \tag{3}$$

In case of perfectly elasticity, the loss component is zero and, consequently, the Poisson's loss factor is also zero. The negative sign in Eq. (2) is related to the compliance nature of the viscoelastic PR. Hence, the absolute value of complex PR can be obtained through

$$|\nu^*(\omega)| = \sqrt{\nu'(\omega)^2 + \nu''(\omega)^2} \tag{4}$$

Besides that, the complex PR can be determined indirectly through any two other viscoelastic functions using the elastic-viscoelastic correspondence principle (EVCP) [32]. The elastic interrelationship between the Young's modulus, E, the shear modulus, G, and the Poisson's ratio, ν is [2]:

$$\nu = \frac{E}{2G} - 1. \tag{5}$$

Thus, the viscoelastic interrelationship in the frequency domain after applying the EVCP into Eq. (5) is

$$\nu^*(\omega) = \frac{E^*(\omega)}{2G^*(\omega)} - 1, \tag{6}$$

where $G^*(\omega)$ is the complex shear modulus and $E^*(\omega)$ is the complex Young's modulus which is simply denoted throughout the paper as complex modulus.

The real and imaginary parts of the complex PR can be written in terms of the real and imaginary parts of the complex modulus ($E'(\omega)$, $E''(\omega)$) and the complex shear modulus ($G'(\omega)$, $G''(\omega)$) as follows

$$\nu'(\omega) = \frac{1}{2}\frac{E'(\omega)G'(\omega) + E''(\omega)G''(\omega)}{[G'(\omega)]^2 + [G''(\omega)]^2} - 1 \tag{7}$$

and

$$\nu''(\omega) = \frac{1}{2}\frac{E'(\omega)G''(\omega) - E''(\omega)G'(\omega)}{[G'(\omega)]^2 + [G''(\omega)]^2}. \tag{8}$$

In principle, once one obtains measurements from $E^*(\omega)$ and $G^*(\omega)$, Eqs. (7) and (8) may be used to obtain estimates of $\nu^*(\omega)$. In this work, they will be used based on the measurements provided by a DMA and the estimates of $\nu^*(\omega)$ will be properly assessed.

3 Experimental Set-Up

3.1 Materials and Test-Specimens Preparation

Two groups of test-specimens were built. The first one was composed of a pure epoxy material. The second one was built by adding a certain amount of flexibilizer. The epoxy resin was Es260 br, a cold-setting one. *Aradur*^{TM} E35, a cycloaliphatic amine based curing agent with low viscosity, was used as hardener for the curing process. Finally, the flexibilizer used was DY 3601. All reagents were purchased from Advanced Vacuum Materials (São Paulo, Brazil).

Firstly, a mixture of epoxy resin and curing agent in a proportion of 45 phr were poured in a becker and were gently mixed using a glass tube. Afterwards, half of the mixture was carefully transferred to a second becker and both beckers were weighted to ensure their equal content. Then, the flexibilizer was added to the first becker in a proportion corresponding to 10% weight of the mixture and this new epoxy system was mixed again to homogenize it. Finally, both beckers were put on an ultrasonic bath for 30 min to avoid the presence of air bubbles. After the bath, the mixtures were poured into the cavities of silicone rubber molds with appropriate dimensions for the DMA tests. The curing process was performed at room temperature for 24 h, followed by a post-curing cycle at 60 °C for 6 h. To avoid residual stresses, the test-specimens were then slowly cooled to room temperature inside the oven.

Furthermore, in order to eliminate variations in material properties caused by processing, all test-specimen were manufactured from only one batch.

3.2 Tests by DMA

Dynamic tests were carried out with a DMA Q800 dynamic mechanical analyzer (produced by TA Instrument Corporation) to measure and investigate the complex modulus $E^*(\omega)$ and the complex shear modulus $G^*(\omega)$. For that purpose, the testing configuration was set in two different operational modes, respectively: single cantilever bending mode and simple shear mode. The former consists of a test-specimen anchored on one end by a stationary clamp and by a moveable clamp on the other which applies a controlled force. The test-specimen's dimensions are 35.00 mm × 12.70 mm × 3.18 mm. The latter mode consists of two equal-size test-specimens sheared between a fixed and a moveable plate. In this situation, the test-specimen's dimensions are 10.0 mm × 10.0 mm × 2.5 mm.

For both operational modes, frequency sweeps from 1 to 100 Hz, measuring 10 points per decade in logarithmic scale, were carried out at seven temperatures from 60 to 90 °C. Isotherms were maintained for 5 min every 5 °C and the heating rate was 2 °C/min.

4 Results and Discussion

Following recommendations provided in [23] to reduce experimental errors, each modality test was performed on three test-specimens of each material. Through the results, the terms pure epoxy and epoxy with flexibilizer will be used extensively and replaced by PE and EWF, respectively.

4.1 *Complex Modulus, $E^*(\omega)$*

Figure 1 shows the real part of complex modulus $E'(\omega)$, also known as storage modulus, as a function of frequency for both materials. Both materials presented the same pattern: storage modulus decreased with the increase in temperature and increased with the increase in frequency. This may be more noticeable at high temperatures and for EWF. This trend is coherent with the classical behavior of polymeric materials [29]. Comparing both materials, PE presented a higher modulus value for all temperature range.

The imaginary part of complex modulus $E''(\omega)$, known as loss modulus, is then represented in Fig. 2. Differences can be observed among the results. For PE, below 80 °C, it increased with the increase in temperature and did not vary with frequency. From 80 °C, it decreased with temperature and increased with frequency. For EWF, on the other hand, the loss modulus presented a similar behavior as the storage modulus.

The thermorheologically simple behavior [7] of these materials was verified through the Cole-Cole Diagram [8] and the Black Space [33] as shown, respectively, in Fig. 3 and Fig. 4. The $E'(\omega)$ versus $E''(\omega)$ and $E''(\omega)/E'(\omega)$ versus $|E^*(\omega)|$ curves in double logarithmic scales at different temperatures coincided in one continuous curve for most points. Points that deviate from the curves are related to measurements at high frequencies and they may be related to resonance phenomena in the DMA [24]. Additionally, both diagrams show a slight temperature dependence of complex modulus of both materials, indicating a need for vertical shifting [27].

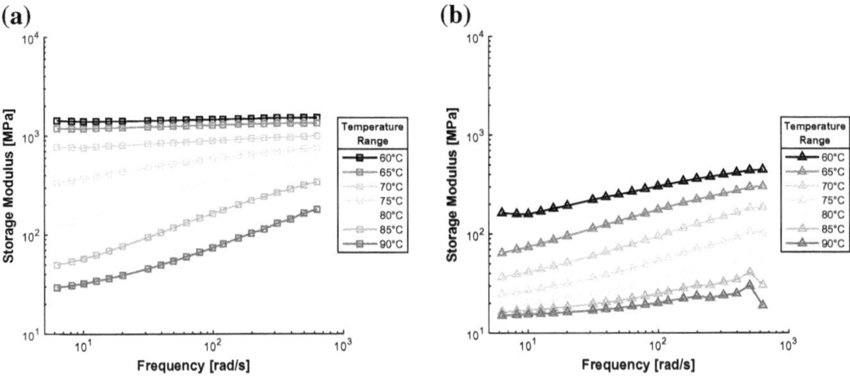

Fig. 1 Storage modulus ($E'(\omega)$) versus frequency (ω). **a** PE **b** EWF

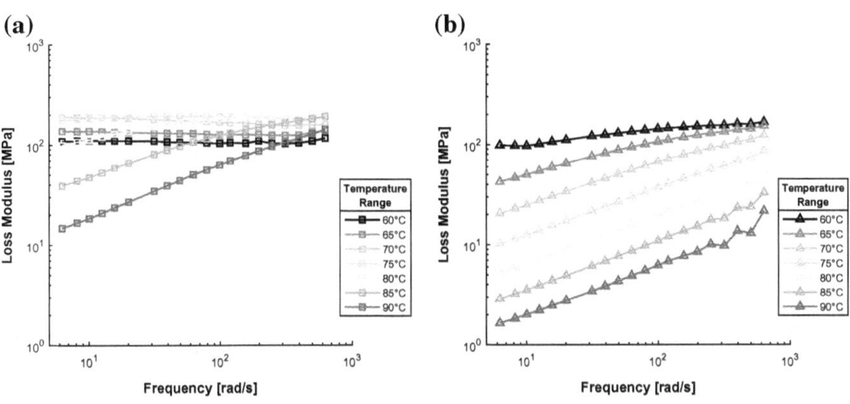

Fig. 2 Loss modulus ($E''(\omega)$) versus frequency (ω). **a** PE **b** EWF

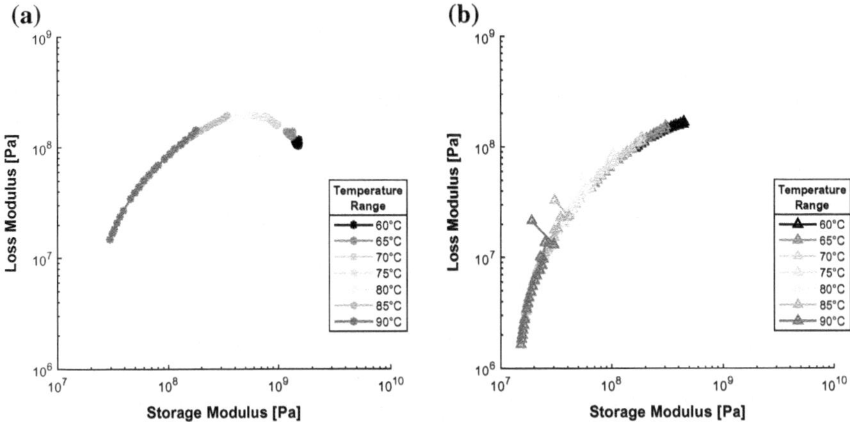

Fig. 3 Cole-cole diagrams: $E''(\omega)$ versus $E'(\omega)$. **a** PE **b** EWF

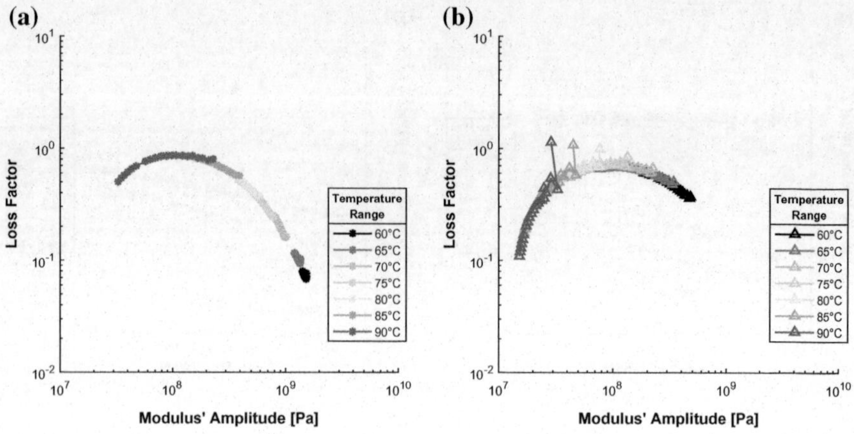

Fig. 4 Black diagrams: $E''(\omega)/E'(\omega)$ versus $|E^*(\omega)|$. **a** PE **b** EWF

Therefore, it is possible to apply the time-temperature superposition principle (TTSP) to characterize the materials on a broader frequency range. Briefly speaking, when assuming that the time-temperature superposition holds, one may write a relation between a viscoelastic property measured at two different temperatures as follows

$$Q^*(\omega;T) = Q^*(a_T \times \omega, T_0) \tag{9}$$

where $Q^*(\omega, T)$ corresponds to some viscoelastic property in the frequency domain at the temperature T, T_0 is a reference temperature and a_T is the thermal shift factor (a_T), also known as horizontal shift coefficient [10].

Concerning the thermal shift factor a_T, one of the most used models is given by the Williams-Landel-Ferry (WLF) equation [34]. Here, the thermal-shift factors were computed for all three test-specimens as a function of the relative temperature and were fitted by the Williams-Landel-Ferry (WLF) equation:

$$\log a_T = -\frac{C_1(T - T_0)}{C_2 + (T - T_0)}, \tag{10}$$

with T_0 as the reference temperature in Kelvin, T as the temperature in Kelvin and C_1 and C_2 as parameters that depend on the material and the reference temperature.

The reference temperature was 348 K (75 °C) and the parameters $C_1 = 19.07$ and $C_2 = 43.17$ K were estimated for PE system, while $C_1 = 15.28$ and $C_2 = 68.41$ K for EWF. These parameters were estimated through the least squares method. From Fig. 5, it can be observed that the thermal shift factor presented high levels of correlation with measured data.

Finally, the master curves were built by using the method described by Rouleau et al. in [27]. They are presented in Figs. 6 and 7. The frequency range is up to approximately 10^6 rad/s. The storage and loss moduli present an asymptotic

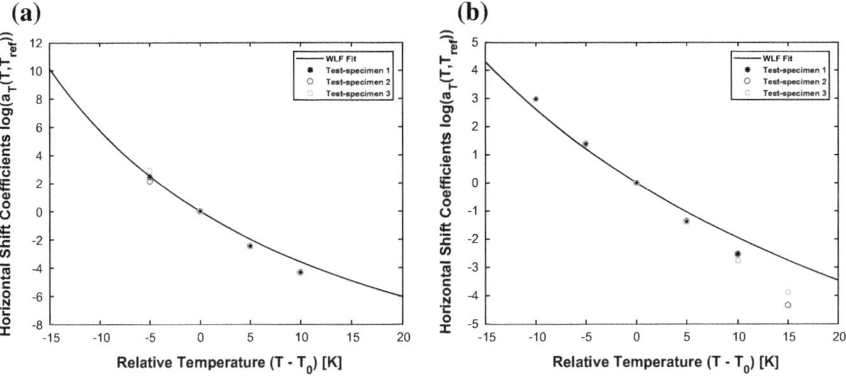

Fig. 5 Thermal shift factors for WLF Equation from $E^*(\omega) - T_0 = 348$ K. **a** PE: $C_1 = 19.07$ and $C_2 = 43.17$ K. **b** EWF: $C_1 = 15.28$ and $C_2 = 68.41$ K

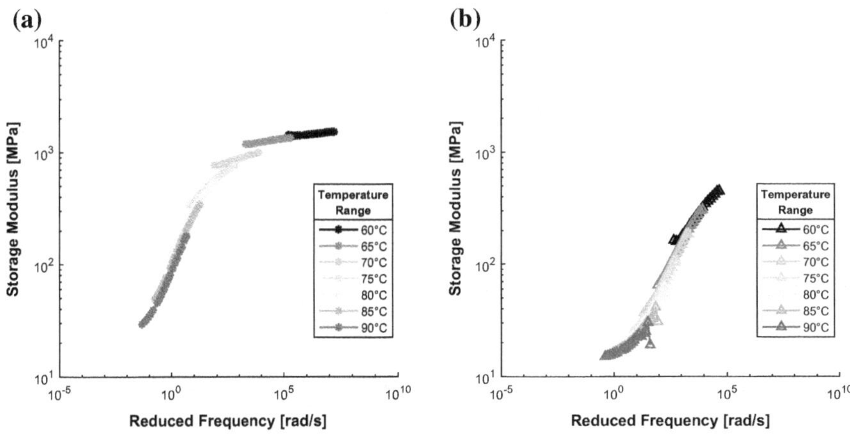

Fig. 6 Master curves for storage modulus $E'(\omega)$. **a** PE **b** EWF

behavior. However, the superposition of these curves makes evident the presence of some discontinuities between them for the highest frequencies. This could probably be circumvented by the use of the vertical shift coefficients as shown by Rouleau et al. [27].

4.2 *Complex Shear Modulus, $G^*(\omega)$*

The real part of the complex shear modulus $G'(\omega)$, known as shear storage modulus, is shown in Fig. 8. The imaginary part $G''(\omega)$, known as shear loss modulus, is depicted in Fig. 9. As for the dependence of the complex shear modulus $G^*(\omega)$ with

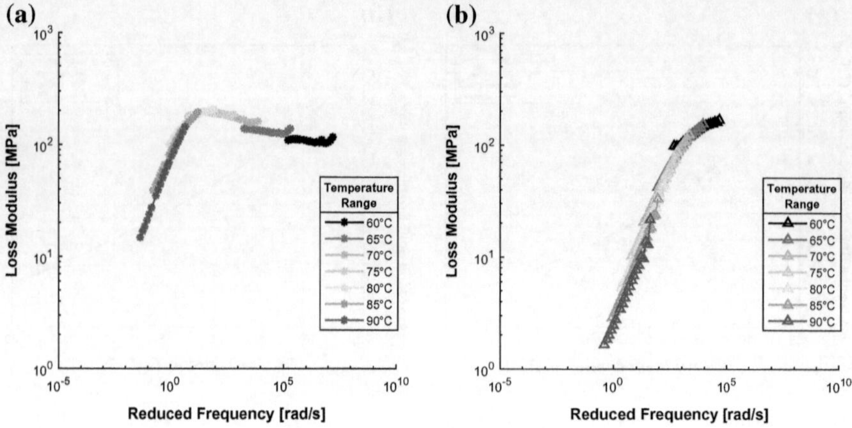

Fig. 7 Master curves for loss modulus $E''(\omega)$. **a** PE **b** EWF

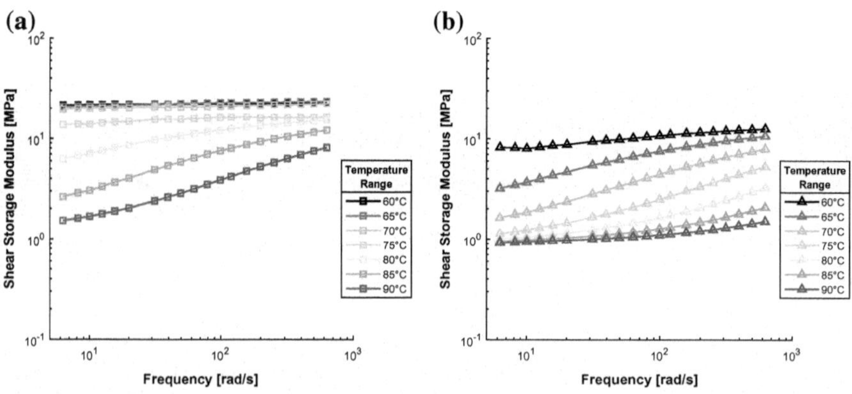

Fig. 8 Shear storage modulus ($G'(\omega)$) versus frequency (ω). **a** PE **b** EWF

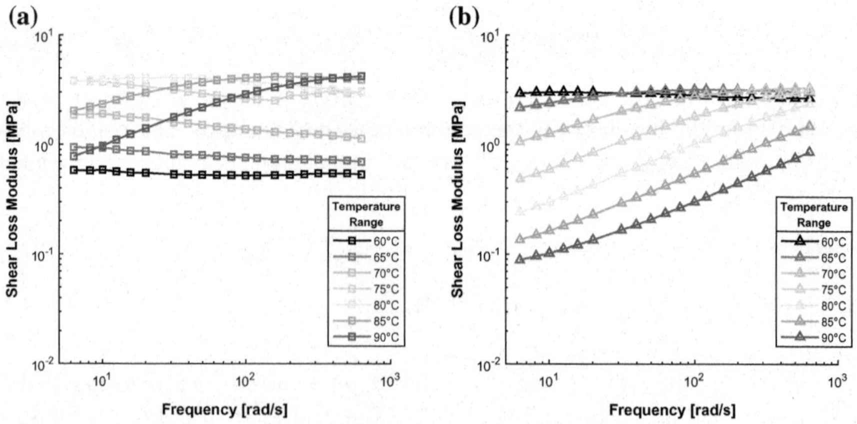

Fig. 9 Shear loss modulus ($G''(\omega)$) versus frequency (ω). **a** PE **b** EWF

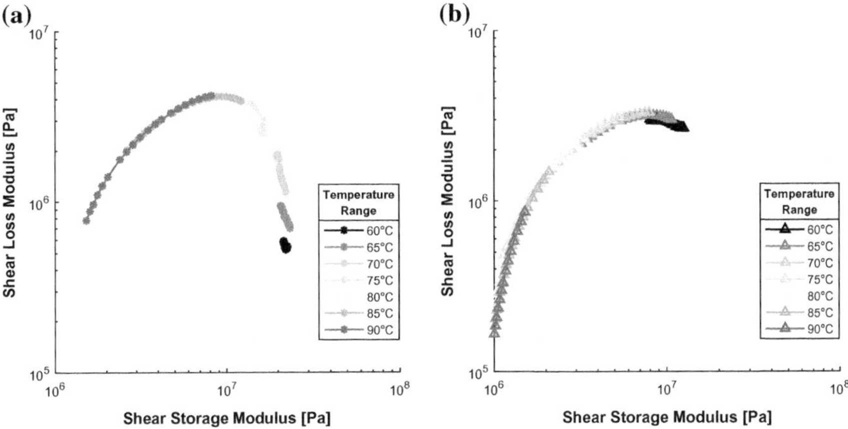

Fig. 10 Cole-cole diagrams: $G''(\omega)$ versus $G'(\omega)$. **a** PE **b** EWF

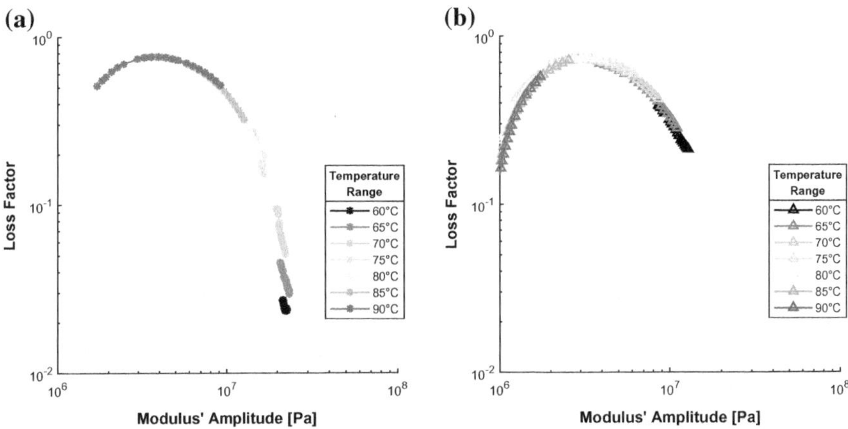

Fig. 11 Black diagrams: $G''(\omega)/G'(\omega)$ versus $|G^*(\omega)|$. **a** PE **b** EWF

temperature and frequency, it presents the same pattern as the one presented by the complex modulus $E^*(\omega)$.

The validity of the TTSP was also assessed with the results of complex shear modulus. Figure 10 shows the Cole-Cole Diagram and Fig. 11, the Black Space. Most points coincided along one single continuous curve, fulfilling the characteristics of the thermorheologically simple material postulate. This validates the hypothesis made when using the results of $E^*(\omega)$.

Once again, a model for the thermal shift factor $a_T(C_1, C_2)$ was calibrated using the complex shear modulus data. Figure 12 presents the measured data and the model built with the estimated parameters. Comparing the results shown in Figs. 5 and 12, one can observe that the models are highly correlated and even the empirical con-

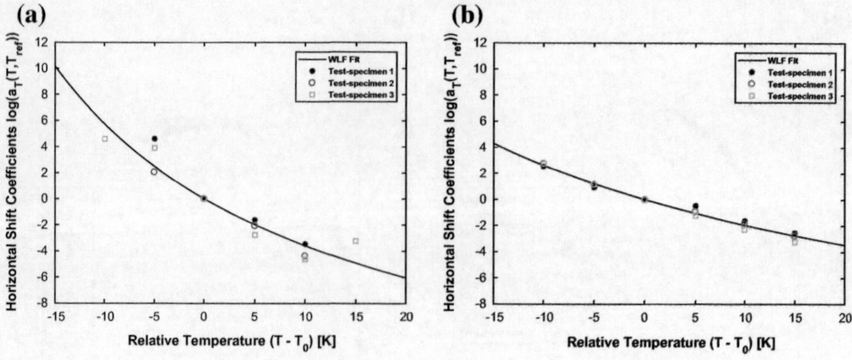

Fig. 12 Thermal shift factors for WLF Equation from $G^*(\omega) - T_0 = 348$ K. **a** PE: $C_1 = 19.07$ and $C_2 = 43.17$ K. **b** EWF: $C_1 = 15.28$ and $C_2 = 68.41$ K

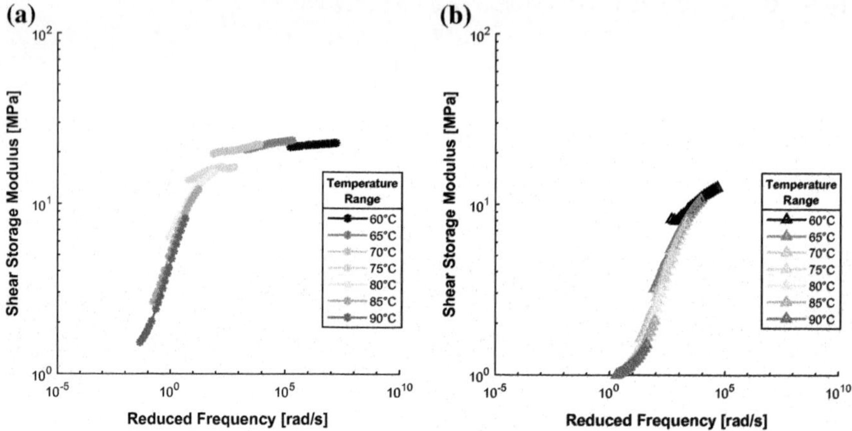

Fig. 13 Master curves for shear storage modulus $G'(\omega)$. **a** PE **b** EWF

stants C_1 and C_2 were found the same. It should be emphasized that Arzoumanidis and Liechti in [3] also obtained similarities between the thermal shift factor estimated using different test modalities.

The master curves were finally obtained at a reference temperature $T_0 = 348$ K (75 °C). Figures 13 and 14 show, respectively, the master curves of shear storage modulus and shear loss modulus. One may notice that they are not so continuous for the PE material. This could probably be circumvented by the use of the vertical shift coefficients.

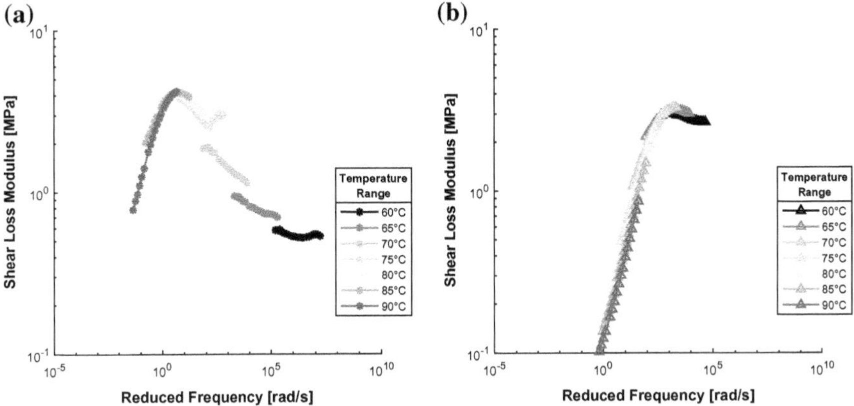

Fig. 14 Master curves for shear loss modulus $G''(\omega)$. **a** PE **b** EWF

4.3 Complex Poisson's Ratio, $\nu^*(\omega)$

After measuring $E^*(\omega)$ and $G^*(\omega)$, the real and imaginary components of complex Poisson's ratio $\nu^*(\omega)$ was, respectively, estimated using Eq. (7) and Eq. (8). However, the results of the real component, $\nu'(\omega)$, were not consistent with the limits of 0.5 and −1 [14].

As the estimates of the complex Poisson's ratio were meaningless from the physical point of view, a further look at the DMA specific experimental apparatus revealed a systematic error due to a scale factor that was also reported in the literature [11, 18, 30]. Firstly, it was observed then that there was a discrepancy in scale with the results of the complex shear modulus.

Therefore, the complex shear modulus measured in this work were compared to the ones reported by Rao et al. in [26] which allowed us to estimate a proper scale factor. It is worth noting that the shift factors for temperatures close to glass transition temperature were quite difficult to identify due to the great variations of the moduli within this range. Figures 15, 16 and 17 show, respectively, the results obtained for the dynamic Poisson's ratio $\nu'(\omega)$, the loss component $\nu''(\omega)$ and the absolute value $|\nu^*(\omega)|$ for each material.

For PE, the dynamic Poisson's ratio and the loss component had different behaviors according to temperature. For temperatures below 75 °C, both components were almost constant over the frequency range. However, for the ones above 75 °C, the dynamic Poisson's ratio increased with the increase in frequency whereas the loss component decreased. For EWF, on the other hand, the dynamic Poisson's ratio increased with the increase in frequency, whereas the loss component decreased. For both materials, the absolute value presents quite similar behavior.

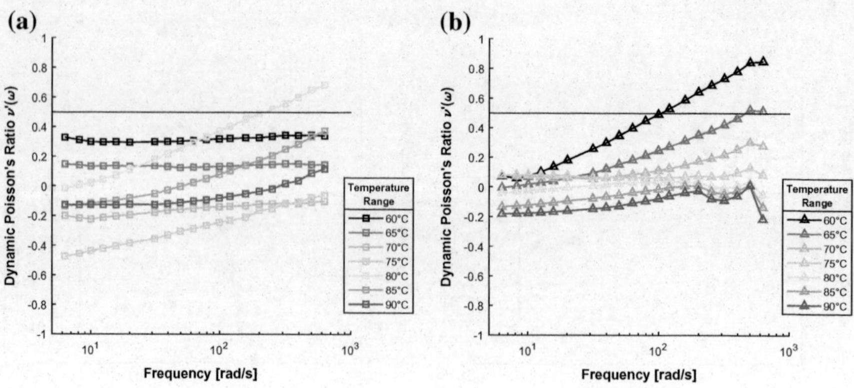

Fig. 15 Dynamic poisson's ratio ($\nu'(\omega)$) versus frequency (ω). **a** PE **b** EWF

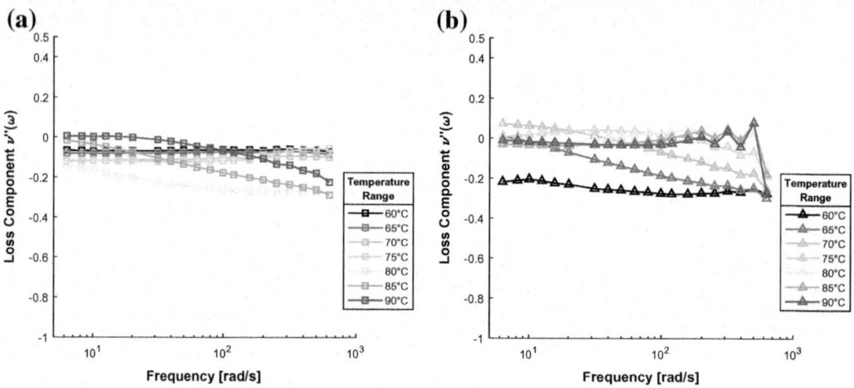

Fig. 16 Loss component ($\nu''(\omega)$) versus frequency (ω). **a** PE **b** EWF

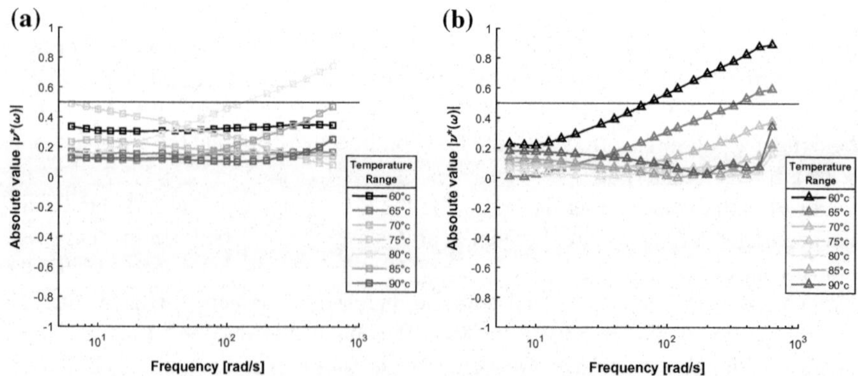

Fig. 17 Absolute value ($|\nu^*(\omega)|$) versus frequency (ω). **a** PE **b** EWF

5 Conclusion

This work proposes an approach to estimate the complex Poisson's ratio through an indirect method. The complex modulus and the complex shear modulus were first determined by performing two different tests using a dynamic mechanical analyzer. Afterwards, the estimates for complex Poisson's ratio are provided by the use of the elastic-viscoelastic correspondence principle.

The complex Poisson's ratio based on the modulus provided from DMA output software proved meaningless from the physical point of view. Therefore, the use of the shear modulus provided by DMA required a correction scale factor that was estimated using measured data provided by works found in the literature.

A key point to be emphasized is that investigations in the literature have been indicating that the viscoelastic properties provided by DMA equipment present great levels of discrepancy in scale depending on the test modality chosen for the material characterization as reported in [11, 18, 30].Therefore, based on these articles found in the literature and on the data analyzed in this work, two main points are worth to be highlighted here: (i) DMA seems to be an efficient tool aimed at examining material properties for quality control, research and development, and also for the establishment of optimum processing conditions; (ii) care should be taken when comparing viscoelastic properties provided directly from DMA software that come from different test modalities inasmuch as these are possibly affected by a biased scale factor.

References

1. Alotta, G., Barrera, O., Cocks, A.C., Paola, M.D.: On the behaviour of a three-dimensional fractional viscoelastic constitutive model. Meccanica 52(9), 2127–2142 (2017). https://doi.org/10.1007/s11012-016-0550-8
2. Armenkas, A.E.: Advanced Mechanics of Materials and Applied Elasticity. CRC Press, Florida (2005)
3. Arzoumanidis, G.A., Liechti, K.M.: Linear viscoelastic property measurement and its significance for some nonlinear viscoelasticity models. Mech. Time-Depend. Mater. **7**, 209–250 (2003). https://doi.org/10.1023/B:MTDM.0000007357.18801.13
4. Bonfiglio, P., Pompoli, F.: A simplified compression test for the estimation of the Poisson's ratio of viscoelastic foams. Polym. Test. **61**, 324–332 (2017). https://doi.org/10.1016/j.polymertesting.2017.05.040
5. Borges, F.C.L., Castello, D.A., Magluta, C., Rochinha, F.A., Roitman, N.: An experimental assessment of internal variables constitutive models for viscoelastic materials. Mech. Syst. Signal Process. 50–51 (2015). https://doi.org/10.1016/j.ymssp.2014.04.023
6. Charpin, L., Sanahuja, J.: Creep and relaxation Poisson's ratio: Back to the foundations of linear viscoelasticity. Application to concrete. Int. J. Solids Struct. 110–111 (2017). https://doi.org/10.1016/j.ijsolstr.2017.02.009
7. Christensen, R.M.: Theory of Viscoelasticity—An Introduction. Academic Press, New York (1982)

8. Dae Han, C., Kim, J.K.: On the use of time-temperature superposition in multicomponent/multiphase polymer systems. Polymer **34**(12), 2533–2539 (1993). https://doi.org/10.1016/0032-3861(93)90585-X
9. Dauvillier, B.S., Feilzer, A.J., De Gee, A.J., Davidson, C.L.: Visco-elastic parameters of dental restorative materials during setting. J. Dent. Res. **79**(3), 818–823 (2000). https://doi.org/10.1177/00220345000790030601
10. Dealy, J., Plazek, D.: Time-temperature superposition—a users guide. Rheol. Bull. **78**, 16–31 (2009)
11. Deng, S., Hou, M., Ye, L.: Temperature-dependent elastic moduli of epoxies measured by DMA and their correlations to mechanical testing data. Polym. Test. **26**(6), 803–813 (2007). https://doi.org/10.1016/j.polymertesting.2007.05.003
12. Estrada-Royval, I.-A., Díaz-Díaz, A.: Post-curing process and visco-elasto-plastic behaviour of two structural adhesives. Int. J. Adhes. Adhes. **61**(9), 9–111 (2015). https://doi.org/10.1016/j.ijadhadh.2015.06.001
13. Graziani, A., Bocci, M., Canestrari, F.: Complex Poisson's ratio of bituminous mixtures: measurement and modeling. Mater. Struct. **47**, 1131–1148 (2014). https://doi.org/10.1617/s11527-013-0117-2
14. Greaves, G.N., Greer, A.L., Lakes, R.S., Rouxel, T.: Poisson's ratio and modern materials. Nat. Mater. **10**(11), 823–837 (2011). https://doi.org/10.1038/nmat3134
15. Hernández, W.P., Castello, D.A., Ritto, T.G.: Uncertainty propagation analysis in laminated structures with viscoelastic core. Comput. Struct. **164**(2), 23–37 (2016). https://doi.org/10.1016/j.compstruc.2015.10.006
16. Hilton, H.: Clarifications of Certain Ambiguities and Failings of Poisson's Ratios in Linear Viscoelasticity. J. Elast. **104**(1), 303–318 (2011). https://doi.org/10.1007/s10659-010-9296-z
17. Hilton, H.: Elastic and Viscoelastic poisson's ratios: the theoretical mechanics perspective. Mater. Sci. Appl. **8**, 291–332 (2017)
18. Kumar, N.P.: Viscoelastic characterization and effective damping of a carbon/polyurethane laminate. Master thesis, Rochester Institute of Technology, New York (2016)
19. Lack, I., Krupa, I., Stach, M., Kuma, A., Juriov, J., Chodk, I.: Thermal lag and its practical consequence in the dynamic mechanical analysis of polymers. Polym. Test. **19**, 775–771 (2000). https://doi.org/10.1016/S0142-9418(99)00046-X
20. Lakes, R., Wineman, A.: On poisson's ratio in linearly viscoelastic solids. J. Elast. **85**(1), 45–63 (2006). https://doi.org/10.1007/s10659-006-9070-4
21. Lakes, R.S.: Viscoelastic measurement techniques. Rev. Sci. Instrum. **75**(4), 797-810 (2004). https://doi.org/10.1063/1.1651639
22. Lee, E.H.: Stress analysis for linear viscoelastic materials. Rheol. Acta. **1**(4-6), 426–430 (1961). https://doi.org/10.1007/BF01989085
23. Menard, K.: Dynamic mechanical analysis—a practical introduction. CRC Press LLC, Florida (2008)
24. Placet, V., Foltête, E.: Is Dynamic Mechanical Analysis (DMA) a non-resonance technique?. EPJ Web Conf. **6**, (2010). https://doi.org/10.1051/epjconf/20100641004
25. Pritz, T.: The Poisson's loss factor of solid viscoelastic materials. J. Sound Vib. **306**(3), 790–802 (2007). https://doi.org/10.1016/j.jsv.2007.06.016
26. Rao, K.V., Dayananda, G.N., Ananthapadmanabha, G.S.: Viscoelastic characterisation of an epoxy based shape memory polymer (SMEP). Indian J. Adv. Chem. Sci. **2**, 64–67 (2014)
27. Rouleau, L., Deu, J.-F., Legay, A., Le Lay, F.: Application of KramersKronig relations to timetemperature superposition for viscoelastic materials. Mech. Mater. **65**(10), 66–75 (2013). https://doi.org/10.1016/j.mechmat.2013.06.001
28. Rouleau, L., Pirk, R., Pluymers, B., Desmet, W.: Characterization and modeling of the viscoelastic behavior of a self-adhesive rubber using dynamic mechanical analysis tests. J. Aerosp. Technol. Manag. **7**(2), 200–208 (2015). https://doi.org/10.5028/jatm.v7i2.474
29. Shaw, M.T., MacKnight, W.J.: Introduction to Polymer Viscoelasticity. Wiley, New York (2005)

30. Swaminathan, G., Shivakumar, K.: A re-examination of DMA testing of polymer matrix composites. J. Reinf. Plast. Comp. **28**(8), 979–994 (2008). https://doi.org/10.1177/0731684407087740
31. Tschoegl, N.W., Knauss, W.G., Emri, I.: Poisson's ratio in linear viscoelasticity a critical review. Mech. Time-Depend. Mater. **6**(1), 3–51 (2002). https://doi.org/10.1023/A:101441150317
32. Tschoegl, N.W.: The Phenomenological Theory of Linear Viscoelastic Behavior: an introduction. Springer, Berlin (1989)
33. Van Gurp, M., Palmen, J.: Time-temperature superposition for polymeric blends. Rheol. Bull. **67**(1), 5–8 (1998)
34. Williams, M.L., Landel, R.F., Ferry, J.D.: The temperature dependence of relaxation mechanisms in amorphous polymers and other glass-forming liquids. J. Am. Chem. Soc. **77**(14), 3701–3707 (1955). https://doi.org/10.1021/ja01619a008
35. Yu, H., Kongsmo, R., Patil, N., He, J., Breiby, D.W., Zhang, Z.: On determining the Poisson's ratio of viscoelastic polymer microparticles using a flat punch test. Int. J. Mech. Sci. **128–129**, 150–158 (2017). https://doi.org/10.1016/j.ijmecsci.2017.04.019

Curvature Effects on Vibrational Power Flow of Smooth Bent Beams

Paulo Martins and Arcanjo Lenzi

Abstract This study discusses an approach to analyze curvature effects on the vibrational powerflow of slender beams. A finite element method (FEM) model was used to calculate transmitted and reflected power via the "propagating wave approach". Previously, the same investigation was addressed with focus on the analytical formulation of curved beams, and an FEM model was programmed using simple two-node straight (Euler-Bernoulli) beam elements. However, the model could simulate in-plane vibrations only, with restrictions to high frequencies (lower than two wavelengths inside the curvatures length). Hence, a new model was proposed, and the curvature parametrization was updated. A novel method to parametrize the curvature in 3D is discussed using quaternions. Results from the updated FEM model and analytical approach were compared for validation. Moreover, the algorithm performed almost exactly like the analytical model, even at high frequencies, which made it suitable to simulate power flow based on the wave approach. The algorithm allows any type of curve configuration to be tested. Curvature effects for in- and out-of-plane vibrations are shown as well. Finally, this work introduces a basis for designing and optimizing slender pipe structures from the perspective of vibration control.

Keywords Curvature · Power flow · Quaternion · FEM · Wave approach

1 Literature Background

The present work mitigates the changes in power flow of vibrational waves while propagating through curved slender beams. The current literature reveals a reasonable amount of work on "smooth bends" and "curves", but the majority only

P. Martins (✉) · A. Lenzi
CTC EMC, LVA, UFSC, Florianópolis, SC CEP 88040-900, Brazil
e-mail: paulo.victor@lva.ufsc.br
URL: http://lva.ufsc.br

A. Lenzi
e-mail: arcanjo@lva.ufsc.br

A. de T. Fleury et al. (eds.), *Proceedings of DINAME 2017*, Lecture Notes in Mechanical Engineering, https://doi.org/10.1007/978-3-319-91217-2_6

addresses static applications (e.g., structural analysis or civil engineering). Therefore, behavior under a dynamic/vibrational perspective is not well explored for curve configurations.

One can highlight studies that aimed at dynamic behavior, such as the work of Walsh and White (see [1]), who studied the vibrational power transmission of waves traveling in the same curves plane. They provided the equations of motion considering different theories: Loves approximations, Flugge's, only rotary inertia, only shear deformation, and Timoshenko (both rotary inertia and shear deformation). They also compared the curved beams wave numbers with that of a straight beam. The latter was also conducted by Lee et al. [2], who linked wave numbers to power flow by determining analytically how real or imaginary wave numbers can change the power transmission. Lee also reported a "displacement ratio", which is a robust approach to understand how curvatures couple both flexural and longitudinal movement. Wu and Lundberg [3] also analyzed the vibration in the plane of curvature; they explained the wave propagation approach in a very detailed manner, which consolidated a basis for the present work. They also proposed a non-dimensional set of parameters for analysis that strengthened understanding on the curvatures behavior. Results for constant frequency and curves parameters were also discussed. For vibrations occurring perpendicular to a curves plane (out-of-plane vibrations), in [4] was investigated the natural frequencies for a continuous bent beam and also provided the governing equations for this type of movement. In the numerical approach, in [5] was provided a semi-infinite straight beam element based on its dynamic stiffness, and in [6] was derived a consistent constant curvature beam element. Both elements are used in finite element analysis.

The next section explains how the curvature will be analyzed, how power from wave constants is obtained, and how the power coefficients are calculated. The third section explains the numerical tool, which is tailored specifically to solve the framework proposed in the second section. The fourth section explains the use of an analytical approach to validate the numerical tool presented in the third section. The fifth section is dedicated to the results and discusses a "map" of curvature effects over power flow. Finally, the last section presents the concluding remarks, as well as some suggestions for future work.

2 Wave Propagation Approach

Consider an arbitrary wave propagating through a semi-infinite media, which may be longitudinal, torsional or flexural (Fig. 1). As this incident wave (superscript I) reaches a curvature (discontinuity), part is reflected (superscript R) and part is transmitted, i.e., passes through and continues (superscript T). The "wave propagation approach" obtains the amplitude constant for each wave type. As this work addresses only the curvature effects, any type of losses or damping are neglected from the structure.

To obtain the amplitude constants of waves, continuity between both straight and curved interfaces must be guaranteed, i.e., all forces and displacements should be the

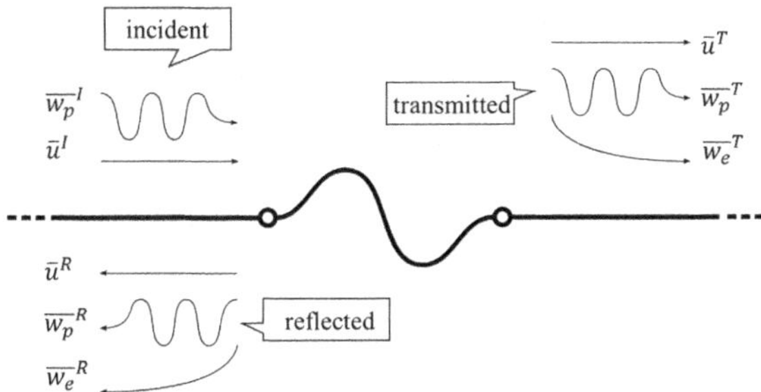

Fig. 1 Diagram of a typical wave approach setup

same at curves ends. Therefore, the formulations of both straight and curved parts are necessary.

First, a straight beams longitudinal wave is known to have two wave numbers (one for each waves traveling direction). However, for a single longitudinal wave propagating through a medium in a single direction, can be written

$$u(x,t) = \bar{u}e^{-j(k_u x+\omega t)}, \tag{1}$$

where $\bar{u}$ is the wave constant, $j = \sqrt{-1}$, k_u is the longitudinal wave number, x is the beams position, ω the angular frequency in rad/s, and t is the time. The other wave solutions can be given in the form

$$\theta_x(x,t) = \bar{\theta}_x e^{-j(k_\theta x+\omega t)}, \tag{2}$$

$$w(x,t) = \bar{w}_p e^{-j(k_w x+\omega t)} + \bar{w}_e e^{-(k_w x+j\omega t)}, \tag{3}$$

where $\bar{\theta}_x$ is the propagating torsional constant, $\bar{w}_p$ is the propagating wave constant and $\bar{w}_e$ is the constant of evanescent part. Similarly, the flexural motion on a perpendicular plane will be the same as before if the beam is symmetric. Otherwise, only the wave number k_w is changed.

The wave numbers for a straight beam are

$$\begin{aligned} k_u &= \omega\sqrt{\rho/E},\ \ k_w = \sqrt{\omega^2\rho A/EI_z}, \\ k_v &= \sqrt{\omega^2\rho A/EI_y},\ \ k_\theta = \omega\sqrt{\rho/G}, \end{aligned} \tag{4}$$

where ρ is the density per length, E is Young's modulus, A is the cross-sections area, I_z and I_y are the areas second moment of inertia on z and y axes, respectively, and G is the shear modulus. Hence, with the configuration shown in Fig. 1, the balanced equations for "in-plane" and "out-of-plane" vibration can be written. Bear in mind that inlet waves should consider the *incident* waves as propagation only and *reflected* waves with both propagating and evanescent parts. Outlet waves should consider both parts as well, but pure *transmission* only (no further reflections).

Given that the four incident constants are known, a total of twelve "straight" wave constants need to be determined for the setup. Note that "in-plane" (and subsequently "out-of-plane") is presented in quotation marks because it is referring to a straight beam, whereas a curved section creates a plane for analysis.

Walsh and White [1] used the propagating wave constants to obtain "in-plane" longitudinal and flexural and "out-of-plane" flexural and torsional powers as follows:

$$
\begin{aligned}
W_u &= \frac{1}{2} EA\omega k_u \bar{u}^2, \quad W_v = EI\omega k_v^3 \bar{v}^2, \\
W_w &= EI\omega k_w^3 \bar{w}^2, \quad W_\theta = \frac{1}{2} GJ\omega k_\theta \bar{\theta}_x^2,
\end{aligned}
\tag{5}
$$

where J is the torsional constant, which can be given by $I_z + I_y$ for a symmetric beam (and rotation axis equals to the centerline).

In the present paper, the concept of "power coefficient" is used to standardize the results as a ratio of transmitted (or reflected) power over the incident power. For example, W_u^T/W_{ip}^I would be a longitudinal transmission coefficient, with $W_{ip}^I = W_u^I + W_w^I$; and W_v^R/W_{op}^I would be an "out-of-plane" reflection coefficient, with $W_{op}^I = W_v^I + W_\theta^I$.

The use of this approach is detailed in the next section, in which the finite element method (FEM) is used to obtain the wave constants and power coefficients.

3 Finite Element Solution

The developed FEM algorithm is suitable to simulate any geometry based on beam-like elements. Moreover, it considers both ends of the geometry as semi-infinite straight beams. Finally, the algorithm provides power coefficients as output.

Common Euler-Bernoulli's straight beam elements are used to assemble the global problems mass and stiffness matrices. The elements of two nodes, and six degrees of freedom per node (three translations and three rotations), can be determined using formulations that are easily found on the Internet.

For completions sake, the work of Zienkwickz and Taylor (see [7]) is recommended, from where K_{el} and M_{el} matrices were given.

Euler's theory of curved beam element for in-plane vibration can be found in the work of Davis and Warburton (see [6]).

The dynamic stiffness of semi-infinite element and force input (to simulate a wave that is propagating through an infinite media) was taken from [5].

3.1 Geometry Crafting

Building curvatures in a single plane is somewhat trivial (a single angle and single radius are sufficient to its parametrization), but thinking about three dimensions is much more complicated. One can work with Eulers angles or Rodrigues' rotation, but all of them leave up to three independent angles of rotation and three radii (one for each plane).

To simplify parametrization, inspiration from *aviation* has been sought, with concepts of an airplane "rolling" (for creating a plane of curve) and "pitching" (for "flying" a curve).

This choice of angles allows a sequence of curved and straight lines to be built using only three known parameters per section: length L, roll angle θ_x, and pitch angle θ_y. The radius of curvature R is obtained as a function of L and θ_y, and it is designated as the "pitching movement". One has chosen for the "rolling movement" to occur instantly, before any other "movement" takes part. For example, if the instruction was $L = 10$ mm, $\theta_x = 30°$ and $\theta_y = 90°$, the hypothetical airplane would be tilted (rolled) 30° from the ground, and start climbing until an arch of 90° with $R \approx 6.37$ mm would be drawn.

To automate this instruction set, the "geometry builder" algorithm used the concept of "quaternions" introduced by Irish mathematician Hamilton [8]. Also another good reference in this matter is the work of Kuipers (see [9]).

4 Analytical Validation

All the analytical formulation for curves and equations of motion for in-plane and out-of-plane vibrations can be found in the works of [1–4]. Thus they will be omitted here for the sake of simplicity.

For an arbitrary curvature of $\Theta_y = 90°$ and $R = 38.2$ mm (as depicted in the left of Fig. 2), frequency range of 0–10 kHz, and prescribing a unitary longitudinal wave, the curves of Fig. 2 can be obtained. The right plot shows the differences between numerical and analytical power coefficients for this setup. The curves represents the coefficients type. The line is longitudinal transmission W_u^T, dashed is flexural (in-plane) transmission W_w^T, dash-dotted is longitudinal reflection W_u^R, and dotted is flexural reflection W_w^R. Figure 2 indicates a maximum difference of 0.05%.

For the same geometry, by prescribing a unitary flexural wave, the differences are approximately 0.06%, showing an almost exact concordance between numerical and analytical analyses for the in-plane vibration.

Unitary flexural wave perpendicular to the plane of curvature (out-of-plane wave) and torsional wave inputs were also tested and the differences between analytical and numerical approaches where consistently low.

The worst case scenario was at $\Theta_y = 30°$ for a prescribed torsional wave ($\bar{\theta}_x^I$), with difference of approximately 6% for frequency until 10 kHz. This discrepancy

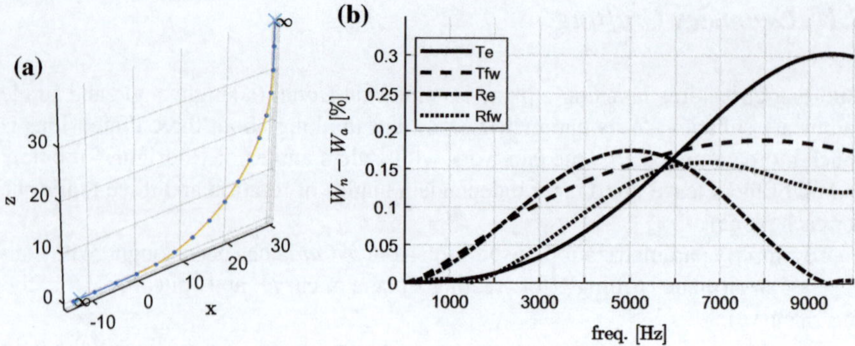

Fig. 2 Arbitrary geometry for testing the tool performance, with $R = 30$ mm and $\Theta_y = 90°$ (**a**). Differences between numerical and analytical power coefficients, with longitudinal wave input (**b**)

is due to a numerical error generated by the straight elements, because they are not designed to fully represent torsion-to-flexural coupling.

Nevertheless, for the frequency range and application type, the numerical approach is considered validated and good for use.

5 Curvature Effects on Vibrational Powerflow

This section provides a more general understanding of which curve configurations are prone to let incoming waves pass through (transmit) the bend. This knowledge will allow a sort of "curvature map" of power flow effects. The sub-sections will address the effects over vibrations occurring in the same curves plane and then focus on an out-of-planes effects.

5.1 In-plane Curvature Effects

This analysis involves three parameters (curvature radius R, angle Θ_y and frequency f), so the coefficient of transmitted power is plotted in a level curve manner. That is, for a given radius R, we chose the horizontal slices to represent the transmitted power coefficient per frequency, while each vertical line represents a given angle Θ_y, from 0° to a full loop (360°). In other words, vertical slices shows transmitted power per angles. So, this will be the standard for all level curve plots herein.

Changes in radius simply shift the level curves to the left or right in frequency, as shown for $R = 30$, 100 and 200 mm (Fig. 3). Therefore, this method is robust in mapping curvature effects on vibrations.

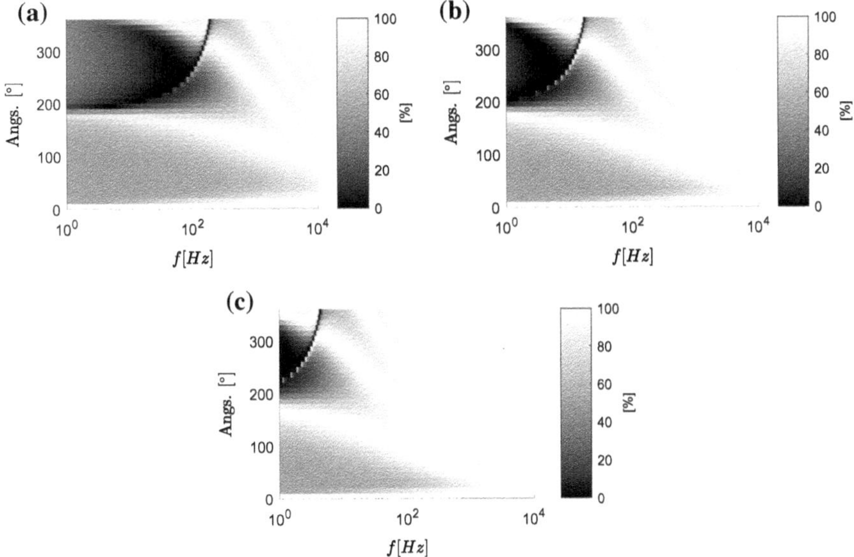

Fig. 3 Sum of all power coefficients of transmitted waves with varying radius. figures shows results for $R = 30$ mm **a**, for $R = 100$ mm **b** and for $R = 200$ mm **c**. These images shows that the "main curve behavior" is maintained

For an arbitrarily chosen radius of 100 mm and prescribing a unitary flexural wave, several repeating regions can be identified over the flexural transmitted power (Fig. 4).

A region of medium to low transmission is present over a wide range of frequency, which occurs for angles $0° < \Theta_y < 150°$. This area is particularly interesting because of its "reflective nature". A high flexural transmission region exists on the plots top-right portion, which reveals that curvatures are negligible for sufficiently small wave length. Meanwhile, the lowest transmission region can be found at the plots top-left area (around $180° < \Theta_y < 350°$). This region is heavily dependent on the flexural wave length being greater than $2\pi R\sqrt{3}$, which is the condition for the first cutoff frequency Ω_1 and will be explained immediately after. A discontinuity can be observed from $\Theta \approx 200°$, which occurs because of a longitudinal transmission predominance (Fig. 5).

Figure 5 shows only longitudinal transmission W_u^T for the same prescribed flexural wave in 3D. In a region of almost no longitudinal transmission located on the plots left, a small peak (≈15%) of longitudinal transmission can be found.

As was observed in [1–3], those three "cutoff" frequencies are due to a curved beams wave numbers. It is known that the wave numbers behave as real numbers before Ω_1 (or in part I of Fig. 6); then, two wave numbers becoming complex at region II, and later turning completely imaginary at III, and finally one of the wave numbers returning to be a real number at IV.

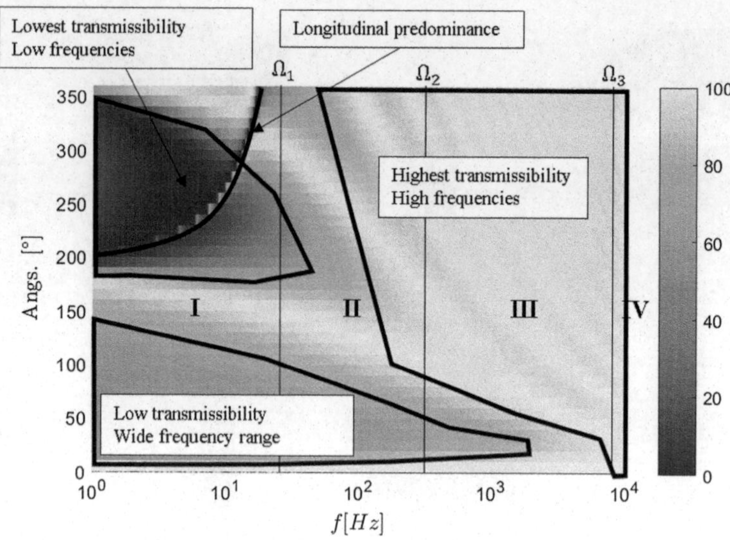

Fig. 4 Mapping of total transmission of power coefficients for a generic curvature type. This shows a single arbitrary curve with a $R = 100$, varying angle from $0° \rightarrow 360°$, while prescribing a flexural wave of unitary amplitude

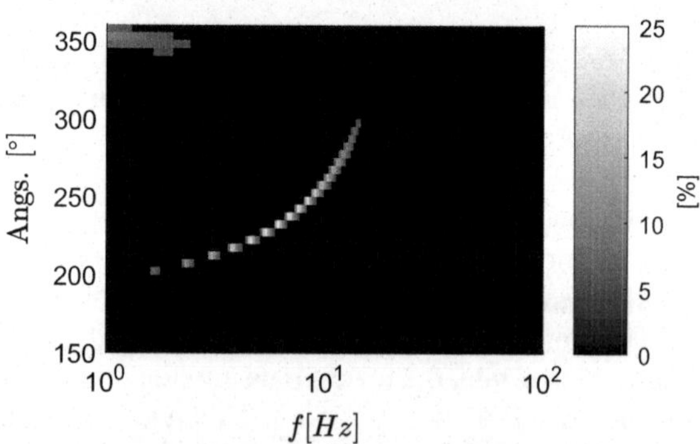

Fig. 5 Longitudinal wave predominance found in region I, with flexural wave input, occurring in angles higher than 180°

Figure 6 shows these curved beams wave numbers (γ_1, γ_2 and γ_3) and compares them with a straight beam wave numbers (γ_u longitudinal and γ_v flexural). Lee et al. [2] also pointed out that Ω_1 and Ω_2 are approximately

$$\Omega_1 \approx \frac{\sqrt{\frac{EI_z}{\rho A}}}{3R^2}, \quad \Omega_2 \approx \frac{\sqrt{\frac{EI_z}{\rho A}}}{R^2}, \tag{6}$$

which leads to the flexural wavelength $\lambda_f \approx 2\pi R\sqrt{3}$ and πR respectively. This last "cutoff frequency" is called the *ring frequency*, and it is expressed as

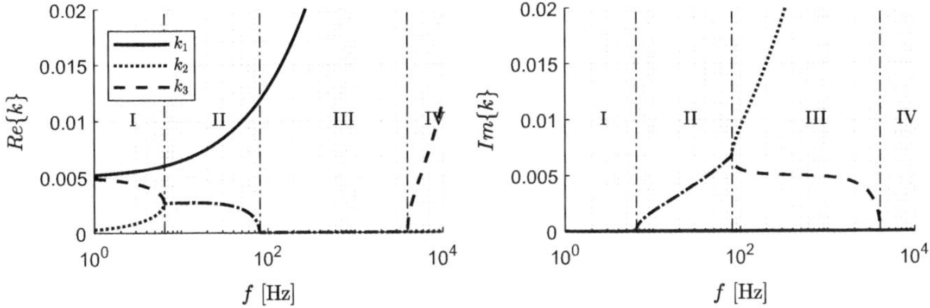

Fig. 6 Wave numbers for in-plane vibration of a curved beam

$$\Omega_3 = \frac{\sqrt{\frac{I_z}{A}}}{R}, \tag{7}$$

which depends only on geometric properties and occurs at a longitudinal wave length $\lambda_l = 2\pi R$.

One can categorize the first medium-low transmission region as $0^\circ < \Theta_y < 150^\circ$ below Ω_3, the very low transmission region as $180^\circ < \Theta_y < 350^\circ$ until mid-frequency of part II (between Ω_1 and Ω_2), and finally, by exclusion, all other regions as high transmission.

For the longitudinal transmission discontinuity, the following formula can be fit as follows:

$$\Omega_d(\Theta_y) \approx 0.9\Omega_1 - \frac{4.3}{\Theta_y^2 + \sqrt{\Theta_y}}, \tag{8}$$

$$\approx \frac{0.3\sqrt{\frac{EI_z}{\rho A}}}{R^2} - \frac{4.3}{\Theta_y^2 + \sqrt{\Theta_y}}, \tag{9}$$

which depend only on the angle Θ_y and Ω_1, because the discontinuity grows asymptotically to it, as shown in the Fig. 4.

This fitted formula can be used to estimate a critical scenario of longitudinal waves, e.g., in an application which this kind of transmission should be avoided by design.

5.2 *Out-of-Plane Curvature Effects*

The level curve plots for out-of-plane vibrations are almost the same of those from in-plane vibrations. The main difference is the coupling between flexural out-of-plane and torsional vibrations. However, given that the wave numbers for this case are also very similar to those in the previous case, all discussion about "transmission regions" is valid by analogy (in- and out-of-plane flexural waves, as well as longitudinal and torsional waves, are similar).

A new approach is necessary to investigate 3D effects, so the following setup is proposed: Let there be a simple curve with angle Θ_y (which is the same "curves angle" Θ_y from before) and length L with an in-plane prescribed flexural wave. This curve will be sliced in the middle ($\Theta_y/2$, therefore $L/2$) and a "rolling" of Θ_x will be applied to the second section.

As this case will be tested for Θ_y from 0° to 360°, Θ_x from 0° to 180°, and frequency f from 0 to 10 kHz, the level curve approach will not suffice because of the extra variable. Therefore, a mean power over frequency is used as a parameter in the form of

$$\bar{W} = \frac{1}{n_f} \sum_{i=1}^{n_f} W. \tag{10}$$

This parameter condenses the results, allowing another "level curve-like" plot. The only disavantage to this approach is the loss of detailed frequency analysis.

To simplify even further for preliminary analysis, the observed transmitted power is a sum of all transmission coefficients

$$W = W_u^T + W_w^T + W_v^T + W_{\theta x}^T, \tag{11}$$

so that details about a waves direction are not overwhelming. The results are shown in Fig. 7, in which x shows the variation for pitch angles and y axis shows the variation in "rolling angles" (Θ_x), for a $R = 6$ mm configuration.

One may immediately detect in Fig. 7 some low transmission areas, which are highlighted by black contours: Region I on the left with curves from 30° to 120° and rolling angles up to about 140° (about 63% of mean power transmission); Region II, a small one with low transmission levels (about 34%), with curves from 180° to 330° but rolling angles only going up to 60°; and Region III, the region of lowest lowest

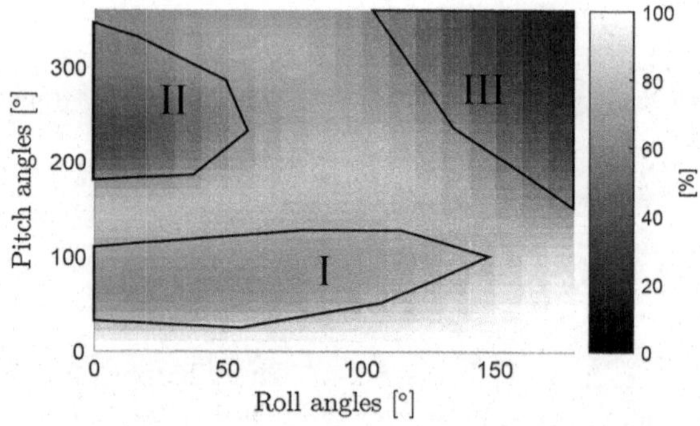

Fig. 7 Condensed mean transmission for out-of-plane flexural vibration acting over a $R = 6$ mm curve

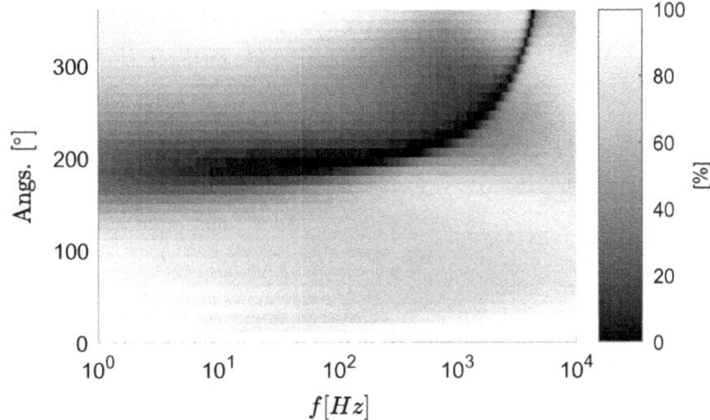

Fig. 8 Power coefficients for a flexural out-of-plane vibration acting over a single $R = 6$ mm curve (without the "rolling aspect")

transmission (reaching below 12%), with curves going from 180° all the way up to a full loop, and rolling angles above 120°.

These regions can be explained by an in-plane level curve map with $R = 6$ mm, resulting in Fig. 8. In- and out-of-plane vibrations are changed in a very similar manner, so it makes clear why regions above 180° are prone to low transmission levels, given that a small radius will increase the cutoff frequencies, especially the first ($\Omega_1 \approx \sqrt{EI_z/\rho A}/3R^2$).

Thus, for a small radius, the best configuration possible is a series of $\Theta_x = 180°$ curves aligned in the same plane.

As the radius increases (now $R = 50$ mm), the low transmission regions for pitch angles $\Theta_y > 180°$ vanishes, as illustrated in Fig. 9.

This phenomenon is due to the low cutoff frequencies, which result from the high transmission region discussed in the previous section. Hence, for 50 mm radius, the "medium-to-low transmission" region appears as the best option, with curve angle Θ_y from 20° to 120° and rolling angles higher than 95° (highlighted by the black line).

When the radius is increased even further, the tendency is to obtain more transmission overall. However, starting from $R = 100$ mm, low power transmission starts to concentrate between $30° < \Theta_y < 60°$ and high rolling angles.

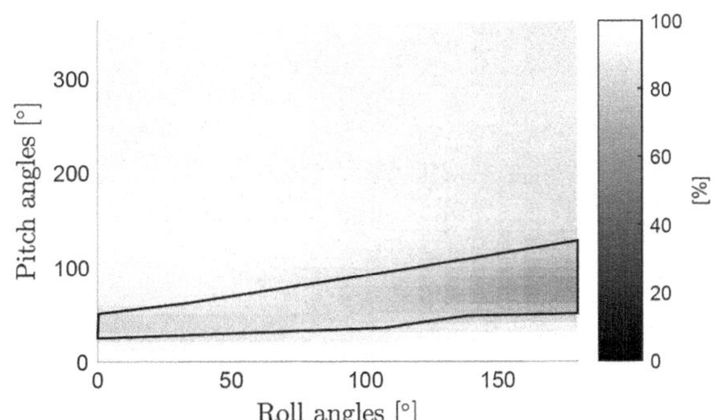

Fig. 9 Condensed mean transmission for out-of-plane flexural vibration acting over a $R = 50$ mm curve. Highlighted area shows the minimum transmission region

In general, from a vibrations perspective, best case scenario is when the structure have a succession of 50° curves, aligned in the same plane. Evidently, for more complex structures, the curves radius should be considered to yield better results.

6 Conclusions

The present work investigated the changes in power flow of vibrational waves, while propagating through smooth bent slender beams. This paper presented the wave propagation approach, as well as the standardization of results using power coefficients.

Subsequently, an FEM program was constructed to simulate the proposed approach. The power coefficients were obtained through this numerical tool. Analytical validation was carried out, with all formulations for both in-plane and out-of-plane vibrations. Differences between the two were discussed.

With the validated FEM tool, the effects of in-plane and out-of-plane vibrations were discussed in detail. Moreover, different setups for obtaining results from each case were proposed. First, longitudinal and torsional waves do not transmit meaningful vibrational power through curves, except for curve angles nearing 0° and for one "discontinuity line" given by $\Omega_d(\Theta_y) \approx 0.9\Omega_1 - 4.3/(\Theta_y^2 + \sqrt{\Theta_y})$.

Both in- and out-of-plane vibrations are similarly changed when their paths have a smooth bend. Longitudinal waves are analogous to torsional ones, and flexural waves on the same plane of a curve are analogous to the plane perpendicular to the curve.

For the effects of in-plane vibration, three major regions were identified: medium-to-low transmission/wide frequency range, very low frequency/low frequency range and high transmission.

Finally, for sequential out-of-plane curves, the radius plays a huge influence over overall transmitted power. If the cutoff frequencies are sufficiently high, two important low-transmission regions exist for curve angles above 180°. Otherwise, the previously mentioned medium-to-low region is highlighted, showing that the lowest transmission values are between 30° and 120° curves.

This study aimed to enlighten the curvature effects on vibrations, for the benefit of engineers to design of new products that use curved slender structures.

References

1. Walsh, S.J., White, R.G.: Vibrational power transmission in curved beams. J. Sound Vib. (Elsevier) **233**(3), 455–488 (2000)
2. Lee, S.K., Mace, B.R., Brennan, M.J.: Wave propagation, reflection and transmission in curved beams. J. Sound Vib. (Elsevier) **306**, 636–656 (2007)
3. Wu, C.M., Lundberg, B.: Reflection and transmission of the energy of harmonic elastic waves in a bent bar. J. Sound Vib. (Elsevier) **190**(4), 645–659 (1996)

4. Wang, T.M., Nettleton, R.H., Keita, B.: Natural frequencies for out-of-plane vibrations of continuous curved beams. J. Sound Vib. (Elsevier) **68**(3), 427–436 (1980)
5. Gavric, L.: Power-flow analysis using infinite beam elements. Le J. de Phys. IV, EDP Sci. **2**(C1), C1–511 (1992)
6. Davis, R., Henshell, R.D., Warburton, G.B.: Constant curvature beam finite elements for in-plane vibration. J. Sound Vib. **25**(4), 561–576 (1972). https://doi.org/10.1016/0022-460X(72)90478-6
7. Zienkwiecz, O.C., Taylor, R.L.: The Finite Element Method for Solid and Structural Mechanics, 6th edn, p. 736. [S.l.]: Butterworth-Heinemann (2005)
8. Hamilton, W.R.: On quaternions, or on a new system of imaginaries in algebra. In: David R. Wilkins (ed.) Philosophical Magazine in 2000. Dublin (1843)
9. Kuipers, J.B.: Quaternions and Rotation Sequences. Princeton Univeristy Press (1999). ISBN 0-691-10298-8

Nonlinear Identification Using Polynomial NARMAX Model and a Stability Analysis of an Aeroelastic System

Raphaela C. M. G. Barbosa, Luiz C. S. Góes, Airton Nabarrete, José M. Balthazar and David F. C. Zúñiga

Abstract This work describes the nonlinear identification applied to an aeroelastic pitch-plunge system using polynomial NARMAX model and a stability analysis. The apparatus is available and consists of a wing typical section with pitch and plunge degrees of freedom. The identification procedure aims to obtain the parameters for the mathematical model including the torsional stiffness as a quadratic polynomial function. The candidate structure to the polynomial model is obtained from discretization of a continuous-time state-space model and the predictions are obtained via the identification procedure using simulated data. The simulation is performed considering the aerodynamics with free stream velocity increased within an established velocity range which includes the flutter phenomenon. In future work, a data acquisition from the experimental apparatus will be performed. The NARMAX model indicates a polynomial function of fourth order for the nonlinearity and a stability analysis, discussed in this work, mapping the nonlinear regions.

Keywords Nonlinear system identification · Polynomial NARMAX model
Aeroelastic system · Stability analysis

1 Introduction

Recent researches exhibited aeroelastic systems presenting phenomena such as limit cycle oscillation (LCO) occurring due to nonlinearities. Many of these studies demonstrate the aeroelasticity of aircraft wings based on typical section model. In other work O'Neil et al. [1] demonstrates the LCO caused by the nonlinearity and considers the free stream velocity as the main parameter to increase oscillations amplitudes.

R. C. M. G. Barbosa (✉) · L. C. S. Góes · A. Nabarrete · J. M. Balthazar · D. F. C. Zúñiga
Instituto Tecnológico de Aeronáutica, São José dos Campos, Brazil
e-mail: machado@ita.br

L. C. S. Góes
e-mail: goes@ita.br

A. de T. Fleury et al. (eds.), *Proceedings of DINAME 2017*, Lecture Notes in Mechanical Engineering, https://doi.org/10.1007/978-3-319-91217-2_7

The stiffness nonlinearity is also estimated through the experiments by O'Neil and Strganac [2], where they proposed an aeroelastic system composed by a rigid wing supported by cubic springs. Indeed, unstable regions associated to LCO are predicted and evidences of internal resonance behavior is verified in the apparatus.

Strganac et al. [3] presents the identification of a limit cycle oscillation (LCO) in an aeroelastic system, concluding that sometimes a nonlinear approach is necessary for designing a control system. Based on this research other control strategies are discussed for the suppression of LCO, for instance the nonlinear adaptive controller. More recently Xu et al. [4] report a nonlinear flutter control law based on state variable feedback that is designed to suppress the LCO to an airfoil system with a hysteresis as nonlinearity.

The researches on nonlinear aeroelastic systems have been stated and the system identification has been performed to identify or adjust the characteristic of the structural nonlinearities (cubic, quadratic, hysteresis or freeplay) using its time history. In this work an identification procedure is applied to the experimental apparatus of a typical section intending the posterior stability analysis based on the model performance. In the model the torsional spring has a nonlinear stiffness that behaves as a fourth order polynomial, and for which the nonlinear effects are investigated in this research. The experimental apparatus is presented in Fig. 1.

Some methodologies for nonlinear identification are described in the literature. Popescu et al. [5] presents the identification the nonlinearities based on nonparametric estimation. On the other hand, the parametric estimation proposed by Kukreja [6] is performed using the polynomial NARMAX (nonlinear autoregressive moving average exogenous) to identify the quadratic nonlinearity from the experimental data of an airfoil without control surface. Additionally, Obeid [7] presented a closed loop feedback control of an airfoil with high angles of attack using a NARMAX model.

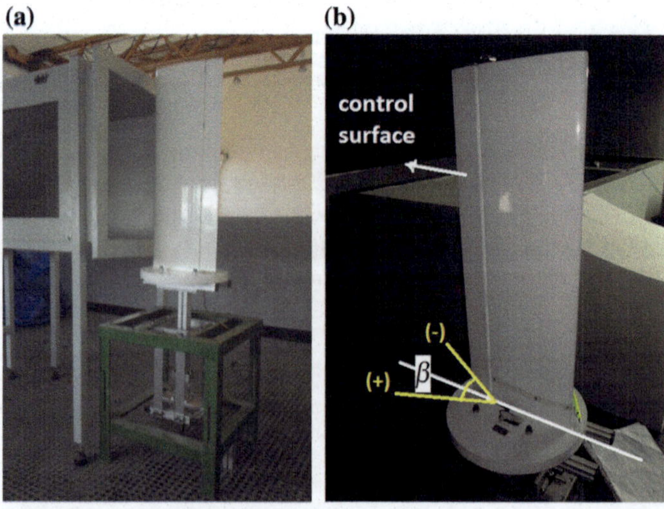

Fig. 1 Airfoil apparatus available at the laboratory

In this way, the nonlinear model is identified from simulated data using a polynomial NARMAX model to capture the static nonlinearity due to torsional spring. The identified model explains the system behavior, e.g., the flutter phenomenon that causes the limit cycle oscillations, useful to the feedback control system design and the evaluation of the airfoil performance. This procedure will be used in a future work with the experimental data acquisition of the experimental apparatus.

2 Nonlinear Aeroelastic Model

The equations of motion of the typical section as depicted in Fig. 2 consider two degrees of freedom as functions of time, plunge $h(t)$ and pitch $\alpha(t)$, for the dynamic response of the system. The trailing edge control surface angle β is considered as a control input of the model.

The coefficients of the dynamic model consist of mass m_T, linear stiffness k_h related to the translation (plunge), moment of inertia I_α and nonlinear torsional stiffness $k_\alpha(\alpha)$ related to the angular displacement of pitch. The inherent damping of the system is modeled as produced exclusively by the deformation of the elastic elements. The damping coefficients are obtained from modal testing.

Aerodynamic lift L and moment M are external efforts applied to the dynamic system. Some coefficients are chosen in accordance with experimental apparatus as shown in Table 1 and some others are defined according to O'Neil [8]. The mathematical model of the aeroelastic system is presented as

$$\begin{bmatrix} m_T & m_T x_\alpha b \\ m_T x_\alpha b & I_\alpha \end{bmatrix} \begin{Bmatrix} \ddot{h} \\ \ddot{\alpha} \end{Bmatrix} + \begin{bmatrix} c_h & 0 \\ 0 & c_\alpha \end{bmatrix} \begin{Bmatrix} \dot{h} \\ \dot{\alpha} \end{Bmatrix} + \begin{bmatrix} k_h & 0 \\ 0 & k_\alpha(\alpha) \end{bmatrix} \begin{Bmatrix} h \\ \alpha \end{Bmatrix} = \begin{bmatrix} -L \\ M \end{bmatrix} \tag{1}$$

The influence of aerodynamic lift and moment on the airfoil is considered under free stream velocity V_∞. Additionally, the model includes the angular position of the control surface β. With some coefficients adopted as referred in Kukreja [6], the quasi-steady aerodynamic lift and moment are evaluated as

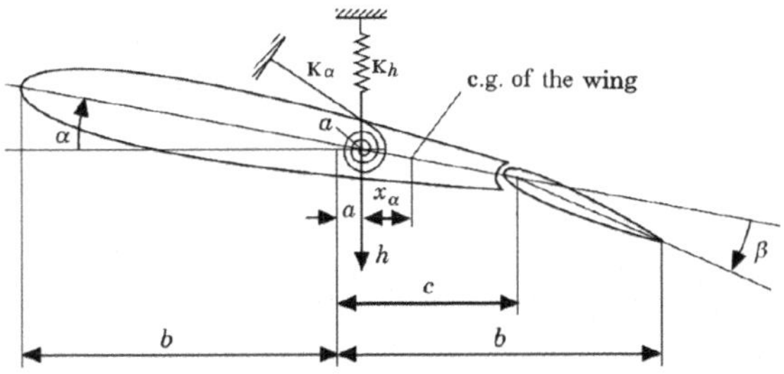

Fig. 2 Typical section model of an airfoil [9]

Table 1 Continuous model coefficients

Coefficient	Value	Coefficient	Value	Coefficient	Value
m_T (kg)	8	ξ_h	0.022	k_{α_0} (N m/rad)	3
I_a (kg m^2)	0.0505	ξ_α	0.03	k_{α_1} (N m/rad^2)	−30
x_a (m)	0.084	c_h (N s/m)	6.6373	k_{α_2} (N m/rad^3)	6600
b (m)	0.1064	c_α (N s/m)	0.2939	k_{α_3} (N m/rad^4)	−21000
a	−0.8	k_h (N/m)	2844.4	k_{α_4} (N m/rad^5)	48000

$$L = \rho U^2 b C_{l\alpha}\left[\alpha + \frac{h}{U} + \left(\frac{1}{2} - a\right) b\frac{\dot\alpha}{U}\right] + \rho U^2 b C_{l\beta}\beta \tag{2}$$

$$M = \rho U^2 b^2 C_{m\alpha}\left[\alpha + \frac{\dot h}{U} + \left(\frac{1}{2} - a\right) b\frac{\dot\alpha}{U}\right] + \rho U^2 b^2 C_{m\beta}\beta$$

The nonlinear torsional stiffness is modeled as a polynomial of fourth order

$$k_\alpha(\alpha, t) = k_{\alpha_0} + k_{\alpha_1}\alpha + k_{\alpha_2}\alpha^2 + k_{\alpha_3}\alpha^3 + k_{\alpha_4}\alpha^4 \tag{3}$$

Rearranging the terms from Eq. 1 the stiffness and damping matrices are obtained including the aeroelastic influence under the free stream velocity as

$$\begin{bmatrix} m_T & m_T x_\alpha b \\ m_T x_\alpha b & I_\alpha \end{bmatrix}\begin{Bmatrix} \ddot h \\ \ddot\alpha \end{Bmatrix} + \begin{bmatrix} c_h & 0 \\ 0 & c_\alpha \end{bmatrix}\begin{Bmatrix} \dot h \\ \dot\alpha \end{Bmatrix} + \begin{bmatrix} \rho U b C_{l\alpha} & \rho U b^2 C_{l\alpha}\left(\frac{1}{2} - a\right) \\ \rho U b^2 C_{m\alpha} & -\rho U b^3 C_{m\alpha}\left(\frac{1}{2} - a\right) \end{bmatrix}\begin{Bmatrix} \dot h \\ \dot\alpha \end{Bmatrix}$$

$$+ \begin{bmatrix} k_h & 0 \\ 0 & k_\alpha(\alpha) \end{bmatrix}\begin{Bmatrix} h \\ \alpha \end{Bmatrix} + \begin{bmatrix} 0 & \rho U^2 b C_{l\alpha} \\ 0 & -\rho U^2 b^2 C_{m\alpha} \end{bmatrix}\begin{Bmatrix} h \\ \alpha \end{Bmatrix} = \begin{bmatrix} -\rho b C_{l\beta} \\ \rho b^2 C_{m\alpha} \end{bmatrix} U^2\beta \tag{4}$$

The constant value of the plunge stiffness and the coefficients of the nonlinear torsional stiffness are informed in the Table 1, in the same way as some other necessary coefficients. The system response is evaluated with constant free stream velocity $V_\infty = 6$ m/s. The plunge h and pitch α responses are based on their initial conditions. Two upper graphics on Fig. 3 show the time history for displacements with the initial conditions $h(0) = 0.1$ m and $\alpha(0) = 0.01$ rad. In these graphics the control surface is located in the neutral angular position, $\beta = 0°$. It is not possible to note the influence of the stiffness nonlinearity considering these responses only. The two graphics below in the same figure show the time history for linear and angular velocities.

In Fig. 3 the nonlinear influence is very difficult to be noted, but it is better observed in Fig. 6. After some calculations the angular moment is showed as a function of the angular displacement α, being the static nonlinearity conveniently modeled as $k_\alpha(\alpha) = 3(1 - 10\alpha + 2200\alpha^2 - 7000\alpha^3 + 16000\alpha^4)$. The time history considers also the free stream velocity $V_\infty = 6$ m/s and a pitch angle of −15° is outlined as initial condition. The angular displacement is set to $\beta = 15°$, considering it

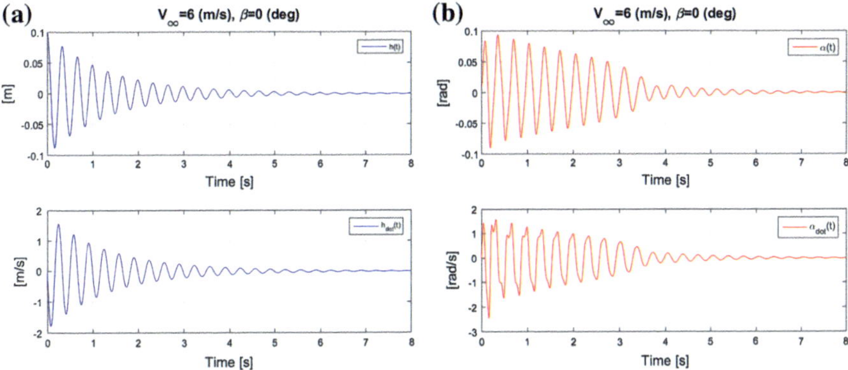

Fig. 3 Time history for initial conditions $h(0) = 0.1$ m and $\alpha(0) = 0.01$ rad: **a** vertical displacement and velocity; **b** angular displacement and velocity

positive or negative, as specified for the maximum nominal angular displacement of the control surface.

3 Nonlinear Identification Using Polynomial NARMAX Model

After obtaining the time history from system simulation, the nonlinear identification procedure is applied to obtain a discrete-time model that represents the system dynamics. The nonlinear system identification is performed considering the measurement noise applied to the linear velocity and angular velocity.

System identification using only input-output experimental data allows for a black box model. In this section the nonlinear identification using a polynomial NARMAX model is applied, but the structure selection is in some sense based on the continuous-time model from the previously section. Therefore the known system, narrow box model, is used and a relationship between a continuous-time model and an identified discrete-time model is possible. Finally the model is not totally unknown and a plunge stiffness estimating can be obtained based on the procedure presented in Kukreja [6].

The nonlinear identification using polynomial NARMAX model consists in the representation of the system output as a polynomial function of $y(n)$, $u(n)$ and $e(n)$ with nonlinearity order l. It follows

$$y(n) = F^l[y(n-1), \ldots, y(n-n_y), u(n-n_u-d), \ldots, u(n-n_u), \\ e(n-1), \ldots, e(n-n_e)] + e(n) \tag{5}$$

The following section presents the structure obtained to represent the system dynamics. The model structure have two outputs, $\dot{h}(n)$ and $\dot{\alpha}(n)$, with delays $n_y = 1$,

$n_u = 1$, $n_e = 1$, dead time $d = 0$, nonlinearity order $l = 5$ corresponding to polynomial of fourth order to represent the torsional stiffness $k_\alpha(\alpha)$. The nonlinear identification deals with a typical least square problem. As so, the nonlinearities appear in the regression matrix and the problem is linear in the parameters. It becomes

$$y(n) = \psi^T(n-1)\hat{\theta} + \xi(n) \tag{6}$$

3.1 Model Structure Selection

The model structure considered to this system is represented by a NARMAX model based in results obtained by Kukreja [6], whose quadratic stiffness type is described in [10]. In order to describe the model used in the identification process, it is mass-normalized as

$$\ddot{\mathbf{x}} + \mathbf{M}^{-1}\mathbf{C}\dot{\mathbf{x}} + \mathbf{M}^{-1}\mathbf{K}\mathbf{x} = \mathbf{M}^{-1}\mathbf{Q} \tag{7}$$

where, $\ddot{\mathbf{x}} = \begin{Bmatrix} \ddot{h} \\ \ddot{\alpha} \end{Bmatrix}$ and $\mathbf{Q} = \begin{Bmatrix} -L \\ M \end{Bmatrix}$.

Rearranging, the continuous-time model is represented as

$$\begin{Bmatrix} \ddot{h} \\ \ddot{\alpha} \end{Bmatrix} = \begin{Bmatrix} -k_1 h - [k_2\mu + pk_\alpha(\alpha)] - c_1\dot{h} - c_2\dot{\alpha} \\ -k_3 h - [k_4\mu + qk_\alpha(\alpha)] - c_3\dot{h} - c_4\dot{\alpha} \end{Bmatrix} \tag{8}$$

The coefficient $\mu = U^2\beta$ and coefficients $k_1, k_2, k_3, k_4, c_1, c_2, c_3, c_4$ are described according to [6] and $p = (mx_\alpha b)/b$ and $q = m/d$.

The discrete-time nonlinear model used in the identification procedure is obtained from discretization of Eq. 8 by applying the Forward Euler Method given by

$$\dot{h}(n) = \dot{h}(n-1) + T\ddot{h}(n-1) \tag{9}$$

$$\dot{\alpha}(n) = \dot{\alpha}(n-1) + T\ddot{\alpha}(n-1)$$

Note that the variable n is the multiple of sample period. The sampling rate of 1 kHz is defined considering the fastest system dynamics and the Nyquist Theorem. The NARMAX structure follows the discrete-time model represented as

$$\begin{aligned} \dot{h}(n) &= \theta_1\dot{h}(n-1) + \theta_2 h(n-1) + \theta_3\alpha(n-1) + \theta_4\alpha(n-1)^2 + \theta_5\alpha(n-1)^3 \\ &\quad + \theta_6\alpha(n-1)^4 + \theta_7\alpha(n-1)^5 + \theta_8\dot{\alpha}(n-1) + \theta_9 u(n-1) + \theta_{10}e_{\dot{h}}(n-1) + e_{\dot{h}}(n) \\ \dot{\alpha}(n) &= \gamma_1\dot{\alpha}(n-1) + \gamma_2 h(n-1) + \gamma_3\alpha(n-1) + \gamma_4\alpha(n-1)^2 + \gamma_5\alpha(n-1)^3 \\ &\quad + \gamma_6\alpha(n-1)^4 + \gamma_7\alpha(n-1)^5 + \gamma_8\dot{h}(n-1) + \gamma_9 u(n-1) + \gamma_{10}e_{\dot{\alpha}}(n-1) + e_{\dot{\alpha}}(n) \end{aligned} \tag{10}$$

All parameters used above are shown in Table 2.

Table 2 Discrete-time model parameters

Model term	Identified	Model term	Identified
θ_1: 0.9986	0.9986	γ_1: 0.9938	0.9938
θ_2: −0.3601	−0.3601	γ_2: 0.5099	0.5105
θ_3: −0.0041	−0.0041	γ_3 : −0.0737	−0.0737
θ_4: 0.0018	0.0019	γ_4 : 0.2006	0.2024
θ_5: −0.3944	−0.4060	γ_5: −44.1227	−44.1026
θ_6: 1.2548	0.9508	γ_6: 140.3904	135.9770
θ_7: −2.8680	21.3053	γ_7: −320.8924	−238.6472
θ_8: −0.0000	−0.0000	γ_8 : −0.0011	−0.0011
θ_9: −0.0019	−0.1110	γ_9 : −0.0035	−0.2021
θ_{10}: −0.9986	−0.9529	γ_{10}: −0.9938	−0.9515

3.2 *Identification of the System with Noise*

A polynomial NARMAX model is used to identify the system with noise. The identification procedure consists on the application of an Extended Least Square Estimator (ELS) as indicated by Aguirre [11]. This procedure basically consists in identifying a NARX model to follow with the construction of a NARMAX structure, including the moving average.

The coefficients used for the system simulation are extracted from the relationship between discrete-time and continuous-time model [6]. In order to prepare the method for future real applications considering the experimental data, the noise is modeled as a Gaussian distribution with covariance $\sigma_{\dot{h}}^2 = 0.04 \times 10^{-5}$ and $\sigma_{\dot{\alpha}}^2 = 0.12 \times 10^{-4}$. The signal to noise ratio is 99.51 dB and 37.60 dB for the outputs $\dot{h}(n)$ e $\dot{\alpha}(n)$, respectively.

The process is excited by the angular displacement of the control surface β, with an aleatory signal having constant levels of 12 s each as depicted in Fig. 4. The ELS procedure is repeated until iteration $k = 10$ and the identified parameters is presented in the Table 2.

To present the model output predictions, by convenience, the same data set is used for identification and validation. The identification data corresponding to 2/3 of the complete data set. The Fig. 5 shows the output predictions of the identified model.

From the identified discrete-time model parameters is possible to return to the continuous-time model following the relationships presented in Eq. 11. The identified stiffness parameters are present in the Table 3.

$$k_{\alpha_0} = \frac{b_2 - k_4\mu}{m/d}, k_{\alpha_1} = \frac{b_3}{m/d}, k_{\alpha_2} = \frac{b_4}{m/d}, k_{\alpha_3} = \frac{b_5}{m/d}, k_{\alpha_4} = \frac{b_6}{m/d} \tag{11}$$

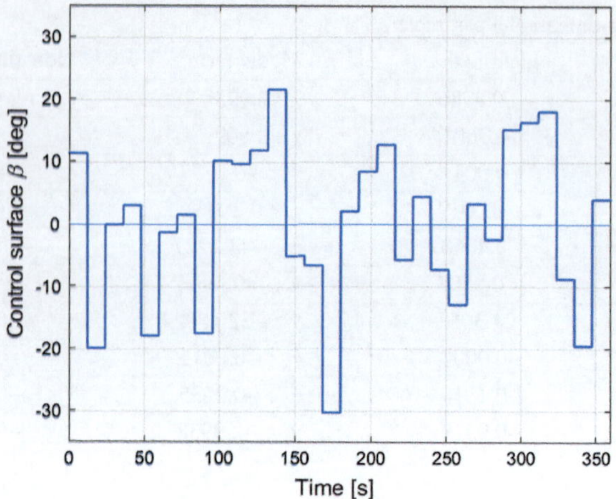

Fig. 4 Excitation signal for system identification

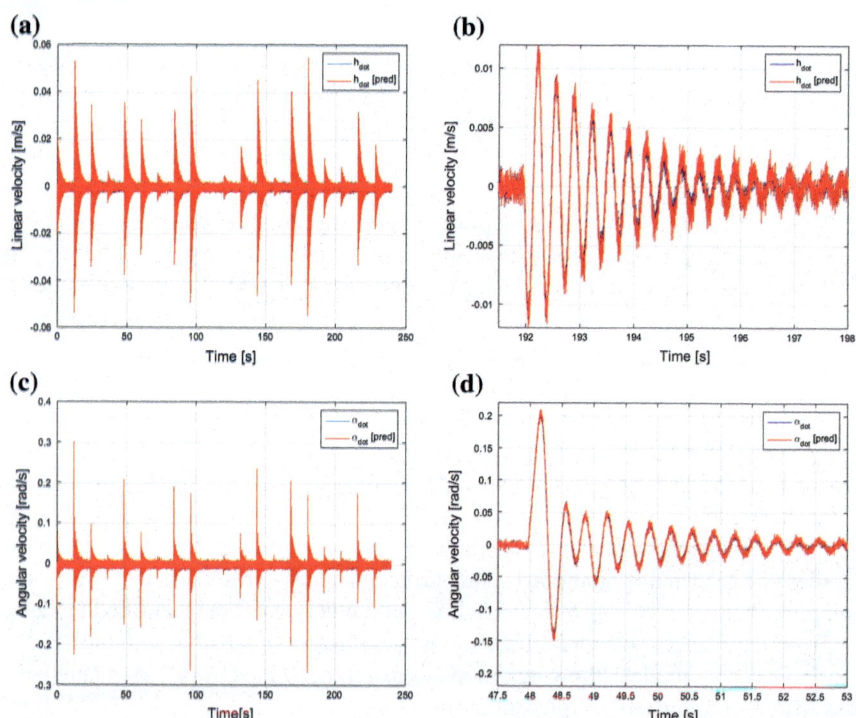

Fig. 5 **a** Output prediction for the linear velocity, $\dot{h}(n)$, **b** enlarged time for linear velocity, **c** output prediction for the angular velocity, $\dot{\alpha}(n)$ and **d** enlarged time for angular velocity

Table 3 Parameters of the nonlinear stiffness

Parameter	Value	Identified
k_{α_0}	3	2.99
k_{α_1}	−30	−30.24
k_{α_2}	6600	6592.35
k_{α_3}	−21000	−20325.51
k_{α_4}	48000	35672.42

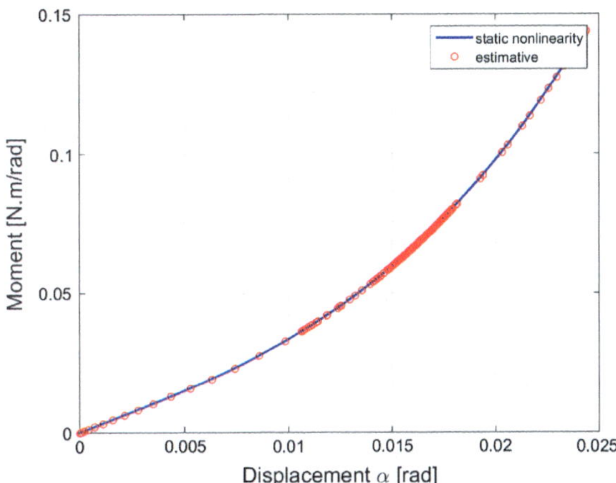

Fig. 6 A representation of the nonlinear behavior: moment as function of the angular displacement

where, $d = m(I_\alpha - mx_\alpha^2 b^2)$ and $k_4 = (-mx_\alpha b^2 \rho C_{l\alpha} - m\rho b^2 C_{m\alpha})/d$. The parameters b_2, b_3, b_4, b_5 and b_6 are obtained from the identified discrete-time model parameters following $b_2 = \gamma_3/(-T), b_3 = \gamma_4/(-T), b_4 = \gamma_5/(-T), b_5 = \gamma_6/(-T)$ and $b_6 = \gamma_7/(-T)$, with sample period T.

The identified parameters of the nonlinear torsional stiffness are also suitable to represent the static nonlinearity present on the system dynamics. Although the parameter k_{α_4} has a very different value in comparison with the true one, the nonlinearity is in the range of interest, between the operational limits for the pitch angle, around 1.5°. The Fig. 6 is obtained with simulated of continuous-time model. The static nonlinear behavior is well captured by the model. The nonlinearity of this system is assumed as symmetric.

4 A Stability Analysis

Stability analysis is important to map the regions where exists stable behavior for open loop operation. In cases where the instability occurs, the solution is the improvement of the control design in order to stabilize the system. One of the goals of this paper is the stability analysis using the identified model. The root locus and the damping factor diagram as function of the free stream velocity and the nonlinear stiffness are presented.

A continuous state-space model may be expressed as follows

$$\dot{\mathbf{x}}(t) = f_\mu[\mathbf{x}(t)] + g[\mathbf{x}(t)]\mu\beta \tag{12}$$

where the state-space variables are defined as

$$\left\{ x_1(t),\ x_2(t),\ x_3(t),\ x_4(t) \right\}^T = \left\{ h(t),\ \theta(t),\ \dot{h}(t),\ \dot{\theta}(t) \right\}^T \tag{13}$$

and consequently their derivatives are

$$\left\{ \dot{x}_1(t),\ \dot{x}_2(t),\ \dot{x}_3(t),\ \dot{x}_4(t) \right\}^T = \left\{ \dot{h}(t),\ \dot{\theta}(t),\ \ddot{h}(t),\ \ddot{\theta}(t) \right\}^T \tag{14}$$

Therefore the dynamic equation may be written as Eq. 15. Note that the term $f_\mu(\mathbf{x})$ depends of the nonlinearity and will compound the Jacobian matrix. The control surface angle is null for the stability analysis.

$$\begin{Bmatrix} \dot{x}_1(t) \\ \dot{x}_2(t) \\ \dot{x}_3(t) \\ \dot{x}_4(t) \end{Bmatrix} = \begin{bmatrix} 0 & 0 & 1 & 0 \\ 0 & 0 & 0 & 1 \\ -k_1 & -[k_2\mu + pk_{x_2}(x_2(t))] & -c_1 & -c_2 \\ -k_3 & -[k_4\mu + qk_{x_2}(x_2(t))] & -c_3 & -c_4 \end{bmatrix} \begin{Bmatrix} x_1(t) \\ x_2(t) \\ x_3(t) \\ x_4(t) \end{Bmatrix} + \begin{bmatrix} 0 \\ 0 \\ g_3 \\ g_4 \end{bmatrix} \mu\beta \tag{15}$$

It can be observed that the term $k_{x_2}(x_2)$ is $k_\alpha(\alpha)$ and due the nonlinearity it is necessary to calculate the Jacobian matrix for evaluating the eigenvalues. Therefore the Jacobian matrix is given by

$$\mathbf{J} = \begin{bmatrix} 0 & 0 & 1 & 0 \\ 0 & 0 & 0 & 1 \\ -k_1 & -[k_2\mu + p(k_{\alpha 0} + k_{\alpha 1}\alpha(t) + k_{\alpha 2}\alpha(t)^2) + k_{\alpha 3}\alpha(t)^3] & -c_1 & -c_2 \\ -k_3 & -[k_4\mu + q(k_{\alpha 0} + k_{\alpha 1}\alpha(t) + k_{\alpha 2}\alpha(t)^2) + k_{\alpha 3}\alpha(t)^3] & -c_3 & -c_4 \end{bmatrix}_{|\alpha=\alpha_{op}} \tag{16}$$

It is possible verify the eigenvalues of the Jacobian matrix depends merely on the state values α. For this system the origin is a equilibrium point, therefore, the Fig. 7a shows the root locus for $\mathbf{x} = \{0, 0, 0, 0\}^T$.

From Fig. 7a, it can be seen that the instability occurs when the eigenvalue results cross the imaginary axis. This occurs for free stream velocity around 15.32 m/s, as shown in the Fig. 7b. In the instability region starts a LCO and the natural frequency of oscillation varies according to the angular displacement, because of the nonlinearity effects $k_\alpha(\alpha)$.

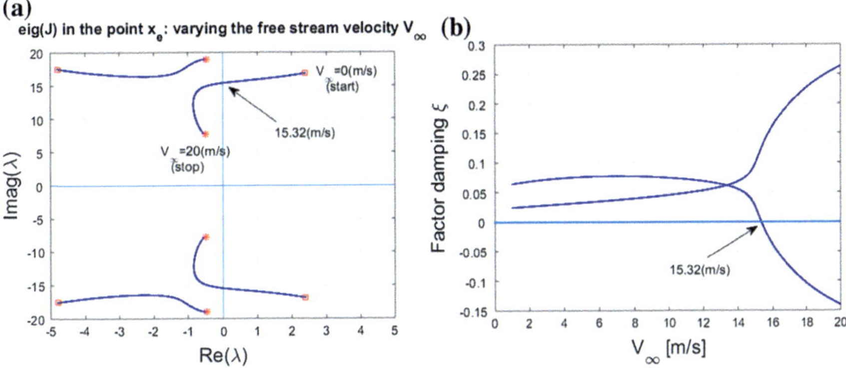

Fig. 7 **a** Root locus for the free stream velocity variation considering the equilibrium point and **b** damping factor as functions of the free stream velocity

4.1 Influence of the Pitch Angle

According to equations of motion it is possible to verify that the nonlinearity affects the stability when a negative damping factor occurs, resulting in unstable behavior. The system becomes unstable due to the nonlinearity as shown in Fig. 8. The figure is obtained varying the system operation point (named x_{op}). In this case, the angular displacement sweep occurred because the torsional stiffness is a function of the α values.

Figure 8 shows six curves from the left to the right, each one corresponding to a free stream velocity of the range $\{6; 6.5; 7; 7.5; 8; 8.5\}$ in m/s. The angular displacement is varying from 0 to 0.07 rad. Note that for reduced values of the free stream velocity, below 7 m/s, the nonlinearity does not affect the system stability. Above

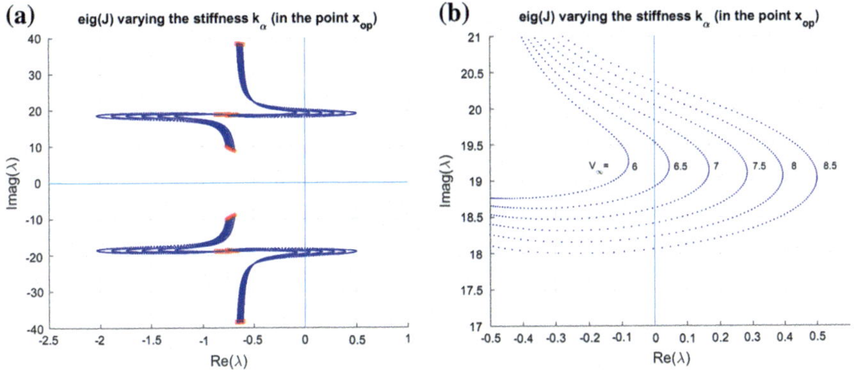

Fig. 8 **a** Root locus for variation in nonlinear stiffness considering six free stream velocities and **b** enlarged region

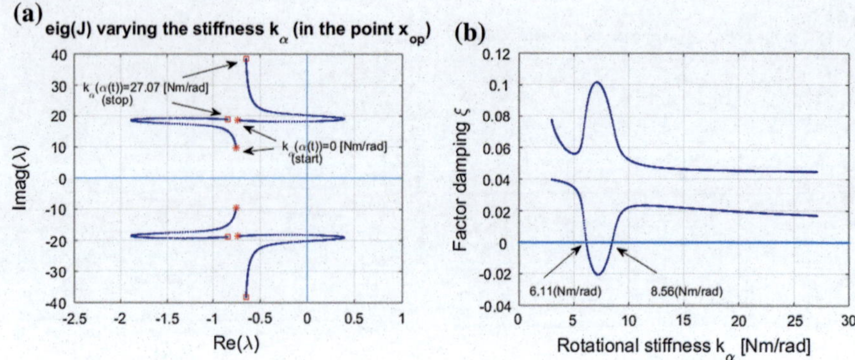

Fig. 9 **a** Root locus for $V_\infty = 8$ m/s varying α between 0 and 0.07 rad, **b** damping factor as function of a nonlinear stiffness

this value, the system is unstable depending on the angular displacement α of the operation point.

The root locus considering the free stream velocity of 8 m/s with the stiffness varying approximately from 0 to 27.07 Nm/rad is shown in Fig. 9. It is observed in Fig. 9b the region of instability for stiffness varying from 6.11 to 8.56 Nm/rad. Through several simulations it is possible to note that the LCO frequency increases as the stiffness rises, considering the stability region.

5 Conclusion

The nonlinear identification applied to an aeroelastic pitch-plunge system is proposed using a polynomial NARMAX model estimated from the extended least square estimator (ELS) just to obtain unbiased estimation even with colored noise in measures. The output predictions and an estimation of the static nonlinearity is presented in order to validate the identified model. The estimates converges quite quickly within ten iterations.

The nonlinear identification results in a suitable nonlinear torsional stiffness. The proposed analysis is validated to be applied with the experimental data from the aeroelastic system. Also, the identified system can be applied to the stability analysis and the control system design with nonlinear approach.

From the stability analysis, the mapping of the unstable regions is outlined based on the eigenvalues of the Jacobian matrix. It concludes that the nonlinearity can lead the system to be dynamically unstable. The aeroelastic system is stable for velocity less than $V_\infty = 15.32$ m/s even with the nonlinearity.

On the other hand, the system becomes unstable with the increase of the pitch angle and the free stream velocity. It is assumed an excursion of control surface with maximum angle of 15°, positive or negative, for normal operational in flight, but larger angles are tested for the identification process. In this work, the system is excited by the control surface from −30° to 30°.

The root-locus is outlined for angular displacement between 0 and 0.07 rad corresponding to maximum angular displacement of 4.01°. The system operating in the unstable region can return to stable region by the action of the control surface leading on eigenvalues crossing the imaginary axis and returning to the left plane.

References

1. O'Neil, T., Gilliatt, H., Strganac, T.W.: Investigations of aeroelastic response for a system with continuous structural nonlinearities. AIAA Meeting Papers on Disc, Paper 96-1390 (1996)
2. O'Neil, T., Strganac, T.W.: Aeroelastic response of a rigid wing supported by nonlinear springs. J. Aircr. **35**(4), 616–622 (1998)
3. Strganac, T.W., Ko, J., Thompson, D.E.: Identification and control of limit cycle oscillations in aeroelastic systems. J. Guidance Control Dyn. **23**(6), 1127–1133 (2000)
4. Xu, X., Gao, Y., Zhang, W.: Aeroelastic dynamic response and control of an aeroelastic system with hysteresis nonlinearities. J. Control Sci. Eng. 10 (2015)
5. Popescu, C., Wong, Y., Lee, B.: System identification for nonlinear aeroelastic models. In: Proceedings of the 46th AIAA/ASME (2005)
6. Kukreja, S.L.: Nonlinear system identification for aeroelastic systems with application to experimental data. In: NASA Technical Memorandum (2008)
7. Obeid, S., Jha, R., Ahmadi, G.: Closed-loop feedback control of flow over a flapped airfoil at high angles of attack using identified NARMAX model. In: Proceedings of the ASME 2012 Fluids Engineering Summer Meeting (2012)
8. O'Neil, T.: Nonlinear aeroelastic response—analyses and experiments. AIAA Meeting Papers on Disc, Paper 96-0014 (1996)
9. Li, D., Guo, S., Xiang, J.: Aeroelastic dynamic response and control of an airfoil section with control surface nonlinearities. J. Sound Vib. **329**, 4756–4771 (2010)
10. Ko, J., Kurdila, A.J., Strganac, T.W.: Nonlinear dynamics and control for a structurally nonlinear aeroelastic system. In: AIAA, p. 13 (1997)
11. Aguirre, L.A.: Introdução a identificação de sistemas: técnicas lineares e não-lineares, Ed. UMFG, p. 730 (2007)
12. Abdelkefi, A., Vasconcellos, R., Marques, F.D., Hajj, M.R.: Modeling and identification of freeplay nonlinearity. J. Sound Vib. **331**, 1898–1907 (2012)
13. Billings, S.A.: Nonlinear System Identification: NARMAX Methods in the Time, Frequency, and Spatio-temporal Domains, p. 574. Wiley, United Kingdom (2013)
14. Dowell, E.: A Modern Course in Aeroelasticity, p. 575. Springer (1989)
15. Farmer, M.G.: A two-degree-of-freedom flutter mount system with low damping for testing rigid wings at different angles of attack. In: NASA Technical Memorandum (1982)
16. Shukla, H.: Nonlinear analysis and control of aeroelastic systems. Doctoral dissertation, Faculty of the Virginia Polytechnic Institute and State University, Blacksburg, Virginia (2016)

Dynamics of Helicopters with DVA Under Structural Uncertainties

Diogo B. P. de Oliveira, João Flávio P. Coelho, Leonardo Sanches and Guilhem Michon

Abstract This paper evaluates the effects and effectiveness of a dynamic vibration absorber (DVA) designed for mitigation of unstable oscillations of helicopters when it is on the ground, i.e.: ground resonance phenomenon. For this purpose, firstly, a parametric analysis shows the stability charts obtained for helicopters without and with DVA as function of the fuselage and rotor damping levels. Also, the present paper is interested on determining the influence of designed DVA devices on the stability robustness of the aircraft once uncertainties are considered in the blade's hinge stiffness. In this sense, μ-analysis is used to predict the smallest stiffness perturbation that leads the helicopter to instability. Indeed, it is not possible to assure the same properties of mechanical components which may alters the effectiveness of the designed DVA. Therefore, the assessment of stability robustness is verified and compared between both aircrafts (without and with DVA). The results showed for both helicopters, if proper combination of fuselage and rotor damping is considered, stability robustness is verified. Also, the inclusion of DVA device on helicopters does not affect the aircraft stability robustness.

Keywords Helicopter ground resonance · Dynamic vibration absorber μ-analysis · Stiffness uncertainty

D. B. P. de Oliveira
Instituto Tecnológico de Aeronáutica, Praça Marechal Eduardo Gomes 50, São José dos Campos, SP, Brazil
e-mail: diogobpoliveira@gmail.com

J. F. P. Coelho · G. Michon
Institut Clemant Ader, ISAE, Université de Toulouse, Avenue Édouard Berlin, 10, Toulouse, France
e-mail: joao-flavio.pafume-coelho@isae.fr

G. Michon
e-mail: guilhem.michon@isae.fr

L. Sanches (✉)
Universidade Federal de Uberlândia, Avenida Joáo Naves de Avila, 2121, Uberlândia, MG, Brazil
e-mail: lsanches@ufu.br

A. de T. Fleury et al. (eds.), *Proceedings of DINAME 2017*, Lecture Notes in Mechanical Engineering, https://doi.org/10.1007/978-3-319-91217-2_8

1 Introduction

The ground resonance is a phenomenon that can occur in helicopters with hinged and hingeless blades and it happens when the fuselage oscillations excite one or more rotor modes, which creates a wobble of the rotor center of gravity. This makes the rotor effective mass couples with the vibration of the fuselage (fuselage mode), which in turn excites the rotor modes. The continuation of this process increases the displacements the rotor and the fuselage experience, leading to the aircraft destruction [2–4].

When it comes to analyzing the helicopter ground resonance instabilities, it is important to study all the relevant parameters of the system and their effects on the overall stability. Wang and Chopra show in their study that when a least stable mode is improved due to blade dissimilarities, the other modes become less stable [15]. Indeed, dynamic systems are susceptible to failure and aging of its components that may appear randomly, compromising its nominal operation, which can lead to damage of the structure. Sanches et al. [10] verified the existence of new instability regions on the ground resonance phenomenon as blade dissimilarities were taken into account.

Since the mechanical properties might evolve in time, it is important to address, in these cases, a robustness analysis namely μ synthesis. The method was often applied to robust control design of dynamical system under perturbations [5, 7]. It is possible to assure the robustness of the controlled system with respect to the perturbations level of any properties [11]. For instance, with the application μ-synthesis, researches could determine the smallest stiffness deviation (from the nominal value) that leads the aircraft to the ground resonance phenomenon [9].

Concerning the vibration isolation in helicopters, several studies were addressed to conceive passive, semi-active and active methodologies. Dynamic vibration absorber (DVA) composes the anti-resonant isolation system (ARIS) applied to mitigation the vibration induced by the rotor in helicopters [6, 8] and also to chatter suppression in turning operations [12, 14].

Generally, the dynamic vibration absorber (DVA) parameters (mass, stiffness, and damping) are of great relevance to mitigate the instabilities or reduce the vibration amplitudes. These parameters might be designed properly according to the primary system. However, since mechanical properties of the aircraft can be modified along its operational life and between two identical elements, the DVA might develop new instabilities regions.

The objective of this work is to assess the influence of the DVA attached to a helicopter on its stability. Firstly, it is verified the damping necessary on the fuselage and on the rotor to mitigate the unstable oscillations. Later, through a robustness analysis, it is determined the minimum blade's stiffness perturbation value that destabilize the aircraft. All these results will be compared with a helicopter without DVA.

2 Mechanical Model

The helicopter studied in this work, shown in Fig. 1 has been simplified according to the model used in [9, 10]. The fuselage, modeled as a rigid body with mass m_f, has one translational movement $x(t)$ along its lateral direction. The spring attached along the x-direction has stiffness K_x and the damping, modeled as a viscous damper, has coefficient C_x. The helicopter at its equilibrium position has the center of mass of the fuselage (point O) coincident with the origin of the inertial reference frame (X_0, Y_0, Z_0).

The rotor system comprises one rigid rotor hub and an assembly of N blades and it revolves at a speed Ω. The blades have mass m_b and moment of inertia I_{zb} around the z-axis (located at its center of mass). The blades have stiffness K_{bk} and damping coefficient C_{bk} at point B, which is where the rotor head and the fuselage are joined by a rigid shaft.

The radius of gyration is defined by the length b and a is related to the rotor eccentricity. For each kth blade there is an in-plane lead-lag motion $\varphi_k(t)$ and an azimuth angle $\psi_k(t) = \Omega t + 2\pi(k-1)/N_b$ with respect to the x-axis. The origin of the rotational reference frame (x, y, z) is parallel to the inertial one and is located at the geometric center of the rotor hub (coincident at point O). Aerodynamic forces on the blades are not taken into account. Such an assumption is quite realistic since the helicopter is on the ground and the vibration induced in ground resonance phenomenon has greater effects than aerodynamic forces generated by the rotor.

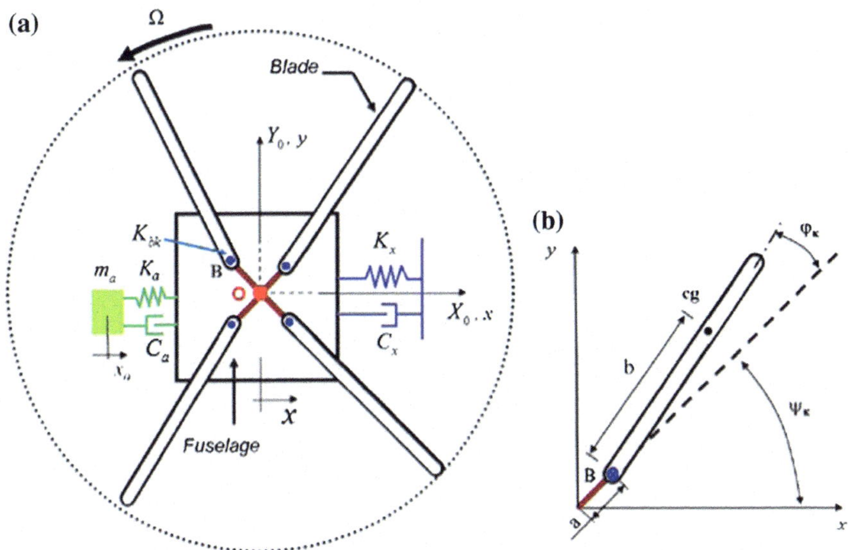

Fig. 1 Sketch of the mechanical model: **a** general view and **b** blade view

The spring-mass-damping vibration system has mass m_a, stiffness K_a and damping coefficient C_a. Note that it translates $x_a(t)$ with respect to the inertial frame. The mechanical system equations of motion are obtained by applying Lagrange's equations of the kinetic (T) and potential (V) energies and the virtual work (δW) of the external forces and moments applied to the system. Sanches et al. [10] derived the energies expressions for the helicopter with hinged blades. The ones used in this work are slightly different for additional terms must be considered concerning the effects of the spring-mass-damping vibration system (SMDVS). The equations of motion in the matrix form are expressed as:

$$\mathbf{M}(t)\ddot{\mathbf{x}}(t) + \mathbf{G}(t)\dot{\mathbf{x}}(t) + \mathbf{K}(t)\mathbf{x}(t) = \mathbf{F_{ext}}(t) \tag{1}$$

with the generalized coordinates $\mathbf{u}(t) = [x(t)\ x_a(t)\ \varphi_1(t)\ \varphi_2(t)\ \varphi_3(t)\ \varphi_4(t)]$. The matrices $\mathbf{M}(t)$, $\mathbf{G}(t)$ and $\mathbf{K}(t)$ correspond to the mass, damping and stiffness matrix, respectively. $\mathbf{F_{ext}}(t)$ is equal to zero since all blades possess the same inertial and geometrical properties. The matrices have periodic terms, as shown by Eq. (2).

$$\mathbf{M}(t) = \begin{bmatrix} 1 & 0 & -r_{m1}\sin(\psi_1) & -r_{m2}\sin(\psi_2) & -r_{m3}\sin(\psi_3) & -r_{m4}\sin(\psi_4) \\ 0 & 1 & 0 & 0 & 0 & 0 \\ -r_{b1}\sin(\psi_1) & 0 & 1 & 0 & 0 & 0 \\ -r_{b2}\sin(\psi_2) & 0 & 0 & 1 & 0 & 0 \\ -r_{b3}\sin(\psi_3) & 0 & 0 & 0 & 1 & 0 \\ -r_{b4}\sin(\psi_4) & 0 & 0 & 0 & 0 & 1 \end{bmatrix} \tag{2a}$$

$$\mathbf{G}(t) = \begin{bmatrix} r_{ca}\,r_{mdva}+r_{cf} & -r_{ca}\,r_{mdva} & -2\Omega r_{m1}\cos(\psi_1) & -2\Omega r_{m2}\cos(\psi_2) & -2\Omega r_{m3}\cos(\psi_3) & -2\Omega r_{m4}\cos(\psi_4) \\ -r_{ca} & r_{ca} & 0 & 0 & 0 & 0 \\ 0 & 0 & r_{cb1} & 0 & 0 & 0 \\ 0 & 0 & 0 & r_{cb2} & 0 & 0 \\ 0 & 0 & 0 & 0 & r_{cb3} & 0 \\ 0 & 0 & 0 & 0 & 0 & r_{cb4} \end{bmatrix} \tag{2b}$$

$$\mathbf{K}(t) = \begin{bmatrix} \omega_a^2 r_{mdva}+\omega_x^2 & -\omega_a^2 r_m dva & \Omega^2 r_{m1}\sin(\psi_1) & \Omega^2 r_{m2}\sin(\psi_2) & \Omega^2 r_{m3}\sin(\psi_3) & \Omega^2 r_{m4}\sin(\psi_4) \\ -\omega_a^2 & \omega_a^2 & 0 & 0 & 0 & 0 \\ 0 & 0 & \Omega^2 r_{a1}^2+\omega_{b1}^2 & 0 & 0 & 0 \\ 0 & 0 & 0 & \Omega^2 r_{a2}^2+\omega_{b2}^2 & 0 & 0 \\ 0 & 0 & 0 & 0 & \Omega^2 r_{a3}^2+\omega_{b3}^2 & 0 \\ 0 & 0 & 0 & 0 & 0 & \Omega^2 r_{a4}^2+\omega_{b4}^2 \end{bmatrix} \tag{2c}$$

$$\mathbf{F_{ext}}(t) = \left[0\ 0\ 0\ 0\ 0\ 0\right]^T \tag{2d}$$

where for $k = 1, \ldots, 4$:

$$r_{mk} = \frac{b\,m_{bk}}{m_f + \sum_{i=1}^{N_b} m_{bi}} \qquad r_{mdva} = \frac{b\,m_a}{m_f + \sum_{i=1}^{N_b} m_{bi}}$$

$$r_{bk} = \frac{b\,m_{bk}}{b^2\,m_{bk} + I_{zbk}} \qquad r_{ak}^2 = a\,r_{bk}$$

$$r_{cf} = \frac{C_x}{m_f + \sum_{i=1}^{N_b} m_{bi}} \qquad r_{cbk} = \frac{C_{bk}}{b^2\,m_{bk} + I_{zbk}}$$

$$r_{ca} = \frac{C_a}{m_a} \qquad \omega_a^2 = \frac{K_a}{m_a}$$

Table 1 Mechanical properties for Helicopter Type 1 and 2

Fuselage	Rotor	DVA
$m_f = 2902.9$ kg	$m_{bk} = 31.9$ kg $a = 0.2$ m $b = 2.5$ m	$m_a = 16.1$ kg
$\omega_x = 6.0\pi$ rad/s	$\omega_{bk} = 3.0\pi$ rad/s	$I_{zbk} = 259$ kg m^2 $\omega_a = 6.02\pi$ rad/s
$C_x = 2284.9$ N s/m	$C_{bk} = 172.8$ N s m/rad	$C_a = 22.67$ N s/m

$$\omega_x^2 = \frac{K_x}{m_f + \sum_{i=1}^{N_b} m_{bi}} \qquad \omega_{bk}^2 = \frac{K_{bk}}{b^2\, m_{bk} + I_{zbk}}$$

3 Helicopter Numerical Data

In this work two helicopters are to be considered: one helicopter with no DVA (Helicopter Type 1) and one helicopter with DVA (Helicopter Type 2). The fuselage and rotor properties are the same for both types. The DVA attached in Helicopter Type 2 has its parameters defined through the application of a differential evolution algorithm to find the optimal DVA parameters (i.e., m_a ω_a), whose variables define the design vector. In this process, the DVA damping coefficient C_a is calculated by considering the optimal damping ratio ζ_{opt} [13], given as:

$$C_a = 2\,\zeta_a\, m_a\, \omega_a \tag{3a}$$

$$\left(\zeta_a\right)_{opt} = \sqrt{\frac{3\mu}{8(1+\mu)^3}} \tag{3b}$$

where $\mu = m_a/m$.

The objective functions adopted for DVA parameters optimization are the exponential growth of the dynamical system and the DVA mass since high mass values are not benefit for aeronautical applications. An analysis of the Pareto distribution (i.e.: m_a vs. ω_a) leads to determining the DVA parameters used in along this work, (see Table 1).

4 Influence of Fuselage and Rotor Damping

According to the literature [2, 10, 15], the ground resonance phenomenon is characterized by the dynamic coupling of the poorly damped cyclic rotor modes with that from the fuselage. The fact of adding a DVA in the fuselage changes its modal char-

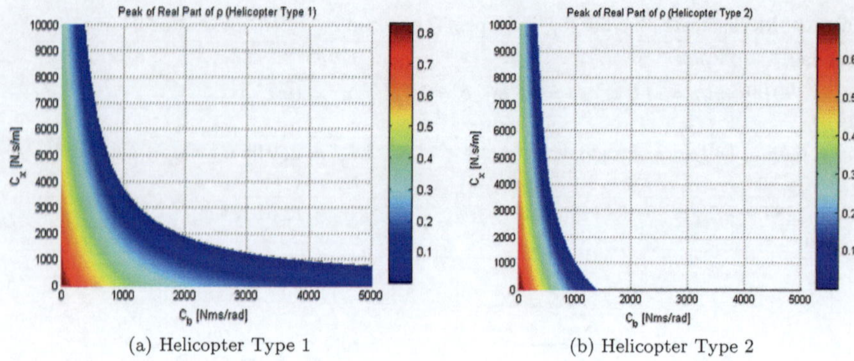

(a) Helicopter Type 1 (b) Helicopter Type 2

Fig. 2 Peak of the real part of eigenvalues

acteristics. The fuselage consists, afterwards, into a dynamic system represented by two lumped mass/spring/damping elements (see Fig. 1, having two resonance frequencies and two modes of vibration. This fact might become helpless the use of DVA on helicopters.

The partial purpose of this work consists to assess the influence of including the SMDVS in the fuselage on the damping level (i.e., at fuselage and rotor) on ground resonance instabilities. For this, Coleman's Transformation is used [1]. It introduces modal coordinates of an isotropic rotor (i.e., all the blades have the same properties) in the equations of motion which were written considering the blades displacement on the rotor rotating frame. By applying the referred transformation to Eq. (1), the equations of motion become time independent.

The stability of such system is guaranteed if the real part of all the eigenvalues ρ are negative over all revolving speeds Ω. Moreover, the maximum value (peak) reached by the real part of the eigenvalue plays a big role on the ground resonance phenomenon, in which it is related to the exponential growth of the helicopter dynamical responses. Figure 2 shows the stability analysis, evidenced by the peaks, for different combinations of C_x and C_b for the helicopter without DVA (Helicopter Type 1) and for helicopter with DVA (Helicopter Type 2). The stability analysis is done for the rotor speed comprised between $0 \leq \Omega \leq 10$ Hz and the damping coefficients between $0 \leq C_x \leq 10{,}000$ N s/m and $0 \leq C_b \leq 5000$ N m s/rad.

The unstable oscillations (colored region, i.e., positive peaks) happen if improperly combination of fuselage and rotor blade damping level is considered. It can be seen that the behavior of those regions are deeply different between both helicopters, especially with respect to the values for C_b. The unstable oscillations for Helicopter Type 2 (i.e., white region) is achieved with a smaller value for C_b ($C_b = 1407$ N m s/rad when $C_x = 0$), when compared to helicopter 1 ($C_b > 5000$ N m s/rad when $C_x = 0$). This means that the DVA is good for the overall system stability, and by applying it to the fuselage, the helicopter can make use of smaller values for damping, which makes the rotor system more compact and less complex. Also, one can see that the stability of the system is more sensitive to variations of C_b, since with small values of that damping even if the values for C_x are high, the system is still unstable. Concerning the peaks, as shown in Fig. 2, the critical portion, i.e., the highest peaks of real part are present in the region which combines small values for both fuselage and rotor blade damping. Although for Helicopter Type 2 the values for peaks are smaller, both helicopters exhibit the same behavior.

5 Stability Robustness Analysis

Dynamic systems are susceptible to parameters variations, which can be caused by failure and aging of its components that may appear randomly, and may compromise its nominal operation. In the case of the helicopter, the blade stiffness may have variations in its natural frequency, affecting the system (helicopter + DVA) dynamics. It may be catastrophic if new instability regions are created. This topic aims in analyzing the stability robustness of the aircraft once blade stiffness perturbations are considered from the nominal values. The analyses are done for both helicopters (Helicopter Type 1 and 2) and later compared, in order to capture the influence of the DVA on the aircraft stability robustness.

For this purpose, since Coleman's Transformation is no more valid once rotor anisotropy is taken into account, the continuous Linear Time Periodic (LTP) system must be considered—Eq. (1). The robustness analysis of LTP systems is considered by transforming it into a LTI system and later applying μ-analysis [5]. In order to apply those methods, one needs first to cast the LTP system into the Linear Fractional Transformation (LFT) form. In order to do so, the lifting procedure is applied to provide a LFT, which leads to uncertainty structures with highly-repeated parameters.

Considering a diagonal matrix $\mathbf{\Delta}(t)$ comprising bounded real unknown parameters which represent the uncertainties:

$$\mathbf{\Delta} = \text{diag}\left[\delta_1,\ \delta_2,\ \ldots,\ \delta_k\right] \tag{4}$$

the system must be nominally stable when $\mathbf{\Delta} = \mathbf{0}$. The primary purpose of the parametric robustness analysis is to find the smallest uncertainties δ_k which destabilize the closed-loop system $\delta(\mathbf{\Delta})$. Floquet's method can be applied to perform this analysis, it is very CPU time-consuming, which is why the μ-analysis, a much faster technique, is used on the LFT model. A comparison between both methods is shown in [9], in which the authors conclude that both yield the same results.

5.1 *Ground Resonance Parametric Analysis*

The uncertainties introduced in the dynamic system in Eq. (1) are related to blade hinge stiffness and are presented as a function of in-plane lead-lag resonance frequency squared. In this work, only uncertainties in the 1st blade are accounted for the robustness analysis. By assuming $\overline{\omega}_{b1}$ as the nominal natural frequency of the blade, the modified frequency ω_{b1} is given by:

$$\omega_{b1}^2 = (1 + \delta_1)\overline{\omega}_{b1} \tag{5}$$

The first robustness analysis is to be done by choosing different combinations of C_b and C_x that gives a specific peak (h) of the real part of the eigenvalues. One admits therefore the same stability level of the aircraft for all set of damping coefficients (C_b, C_x) considered. In order to do so, first the curve that gives combinations of both damping that yield the peak $h = -0.5$ (i.e., the maximum real part of the eigenvalue) has to be found. Afterwards, for each pair of (C_b, C_x), the μ-analysis determines the minimal perturbation δ_1 that destabilizes the Helicopter Types 1 and 2, accordingly to Tables 2 and 3, respectively. Note that the analysis was carried out at four different rotor speeds $\Omega = [2, 4, 6, 8]$ Hz. It is important to note from Tables 2 and 3 the minimal perturbation δ_1 that leads the helicopter unstable remains practically constant at a given rotor speed and for different set of (C_b, C_x). Also, the result does not change when DVA is considered. Moreover, the perturbation δ_1 evolves as the rotor speed increases, which means the system is less robust in low rotor speeds.

By taking into account Eq. (5), one can note that the system is extremely robust once the perturbations less than −1 lead to negative blade natural frequencies. This means the helicopter can lose one of its rotor blade stiffness and still be a stable system with those combinations of damping at these rotating speed. Also, considering both helicopters have the same stability level ($h = -0.5$) for all set of damping coefficients, the fact of adding the DVA on the system does not change the aircraft robustness. This is concluded since the same perturbations (δ_1) are obtained for HT1 and HT2 systems. From a practical point of view, this fact means only positive contritions of the DVA for the helicopter.

Table 2 Perturbations δ_1 for Helicopter Type 1

C_b (N m s/rad)	C_x (N s/m)	$\Omega = 2$ Hz	$\Omega = 4$ Hz	$\Omega = 6$ Hz	$\Omega = 8$ Hz
		δ_1	δ_1	δ_1	δ_1
1290	7688.5636	−1.0586	−1.2821	−1.6038	−2.0627
1668	6261.2820	−1.0586	−1.2827	−1.6039	−2.0627
2047	5515.3003	−1.0586	−1.2829	−1.6040	−2.0628
2426	5049.8542	−1.0586	−1.2830	−1.6040	−2.0628
2805	4741.7484	−1.0586	−1.2831	−1.6040	−2.0628
3184	4518.1112	−1.0586	−1.2832	−1.6040	−2.0628
3563	4344.1412	−1.0586	−1.2832	−1.6040	−2.0628
3942	4217.9519	−1.0586	−1.2832	−1.6041	−2.0628
4321	4119.8877	−1.0586	−1.2832	−1.6041	−2.0628
4700	4025.2589	−1.0586	−1.2833	−1.6041	−2.0628

Table 3 Perturbations δ_1 for Helicopter Type 2

C_b (N m s/rad)	C_x (N s/m)	$\Omega = 2$ Hz	$\Omega = 4$ Hz	$\Omega = 6$ Hz	$\Omega = 8$ Hz
		δ_1	δ_1	δ_1	δ_1
1131	7031.7050	−1.0586	−1.2828	−1.6040	−2.0628
1533	5126.2450	−1.0586	−1.2835	−1.6041	−2.0628
1935	4218.1842	−1.0585	−1.2837	−1.6042	−2.0629
2337	3733.5303	−1.0585	−1.2838	−1.6042	−2.0629
2740	3386.5867	−1.0585	−1.2839	−1.6042	−2.0629
3142	3167.5168	−1.0585	−1.2839	−1.6042	−2.0629
3544	3016.2318	−1.0585	−1.2840	−1.6042	−2.0629
3946	2859.3988	−1.0585	−1.2840	−1.6042	−2.0629
348	2803.9668	−1.0585	−1.2840	−1.6042	−2.0629
4750	2651.9956	−1.0585	−1.2840	−1.6042	−2.0629

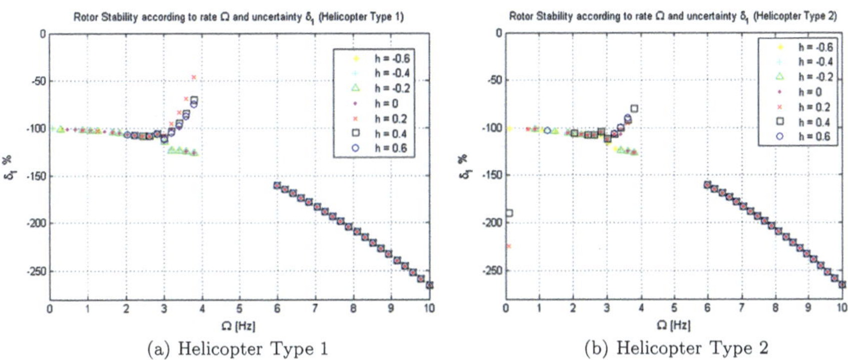

(a) Helicopter Type 1 (b) Helicopter Type 2

Fig. 3 Rotor stability according to rotor speed Ω and perturbations δ_1

The previous analysis for both Helicopter Type 1 and Helicopter Type 2 was done considering only the peak of the maximum real part as $h = -0.5$. However, it is important to assess the different perturbations related to all peak levels (h). As seen before, the perturbations are practically constant along the curve for a specific peak of real part, which means that every combination of fuselage and rotor damping that lies in that curve will give the same perturbation as result. With that in mind, it is possible to analyze the minimum destabilizing perturbations for different h by choosing one set of damping coefficients. Here, the peaks chosen are −0.6, −0.4, −0.2, 0, 0.2, 0.4 and 0.6. In this case, the perturbations were evaluated in the range (i.e., $0 \leq \Omega \leq 10$ Hz) in which the helicopter is on the ground. Figure 3 shows the minimum perturbations at different rotor speed allowed to guarantee stability for helicopters HT1 and HT2.

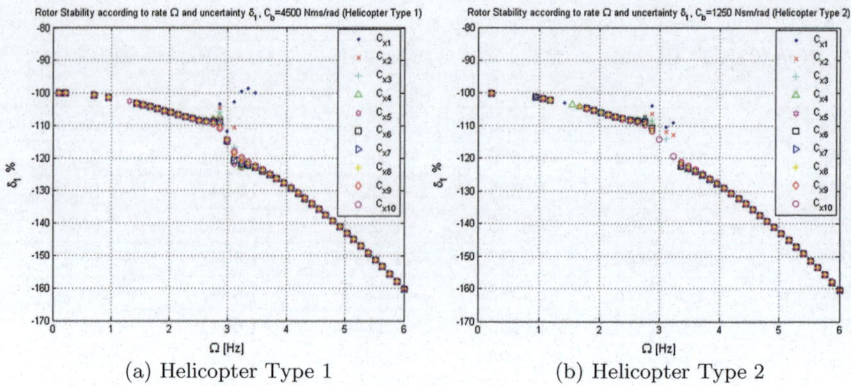

(a) Helicopter Type 1 (b) Helicopter Type 2

Fig. 4 The influence of fuselage damping on the rotor stability sensitivity due to blade stiffness asymmetry: a) Helicopter Type 1 $C_b = 4500$ N s/m; b) Helicopter Type 2 $C_b = 1250$ N s/m

It is worth mentioning that the μ-analysis method can only be applied when the system is nominally stable. Therefore, the analysis is not performed in the range $4 \leq \Omega \leq 6$ Hz since, when $h > 0$, the system is unstable. Also, in order to be able to compare all the results, the analysis was not performed in that region for $h < 0$.

One can see from Fig. 3a that in the region $0 \leq \Omega \leq 3$ Hz, the perturbations are practically the same for all h, with some small differences. This means that the peak of the real part has almost no influence in the robustness of the system. However, in the range of $3 \leq \Omega \leq 4$ Hz, it is possible to see that the perturbations are different between positive and negative h values. In fact, the higher the peak (from negative to positive h), the less robust the system is, since the perturbations are bigger. In the region $6 \leq \Omega \leq 10$ Hz the perturbations are exactly the same for all values of h, meaning that the peak has absolutely no effect in the robustness of the system. The same conclusions can be made for Helicopter Type 2 in Fig. 3b. Also, by comparing the results for Helicopter Type 1 and Helicopter Type 2 in the region $3 \leq \Omega \leq 4$ Hz, it is possible to see that the perturbations for the helicopter with DVA are greater than the ones for the helicopter without DVA, which means that Helicopter Type 2 is more robust than Helicopter Type 1 in that region.

The robustness analysis made for different values of peak of real part does not assess the influence of fuselage and rotor blade damping in the robustness of the system, since for each value of h a different set of C_b and C_x is considered. Therefore, robustness analyses are carried out by assuming isolated variations at fuselage damping coefficients (see Fig. 4) and at blade damping coefficients (see Fig. 5), respectively.

In order to evaluate the influence of C_x in helicopters robustness in Fig. 4, the rotor blade damping was considered fixed at $C_b = 4500$ N m s/rad for Helicopter Type 1 and $C_b = 1250$ N m s/rad for Helicopter Type 2 and the fuselage damping was varied from 900 N s/m $\leq C_x \leq$ 9000 N s/m with increments of 900 N s/m. For ease to see

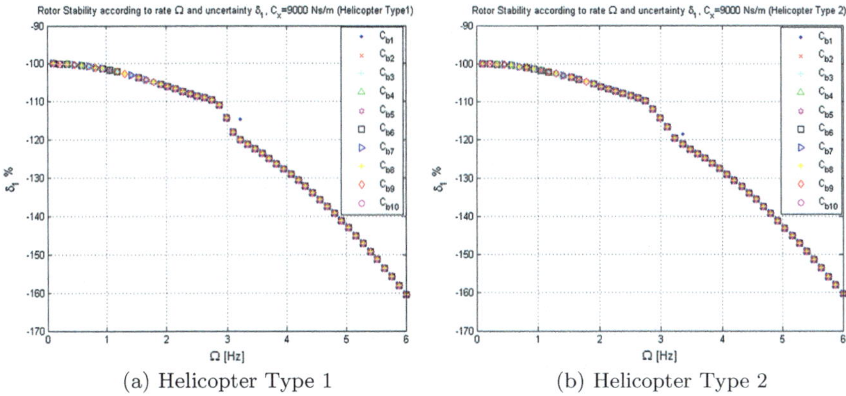

(a) Helicopter Type 1

(b) Helicopter Type 2

Fig. 5 The influence of blade damping on the rotor stability sensitivity due to blade stiffness asymmetry: a) Helicopter Type 1 $C_b = 4500$ N s/m; b) Helicopter Type 2 $C_b = 1250$ N s/m

the results, for both cases the fuselage damping were given subscript index in which $C_{x1} = 900$ N s/m, $C_{x2} = 1800$ N s/m, and so on until $C_{x10} = 9000$ N s/m.

One can see from Fig. 4 that the perturbations are practically the same in all the range of rotor speed, with the exception being in the region around $\Omega = 3$ Hz. At this region one may conclude that the helicopter robustness is sensitive to the fuselage damping, specially concerning low values of C_x that yield higher perturbations values. Also, with the exception of that region, both helicopters have the same values for perturbation, meaning that the DVA has no effect on the robustness of the system, which corroborates with the conclusions made with the analysis of the influence of the peak of real part.

Now the influence of rotor blade damping is evaluated in Fig. 5. For both helicopters the fuselage damping is fixed at $C_x = 9000$ N s/m and the rotor blade damping is varied from 500 N m s/rad $\leq C_b \leq$ 4500 N m s/rad. One can see from the results that the perturbations are exactly the same in all the range of rotor speed, which means that the rotor blade damping has no influence on the robustness of both systems. Also, both helicopters have the same values for perturbation, which once again corroborates with the previous conclusions.

6 Conclusion

Electrical or mechanical devices might be attached to the fuselage of helicopters, which may be modeled as spring-mass-damping vibration systems and can interact with the helicopter ground oscillations, which might alter the stability characteristics of the helicopter. The influence of the SMDVS parameters added to the fuselage of a helicopter were under investigation in the present paper. Firstly, the influence of fuselage and rotor blade damping, specially their combination, was evaluated. The results showed that in order to vanish the instabilities due to ground resonance, the system should have the proper combination of damping. Also, it was possible to see that the addition of the DVA is good for the overall system stability, since by applying it to the fuselage, the helicopter can make use of smaller values for damping, which makes the system more compact and less complex.

Lastly, a stability robustness analysis of the entire system was performed by changing the stiffness properties of one rotor blade for the helicopters with and without DVA to assess how the system handles those variations. The results showed that for both helicopters, as the rotor speed increases, the system become more robust. It was seen that with the proper combination of fuselage and rotor blade damping, the system can be so robust as to lose one of its rotor blade stiffness and still remain stable. Also, the analysis showed that in the region around the natural frequency of the fuselage, the robustness of the system is sensitive to the fuselage damping, since different values of it yield different values of perturbations.

Acknowledgements The authors would like to thank the French and Brazilian funding provided by CNPq (Scholarship IC-CNPQ2015-0139 and Project 460207/2014-8), through INCT-EIE and FAPEMIG.

References

1. Bir, G.: Multiblade coordinate transformation and its application to wind turbine analysis. In: ASME Wind Energy Symposium, pp. 1–15 (2008)
2. Coleman, R.P., Feingold, A.M.: Theory of self-excited mechanical oscillations of helicopter rotors with hinged blades. NACA Report No. 1351 (1957)
3. Ganiev, R.F., Pavlov, I.G.: The theory of ground resonance of helicopters. Int. Appl. Mech. **9**(5), 505–510 (1973)
4. Johnson, W.: Helicopter theory. Courier Corporation (2012)
5. Kalender, S., Flashner, H.: Control design and robustness analysis of linear time-periodic systems. J. Comput. Nonlinear Dyn. **3**(4) (2008)
6. Konstanzer, P., Enenkl, B., Aubourg, P., Cranga, P.: Recent advances in Eurocopter's passive and active vibration control. In: Annual Forum Proceedings-American Helicopter Society, vol. 64, p. 854. American Helicopter Society, Inc (2008)
7. Lauridsen, J., Santos, I.: Design of robust AMB controllers for rotors subjected to varying and uncertain seal forces. Mech. Eng. J. 16-00618 (2017)
8. Narayana Rao, K.S., Pranesh, B., Ravindranath, R.: Design, development and validation of an anti-resonant isolation. In: 24th European Rotorcraft Forum, Marseille, France (1998)
9. Sanches, L., Alazard, D., Michon, G., Berlioz, A.: Robustness analysis of helicopter ground resonance with parametric uncertainties in blade properties. J. Guid. Control Dyn. (2013)
10. Sanches, L., Michon, G., Berlioz, A., Alazard, D.: Parametrically excited helicopter ground resonance dynamics with high blade asymmetries. J. Sound Vib. **331**(16), 3897–3913 (2012)
11. Sawicki, J.T., Madden, R.: Identification of missing dynamics in rotor systems using robust control theory approach. In: Vibration Problems ICOVP 2011, pp. 581–587. Springer (2011)
12. Sims, N.D.: Vibration absorbers for chatter suppression: a new analytical tuning methodology. J. Sound Vib. **301**(3), 592–607 (2007)
13. Steffen Jr, V., Rade, D.: Dynamic vibration absorber. In: Encyclopedia of Vibration, pp. 9–26 (2001)

14. Tarng, Y., Kao, J., Lee, E.: Chatter suppression in turning operations with a tuned vibration absorber. J. Mater. Process. Technol. **105**(1), 55–60 (2000)
15. Wang, J., Chopra, I.: Dynamics of helicopters in ground resonance with and without blade dissimilarities. In: Dynamics Specialists Conference, p. 2108 (1992)

Cable Dynamic Modeling and Applications in Three-Dimensional Space

Sebastião C. P. Gomes, Elisane B. Zanela and Adriana E. L. Pereira

Abstract Dynamic modeling of cables in three-dimensional space is a problem with great difficulty and complexity. This article discusses a new dynamic modeling formalism, including applications in the underwater environment. It is assumed that the cable is formed by rigid links connected by elastic fictitious joints, allowing elevation, azimuth and torsion movements. Algorithms have been developed to automatically generate the dynamic model for any number of links selected for the discrete approximation of the flexible structure. Three practical situations are tested: cable out of the water with free terminal load; underwater considering dynamics with or without ocean current; with terminal load fixed to the seabed. Constraint forces obtained through proportional and derivative control were applied to the terminal load to fix it to the seabed. The algorithms were determined from the Euler-Lagrange formalism, and in all situations the simulations showed physically consistent results.

Keywords Cable · Dynamic · Modeling · Algorithms · Automatic generation model · Underwater applications

1 Introduction

Cable dynamic modeling is a complicated task because of its complexity, especially in the case of movement in three dimensional space. Many applications involving cable dynamics occur in the underwater environment: risers, mooring lines, towing

S. C. P. Gomes (✉) · E. B. Zanela · A. E. L. Pereira
Universidade Federal do Rio Grande FURG, Av. Itália, km 8,
Rio Grande 96201-900, Brazil
e-mail: sebastiaogomes@furg.br; scpgomes@gmail.com

E. B. Zanela
e-mail: laplaciano2000@bol.com.br

A. E. L. Pereira
e-mail: adrianapereira@furg.br

A. de T. Fleury et al. (eds.), *Proceedings of DINAME 2017*, Lecture Notes in Mechanical Engineering, https://doi.org/10.1007/978-3-319-91217-2_9

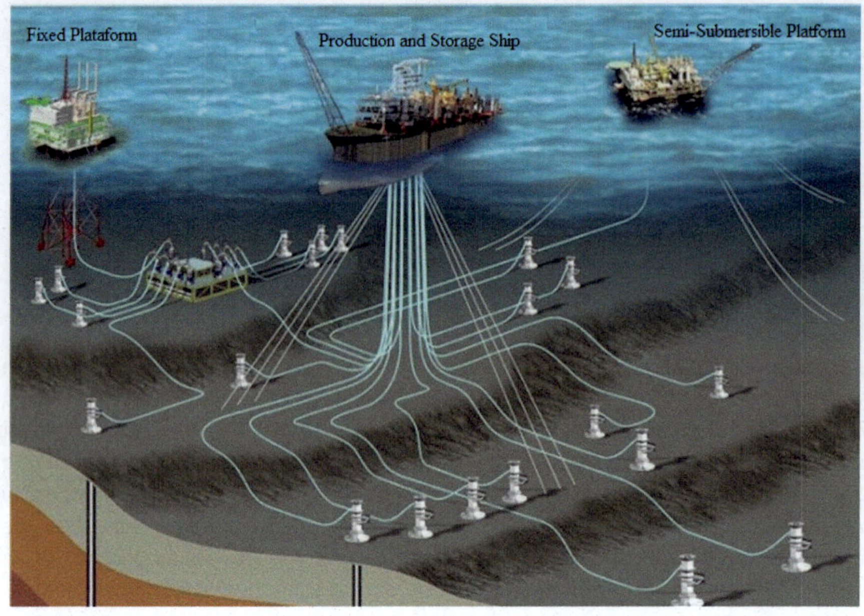

Fig. 1 Floating system of oil production (adapted from: http://diariodopresal.wordpress.com/petroleo-e-gas)

cables, etc., can be examples of offshore oil industry applications (see Fig. 1). Most studies found in the literature address the modeling of these structures using Finite Element Methods [1, 2]. Sun et al. [3] introduced a finite element method to modeling a cable towed body.

Some authors have developed their works performing a cable static analysis [4, 5], using the method of finite differences. Discrete approach was used specially in static analysis: Raman-Nair and Williams [6] have used a discrete model to reproduce structural forces acting into a flexible marine riser under effects of flow and pressure of fluid within the riser; Zhu et al. [7] proposed a discrete model to determine the forces that an umbilical cable exerts on a ROV (Remotely Operated Vehicle).

When the discrete formalism is used in dynamic modeling, usually lumped mass approach is applied, considering the dynamics evolving in a single plane [8].

Finite differences are widely used in cable modeling. Matulea et al. [9] used finite differences to determine the riser's static equilibrium configuration. Lee et al. [10] applied finite differences method with lumped mass to model a flexible pipe.

In short, most of the articles that deal with cables treat the problem as restricted to a single plane using finite elements or finite differences for the dynamic model.

Other works focus interest in static analysis of axial forces on the cable. Gobat in his thesis provides details about these main methods used in cable dynamics [11].

This paper introduces a new method to automatically obtain vectors and matrices elements of a cable dynamic model. We use a discrete formalism to represent the continuous flexibility from a chain of rigid links connected by fictitious elastic joints. Each joint allows three elastic movements: elevation, azimuth and torsion. We take as a basis the work of Gomes et al. [12], where a discrete formalism was used to model a robotic manipulator with a single flexible link. In that case, the flexibility occurred on a single plane and each joint had one degree of freedom. Based on this work, Pereira et al. [13] developed the analytical modeling of a cable considering three links, but with spatial flexibility, i.e. the discrete formalism was used without the motion being restricted to a single plane. In Gomes et al. [14] algorithms were presented to automatically generate the dynamic model, for any number of links to be used in discrete approximation of the continuous cable flexibility. The present article shows this theory in three possible applications, considering also underwater currents acting as external efforts. Automatic retrieval models are very important due to the great complexity of the equations that turn unfeasible obtain these models manually through the application of the Euler-Lagrange equations.

2 Dynamic Modeling

In this work it is considered a cylindrical cable with constant radius, fixed at one extremity (fixed base) and free at the other, where there is a terminal load m_c. The basic principle of this modeling theory is to approximate the continuous flexibility by a discrete equivalent one, consisting of rigid links connected by flexible fictitious joints, as showed in Fig. 2. Each fictitious elastic joint allows three movements: elevation (θ_{ie}), azimuth (θ_{ia}) and torsion (θ_{iT}), $i = 1, \ldots, n$. Therefore, this dynamic system has $3n$ degrees of freedom when considering n links. In each fictitious joint is positioned a reference frame, as shown in Fig. 3 for the first two systems. The first is an inertial system (X_0 Y_0 Z_0). It was adopted the following convention for reference systems: all Z axes point to the center of the Earth and thus, the XY axes form horizontal planes. The Y_i axes are parallel to the projection of the link i on the $X_{i-1}Y_{i-1}$ plane, as showed in Figs. 3 and 4. For instance, Y_1 is parallel to r in Fig. 3. Figure 4 also shows the three angular positions coordinates of the first joint and the three others of the second fictitious joint. As all links are rigid, torsion motions are considered as rotations about the longitudinal axis of the links.

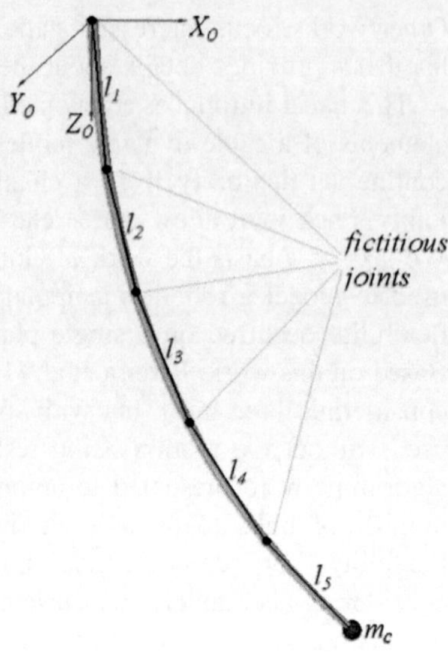

Fig. 2 Continuous flexibility and its discrete approximation

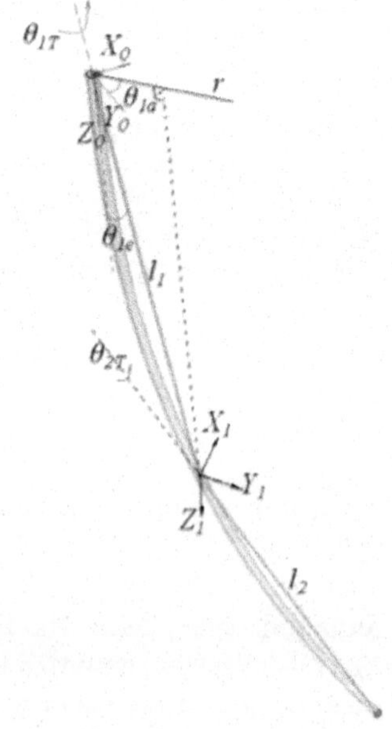

Fig. 3 Angular coordinates of the first joint

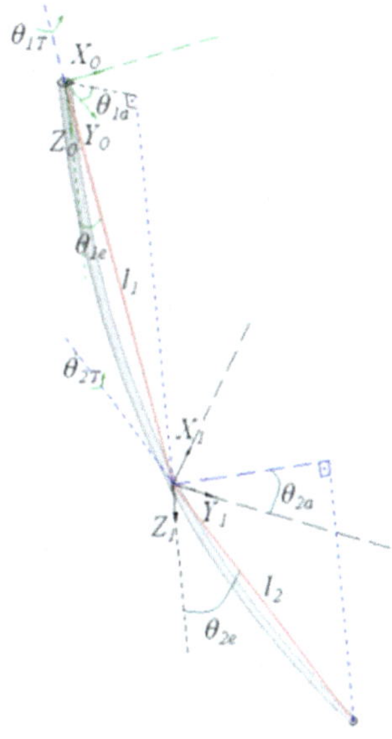

Fig. 4 The first two reference systems and its angular coordinates

It is simple to find a homogeneous transformation matrix between two consecutive reference systems. For example, the homogeneous matrix that relates $X_0Y_0Z_0$ and $X_1Y_1Z_1$ systems has the form:

$$H_{01} = \begin{bmatrix} \cos\theta_{1a} & \sin\theta_{1a} & 0 & l_1 \sin\theta_{1e} \sin\theta_{1a} \\ -\sin\theta_{1a} & \cos\theta_{1a} & 0 & l_1 \sin\theta_{1e} \cos\theta_{1a} \\ 0 & 0 & 1 & l_1 \cos\theta_{1e} \\ 0 & 0 & 0 & 1 \end{bmatrix} \tag{1}$$

The products between successive homogeneous matrices generate another homogeneous matrix that can relate any mobile reference system to the base inertial system of the structure. Thus, the spatial position of the center of mass of any link in the inertial frame may be determined as functions of the lengths of the links and the angular position coordinates, as specified below $(k = 1, \ldots, n)$:

$$\begin{cases} x_k = \frac{l_k}{2} \sin\theta_{ke} \sin\left(\sum\limits_{i=1}^{k} \theta_{ia}\right) + \sum\limits_{j=1}^{k-1} \left[l_j \sin\theta_{je} \sin\left(\sum\limits_{i=1}^{j} \theta_{ia}\right)\right] \\ y_k = \frac{l_k}{2} \sin\theta_{ke} \cos\left(\sum\limits_{i=1}^{k} \theta_{ia}\right) + \sum\limits_{j=1}^{k-1} \left[l_j \sin\theta_{je} \cos\left(\sum\limits_{i=1}^{j} \theta_{ia}\right)\right] \\ z_k = \frac{l_k}{2} \cos\theta_{ke} + \sum\limits_{j=1}^{k-1} l_j \cos\theta_{je} \end{cases} \tag{2}$$

Arising from the same formalism, spatial coordinates of the terminal load (written in the inertial frame) have the form:

$$\begin{cases} x_c = \sum_{j=1}^{n} \left[l_j \sin\theta_{je} \sin\left(\sum_{i=1}^{j} \theta_{ia} \right) \right] \\ y_c = \sum_{j=1}^{n} \left[l_j \sin\theta_{je} \cos\left(\sum_{i=1}^{j} \theta_{ia} \right) \right] \\ z_c = \sum_{j=1}^{n} l_j \cos\theta_{je} \end{cases} \tag{3}$$

The time derivatives of the Eqs. (2) and (3) can be easily obtained and thus the Lagrangian of the system can be obtained only as function of the angular position and velocity coordinates (see [14] for more details). Each new link considered in the discrete approach means three degrees of freedom more in the model. Thus, when n links are considered, the dynamic model will have $3n$ degrees of freedom. The application of the $3n$ Euler-Lagrange equations allows finding the dynamic model in the form:

$$I(\vec{\theta})\ddot{\vec{\theta}} + C\dot{\vec{\theta}} + K\vec{\theta} + \vec{f}(\vec{\theta}, \dot{\vec{\theta}}) + \vec{G}(\vec{\theta}) = \vec{\tau} \tag{4}$$

where $\vec{\theta} = [\theta_{1e}\,\theta_{2e}\ldots\theta_{ne}\,\theta_{1a}\,\theta_{2a}\ldots\theta_{na}\,\theta_{1T}\,\theta_{1T}\ldots\theta_{nT}]^T$ is the angular position vector, θ_{ie}, θ_{ia} and θ_{iT} are elevation, azimuth and torsion angles of the link $i\ (i = 1, \ldots, n), I(\vec{\theta})$ is the inertia matrix, C is the friction coefficient matrix, K is the elastic constant matrix, $\vec{f}(\vec{\theta}, \dot{\vec{\theta}})$ is the Coriolis-centrifugal vector, $\vec{G}(\vec{\theta})$ is the gravitational vector and $\vec{\tau}$ is the external torques vector. All matrices are $(3n \times 3n)$ and vectors are $(3n \times 1)$. It is important to explain that the matrices of the friction coefficients and elastic constants have the same configuration and also, all matrices of the dynamic models are symmetric.

3 Generic Algorithm

Equation (4) was manually developed considering 1, 2, 3 and 4 links and so it was possible to identify growth patterns for matrices and vectors of the dynamic model, shown below in the form of algorithms that can automatically generate the vectors and matrices of the model, for any adopted number of links.

3.1 Algorithms for the Elements of Inertia Matrix

To facilitate understanding, the inertia matrix I is represented through nine sub-matrices, as indicated below, where indices e, a and t mean elevation, azimuth and torsion, respectively.

$$I = \begin{bmatrix} I_e & N_e & T_e \\ I_a & N_a & T_a \\ I_t & N_t & T_t \end{bmatrix} \tag{5}$$

I is symmetric and thus, $I_a = N_e^T; I_t = T_e^T; N_t = T_a^T$. As there is a dynamic decoupling of the torsion movement with respect to elevation and azimuth motions, submatrices T_e and T_a are null and T_t is diagonal. Elements of T_t are constant and equivalent to $I_{iT} = (m_i/2)r_i^2$, where m_i and r_i are mass and radius of the link i, respectively, with $i = 1, \ldots, n$. Due to the symmetry of the inertia matrix the interest is to determine the rules for the automatic generation of I_e, N_e and N_a.

The submatrix I_e can be generated from the following algorithm:

$for\ i = 1{:}n,$

$\quad for\ j = i+1{:}n,$

$$I_e(i,j) = l_i l_j \left(\frac{m_j}{2} + m_c + \sum_{k=j+1}^{n} m_k\right)\left[\cos\theta_{ie}\cos\theta_{je}\cos\left(\sum_{k=i+1}^{j}\theta_{ka}\right) + \sin\theta_{ie}\sin\theta_{je}\right]; \tag{6}$$

$\quad end,$

$$I_e(i,i) = l_i^2\left[\frac{m_i}{4} + m_c + \sum_{k=i+1}^{n}(m_k)\right] + I_{ie};$$

$end,$

As I_e is symmetric, $I_e(j,i) = I_e(i,j) \cdot I_{ie}$ is the elevation rotational inertia moment of the link i.

The submatrix N_e can be generated from the following algorithm:

```
for i = 1:n,
  for j = 1:n,
    if i ≠ j,
      S = 0;
      for k = j:n,
        if k ≠ i,
          if i > k, μ = 1; else μ = −1;   endif,
          v = [i   k];
```

$$S = S + \mu l_k \left[\frac{m_\alpha}{2} + m_c + \sum_{g=max(v)+1}^{n} m_g \right] cos(\theta_{ie}) sin(\theta_{ke}) sin\left(\sum_{g=min(v)+1}^{max(v)} \theta_{ga} \right);$$

```
        endif,
      end,
      N_e(i, j) = l_i S;
    endif,
  end,
end,
for i = 1:n − 1,
  N_e(i, i) = N_e(i, i + 1);
end,
N_e(n, n) = 0;
```

(7)

The submatrix N_a can be generated from the following algorithm:

```
for i = 1:n,
  for j = i:n,
    S_1 = 0;
    for k = j:n,
          S_1 = S_1 + l_k^2 [ m_k/4 + m_c + sum_{h=k+1}^{n} m_h ] sin^2(θ_ke);
    end,
    S_2 = 0;
    v = [i j];
    for k = min(v) : n − 1,
      S_3 = 0;
      for g = j:n,
        if g > k,
          σ = 1;
          if (i = 1 and j = 1) or k ≥ j,
                σ = 2;
          endif,
            S_3 = S_3
                  + σ l_g [ m_g/2 + m_c
                  + sum_{h=g+1}^{n} m_h ] sin(θ_ke) sin(θ_ge) cos( sum_{h=k+1}^{g} θ_ha );
        endif,
      end,
      S_2 = S_2 + l_k S_3;
    end,
    if (i = n) and (j = n),
      S_2 = 0;
    endif,
    N_a(i,j) = S_1 + S_2;
    N_a(j,i) = N_a(i,j);
    if i = j,
      N_a(i,i) = N_a(i,i) + I_ia;
    endif,
  end,
end,
```
(8)

I_{ia} is the azimuth rotational inertia moment of the link i.

The algorithm for generation of Coriolis-centrifugal vector can be seen in [15].

3.2 Gravitational Vector, Friction and Elastic Coefficients Matrix Generation

Observing the equations of the dynamic model for cases 1, 2, 3 and 4 links, one realizes that the generation of the first n elements of the gravitational vector obeys the following rule:

$$G_i = l_i\left(\frac{m_i}{2} + \sum_{k=i+1}^{n} m_k\right) g \sin\theta_{ie} \tag{9}$$

with $i = 1, \ldots, n$. The others elements for $i = n+1, \ldots, 3n$ are nulls. Torques caused by buoyancy forces have the same structures as in Eq. (9), but with a negative sign (gravitational and buoyancy forces acting in opposite senses). As the geometry of the links is known, masses of fluid equivalent to the volume of each link are also known, so that Eq. (9) can easily be adapted to generate buoyancy torques.

As explained previously, the matrices of friction coefficients and elastic constants have the same generation rule, showed at the following algorithm (for the elastic constants matrix):

```
for i = 1:n,  for j = 1:n,  k(i,j) = 0;  end,  end,
    for i = 1:n,
        for j = 1:n,
            if (i = j) and (i < n),
                k(i,j) = k_ie + k_(i+1)e;
            endif,
            if (i = j) and (i = n),
                k(i,j) = k_ie;                                  (10)
            endif,
            if j = i + 1,
                k(i,j) = -k_je;
                k(j,i) = k(i,j);
            endif,
        end,
    end,
```

$k_{ie}(i = 1, \ldots, n)$ is the elevation elastic constant of the joint i. Algorithm (10) generates the elevation submatrix K_e. An identical rule is used to generate the azimuth submatrix K_a, considering, in this case, the azimuth elastic constant $k_{ia}(i = 1, \ldots, n)$. The same generation rule is also used to the torsion matrix K_T, considering $k_{iT}(i = 1, \ldots, n)$. The complete elastic constant matrix K is $(3n \times 3n)$, as well as the friction coefficient matrix C, and both are written in the form:

$$K = \begin{bmatrix} K_e & 0 & 0 \\ 0 & K_a & 0 \\ 0 & 0 & K_T \end{bmatrix}; C = \begin{bmatrix} C_e & 0 & 0 \\ 0 & C_a & 0 \\ 0 & 0 & C_T \end{bmatrix} \quad (11)$$

4 Simulation Results

Simulations were performed considering two application situations in the underwater environment. In both situations, the cable has one end fixed to a structure at the water surface. Its other end is free (or fixed to the seabed) and simulations are performed with or without considering ocean current. The hydrodynamic drag is modeled in a simple way, where the drag force is proportional to the square of the relative velocity between the structure and the water. Table 1 shows physical parameters used to perform the simulations (all others parameters can be seen in [15]).

The first simulation shows the cable in free fall from an initial spatial configuration in the underwater environment. Figure 5 shows a sequence of frames with the cable spatial configuration, from zero to 22 s every 2 s (from left to right). The cable terminal load appears in red in the animations with the simulation results. As explained before, it was used a simple model for the hydrodynamic drag, proportional to the square of the relative velocity between the fluid and the structure. This external effort was primarily responsible for the slow cable movement in free fall, seen in Fig. 5. Figure 6 shows results of a simulation similar to the previous one, but considering the cable out of the water. The frames also are from zero to 22 s every 2 s. In this case, cable's dynamic is obviously faster. Figure 7 shows the two cases simultaneously, each with spatial configurations on the same graph. The three-dimensional animations of the spatial configuration of the cable (Figs. 5, 6 and 7) allow showing that the simulation results give a great sense of physical reality. Figures 5 and 6 only show evolutions of the spatial configuration of the cable, while Fig. 7 shows these evolutions in the same graph, making it possible to visualize the dimensional scale in *m*.

Table 1 Principal physical parameters used in simulations

Parameters	Numerical value	Physical meaning
L_c	1200 m	Cable length
n	32; 24	Number of links
r_i	0.01 m	Radius of each link (constant)
m_e	7850 kg/m^3	Cable specific mass
m_c	600 kg	Terminal load mass

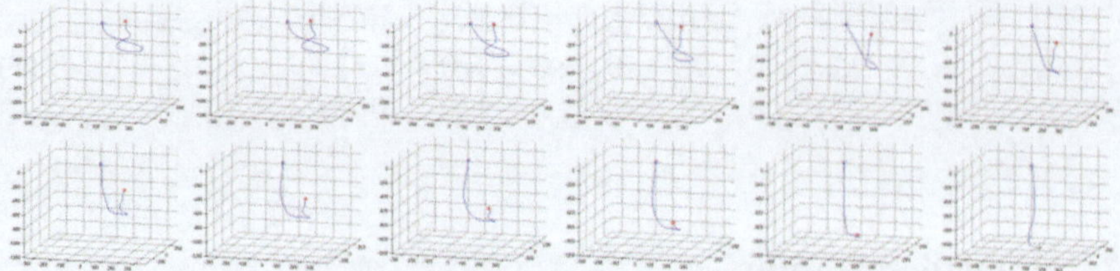

Fig. 5 Free fall simulation, underwater, from zero to 22 s, frames every 2 s

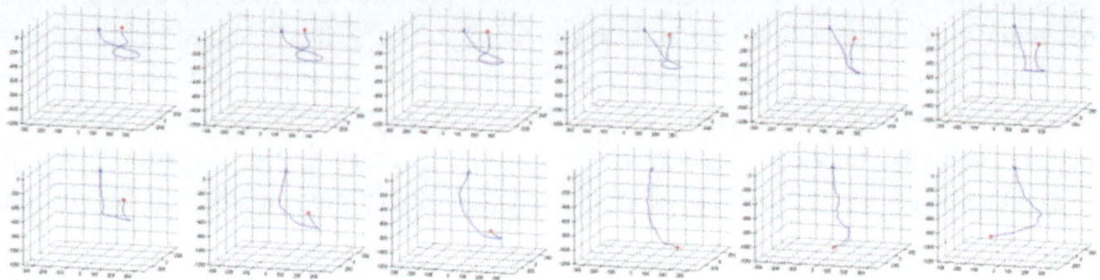

Fig. 6 Free fall simulation, out of water, from zero to 22 s, frames every 2 s

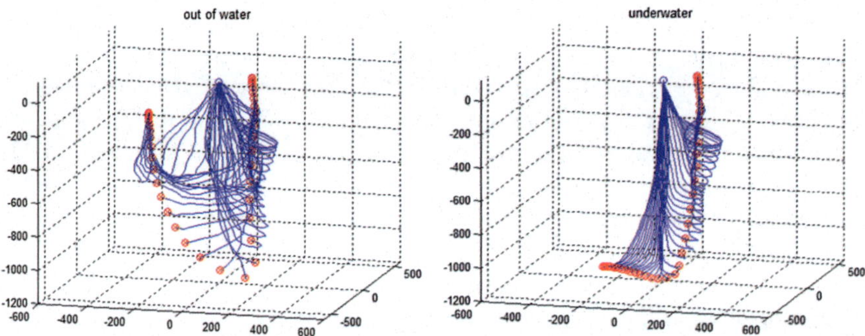

Fig. 7 Free fall simulations, out of water and underwater, from zero to 40 s, frames every 1 s

Figure 8 shows 36 s of simulation with frames every 2 s, considering an ocean current acting on the cable with speed of 3.5 m/s. Drag forces was considered proportional to the square of the relative velocities between fluid and structure. The cable is initially at rest in its vertical position and with free terminal load. The ocean current starts from zero initial time and then imposes a significant dynamic disturbance on the cable. If the cable terminal load is a remotely operated vehicle (ROV), the ocean current can to induce dynamic disturbances to the cable that would be transmitted to the vehicle, thus hindering the performance of any control strategy.

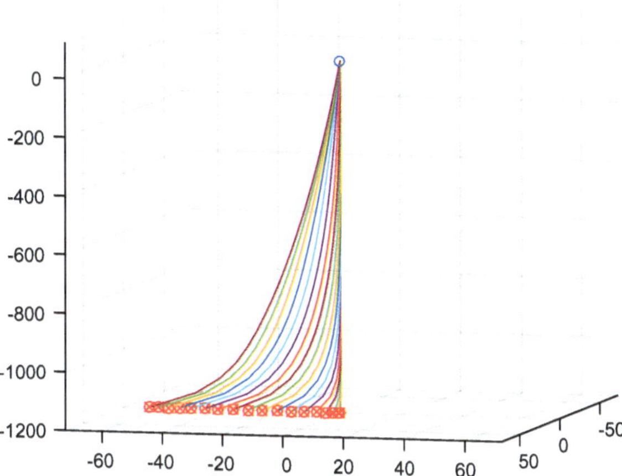

Fig. 8 Cable under the action of underwater current, from 0 to 36 s, frames every 2 s

The following simulation was performed considering the terminal load fixed to the seabed. The cable was placed in an initial spatial configuration (first frame of Fig. 10). The position of the terminal load at this initial configuration has been taken as a reference to a Proportional and Derivative (PD) control, applied to maintain the terminal load fixed in this reference (constraint forces). These forces were applied in the CM of the terminal load and in the three dimensions, generating torques in the last joint, which enter as external torques in Eq. (1). Figure 9 shows the elevation angles and Fig. 10 shows the cable spatial configuration, with frames every 2 s. It is observed that the control was effective in maintaining the terminal load in its original initial position, which could be on the seabed. This simulation is in a single plan, so that all azimuth angles are zero. We verified that the cable stabilization trend in search of its final spatial configuration (catenary static equilibrium), which is fully achieved after 180 s.

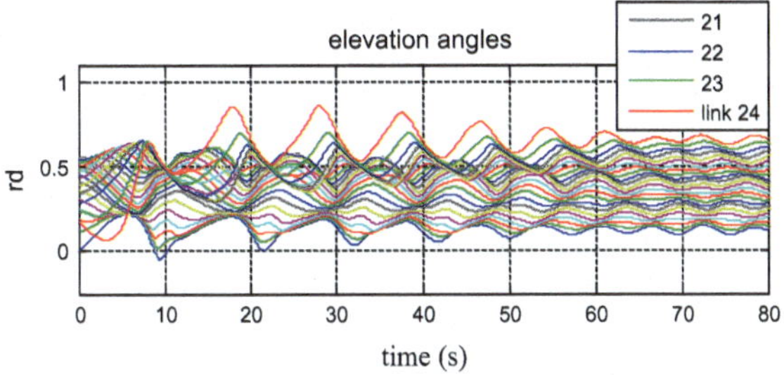

Fig. 9 Elevation angles, terminal load fixed to the seabed, considering 24 links

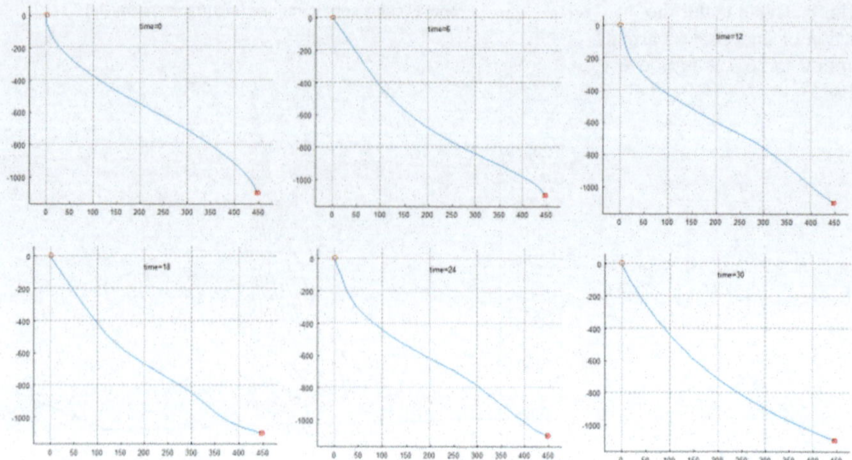

Fig. 10 Frames showing spatial cable configuration considering the terminal load fixed to the seabed (from zero to 30 s, frames every 6 s, 24 links)

5 Conclusions

Nature follows growth patterns in their phenomena, but we cannot always identify them. The discrete approach proposed in the modeling formalism approximates the real continuous flexibility when increasing the number of links. Each new added link increases in three degrees of freedom the cable dynamics and its equations grow considerably in size and complexity. It was possible to identify patterns of this growth, allowing to the proposition of algorithms that automatically generate dynamic models for any number of links considered in discrete approximation.

The simulations were chosen in order to know a priori what should be the dynamic behavior of the cable and, in all cases the results were in agreement with the physically expected. Torsion motion was considered in cable modeling because in future applications, there is an interest in having a ROV as a cable terminal load. In this case, forces applied to the ROV can generate torsion in the cable. It has been specially developed software that performs three-dimensional animations of the cable's spatial configuration, for better visualization and analysis of simulation results. Three-dimensional animations allowed us to identify a great sense of physical reality. Increasing the number of links implies a better discrete approximation of the continuous flexibility. It was observed that over forty links the discrete model closely matches the continuous flexibility.

In future works it is planned to build an experimental support sensed by digital cameras to validate simulations and physical parameters identification strategies.

References

1. Srinil, N., Rega, G., Chucheepsakul, S.: Two to one resonant multi-modal dynamics of horizontal/inclined cables. Part I: Theoretical formulation and model validation. Nonlinear Dyn. **48**, 231–252 (2007)
2. Yoon, J.W., Park, T.W., Yim, H.J.: Fatigue life prediction of a cable harness in an industrial robot using dynamic simulation. J. Mech. Sci. Technol. **22**(3), 484–489 (2008). ISSN 1738-494X
3. Sun, F.J., Zhu, Z.H., LaRosa, M.: Dynamic modeling of cable towed body using nodal position finite element method. Ocean Eng. **38**, 529–540 (2011)
4. Hover, F.S., Grosenbaugh, M.A., Triantafyllou, M.S.: Calculation of dynamic motions and tensions in towed underwater cables. IEEE J. Ocean. Eng. **19**(3) (1994)
5. Wang, F., Huang, G., Deng, D.: Steady state analysis of towed marine cables. J. Shanghai Jiaotong Univ. (Sci.) **13**(2), 239–244 (2008)
6. Raman-Nair, W., Williams, C.D.: Vortex-induced response of a long flexible marine riser in a shear current. In: International Symposium on Technology of Ultra Deep Ocean Engineering, Tokyo, Japan, 1–2 Feb 2005
7. Zhu, K.-Q., Zhu, H.-Y., Zhang, Y.-S., Gao, J.: A multi-body space-coupled motion simulation for a deep-sea tethered remotely operated vehicle. J. Hydrodyn. **20**(2), 210–215 (2008)
8. Hall, M., Goupee, A.: Validation of a lumped-mass mooring line model with DeepCwind semisubmersible model test data. Ocean Eng. **104**, 590–603 (2015)
9. Matulea, I., Ştefan, D., Vladescu, D., Barbu, C., Coelho, A.M.G.: A novel numerical approach to the dynamics analysis of marine cables. Int. J. Appl. Sci. Technol. **4**(2) (2014)
10. Lee, C., Hong, S., Kim, H., Kim, S.: A comparative study on effective dynamic modeling methods for flexible pipe. J. Mech. Sci. Technol. **29**(7), 2721–2727 (2015)
11. Gobat, J.I.: The dynamics of geometrically compliant mooring systems. Ph.D. thesis, Massachusetts Institute of Technology (2000)
12. Gomes, S.C.P., Rosa, V.S., Albertini, B.C.: Active control to flexible manipulators. IEEE/ASME Trans. Mechatron. **11**(1), 75–83 (2006). USA
13. Pereira, A.E.L., Gomes, S.C.P., Bortoli, A.L.: A new formalism for the dynamic modeling of cables. Math. Comput. Model. Dyn. Syst. **1**, 1–14 (2012)
14. Gomes, S.C.P., Zanela, B.E., Pereira, A.E.L.: Automatic generation of dynamic models of cables. Ocean Eng. **121**, 559–571 (2016). https://doi.org/10.1016/j.oceaneng.2016.05.041
15. Gomes, S.C.P., Zanela, E.B., Pereira, A.E.L.: Cable dynamics modeling and applications. In: XVII International Symposium on Dynamic Problems of Mechanics, DINAME, Brazilian Society of Mechanical Sciences and Engineering—ABCM (2017)

Optimization of the Fundamental Frequency of Mechanical Structures by Using the Bidirectional Evolutionary Structural Method

Ederval de Souza Lisboa, João Baptista Dias Moreira, Emanuel Moutinho Cesconeto and Walter Jesus Paucar Casas

Abstract A hierarchical structure is a structure that can be described by different characteristic lengths, and is such that its layout in the smaller scale (microscale) affects its behavior in the bigger scale (macroscale). Each hierarchical level is treated as a continuous medium composed of one or more materials. The simultaneous design of multiphase composite structures aims at finding the optimal distribution of materials such that one or more structural parameters are maximized (or minimized). In this work, the Bi-directional Evolutionary Structural Optimization method, BESO hereinafter, is applied to the maximization of the fundamental frequency of a structure subjected to a constraint on the total volume of materials used. Numerical experiments are made in order to validate the implementation and confirm the efficacy of the method in optimizing the topology of the structure.

Keywords Topology optimization · Hierarchical structure · BESO Homogenization

---

E. de Souza Lisboa (✉) · J. B. D. Moreira · E. M. Cesconeto · W. J. P. Casas
Mechanical Engineering Department, Federal University of Rio Grande do Sul, Porto Alegre 90050-170, Brazil
e-mail: ederval.lisboa@yahoo.com.br

J. B. D. Moreira
e-mail: baptista.moreira@ufrgs.br

E. M. Cesconeto
e-mail: emanuelemc@hotmail.com

W. J. P. Casas
e-mail: walter.paucar.casas@ufrgs.br

A. de T. Fleury et al. (eds.), *Proceedings of DINAME 2017*, Lecture Notes in Mechanical Engineering, https://doi.org/10.1007/978-3-319-91217-2_10

1 Introduction

Hierarchical structures are structures that can be described in varied scales with different characteristic lengths, so that a structural element in a certain scale is composed by periodic substructures in a smaller scale. Each hierarchical level may be treated as a continuous medium, where the material properties on the macro scale are evaluated through a homogenization process considering the micro structure's materials and layout. The design of composite materials aims at finding the distribution of the base material in the microstructure by using topology optimization techniques in order to reach desired material properties.

The simultaneous design with multiple phases of material and composite structures is still limited. For instance, on the design of composite structures, Huang and Xie [1] use the BESO method to solve the problem of minimum compliance of structures with multiple materials, while Huang and Xie [2] study the problem of maximizing the frequency of multimaterial structures, this problem is in force in works by Xu and Xie [3], Liu et al. [4] and Long et al. [5], among others.

Zuo et al. [6] highlight that fundamental frequency optimization is important in engineering, given that it is desirable to modify the natural frequencies spectrum distribution so as to avoid any resonance situation in a larger interval of excitation frequencies. As the natural frequency is associated with the performance of the macrostructure, microscale optimization is seldom studied.

In this context, the objective of this work is to develop an approach of simultaneous design, optimizing the fundamental frequency of the system through the BESO method. The design variables describe the distribution of material in both macro and micro levels. In this approach the physical properties of both materials used in the macroscale optimization are composed via homogenization of the microstructures, and these in turn are composed by isotropic materials. These microstructures are also optimized, so as to approach the ideal physical properties for a given structure.

2 Theoretical Background

2.1 *BESO Method for Fundamental Frequency Optimization*

The modal behavior of an undamped system can be analyzed by means of (1).

$$\left(\mathbf{K}-\omega^{2}\mathbf{M}\right)\mathbf{u}=\mathbf{0} \tag{1}$$

The k-th natural frequency ω_k and its corresponding mode $\mathbf{u}_k$ are related through Rayleigh's quotient given by (2).

$$\omega_k^2 = \frac{\mathbf{u}_k^T \mathbf{K} \mathbf{u}_k}{\mathbf{u}_k^T \mathbf{M} \mathbf{u}_k} \tag{2}$$

Consider the objective of maximizing the k-th natural frequency of a structure described by a discrete mesh using a predetermined volume of material. To every element in the meshed domain is assigned a value of 1 or $x_{\min}$ indicating the presence or absence of material, respectively. In this context, the problem may be presented as in (3) and (4) according to Huang and Xie [2]:

$$\text{Maximize:} f(\mathbf{x}) = \omega_k \tag{3}$$

$$\begin{gathered} \text{Subject to: } V* - \sum_{i=1}^{N} x_i V_i = 0 \\ (x_i = x_{min} \text{ or } 1) \end{gathered} \tag{4}$$

$V*$ is the prescribed volume fraction, meaning the ratio between the volume that the structure should occupy and the total volume of the domain. N is the number of elements in the domain, and V_i is the fraction of the total volume that the i-th element occupies. The binary design variable x_i indicates if the structure occupies the i-th element, and the small value x_{min} (e.g. 10^{-6}) corresponds to a region in the domain with no material.

2.2 *Interpolation Scheme for the Material Properties*

For the solid-void design, the material's density and Young modulus are functions of the design variable. Given that with the SIMP (Solid Isotropic Material with Penalization) method the ratio between elementary mass and stiffness is extremely high for small x_i values (e.g. after applying power law penalization to the stiffness and linear interpolation to the density), artificial localized modes appear in regions with low density. Huang et al. [7] proposed an alternative interpolation scheme where the mass/stiffness ratio is kept constant, as given by Eqs. (5) and (6):

$$\rho(x_{min}) = x_{min} \rho^1 \tag{5}$$

$$E(x_{min}) = x_{min} E^1 \tag{6}$$

A better interpolation scheme takes the explicit form shown in Eqs. (7) and (8), according to Huang and Xie [2]:

$$\rho(x_i) = x_i \rho^1 \tag{7}$$

$$E(x_i) = \left[\frac{x_{min} - x_{min}^P}{1 - x_{min}^P}\left(1 - x_i^P\right) + x_i^P\right] E^1 \tag{8}$$

P is called the penalty exponent and x_i is 1 if the i-th element is composed of the corresponding material, 0 if otherwise. Further details may be find in [7].

2.3 *Frequency Optimization in Hierarchical Structures*

The objective function for dynamic problems in multiscale structures can be written as shown in Eqs. (9) through (14), according to Zuo et al. [6]:

$$\begin{aligned} &\text{Find: } \mathbf{x} = \left\{\mathbf{x}^{mac}, \mathbf{x}^{mic,1}, \mathbf{x}^{mic,2}\right\} \\ &\left(x_i^{mac}, x_i^{mic,1}, x_i^{mic,2} = x_{min} \text{ or } 1\right) \end{aligned} \tag{9}$$

$$\text{Maximize:} f(\mathbf{x}) = \omega_k \tag{10}$$

$$\text{Subject to: } \left(\mathbf{K} - \omega_k^2 \mathbf{M}\right)\mathbf{u}_k = \mathbf{0} \tag{11}$$

$$\sum_{i=1}^{M} x_i^{mac} V_i^{mac} = V^{mac} \tag{12}$$

$$\sum_{j=1}^{N} x_j^{mic,1} V_j^{mic,1} = V^{mic,1} \tag{13}$$

$$\sum_{j=1}^{N} x_j^{mic,2} V_j^{mic,2} = V^{mic,2} \tag{14}$$

V is the volume, and the superscript *mac* represents the macromodel and *mic* represents the micromodel of the structure. The subscripts i and j correspond to the i-th and j-th element of the macromodel and the micromodel, respectively.

In this problem the vector $\mathbf{x}$ is composed by $\mathbf{x}^{mac}$, which describes the macromodel layout, and $\mathbf{x}^{mic,1}$ and $\mathbf{x}^{mic,2}$, which describe the micromodels corresponding to each of the phases present. This means that x_i^{mac} can assume a value of 0, in which case the properties for the i-th element are given by the homogenization of the microstructure defined by $\mathbf{x}^{mic,1}$, or a value of 1, in which case the microstructure defined by $\mathbf{x}^{mic,2}$ is used instead.

For both macro and micromodels the design variable takes binary values corresponding to the presence or absence of a certain phase. As the design variables in the two micromodels vary over the same domain, from now on they will be called only x^{mic}, requiring a distinction between $x^{mic,1}$ or $x^{mic,2}$ when necessary.

ω_k represents the k-th natural frequency associated with the structure. Equation (12) describes the volume constraint on the macromodel, which controls the material distribution on the macroscale. V^{mac} correspond to the volume fraction of the predefined macro phase, where V_i^{mac} is the fraction of the total domain occupied by the i-th element in the macro domain. Similarly, Eqs. (13) and (14) describe volume constraints on the microscale of the first and second phases, respectively, *i.e.*, $V_j^{mic,1}$ is the volume of the j-th element in the first micromodel and $V^{mic,1}$ is the prescribed fraction of volume of the first phase in the micromodel; $V_j^{mic,2}$ and $V^{mic,2}$ have analogous meanings but for the second micromodel.

2.4 Sensitivity Analysis

The element mass matrix $\mathbf{m}$ and the element stiffness matrix $\mathbf{k}$, both of the macromodel, and the material constitutive matrix $\mathbf{D}$ of the micromodel, are defined respectively by Eqs. (15), (16) and (17), as shown by Zuo et al. [6], being necessary components for the sensitivity analysis. The penalized relations for the stiffness and constitutive matrix come from the SIMP model and are valid for both macro and micro levels of the structure.

$$\mathbf{m}\left(x_i^{mac}\right) = x_i^{mac}\mathbf{m}_i^1 + \left[1 - x_i^{mac}\right]\mathbf{m}_i^2 \tag{15}$$

$$\mathbf{k}\left(x_i^{mac}\right) = \left(x_i^{mac}\right)^P\mathbf{k}_i^1 + \left[1 - \left(x_i^{mac}\right)^P\right]\mathbf{k}_i^2 \tag{16}$$

$$\mathbf{D}\left(x_j^{mic}\right) = \left(x_j^{mic}\right)^P\mathbf{D}_j^1 + \left[1 - \left(x_j^{mic}\right)^P\right]\mathbf{D}_j^2 \tag{17}$$

The derivatives for the global mass matrix $\mathbf{M}$ and stiffness matrix $\mathbf{K}$, and for the material constitutive matrix $\mathbf{D}$ of the micromodel are obtained through Eqs. (18), (19) and (20), respectively. According to Zuo et al. [6]:

$$\frac{\partial \mathbf{M}}{\partial x_i^{mac}} = \mathbf{m}_i^1 - \mathbf{m}_i^2 \tag{18}$$

$$\frac{\partial \mathbf{K}}{\partial x_i^{mac}} = P\left(x_i^{mac}\right)^{P-1}\left(\mathbf{k}_i^1 - \mathbf{k}_i^2\right) \tag{19}$$

$$\frac{\partial \mathbf{D}}{\partial x_j^{mic}} = P\left(x_j^{mic}\right)^{P-1}\left(\mathbf{D}_j^1 - \mathbf{D}_j^2\right) \tag{20}$$

2.5 Macroscale Sensitivity Analysis

Deriving the k-th natural frequency from Rayleigh's quotient (2) with respect to the i-th design variable in the macrolevel, normalizing the eigenvectors with respect to the mass matrix, and using the definition of the derivatives given by Eqs. (18) and (19), the sensitivity of the k-th natural frequency for the macromodel is obtained in Eq. (21), as shown by Zuo et al. [6]:

$$\alpha_i^{mac} = \frac{\partial \omega_k}{\partial x_i^{mac}} = \frac{1}{2\omega_k}\mathbf{u}_k^{\mathrm{T}}\left[P\left(x_i^{mac}\right)^{P-1}\left(\mathbf{k}_i^1 - \mathbf{k}_i^2\right) - \omega_k^2\left(\mathbf{m}_i^1 - \mathbf{m}_i^2\right)\right]\mathbf{u}_k \tag{21}$$

2.6 Microscale Sensitivity Analysis

The micromodel describes the microstructure of the phases present in the macromodel. This microstrucuture is then homogenized to get the effective material properties used in the macrostructure. The homogenized matrix $\mathbf{D}^{\mathrm{H}}$ is calculated according to Eq. (22), over the domain Y of the base cell described by the variable x^{mic}. The procedure used to calculate $\mathbf{D}^{\mathrm{H}}$ is explained in a series of papers by Hassani and Hinton [8–10], and a detailed computational implementation may be found, for instance, in Andreassen and Andreasen [11].

$$\mathbf{D}^{\mathrm{H}} = \frac{1}{|Y|}\int_Y \mathbf{D}(\mathbf{I} - \mathbf{bu})dY \tag{22}$$

$\mathbf{D}$ represents the constitutive matrix of the material in the microstructure, $\mathbf{I}$ is identity matrix, $\mathbf{b}$ is strain matrix in the micromodel and $\mathbf{u}$ the displacement field.

The stiffness matrix $\mathbf{k}_i$ may be calculated according to Eq. (23), assuming a 2D domain, by imposition of a periodic boundary condition, where the displacement fields $\mathbf{u}$ are chosen so that the strain $\mathbf{b}$ is uniform $[1,0,0]^{\mathrm{T}}$, $[0,1,0]^{\mathrm{T}}$ and $[0,0,1]^{\mathrm{T}}$:

$$\mathbf{k}_i = \int_{V_i} \mathbf{B}^{\mathrm{T}}\mathbf{D}^{\mathrm{H}}\mathbf{B}dV_i \tag{23}$$

$\mathbf{B}$ represents the strain matrix and V_i the i-th element volume.

Differentiating the k-th natural frequency from Eq. (2) with respect to the j-th variable of the micromodel, Eq. (24) is obtained:

$$\frac{\partial \omega_k}{\partial x_j^{mic}} = \frac{1}{2\omega_k}\left[\mathbf{u}_k^{\mathrm{T}}\left(\frac{\partial \mathbf{K}}{\partial x_j^{mic}} - \omega_k^2\frac{\partial \mathbf{M}}{\partial x_j^{mic}}\right)\mathbf{u}_k\right] \tag{24}$$

In a finite element analysis the global stiffness matrix **K** is composed from the element stiffness matrices, and the global mass matrix **M** is composed from the element mass matrices. The sensitivity of the frequency with respect to the microstructure design variable was developed by Zuo et al. [6], as given by Eq. (25):

$$\alpha_j^{mic} = \frac{\partial \omega_k}{\partial x_j^{mic}} = \frac{1}{2\omega_k} \sum_{i=1}^{M} \mathbf{u}_{k,i}^{\mathrm{T}} \int_{V_i} \mathbf{B}^{\mathrm{T}} \frac{\partial \mathbf{D}^{\mathrm{H}}}{\partial x_j^{mic}} \mathbf{B} dV_i \mathbf{u}_{k,i} \tag{25}$$

The sensitivity of the homogenized matrix can be obtained from Eqs. (20) and (22), obtaining Eq. (26):

$$\begin{aligned} \frac{\partial \mathbf{D}^{\mathrm{H}}}{\partial x_j^{mic}} &= \frac{1}{|Y|} \int_{Y} (\mathbf{I} - \mathbf{bu})^{\mathrm{T}} \frac{\partial \mathbf{D}}{\partial x_j^{mic}} (\mathbf{I} - \mathbf{bu}) dY \\ &= \frac{P\left(x_j^{mic}\right)^{P-1}}{|Y|} \int_{Y} (\mathbf{I} - \mathbf{bu})^{\mathrm{T}} \left(\mathbf{D}_j^1 - \mathbf{D}_j^2\right) (\mathbf{I} - \mathbf{bu}) dY \end{aligned} \tag{26}$$

Thus, the sensitivity of the fundamental frequency with respect to the microstructure design variables is found by substituting Eq. (25) into Eq. (26), producing Eq. (27):

$$\alpha_j^{mic} = \frac{P\left(x_j^{mic}\right)^{P-1}}{2\omega_k |Y|} \sum_{i=1}^{M} \mathbf{u}_{k,i}^{\mathrm{T}} \left\{ \int_{V_i} \mathbf{B}^{\mathrm{T}} \left[\int_{Y} (\mathbf{I} - \mathbf{bu})^{\mathrm{T}} \left(\mathbf{D}_j^1 - \mathbf{D}_j^2\right) (\mathbf{I} - \mathbf{bu}) dY \right] \mathbf{B} dV_i \right\} \mathbf{u}_{k,i} \tag{27}$$

To minimize oscillatory effects that occur during the optimization process, where the same elements are repeatedly added and removed, a filtering technique is applied. The averaged sensitivity with respect to time is given in Eq. (28), where q is the iteration numerical index:

$$\tilde{\alpha} = \frac{1}{2}\left(\alpha_i^q + \alpha_i^{q-1}\right) \tag{28}$$

2.7 Convergence Criteria

The evolution ratio *ER* is an algorithm parameter, and its importance resides in controlling the volume variation between iterations, given in Eq. (29) for the macromodel and in Eq. (30) for the micromodel:

$$V^{mac,q} = V^{mac,q-1}(1 \pm ER^{mac}) \tag{29}$$

$$V^{mic,q} = V^{mic,q-1}(1 \pm ER^{mic}) \tag{30}$$

$V^{mac,q}$ is the value of the volume in the q-th iteration, while $V^{mac,q-1}$ is defined in the previous iteration ($q - 1$). $V^{mic,q}$ and $V^{mic,q-1}$ are defined in the same way.

After the final volume fractions are reached the element addition/removal process continues until the variation in the objective function for consecutive iterations is lower than a certain threshold value. The considered variation of ω_k takes into account the last N iterations, as shown in Eq. (31):

$$\tau = \frac{\sum_{i=1}^{N} \omega_k^{q-i+1} - \omega_k^{q-N-i+1}}{\sum_{i=1}^{N} \omega_k^{q-i+1}} \leq \tau^* \tag{31}$$

τ represents the relative variation of the objective function, τ^* is the tolerance or threshold and N is a predefined number of iterations. When this condition is satisfied the system is considered to have converged and the optimization process is stopped.

2.8 Flowchart

The flowchart in Fig. 1 shows the implementation of the BESO algorithm.

3 Results and Discussion

3.1 BESO Method for Fundamental Frequency Optimization

A program was developed by using the methodology described in this section. It was then tested in order to evaluate the accuracy of the implementation and to mitigate numerical instabilities. For the finite element model four-node quadrilateral plane strain elements were used. The results of the optimization were compared with reference problems found in the literature.

Problem 1 Problem 1 is a two-material beam clamped at both ends with a lumped mass $M = 1.4 \times 10^{-5}$ kg on its center. The dimensions are 0.14 m $\times$ 0.02 m, and the domain is discretized with 280 $\times$ 40 elements (see Fig. 2).

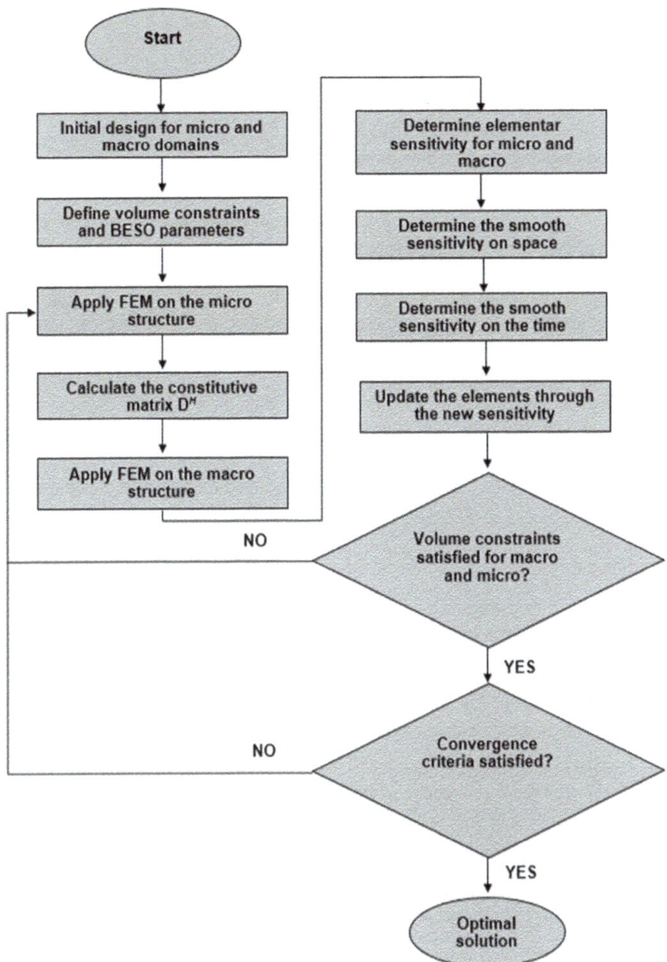

Fig. 1 General scheme of the BESO method applied

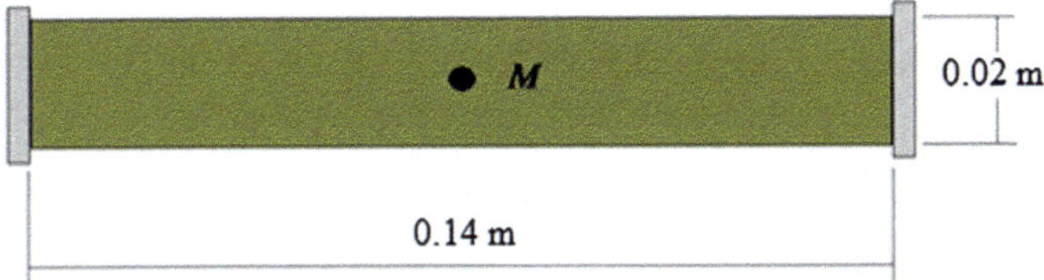

Fig. 2 Problem 1, two-material beam with lumped mass and clamped at both ends

Table 1 Material properties for Problem 1

Properties	Material 1	Material 2
Young modulus (E) (N/cm^2)	100	20
Poisson ratio (ν)	0.3	0.3
Mass density (ρ) (kg/cm^3)	10^{-6}	10^{-7}

Fig. 3 Problem 1, results obtained in this work

Table 2 Natural frequency for Problem 1 (rad/s)

Author	ω_1	ω_2
Huang and Xie [2]	37.1	–
This work	38.2	115

The objective is to maximize the fundamental frequency, where each material occupies half of the domain. Table 1 gives the properties of both materials.

The input data for the BESO method are the evolutionary rate $ER = 2\%$, the maximum addition ratio $AR_{max} = 2\%$, $\tau = 0.1\%$, $V^{mac} = 50\%$, $V^{mic,1} = 100\%$, $V^{mic,2} = 100\%$, filtering radius $r_{min} = 0.0015$ m and penalty factor $P = 3$.

The results of the optimization are shown in Fig. 3, where black regions represent Material 1 and yellow regions represent Material 2. Results of the BESO optimization implemented of this work were compared to those obtained by Huang et al. [7]. Table 2 shows the fundamental frequencies.

Problem 2 Problem 2 considered the same beam shown in Fig. 2, but this time taking the softer material density and stiffness close to zero to emulate voids. The domain is discretized with 280 $\times$ 40 square elements with unitary sides, similar to Problem 1. The beam is composed of a material whose properties are shown in Table 3. The input data are: evolutionary rate $ER = 2\%$, addition ratio $AR_{max} = 2\%$, $\tau = 0.01\%$, $V^{mac} = 50\%$, $V^{mic,1} = 100\%$, $V^{mic,2} = 0\%$, filtering radius $r_{min} =$ 0.0015 m and penalty factor $P = 3$.

The result obtained in this work is shown in Fig. 4, where the material is colored black, while the simulated void is shown as white. Table 4 compares the obtained frequencies.

Table 3 Material properties for Problem 2

Properties	Material
Young modulus (E)	10 N/cm^2
Poisson ratio (ν)	0.3
Mass density (ρ)	1 kg/cm^3

Fig. 4 Problem 2, final optimized structure obtained in this work

Table 4 Natural frequency for Problem 2 (rad/s)

Author	ω_1	ω_2	Iterations
Huang and Xie [2]	33.7	–	55
This work	34.7	104.3	47

Fig. 5 Problem 3, two-material beam clamped at both ends

Problem 3 Problem 3 aims to optimize the two material beam clamped at both ends, as illustrated in Fig. 5. The domain was discretized with 80 × 40 square elements of unitary dimensions. The objective is to maximize the fundamental frequency where each material occupies half of the domain. Table 5 shows properties of both materials.

The input data are: evolutionary rate $ER = 1\%$, addition ratio $AR_{max} = 1\%$, $\tau = 0.01\%$, $V^{mac} = 50\%$, $V^{mic,1} = 100\%$, $V^{mic,2} = 100\%$, filtering radius $r_{min} = 0.0015$ and penalty factor $P = 3$.

The final structure found in this case is shown in Fig. 6, where the stiff material is shown in black while the softer material is shown in yellow. Table 6 compares the obtained frequencies.

Table 5 Material properties for Problem 3

Properties	Material 1	Material 2
Young modulus (E) (N/cm^2)	1	0.2
Poisson ratio (ν)	0.3	0.3
Mass density (ρ) (kg/cm^3)	1	2

Fig. 6 Problem 3, results obtained in this work

Table 6 Natural frequency for Problem 3 (rad/s)

Author	ω_1	ω_2
Zuo et al. [6]	0.011317	–
This work	0.011246	0.019632

4 Conclusions

The comparison shows that the BESO algorithm implemented in this work was capable of optimizing the fundamental natural frequency in different structures. The algorithm makes use of its own FEM module, thus being independent of commercial FEM solvers.

The averaged sensitivity used in this work enabled to optimize the fundamental natural frequency in agreement to most elaborated proposals of the literature, e.g. the fundamental natural frequency obtained in this work was higher in 2.96, 2.97 and 8.21% in Problems 1, 2 and 3, when compared with values obtained by other authors.

Acknowledgements The research described in this paper was financially supported by the National Council of Scientific and Technological Development CNPq-Brazil (grants 140081/2015-1, 148895/2016-6 and 148887/2016-3) from the Ministry of Science, Technology and Innovation.

References

1. Huang, X., Xie, Y.M.: Bi-directional evolutionary topology optimization of continuum structures with one or multiple materials. Comput. Mech. **43**, 393–401 (2009)
2. Huang, X., Xie, Y.M.: Evolutionary Topology Optimization of Continuum Structures: Methods and Applications. Wiley, Chichester (2010)
3. Xu, B., Xie, M.: Concurrent design of composite macrostructure and cellular microstructure under random excitations. Compos. Struct. **123**, 65–77 (2015)
4. Liu, Q., Chan, R., Huang, X.: Concurrent topology optimization of macrostructures and material microstructures for natural frequency. Mater. Des. **106**, 380–390 (2016)

5. Long, K., Han, D., Gu, X.: Concurrent topology optimization of composite macrostructure and microstructure constructed by constituent phases of distinct Poisson's ratios for maximum frequency. Comput. Mater. Sci. **129**, 194–201 (2017)
6. Zuo, Z.H., Huang, X., Rong, J.H., Xie, Y.M.: Muti-scale design of composite materials and structures for maximum natural frequencies. Mater. Des. **51**, 1023–1034 (2013)
7. Huang, X., Zuo, Z.H., Xie, Y.M.: Evolutionary topological optimization of vibrating continuum structures for natural frequencies. Comput. Struct. **88**, 357–364 (2010)
8. Hassani, B., Hinton, E.: A review of homogenization and topology optimization I—homogenization theory for media with periodic structure. Comput. Struct. **69**, 707–717 (1998)
9. Hassani, B., Hinton, E.: A review of homogenization and topology optimization II—analytical and numerical solution of homogenization. Comput. Struct. **69**, 719–738 (1998)
10. Hassani, B., Hinton, E.: A review of homogenization and topology optimization III—topology optimization using optimality criteria. Comput. Struct. **69**, 739–756 (1998)
11. Andreassen, E., Andreasen, C.S.: How to determine composite material properties using numerical homogenization. Comput. Mater. Sci. **83**, 488–495 (2014)

Operational Modal Parameters Identification Using the ARMAV Model

Heraldo N. Cambraia, Leonardo M. L. Contini and Paulo R. G. Kurka

Abstract Applied system identification is an important issue in science and engineering. Experimental modal analysis is used to describe the dynamical behavior of structures, in general, for a given set of input and output data. This article deals with multidimensional modal parameters identification valid for output-only data—operational modal analysis (OMA). This approach is interesting when the input is not known or difficult to be measured. A linear, time-invariant and finite dimensional mechanical system is considered, which is described mathematically by an autoregressive-moving-average-vector (ARMAV) model, excited by unknown operating forces assumed to be a white Gaussian process—a persistent excitation. The focus of the study is, both, theoretical and practical aspects, of the use of the ARMAV model in OMA. Specifically, it discusses the need of using an output-vector as reference for output-only parameters identification scheme. The model order is identified by inspection of the most significant singular values of a block Hankel matrix derived directly from the formulation of the model. The AR parameters matrices of the ARMAV model, contained in a companion matrix, are determined via least-squares technique. Natural frequencies, damping factors and modal shapes are identified by means of eigenvalues and eigenvectors of that companion matrix. Examples using computational simulated data are presented.

Keywords Operational modal analysis · ARMAV · Least squares approach

---

H. N. Cambraia (✉) · L. M. L. Contini
Paraná Federal University, DEMEC, Curitiba, Brazil
e-mail: heraldo@ufsc.br

P. R. G. Kurka
Campinas State University, DEM, Campinas SP, Brazil

A. de T. Fleury et al. (eds.), *Proceedings of DINAME 2017*, Lecture Notes in Mechanical Engineering, https://doi.org/10.1007/978-3-319-91217-2_11

1 Introduction

Mathematical modeling is an analytical approach used to describe the dynamic behavior of a natural phenomenon based on physical laws. System identification is an experimental approach, where parametric models are fitted from measured data [1]. Both approaches are important in system analysis, design and control problems.

Modal parameters can be used in analysis, modal updating via finite elements, damage detection and control. Modal parameters identification techniques, in the time domain, are classically based on the information contained in the impulse response functions (IRF) or in the input-output relationship [2]. In general terms, a modal identification test is conducted under certain laboratory conditions, where the structure is fixed to a test bench and hammers or actuators are used to produce controlled types of input forces, which are required to match a linear time-invariant mathematical model, covering a certain frequency range of interest. However, in many applications, the real operating conditions may differ significantly from those applied during the modal tests, where the input forces are not known, or just impossible to be measured. Parameters identification based on the knowledge of output-only responses, without using excitation information, is known as operational modal analysis (OMA) [3]. The subject is of actual scientific and industrial interest in mechanical and civil engineering opening a way for damage detection and structural health analysis [1, 4–6].

OMA is present in several practical engineering applications. Lardies and Ta [5] have used OMA to assess the structural health and damage detection of stay cables in cable stayed bridges.

Vu et al. [6] proposed a method for the automatic identification procedure to discriminate physical modes from spurious ones using a multivariate autoregressive (AR) model whose parameters are estimated via a least squares (LS) method. Zaghbani and Songmene [7] proposed a methodology based on OMA to compare the modal parameters of machine tools, demonstrating how OMA can be industrially exploited. Rainieri and Fabbrocino [8] present a literature review on automated operational modal-based damage detection for civil engineering structures. Ramos et al. [9] performed structural identification of monuments in Portugal by OMA to assess damage by means of vibration signature.

According to Peeters and Roeck [10], there are many methods used to perform the OMA parameters identification. Formally, for a completely unknown input, it can be assumed that the system is excited by a white Gaussian process known as a persistent excitation. A multivariate linear time-invariant autoregressive-moving-average-vector (ARMAV) model can be used to fit the data, adopting a least squares, maximum likelihood or prediction method as optimization criterion to calculate the model's parameters [1, 3]. Maximum likelihood optimization procedure leads to a highly non-linear minimization problem in order to calculate the parameters of the model. The solution of such a problem has a very high computational cost, especially for the multivariable parameters case.

The focus of the present paper is on, both, theoretical and practical aspects of the use of the ARMAV model in OMA. Despite OMA parameters identification is a well documented subject, some problems remains to be studied. Specifically, the need of using a output vector as reference for an output only parameters identification scheme. Practical aspect consists in the computational implementation of an ARMAV model identification algorithm.

In the present technique, the OMA parameters are identified from eigen decomposition of a companion matrix that contains the AR coefficients of the ARMAV model obtained via least squares optimization of a block Hankel matrix formed by correlation matrices between output measured data. The problem with the least squares approach is the adopting an initial over parametrization of the model order resulting in a number of spurious numerical modes that must be separated from the true modes of the system. The correct order of the ARMAV model is identified via inspection of the more significant singular values of the block Hankel matrix above mentioned, using singular value decomposition (SVD).

The performance of the presented technique is demonstrated using data generated by means of computational simulation. Impulse responses (which have a type of self-reference given by their impulsive force) and input-output data without using excitation information are considered.

The paper is organized as follows: Next section present the multivariate ARMAV model. The algorithm is then introduced. An application based on simulation using data from mechanical system is discussed. Finally, it is brings the main conclusions of the work.

2 The ARMAV Model

The autoregressive-moving-average-vector (ARMAV) model is largely used in multivariate system identification [3]. ARMAV model can represent a multivariate time series from a linear time-invariant dynamical system by means of a multivariate difference equation as,

$$\mathbf{y}(k+p) - \sum_{i=1}^{p} \boldsymbol{\alpha}_i \mathbf{y}(k+p-i) = \sum_{i=1}^{q} \boldsymbol{\beta}_i \mathbf{e}(k+q-i) \tag{1}$$

where $\mathbf{y}(k) = \{y_1(k) \ \ldots \ y_m(k)\}^T$ are the $m \times 1$ vectors representing the measurements of m outputs variables of the system at discrete time $k\Delta t$, with the superscript "T" denoting vector transposition. The vector $\mathbf{e}(k) = \{e_1(k) \ \ldots \ e_m(k)\}^T$ is a non-observable stochastic $m \times 1$ vector process of with zero mean and nonsingular $m \times m$ covariance matrix Σ, representing the extraneous noise contained in the measurements. The limits p and q represent, respectively, the orders of the autoregressive (AR) and moving-average (MA) matrix parameters. The generic scalar elements $\boldsymbol{\alpha}_i$'s and $\boldsymbol{\beta}_i$'s are, respectively, the p AR and the q MA parameter matrices of dimension $m \times m$.

Multivariable ARMAV model described by Eq. (1) can be converted to following first order difference equations as,

$$\mathbf{Y}(k+1) = \boldsymbol{\alpha}\, \mathbf{Y}(k) + \boldsymbol{\beta}\, \mathbf{E}(k) \tag{2}$$

where the $mp \times 1$ vectors are,

$$\mathbf{Y}(k) = \left\{ \mathbf{y}^T(k) \quad \mathbf{y}^T(k+1) \quad \cdots \quad \mathbf{y}^T(k+p-1) \right\}^T \tag{3}$$

$$\mathbf{E}(k) = \left\{ \mathbf{e}^T(k) \quad \mathbf{e}^T(k+1) \quad \cdots \quad \mathbf{e}^T(k+q-1) \right\}^T \tag{4}$$

and the AR parameters are contained in the following $mp \times mp$ companion matrix as,

$$\boldsymbol{\alpha} = \begin{bmatrix} \mathbf{0} & \mathbf{I} & \mathbf{0} & \cdots & \mathbf{0} \\ \mathbf{0} & \mathbf{0} & \mathbf{I} & \cdots & \mathbf{0} \\ \vdots & \vdots & \vdots & \ddots & \vdots \\ \mathbf{0} & \mathbf{0} & \mathbf{0} & \mathbf{0} & \mathbf{I} \\ \boldsymbol{\alpha}_p & \boldsymbol{\alpha}_{p-1} & \boldsymbol{\alpha}_{p-2} & \cdots & \boldsymbol{\alpha}_1 \end{bmatrix} \tag{5}$$

and the MA parameters are contained in the $mp \times mq$ matrix as,

$$\boldsymbol{\beta} = \begin{bmatrix} \mathbf{0} & \mathbf{0} & \mathbf{0} & \cdots & \mathbf{0} \\ \mathbf{0} & \mathbf{0} & \mathbf{0} & \cdots & \mathbf{0} \\ \vdots & \vdots & \vdots & \ddots & \vdots \\ \mathbf{0} & \mathbf{0} & \mathbf{0} & \mathbf{0} & \mathbf{0} \\ \boldsymbol{\beta}_q & \boldsymbol{\beta}_{q-1} & \boldsymbol{\beta}_{q-2} & \cdots & \boldsymbol{\beta}_1 \end{bmatrix} \tag{6}$$

Let's define the following vectors,

$$\mathbf{Y}^{fut}(k) = \left\{ \mathbf{y}^T(k) \quad \mathbf{y}^T(k+1) \quad \cdots \quad \mathbf{y}^T(k+p-1) \right\}^T \tag{7}$$

$$\mathbf{Y}^{pas}(k) = \left\{ \mathbf{y}^T(k) \quad \mathbf{y}^T(k-1) \quad \cdots \quad \mathbf{y}^T(k-s+1) \right\}^T \tag{8}$$

$$\mathbf{E}^{fut}(k) = \left\{ \mathbf{e}^T(k) \quad \mathbf{e}^T(k+1) \quad \cdots \quad \mathbf{e}^T(k+p-1) \right\}^T \tag{9}$$

where $\mathbf{Y}^{fut}$ and $\mathbf{Y}^{pas}$ are, respectively, $mp \times 1$ and $ms \times 1$ vectors and $\mathbf{E}^{fut}$ is $mp \times 1$, with the superscripts *fut* and *pas* denoting, respectively, future and past data.

Now, for a quantity of measured data of N_p points, post-multiplying Eq. (2) by $\mathbf{Y}^{pas\,T}(k-1)$ and taking the expectation values and assuming that the process $\mathbf{E}^{fut}$ and $\mathbf{Y}^{pas}$ are uncorrelated, i.e., $E\left[\mathbf{E}^{fut}(k)\mathbf{Y}^{pasT}(k-1)\right] = \mathbf{0}$, results in,

$$E\left[\mathbf{Y}^{fut}(k+1)\ \mathbf{Y}^{pasT}(k-1)\right] = \boldsymbol{\alpha}\ E\left[\mathbf{Y}^{fut}(k)\ \mathbf{Y}^{pasT}(k-1)\right] \tag{10}$$

where E means the expectation operation.

Let's define the following matrices,

$$\mathbf{H}^{(1)} = E\left[\mathbf{Y}^{fut}(k)\ \mathbf{Y}^{pasT}(k-1)\right] = \begin{bmatrix} \mathbf{R}_1 & \mathbf{R}_2 & \cdots & \mathbf{R}_s \\ \mathbf{R}_2 & \mathbf{R}_3 & \cdots & \mathbf{R}_{s+1} \\ \vdots & \vdots & \ddots & \vdots \\ \mathbf{R}_p & \mathbf{R}_{p+1} & \cdots & \mathbf{R}_{p+s-1} \end{bmatrix} \tag{11}$$

and

$$\mathbf{H}^{(2)} = E\left[\mathbf{Y}^{fut}(k+1)\ \mathbf{Y}^{pas^T}(k-1)\right] = \begin{bmatrix} \mathbf{R}_2 & \mathbf{R}_3 & \cdots & \mathbf{R}_{s+1} \\ \mathbf{R}_3 & \mathbf{R}_4 & \cdots & \mathbf{R}_{s+2} \\ \vdots & \vdots & \ddots & \vdots \\ \mathbf{R}_{p+1} & \mathbf{R}_{p+2} & \cdots & \mathbf{R}_{p+s} \end{bmatrix} \tag{12}$$

as block Hankel matrices of dimension $mp \times mr$ formed by $m \times m$ covariance matrices defined as,

$$\mathbf{R}_i = E\left[\mathbf{y}(k)\ \mathbf{y}^T(k-i)\right] \tag{13}$$

where i means the correspondent lag of $\mathbf{R}_i$, for a quantity of lags used to build the matrices $\mathbf{H}^{(1)}$ and $\mathbf{H}^{(2)}$ equal to $N_{lags} = p + s$.

Equation (10) can be rewritten as,

$$\mathbf{H}^{(2)} = \boldsymbol{\alpha}\ \mathbf{H}^{(1)} \tag{14}$$

Assuming matrix $\mathbf{H}^{(1)}$ to be nonsingular, it follows that the companion matrix $\boldsymbol{\alpha}$ can be calculated by solving the overdetermined system of linear equation as,

$$\boldsymbol{\alpha} = \mathbf{H}^{(2)}\mathbf{H}^{(1)T}\left(\mathbf{H}^{(1)}\mathbf{H}^{(1)T}\right)^{-1} = \mathbf{H}^{(2)}\mathbf{H}^{(1)+} \tag{15}$$

where $\mathbf{H}^{(1)+} = \mathbf{H}^{(1)T}\left(\mathbf{H}^{(1)}\mathbf{H}^{(1)T}\right)^{-1}$ denotes the Moore-Penrose pseudo inverse of $\mathbf{H}^{(1)}$.

3 Reference-Vectors for OMA Identification Scheme

Theoretically, in classical modal analysis, the impulsive or white Gaussian forces, used as excitation for, respectively, IRF's or input-output modal tests, have constant spectrum. These signals work as a type of reference in the modal parameters identification, for a certain frequency range of interest. However, in OMA parameters identification, where the input forces are not known, it is important to define some coordinates of reference to calculate the modal parameters of the system. The output vector of dimension $m \times 1$ is defined as,

$$\mathbf{y}(k) = \begin{Bmatrix} \mathbf{y}_r(k) \\ \mathbf{y}_{nr}(k) \end{Bmatrix} \tag{16}$$

where $\mathbf{y}_r(k)$ is the reference-output vector of dimension $r \times 1$. The vector $\mathbf{y}_{nr}(k)$ of dimension $(m-r) \times 1$ represents the part of non-referenced of the output vector $\mathbf{y}(k)$. The relation between $\mathbf{y}_r(k)$ and $\mathbf{y}_{nr}(k)$ is given by,

$$\mathbf{y}_r(k) = \mathbf{L}\,\mathbf{y}(k) \tag{17}$$

with $\mathbf{L} = [\mathbf{I}_r \quad \mathbf{0}]$ of dimension $r \times m$.

In OMA parameters identification, the non-referenced covariance matrices defined by Eq. (13) must be substituted by referenced-covariance matrices between the complete output vector $\mathbf{y}(k)$ and the reference-output-vector $\mathbf{y}_r(k)$ defined as,

$$\mathbf{R}_i^r = E\left[\mathbf{y}(k)\ \mathbf{y}_r^T(k-i)\right] = \mathbf{R}_i\,\mathbf{L}^T = E\left[\mathbf{y}(k)\ \mathbf{y}^T(k-i)\right]\mathbf{L}^T \tag{18}$$

For the example, in the case of only one reference as the jth variable $y_j(k)$, the referenced-covariance matrix $\mathbf{R}_i^r$ becomes,

$$\mathbf{R}_i^j = \mathbf{R}_i\mathbf{L}^T = \begin{bmatrix} \mathbf{R}_i(1,1) & \cdots & \mathbf{R}_i(1,j-1) & \mathbf{R}_i(1,j) & \cdots & \mathbf{R}_i(1,m) \\ \mathbf{R}_i(2,1) & \cdots & \mathbf{R}_i(2,j-1) & \mathbf{R}_i(2,j) & \cdots & \mathbf{R}_i(2,m) \\ \vdots & \vdots & \vdots & \vdots & \ddots & \vdots \\ \mathbf{R}_i(m,1) & \cdots & \mathbf{R}_i(m,j-1) & \mathbf{R}_i(m,j) & \cdots & \mathbf{R}_i(m,m) \end{bmatrix} \begin{Bmatrix} 0 \\ \vdots \\ 0 \\ 1 \\ \vdots \\ 0 \end{Bmatrix} = \begin{Bmatrix} \mathbf{R}_i(1,j) \\ \mathbf{R}_i(2,j) \\ \vdots \\ \mathbf{R}_i(m,j) \end{Bmatrix} \tag{19}$$

The above equation shows how referenced covariance matrix $\mathbf{R}_i^r$ can be obtained from non-reference covariance matrix $\mathbf{R}_i$.

4 Modal Parameters Identification

The input-output relationship, based on Eq. (1), can be written as,

$$\mathbf{y}(k+p) - \sum_{i=1}^{p} \boldsymbol{\alpha}_i \mathbf{y}(k+p-i) = \mathbf{u}(k) \tag{20}$$

where $\mathbf{u}(k)$ denotes the $m \times 1$ input-vector related to the external forces applied to the system.

In order to obtain a scheme to estimate the modal parameters of the mechanical system, the z-transform is applied to both sides of Eq. (20) giving the following equations,

$$\left[z^p\mathbf{I} - z^{p-1}\boldsymbol{\alpha}_1 - z^{p-2}\boldsymbol{\alpha}_2 - \cdots - \boldsymbol{\alpha}_p\right]\mathbf{Y}(z) = \mathbf{U}(z) \tag{21}$$

$$\left[z^p\mathbf{I} - z^{p-1}\boldsymbol{\alpha}_1 - z^{p-2}\boldsymbol{\alpha}_2 - \cdots - z\,\boldsymbol{\alpha}_{p-1} - \boldsymbol{\alpha}_p\right]\mathbf{H}(z) = \mathbf{I} \tag{22}$$

where $\mathbf{Y}(z)$ and $\mathbf{U}(z)$ are, respectively, the z-transform of $\mathbf{y}(k)$ e $\mathbf{u}(k)$ and $\mathbf{H}(z)$ is the $m \times m$ transfer function between $\mathbf{Y}(z)$ and $\mathbf{U}(z)$ in the z-domain.

Equation (22) can be re-written in terms of a companion matrix $\boldsymbol{\alpha}$ as,

$$\left\{\begin{bmatrix} z\mathbf{I} & \mathbf{0} & \cdots & \mathbf{0} & \mathbf{0} \\ \mathbf{0} & z\mathbf{I} & \cdots & \mathbf{0} & \mathbf{0} \\ \vdots & \vdots & \ddots & \vdots & \vdots \\ \mathbf{0} & \mathbf{0} & \cdots & z\mathbf{I} & \mathbf{0} \\ \mathbf{0} & \mathbf{0} & \cdots & \mathbf{0} & z\mathbf{I} \end{bmatrix} - \begin{bmatrix} \mathbf{0} & \mathbf{I} & \cdots & \mathbf{0} & \mathbf{0} \\ \mathbf{0} & \mathbf{0} & \cdots & \mathbf{0} & \mathbf{0} \\ \vdots & \vdots & \ddots & \vdots & \vdots \\ \mathbf{0} & \mathbf{0} & \cdots & \mathbf{0} & \mathbf{I} \\ \boldsymbol{\alpha}_p & \boldsymbol{\alpha}_{p-1} & \cdots & \boldsymbol{\alpha}_2 & \boldsymbol{\alpha}_1 \end{bmatrix}\right\} \begin{Bmatrix} \mathbf{I} \\ z\mathbf{I} \\ \vdots \\ z^{p-2}\mathbf{I} \\ z^{p-1}\mathbf{I} \end{Bmatrix} \mathbf{H}(z) = \begin{Bmatrix} \mathbf{0} \\ \mathbf{0} \\ \vdots \\ \mathbf{0} \\ \mathbf{I} \end{Bmatrix} \tag{23}$$

The above equation can be written in a more compactly form as,

$$[z\mathbf{I} - \boldsymbol{\alpha}]\tilde{\mathbf{I}}_z\mathbf{H}(z) = \tilde{\mathbf{B}}(z) \tag{24}$$

where

$$\tilde{\mathbf{I}}_z = \begin{Bmatrix} \mathbf{I} \\ z\mathbf{I} \\ \vdots \\ z^{p-2}\mathbf{I} \\ z^{p-1}\mathbf{I} \end{Bmatrix} \quad \text{and} \quad \tilde{\mathbf{B}}(z) = \begin{Bmatrix} \mathbf{0} \\ \mathbf{0} \\ \vdots \\ \mathbf{0} \\ \mathbf{I} \end{Bmatrix} \tag{25}$$

are $pm \times m$ matrices.

The eigenvalue problem of companion matrix $\boldsymbol{\alpha}$ can be written, from Eq. (24), as,

$$\left[z_j\mathbf{I} - \boldsymbol{\alpha}\right]\tilde{\varphi}_j = \mathbf{0} \tag{26}$$

which leads to the calculation of a quantity of mp z-poles z_j's, where $mp - n$ of then are computational poles and may be separated from the identification process.

The minimal order of the model n can be identified by inspection of the more significants singular values of matrix $\mathbf{H}^{(1)}$.

In general, mechanical systems are modeled in continuous-time in nature using for example Newton's second law. The relation between the continuous-time poles λ_l's and the discrete-time poles z_l's is given by,

$$\lambda_j = \frac{\ln(z_j)}{\Delta t} \quad \text{with} \quad j = 1:n \tag{27}$$

where Δt is the sampling time interval.

The natural frequencies ω_j and modal damping ξ_j, for the case of underdamped vibratory systems, are estimated from λ_j, respectively, according to,

$$\omega_j = |\lambda_j| \quad \text{and} \quad \varsigma_j = \frac{\mathrm{Re}(\lambda_j)}{\omega_j} \tag{28}$$

where symbol || denotes absolute value.

Finally, the *mp* eigenvectors ϕ_j of the companion matrix $\boldsymbol{\alpha}$, from Eq. (26), can be used to estimate the mode-shapes ϕ_j of the mechanical system using the following relation [2],

$$\phi_j = \mathbf{I}_z \phi_j \tag{29}$$

where ϕ_j is identified as,

$$\phi_j = (\mathbf{I}_z^T \mathbf{I}_z)^{-1} \mathbf{I}_z^T \phi_j \tag{30}$$

where ϕ_j is a $pm \times 1$ vector and ϕ_j is a $m \times 1$ mode-shape vector.

5 The ARMAV Algorithm

The ARMAV algorithm for OMA parameters identification consists in the following steps:

(1) Calculation of the matrices $\mathbf{H}^{(1)}$ and $\mathbf{H}^{(2)}$, as Eqs. (11) and (12), for a total of a number of lags equal to $N_{lags} = p + s$, using referenced-covariance-matrices $\mathbf{R}_i^r$ obtained by Eq. (18) for a quantity of N_p measured data,

(2) Calculation of the companion matrix $\boldsymbol{\alpha}$ that contains the matrices parameters of the ARMAV model by Eq. (15),

(3) Calculation of a quantity of *mp* eigenvalues z_i's and the associated *mp* eigenvectors ϕ_i of the companion matrix $\boldsymbol{\alpha}$, with $i = 1, \ldots, mp$. The poles *mp* λ_i's of the mechanical system described in continuous time are calculated according to Eq. (27). For underdamped systems, the natural frequencies ω_j and damping factors ξ_j are calculated by Eq. (28),

(4) The mode-shape vectors and ϕ_j are obtained using Eq. (30),
(5) The minimum order of the system n can be obtained by means of inspection of the number of repeated poles identified by the former step or by the number of significant singular values of Hankel block matriz $\mathbf{H}^{(1)}$.

6 Examples of Application

In order to show the capabilities of the present OMA parameters identification technique using the ARMAV model, a SIMO numerical experiment is conducted. The collection of impulse responses and output-only data are obtained by numerical simulation from the five degrees of freedom mass-spring oscillator without damping, as shown in Fig. 1.

6.1 OMA Parameters Identification Using IRF's

Data of a SIMO, 1-input and 5-outputs, test are then numerically simulated with the unit impulse force acting in block 1. In the present test, it is adopted a number of 2000 data samples for each term $h_{ij}(k)$ for a quantity of 500 lags to build the covariance matrix $\mathbf{R}_i^1$ and the order of AR part of the model $p=4$, resulting in a pair of matrices $\mathbf{H}^{(1)}$ and $\mathbf{H}^{(2)}$ both of dimension 20×496. The time sampling interval Δt used is 0.025 s. Table 1 shows the exact and identified modal parameters.

The order of the system is identified to be equal to 10 by inspection of most significant singular values of matrix $\mathbf{H}^{(1)}$ is shown in Fig. (2). Based on this criterion, it is adopted $p=2$ in the identification process resulting the five modes and modal identified parameters present in the Table 1.

Figure 3 shows the five identified mode shapes associated to five natural frequencies as compared to exact modes derived from numerical simulation.

Theoretically, it is important to note that the modal parameters identification using IRF's data, using the present method, does not require the use of reference-vectors in the identification scheme, as discussed in previous section. This type of data has their own references due to the impulsive forces as integrant part of calculation of IRF's.

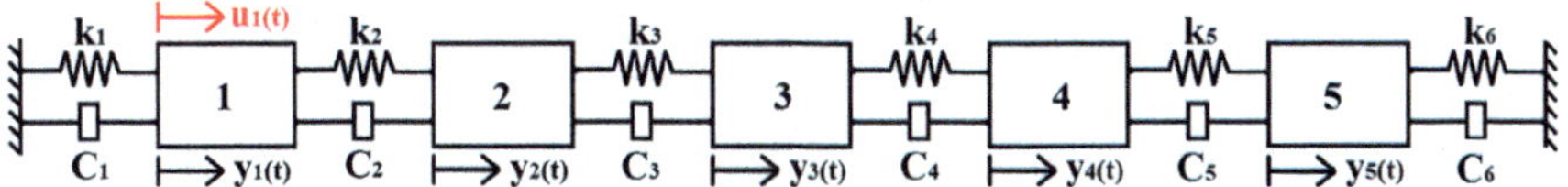

Fig. 1 Five degrees of freedom oscillator system

Table 1 Exact and identified modal parameters

Mode number	Exact natural frequency (Hz)	Identified natural frequency (Hz)	Error (%)
1	5.2105	5.2105	0
2	10.0658	10.0658	0
3	14.2353	14.2353	0
4	17.4346	17.4346	0
5	19.4457	19.4459	0.001

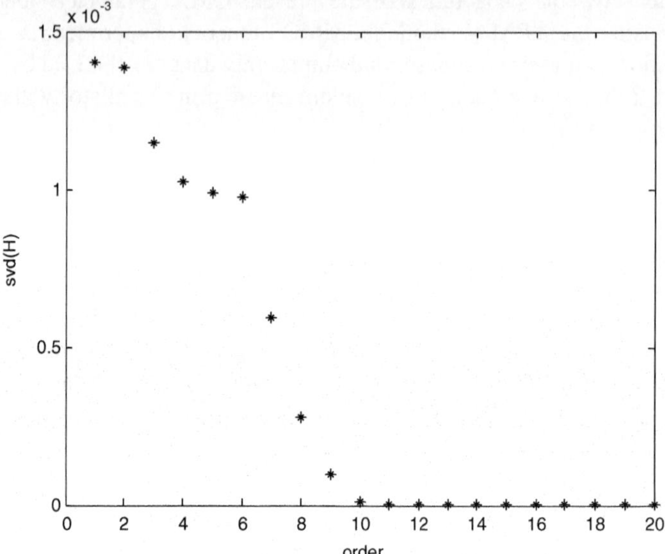

Fig. 2 Singular values of matrix $\mathbf{H}^1$

6.2 OMA Parameters Identification Using Output-Only Data

A SIMO 1-input and 5-outputs test is shown. Adopting as input $u_1(k)$, a mean zero white Gaussian noise signal with amplitude equal to 10 N, that acts on the blocks 1, the responses $\mathbf{y}(k)$'s are obtained by evaluating the following sum of convolution,

$$y_i(k) = \sum_{s=0}^{N_p-1} h_{i1}(s)\, u_1(k-s) \quad i=1,\ldots,5 \tag{31}$$

In the present test, it is adopted a number of 2000 data samples for each term $y_i(k)$ for a quantity of 500 lags to build the covariance matrix $\mathbf{R}_i^1$ and the order of

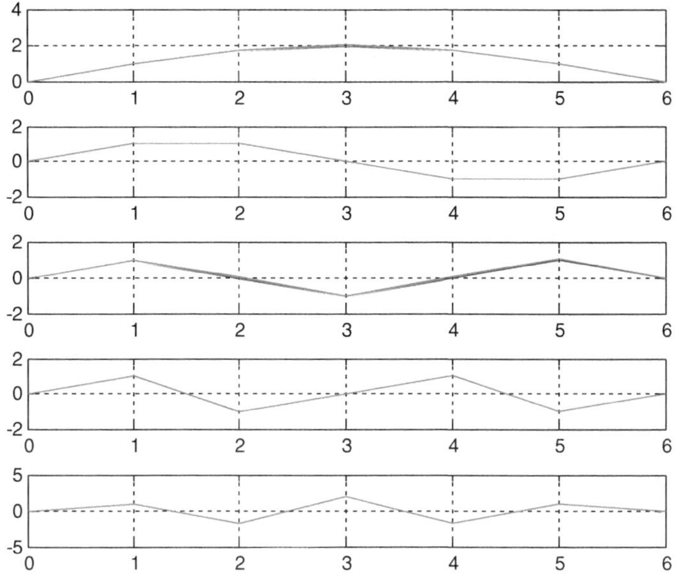

Fig. 3 Exact (green) and identified (blue) mode shapes

AR parameters of the model $p=4$, resulting in a pair of matrices $\mathbf{H}^{(1)}$ and $\mathbf{H}^{(2)}$ both of dimension 20×496. The time sampling interval Δt used is 0.025 s.

The order of the system is identified to be equal to 10 by inspection of most significant singular values of matrix $\mathbf{H}^{(1)}$ shown in Fig. 4. Based on this criterion,

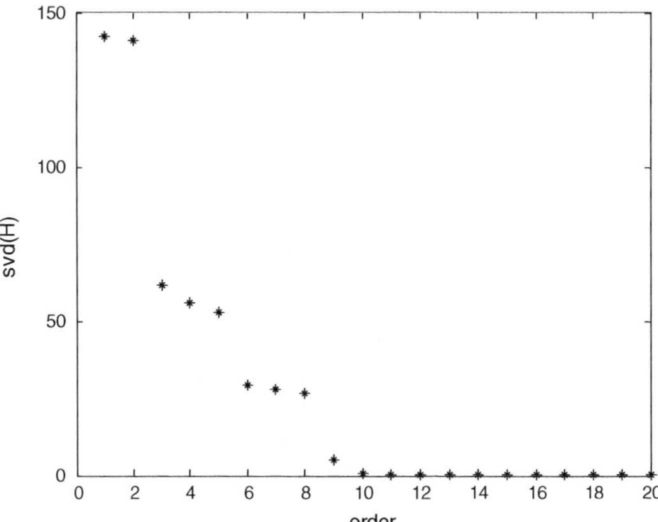

Fig. 4 Singular values of matrix $\mathbf{H}^{1}$

Table 2 Exact and identified modal parameters

Mode number	Exact natural frequency (Hz)	Identified natural frequency (Hz)	Error (%)
1	5.2105	5.2049	0.1074
2	10.0658	10.0713	0.0546
3	14.2353	14.2413	0.0415
4	17.4346	17.4381	0.0207
5	19.4457	19.4479	0.0113

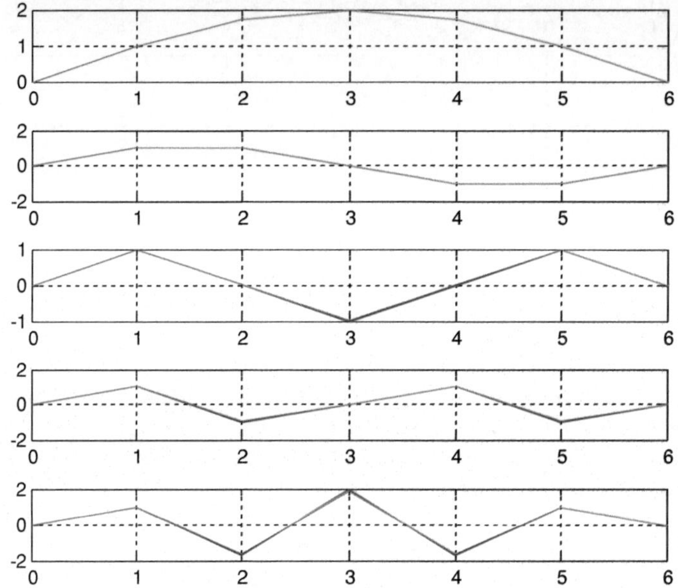

Fig. 5 Exact (green) and identified (blue) mode shapes

the order of AR parameters is changed to $p = 2$ resulting in a total of five identified modes. The exact and identified parameters present in the Table 2.

Figure 5 shows the five identified mode shapes associated to five first natural frequencies as compared to exact modes derived from numerical simulation.

7 Conclusion

OMA is a very attractive field in mechanical and civil engineering and several techniques has been proposed in the literature. The present OMA parameter identification, method based on ARMAV model, is very simple, has robust numerical properties and relatively low computational cost, using only linear algebra manipulations. The tests based on numerical simulated data show that the presented method can be regarded as a way to perform the modal identification—natural frequencies, damping factors and associated mode shapes. The need of an output-vector as reference in the output-only parameters identification is highlighted. The present paper encourages a future implementation of the present algorithm using a more precise (accurate) optimization technique for the parameters identification using, for example, the maximum likelihood technique.

References

1. Söderström, T., Stoica P.: System Identification. Prentice Hall (1989)
2. Maia, N.M.M., Silva, J.M.M.: Theoretical and Experimental Modal Analysis. Research Studies Press (1997)
3. Lardies, J.: Modal parameter identification based on ARMAV and state-space approaches. Arch. Appl. Mech. **80**, 335–352 (2010)
4. Mevel, L., Hermans, L., Der Auweraer, Van: Application of a subspace-based fault detection method in industrial structures. Mech. Syst. Sig. Process. **13**(6), 823–838 (1999)
5. Lardies, J., Ta, M.N.: Modal parameter identification of stay cables from output-only measurements. Mech. Syst. Sig. Process. **25**, 133–150 (2011)
6. Vu, V.H., Thomas, M., Lafleur, F., Marcouiller, L.: Towards an automatic spectral and modal identification from operational modal analysis. J. Sound Vib. **332**, 213–227 (2013)
7. Zaghbani, I., Songmene, V.: Estimation of machine tool dynamic parameters during machining operation through operational modal analysis. Int. J. Mach. Tools Manuf. **49**, 947–957 (2009)
8. Rainieri, C., Fabbrocino, G.: Automated output-only dynamic identification of civil engineering structures. Mech. Syst. Sig. Process. **24**, 678–695 (2010)
9. Ramos, L.F., Marques, L., Lourenço, P.B., De Roeck, G., Campos-Costa, A., Roque, J.: Monitoring historical masonry structures with operational modal analysis: two case studies. Mech. Syst. Sig. Process. **24**, 1291–1305 (2010)
10. Peeters, B., De Roeck, G.: Referenced-based stochastic subspace identification for output-only modal analysis. Mech. Syst. Sig. Anal. **13**(6), 855–878 (1999)

Dynamic Behavior and Optimization of Tow Steered Composite Plates

T. A. M. Guimarães, D. A. Pereira and D. A. Rade

Abstract In the last years, many techniques and procedures have been employed to optimize traditional composite laminates, which can be classified as constant-stiffness composite laminates (CSCL), since the local stiffness is independent on the position over the laminate. On the other hand, recent advances in manufacturing processes now enable to explore non conventional designs. In particular, the development of automatic fiber placement allows the realization of variable stiffness composite laminates (VSCL), in which the local stiffness varies over the laminated as intended by the designer. In practice, VSCL can be achieved by making the fibers follow curvilinear trajectories over the plies (tow steering), or varying the matrix/fiber fraction over the laminate. Some authors have explored the benefits of VSCL to improve the performance of composite laminates in terms of stress distributions, static deformations, buckling, dynamic behavior and aeroelastic stability. In this context, this work proposes a strategy to optimize tow steered rectangular plates by controlling the angles that define the fiber trajectories. These latter are described by Lagrange polynomials of different orders, and two different sets of boundary conditions are considered. A structural model based on the Ritz method, combined with the classical lamination theory to model the composite laminate are used. The plate is considered thin, being modeled based on Kirchhoffs hypotheses. The equations of motion are obtained from Lagrange equations. The proposed model is validated by comparing natural frequencies and mode shapes with the counterparts obtained by using Nastran finite element software. The model is also validated by using experimental results obtained from a tow steered plate manufactured by the automatic fiber placement. A convergence analysis is carried-out to determine the number of functions in the Ritz basis necessary to ensure convergence of the semi-analytical model. A differential evolution (DE) algorithm is used to maximize the first natural

T. A. M. Guimarães (✉)
School of Mechanical Engineering, UFU - Federal University of Uberlândia, Uberlândia, Brazil
e-mail: thiagoamg@ufu.br

D. A. Pereira · D. A. Rade (✉)
Division of Mechanical Engineering, ITA - Technological Institute of Aeronautics, São José dos Campos, Brazil
e-mail: rade@ita.br

A. de T. Fleury et al. (eds.), *Proceedings of DINAME 2017*, Lecture Notes in Mechanical Engineering, https://doi.org/10.1007/978-3-319-91217-2_12

frequency by finding the optimal fiber placement, defined by controlling the interpolation points of Lagrange polynomials of different orders. The results show the possibility of increasing the value of the fundamental frequency for various orders of the interpolation polynomials. However, as this order increases, the fiber paths become more complex, which brings about challenges to manufacturing process. For all simulated conditions, one notices the benefits of VSCL in terms of the vibration behavior, which leads to conclude that tow steering can indeed be used to cope with practical design goals such as to avoid resonances in a specific range of excitation frequency, or to increase the aeroelastic stability margin.

Keywords Variable stiffness composites · Optimization · Plate vibrations · Tow steering

1 Introduction

Traditionally, composite laminates are produced by stacking plies in each of which the fibers are disposed in a predefined direction. In this case, one obtains constant-stiffness composite laminates (CSCL), as the local stiffness is independent from the position over the plate surface. More recently, the so-called variable-stiffness composite laminates (VSCL) are being explored due to the possibility they offer to control the spatial stiffness distribution, which enables to improve the design under different performance criteria.

As a result of the development of advanced manufacturing processes, such as automated fiber placement (AFP), VSCL can be obtained with variable tow angle plies, in which case the fibers follow curvilinear paths [1], or by varying the proportion fiber/matrix over the plate [2].

Many studies reported in the literature have demonstrated that the use of VSCL can improve design characteristics in comparison with traditional composites, without weight penalties [3]. In particular, a number of works have demonstrated the benefits in terms of static structural behavior [2, 4–6]. Other studies are devoted to the vibrational characteristics [7, 8], and aeroelastic behavior [9, 10].

The optimization of CSCL generally consists in finding an optimal combination of orientation angles of the plies in which the fibers are deposited following straight trajectories, to comply with different design purposes [11–13]. In this case, the ply orientation angles are used as design variables. On the other hand, in the case of VSCL, besides the ply orientations, a set of parameters defining the fiber trajectories over the plies can be used as design variables. This certainly provides broader design spaces at the expense of a higher computational cost engendered by larger numbers of design variables. Additionally, manufacturing constraints must be properly dealt with [14].

This paper proposes an approach for the optimal design of tow steered composite rectangular plates, the goal of which is to maximize the fundamental frequency. In the next section, the dynamic model is first derived by using a semi-analytical approach based on the combination of the Classical Lamination Theory, adapted to take into account parameterized fiber trajectories, with the Rayleigh-Ritz (Assumed-Modes) method. Next, after defining the configurations of interest, the semi-analytical model is validated by comparing natural frequencies and mode shapes with the counterparts obtained from standard finite element modeling. The optimization results obtained by using a Differential Evolution (DE) algorithm are presented and discussed, showing the effectiveness of the suggested strategy to optimize the dynamic behavior of tow steered composite plates.

2 Dynamic Model

The i-th ply of a rectangular tow steered plate of dimensions $a \times b$ is depicted in Fig. 1, in which θ indicates the local orientation angle of a generic individual fiber.

Assuming that the laminate is sufficiently thin, it is modeled according to the Classical Lamination Theory (CLT), and the transverse displacements $w(x, y, t)$ are assumed to be constant through the plate thickness. Moreover, each ply is assumed to be in plane stress state. The displacement field is represented as:

$$
\begin{aligned}
u(x, y, z, t) &= u_0(x, y, t) - z\frac{\partial w_0(x, y, t)}{\partial x} \\
v(x, y, z, t) &= v_0(x, y, t) - z\frac{\partial w_0(x, y, t)}{\partial y} \\
w(x, y, z, t) &= w_0(x, y, t)
\end{aligned}
\tag{1}
$$

where (u_0, v_0, w_0) are the displacements on the midplane.

Another assumption adopted in this study is that damping is neglected.

The strain components are expressed as:

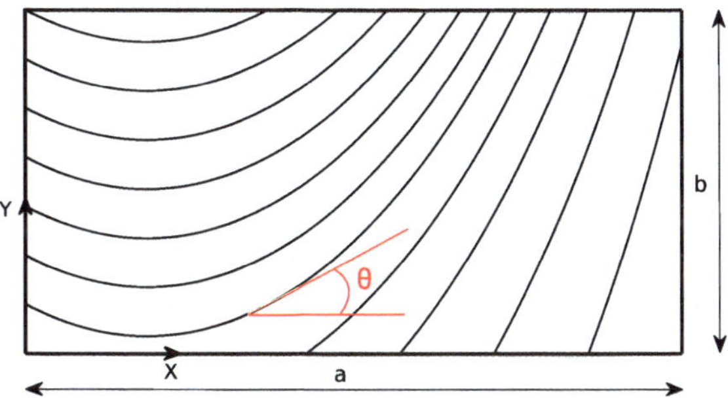

Fig. 1 Illustration of a typical ply of a tow-steered laminate

$$\epsilon = \begin{Bmatrix} \epsilon_x \\ \epsilon_y \\ \gamma_{xy} \end{Bmatrix} = \begin{Bmatrix} \frac{\partial u_0(x,y,t)}{\partial x} - z\frac{\partial^2 w_0(x,y,t)}{\partial x^2} \\ \frac{\partial v_0(x,y,t)}{\partial y} - z\frac{\partial^2 w_0(x,y,t)}{\partial y^2} \\ \frac{\partial u_0(x,y,t)}{\partial y} + \frac{\partial v_0(x,y,t)}{\partial x} - 2z\frac{\partial^2 w_0(x,y,t)}{\partial y \partial x} \end{Bmatrix} \tag{2}$$

where the linear membrane and out of plane terms are denoted as:

$$\epsilon^{\mathbf{L}} = \begin{Bmatrix} \epsilon_x^L \\ \epsilon_y^L \\ \gamma_{xy}^L \end{Bmatrix} = \begin{Bmatrix} \frac{\partial u_0(x,y,t)}{\partial x} \\ \frac{\partial v_0(x,y,t)}{\partial y} \\ \frac{\partial u_0(x,y,t)}{\partial y} + \frac{\partial v_0(x,y,t)}{\partial x} \end{Bmatrix} \tag{3}$$

$$\kappa = \begin{Bmatrix} \kappa_x \\ \kappa_y \\ \kappa_{xy} \end{Bmatrix} = -\begin{Bmatrix} \frac{\partial^2 w_0(x,y,t)}{\partial x^2} \\ \frac{\partial^2 w_0(x,y,t)}{\partial y^2} \\ 2\frac{\partial^2 w_0(x,y,t)}{\partial x \partial y} \end{Bmatrix} \tag{4}$$

Then, the forces **Q** and moments **Bm**, with the assumptions of CLT, are:

$$\begin{Bmatrix} \mathbf{Q} \\ \mathbf{Bm} \end{Bmatrix} = \begin{bmatrix} \mathbf{A} & \mathbf{B} \\ \mathbf{B}^T & \mathbf{D} \end{bmatrix} \begin{Bmatrix} \epsilon^{\mathbf{L}} \\ \kappa \end{Bmatrix} \tag{5}$$

in which, the membrane, membrane-bending and bending stiffness are given by the matrices **A**, **B** and **D**, respectively. In the case of tow steered plates, these matrices must be computed accounting for the fact that the ply angle varies over the plate, according to Lagrange polynomials [14]:

$$\theta_i(x, y) = \Phi_i + \sum_{i=0}^{M-1} \sum_{j=0}^{N-1} \theta_{mn} \cdot \prod_{\substack{m=0 \\ m \neq i}}^{M-1} \frac{x - x_i}{x_m - x_i} \cdot \prod_{\substack{n=0 \\ n \neq j}}^{N-1} \frac{y - y_j}{y_n - y_j} \tag{6}$$

where Φ_i is the reference ply angle and θ_{mn} are the control angles in the reference points (x_m, y_n), as depicted in Fig. 2.

Considering symmetric laminates, $\mathbf{B} = \mathbf{0}$ and matrices **A** and **D** can be formulated in terms of lamination parameters (V_i and W_i) and invariants defined as [15]

$$(V_1, V_2, V_3, V_4)(x, y) = \frac{1}{h} \int_{-h/2}^{h/2} (\cos(2\theta), \sin(2\theta), \cos(4\theta), \sin(4\theta))dz \tag{7}$$

$$(W_1, W_2, W_3, W_4)(x, y) = \frac{12}{h^3} \int_{-h/2}^{h/2} z^2(\cos(2\theta), \sin(2\theta), \cos(4\theta), \sin(4\theta))dz \tag{8}$$

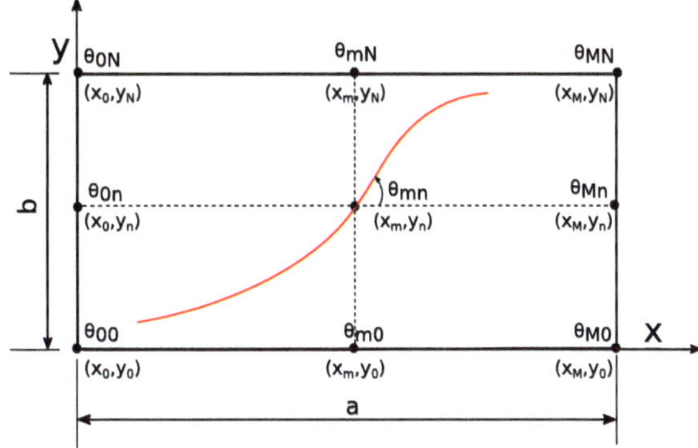

Fig. 2 Curvilinear fiber orientation by Lagrange polynomials. (Adapted from [14])

$$\Gamma_0 = \begin{bmatrix} I_1 & I_4 & 0 \\ I_4 & I_1 & 0 \\ 0 & 0 & I_5 \end{bmatrix}, \quad \Gamma_1 = \begin{bmatrix} I_2 & 0 & 0 \\ 0 & -I_2 & 0 \\ 0 & 0 & 0 \end{bmatrix}, \quad \Gamma_2 = \begin{bmatrix} 0 & 0 & I_2/2 \\ 0 & 0 & I_2/2 \\ I_2 & I_2 & 0 \end{bmatrix},$$
$$\Gamma_3 = \begin{bmatrix} I_3 & -I_3 & 0 \\ -I_3 & I_3 & 0 \\ 0 & 0 & -I_3 \end{bmatrix}, \quad \Gamma_4 = \begin{bmatrix} 0 & 0 & I_3 \\ 0 & 0 & -I_3 \\ I_3 & -I_3 & 0 \end{bmatrix}, \tag{9}$$

where h is the laminate total thickness.

The expression for I_1 to I_5 are omitted here and can be found in [16], as function of the lamina properties.

Hence, matrices **A** and **D** are finally expressed in the forms:

$$\begin{bmatrix} A_{11}(x,y) & A_{12}(x,y) & A_{16}(x,y) \\ A_{12}(x,y) & A_{22}(x,y) & A_{26}(x,y) \\ A_{16}(x,y) & A_{26}(x,y) & A_{66}(x,y) \end{bmatrix} = h(\Gamma_0 + \Gamma_1 V_1(x,y) + \Gamma_2 V_2(x,y) + \Gamma_3 V_3(x,y) + \Gamma_4 V_4(x,y)) \tag{10}$$

$$\begin{bmatrix} D_{11}(x,y) & D_{12}(x,y) & D_{16}(x,y) \\ D_{12}(x,y) & D_{22}(x,y) & D_{26}(x,y) \\ D_{16}(x,y) & D_{26}(x,y) & D_{66}(x,y) \end{bmatrix} = \frac{h^3}{12}(\Gamma_0 + \Gamma_1 W_1(x,y) + \Gamma_2 W_2(x,y) + \Gamma_3 W_3(x,y) + \Gamma_4 W_4(x,y)) \tag{11}$$

From Eqs. (11) and (12) it becomes apparent that, for the tow steered laminates, the stiffness matrices vary according to the position (x, y), which justifies the denomination "variable stiffness composite laminate".

The strain energy of the composite plate can be expressed by:

$$U(x,y) = \frac{1}{2}\int_0^a \int_0^b \left(\epsilon^L\right)^T \mathbf{A}\epsilon^L dxdy + \frac{1}{2}\int_0^a \int_0^b \kappa^T \mathbf{D}\kappa dxdy. \tag{12}$$

In the present study, in-plane effects are neglected in favor of transverse effects; hence, the first term on the right-hand-side of Eq. (12) is neglected.

Based on Kirchhoff's Theory hypotheses, the kinetic energy of the laminate is found to be expressed as [17].

$$T = \frac{1}{2}\int_0^a \int_0^b \int_{-h/2}^{h/2} \rho \left(\frac{\partial w_0}{\partial t}\right)^2 dxdydz \tag{13}$$

According to Rayleigh-Ritz method, the transverse displacement field $w_0(x, y, t)$ is approximated by using Legendre polynomials, as follows [14]:

$$\begin{aligned} L_n(x) &= \frac{1}{2^n}\sum_{k=0}^{K}(-1)^k \frac{(2n-2k)!}{k!(n-k)!(n-2k)!} x^{n-2k} \\ K &= \begin{cases} \frac{n}{2} & (i = 0, 2, 4, \ldots) \\ \frac{n-1}{2} & (i = 1, 3, 5, \ldots) \end{cases} \end{aligned} \tag{14}$$

Therefore, $w_0(x, y, t)$ is approximated in terms of dimensionless coordinates ($\zeta = x/a$ and $\eta = y/b$) as follows:

$$w_0(\zeta, \eta, t) = (\zeta^2 - \zeta)^c(\eta^2 - \eta)^c \sum_{m_0}^{m_{max}} \sum_{n_0}^{n_{max}} q_{mn}(t)L_m(\zeta)L_n(\eta), \tag{15}$$

or:

$$w_0(\zeta, \eta, t) = \mathbf{N}(\zeta, \eta)\mathbf{q} \tag{16}$$

It should be noticed that parameter c enables to define approximations for different boundary conditions, $c = 0, 1, 2$ corresponding to free, simply-supported and clamped edges, respectively.

The virtual work of transverse distributed loads $p(x, y, t)$ and transverse concentrated forces $P_i(x_i, y_i)$ can be expressed as:

$$\delta W = \int_0^a \int_0^b p(x, y, t)\delta w_0(x, y, t)dxdy + \sum_i P_i\delta(x - x_i)\delta(y - y_i)\delta w_0(x_i, y_i, t) \tag{17}$$

By applying Lagrange equations, the equations of motion are found in the form:

$$\mathbf{M}\ddot{\mathbf{q}}(t) + \mathbf{K}\mathbf{q}(t) = \mathbf{Q}(t) \tag{18}$$

with the following associated eigenvalue problem:

$$\left(\mathbf{K} - \omega^2\mathbf{M}\right)\bar{\mathbf{q}} = \mathbf{0} \tag{19}$$

3 Model Description and Validation

A plate whose properties are given in Table 1 has been chosen for validating the semi-analytical model developed based on the formulation presented in the previous section. The procedure adopted for this consists in comparing the values of the natural frequencies and computing the values of MAC (Modal Assurance Criterion) between the natural mode shapes provided by the semi-analytical model and the counterparts obtained by using the commercial finite element software NASTRAN®. This validation is considered to be important since the accuracy of the solutions obtained from Rayleigh-Ritz approximations is strongly dependent on the choice of type and number of approximation functions kept in the expansion given by Eq. 15. A convergence analysis led to conclude that $m = 8$, $n = 8$ are satisfactory for all the test scenarios considered.

The test scenarios of interest include non-steered (N-ST) and steered (ST) configurations, and three different boundary conditions (FFFF: free edges; SSSS: simply-supported edges; CCCC: clamped edges). For all the configurations, the plate is assumed to be composed of 8 plies, with stacking orientations [0° 45° −45° 90°]s. For the N-ST configurations, a second-order Lagrange polynomial (Eq. 6) is used to describe the fiber trajectories over the plies.

Tables 2 and 3 enable to compare the values of the natural frequencies of the N-ST and ST configurations for the three boundary conditions considered. Additionally, Fig. 3 depicts the MAC values computed for the SSSS configuration.

Table 1 Material properties and plate dimensions

Property	Value	Property	Value
E_1	129500 MPa	Length, a	400 mm
E_2	9370 MPa	Width, b	300 mm
G_{12}	5240 MPa	Density, ρ_0	1500 kg/m^3
μ_{12}	0.38	Ply thickness, t	0.19 mm

Table 2 Values of natural frequencies for the N-ST plates

FFFF			SSSS			CCCC		
R-Ritz (Hz)	FE (Hz)	Deviation (%)	R-Ritz (Hz)	FE (Hz)	Deviation (%)	R-Ritz (Hz)	FE (Hz)	Deviation (%)
41.65	41.31	0.83	64.36	63.88	0.75	117.89	116.82	0.92
63.77	62.96	1.28	153.82	152.47	0.88	230.42	227.77	1.17
81.23	80.54	0.85	168.59	166.97	0.97	250.19	247.07	1.26
103.54	101.76	1.75	253.56	248.68	1.96	349.30	340.78	2.50
118.75	116.76	1.70	313.31	309.94	1.09	407.58	412.92	1.29
192.67	188.13	2.41	336.30	333.02	0.98	418.56	444.20	5.77

Table 3 Values of natural frequencies for the ST plates

FFFF			SSSS			CCCC		
Ritz (Hz)	Nastran (Hz)	Error (%)	Ritz (Hz)	Nastran (Hz)	Error (%)	Ritz (Hz)	Nastran (Hz)	Error (%)
41.53	41.19	0.82	65.90	65.38	0.80	118.16	116.94	1.04
64.65	63.86	1.24	152.82	151.55	0.84	227.87	225.18	1.19
82.46	81.65	0.99	172.28	170.43	1.09	253.84	250.14	1.48
105.47	103.81	1.61	251.61	248.68	1.18	342.66	334.51	2.44
118.60	116.51	1.80	323.15	319.74	1.07	432.24	425.86	1.50
196.31	191.42	2.55	334.67	330.67	1.21	446.60	439.84	1.54

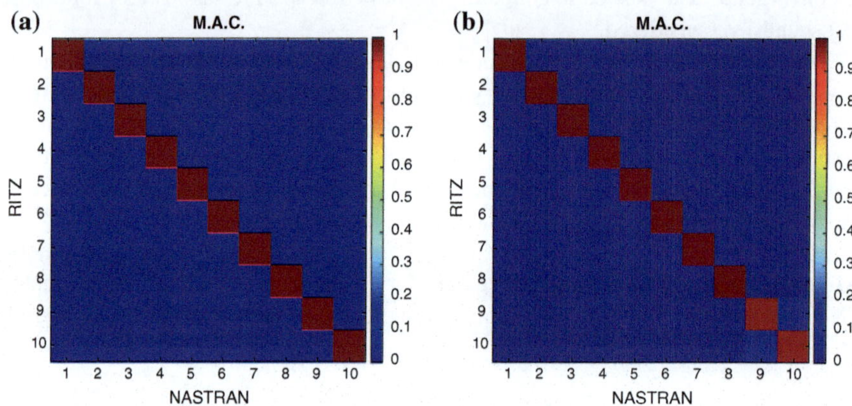

Fig. 3 Graphical representation of the values of the MAC computed between semi-analytical and FE models **a** N-ST configuration; **b** ST configuration

The results presented above enable to conclude that the natural frequencies and mode shapes obtained from the semi-analytical model are in satisfactory agreement with the counterparts obtained from FE modeling, which leads to consider the semi-analytical model as validated and adequate for optimization of the dynamic behavior of tow steered composite plates. It should be mentioned that similar validation has been made for other orders of the polynomials used to represent the fiber trajectiories (results not shown here).

4 Experimental Validation

The numerical analysis concerning the first and second natural frequencies and vibration modal shapes were validated by confrontation with experimental results. For such purpose, a tow steered plate was manufactured by using an automated fiber

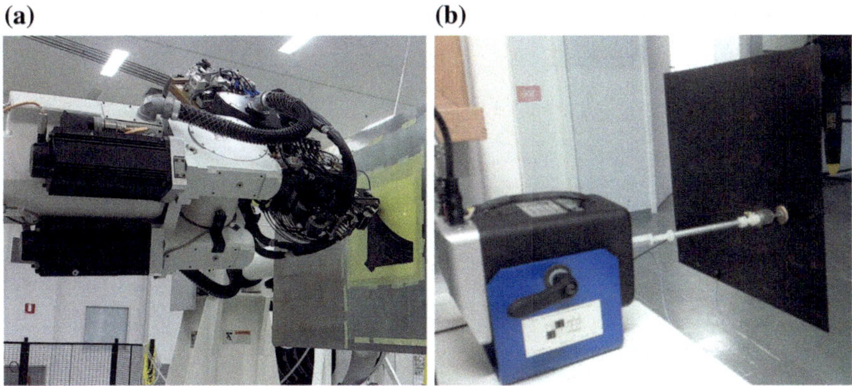

Fig. 4 **a** Automated fiber placement and **b** experimental set-up

placement machine (AFP), which is settled in the Lightweight Structures Laboratory of Institute for Technological Research of the State of São Paulo, shown in Fig. 4a. Each course of the AFP had from 4 up to 12 tows. In this work, each fiber had 3.175 mm width and 8 tows were used, totalizing 25.4 mm width per course. Figure 4b shows the experimental setup used for vibration testing, for which the plate was suspended and excited horizontally with an electromagnetic mini-shaker (PCB model K2007E1); the excitation force (a random signal with 0–200 Hz frequency band) was measured with a piezoelectric force sensor (Bruel & Kjaer model LW34448). The response measurements were performed using a mini-accelerometer (PCB model 352C22). To perform data acquisition and processing, including experimental modal analysis, the responses were measured over a mesh consisting of 7 points in the horizontal direction and 6 points in the vertical direction. The measured frequency response functions (FRF) were used to identify the natural frequencies and natural modes shapes using MEscope® software.

The material used to manufacture the plate was HexPly® M21 from HEXCEL®, which is a continuous, high performance carbon fiber. The plate has the stacking sequence [0° 0° 90°]s, resulting in a laminate with the final thickness of 1.08 mm. As depicted in Fig. 5, the ST plate was manufactured with fiber trajectories having constant radius of 605 mm with T_0 and T_1 equal to 0° and −32°, respectively. This strategy was adopted to avoid manufacturing-induced defects, as discussed in [18].

The AFP equipment has an integrated software that simulates the manufacturing process. Figure 6a, b show the AFP simulation for the 0° and 90° plies, respectively. Table 4 presents the material properties and plate geometry used in the numerical analysis. Figure 7 presents the numerical and experimental FRFs of the tow steered plate, on which the values of the first and second natural frequencies are indicated. The relative difference between experimental and numerical values are 1.42% and 0.03% for the first and second natural frequency, respectively. Figure 8 depicts the experimental and numerical mode shapes corresponding to the first and second

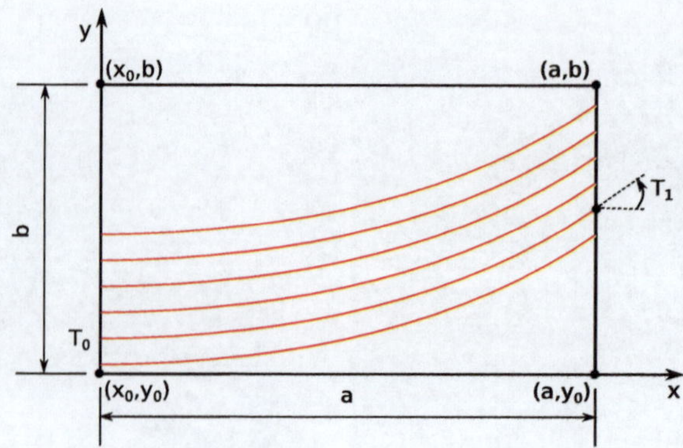

Fig. 5 Fiber trajectories of constant radii

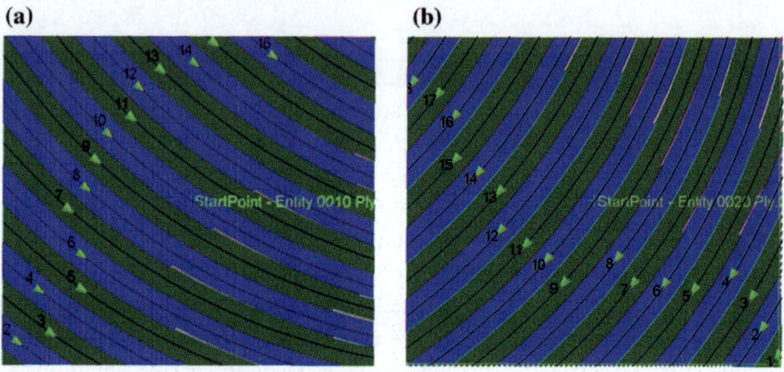

Fig. 6 Simulation of tow steered plies. **a**: 0°; **b**: 90°

Table 4 Material property and plate geometry

Property	Value	Property	Value
E_1	160 GPa	Length, a	320 mm
E_2	28 GPa	Width, b	270 mm
G_{12}	5.6 GPa	Density, ρ_0	1620 kg/m^3
μ_{12}	0.31	Ply thickness, t	0.18 mm

natural frequencies. One can infer that the Rayleigh-Ritz method is capable of representing quite well the natural frequencies and mode shapes of the experimentally tested tow steered plate. However, since damping is not included in the theoretical model, the FRF amplitudes are not well represented.

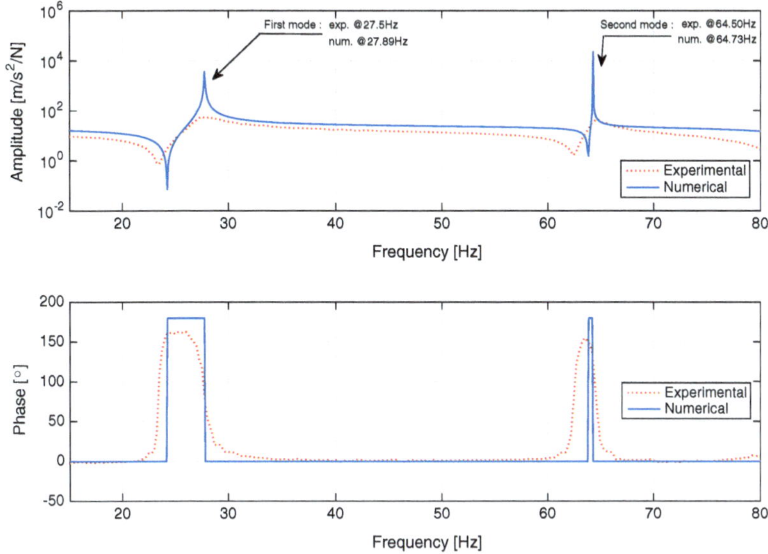

Fig. 7 Numerical and experimental frequency response functions

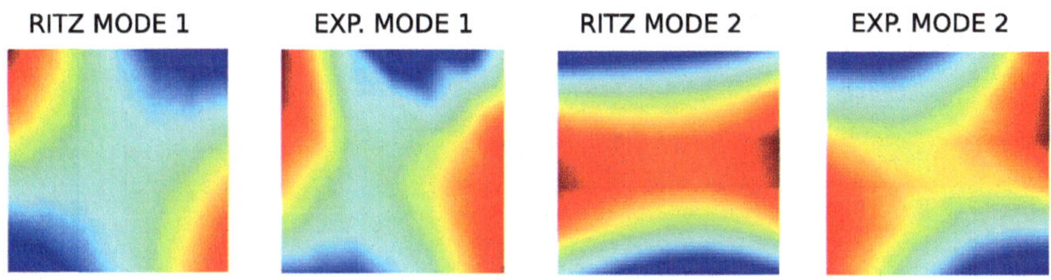

Fig. 8 Numerical and experimental mode shapes

5 Optimization Procedure

The interest here is to maximize the value of the fundamental frequency of the laminate by choosing appropriate fiber trajectories. Hence, an optimization problem is formulated as follows:

$$\begin{aligned} &\textit{Maximize}: \ \omega_1(\theta_{mn}) \\ &\textit{Design variables}: \ \theta_{mn}(\text{see Table 5}); \\ &\textit{Suject to}: \ \theta_{mn} \in [-90°; 90°]; \end{aligned}$$

To solve the optimization problem, a Differential Evolution (DE) algorithm was used considering suitable generations and populations, and performing five optimization runs for each configuration. Table 5 indicates the four optimization setups

Table 5 Definition of the optimization setups

Configuration	Design variables	Variable type	Constraints
O1	$\begin{bmatrix}\theta_0 & \theta_1\end{bmatrix}$	Continuous	[−90°; 90°]
O2	$\begin{bmatrix}\theta_{01} & \theta_{11} \\ \theta_{00} & \theta_{10}\end{bmatrix}$	Continuous	[−90°; 90°]
O3	$\begin{bmatrix}\theta_{02} & \theta_{12} & \theta_{22} \\ \theta_{01} & \theta_{11} & \theta_{21} \\ \theta_{00} & \theta_{10} & \theta_{20}\end{bmatrix}$	Continuous	[−90°; 90°]
O4	$\begin{bmatrix}\theta_{03} & \theta_{13} & \theta_{23} & \theta_{33} \\ \theta_{02} & \theta_{12} & \theta_{22} & \theta_{32} \\ \theta_{01} & \theta_{11} & \theta_{21} & \theta_{31} \\ \theta_{00} & \theta_{10} & \theta_{20} & \theta_{30}\end{bmatrix}$	Continuous	[−90°; 90°]

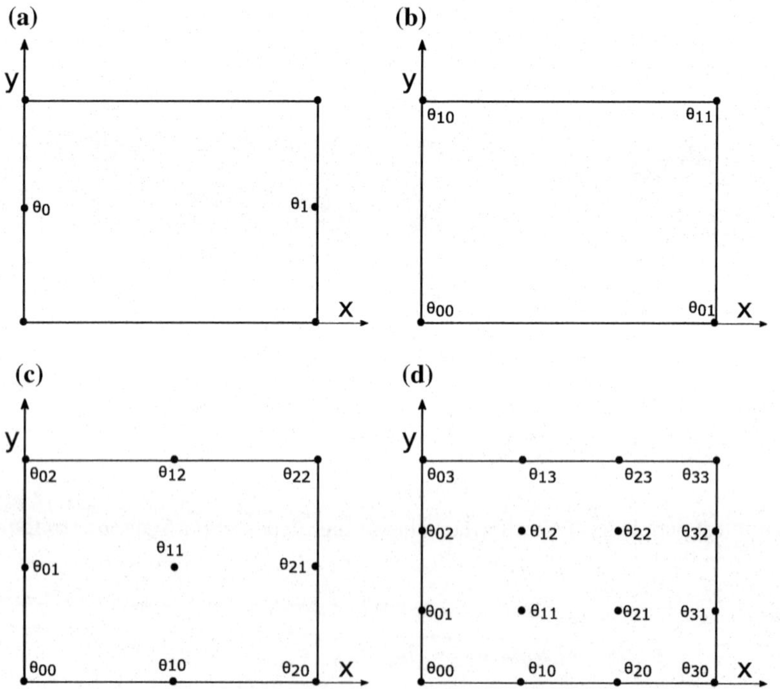

Fig. 9 Illustration of the design variables (control tow angles) for each of the optimization setups (**a** O1; **b** O2; **c** O3; **d** O4)

defined for the problem, which differ from each other by the number of design variables, which, on its turn, is defined by the degree of the polynomial used to represent the fiber trajectories (see Eq. 7). In Fig. 9, the interpretation of the design variables as control tow angles is clarified for each of the four optimization setups.

6 Optimization Results

The optimization results obtained for each boundary condition considered are shown in Table 6. For the purpose of comparison, the N-ST plate is adopted as the baseline configuration, for which the values of the fundamental natural frequencies are 41.65 Hz, 64.36 Hz and 117.89 Hz for the free, simply-supported and clamped configurations, respectively.

Figure 10 depicts the optimal fiber trajectories obtained for the plies oriented at [0° 45° −45° 90°]s, for each boundary condition considered.

It can be seen in Fig. 10 that there is a direct relation between the path complexity (higher order polynomials) and the attainable increase of the fundamental frequency. In addition, this increase depends on the boundary conditions, being more significant for the more constrained configurations. This fact can be explained by the different distributions of the modal strain energies induced by the constraints.

As an additional evidence of the influence of fiber steering on the dynamic behavior of the plate, Fig. 11 enables to compare the amplitudes of typical frequency response functions for the optimal N-ST and ST plates, for the simply-supported and clamped boundary conditions. In these cases the point of excitation is (x = a/3, y = b/3) and the measurement point is (x = a/2 and y = b/2).

7 Conclusions

A procedure has been developed and evaluated for the modeling and optimization of the dynamic behavior of tow steered composite laminates. It has been shown that

Table 6 Optimization results

	FFFF			SSSS			CCCC		
	Baseline (Hz)	Optimal (Hz)	Incr. (%)	Baseline (Hz)	Optimal (Hz)	Incr. (%)	Baseline (Hz)	Optimal (Hz)	Incr. (%)
O1	41.65	45.15	8.40	64.36	78.53	22.02	117.89	153.64	30.32
O2		47.24	13.42		79.50	23.52		154.03	30.66
O3		47.49	14.02		82.26	27.81		160.06	35.77
O4		48.60	16.69		82.59	28.33		160.40	36.06

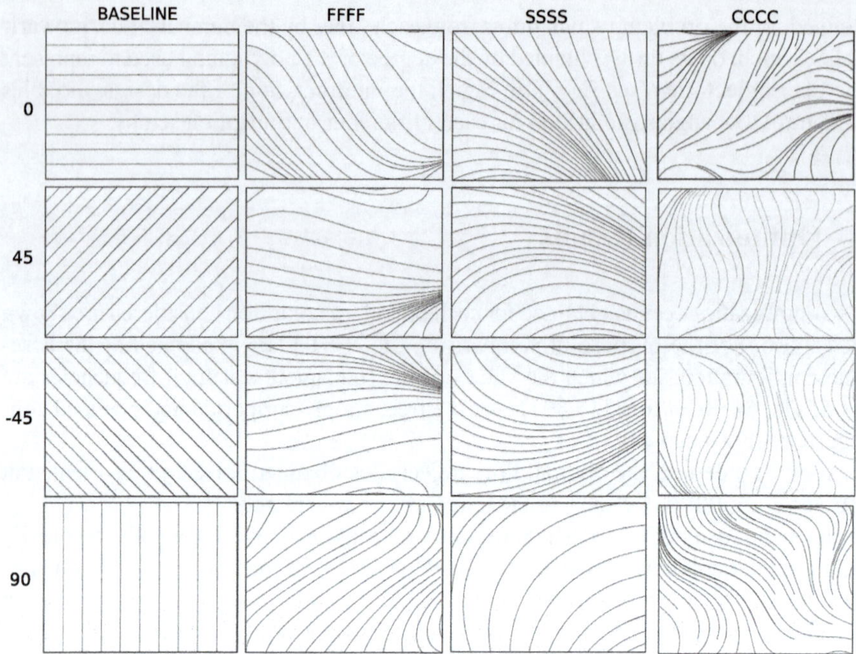

Fig. 10 Illustration of the fiber trajectories obtained for the optimized configurations

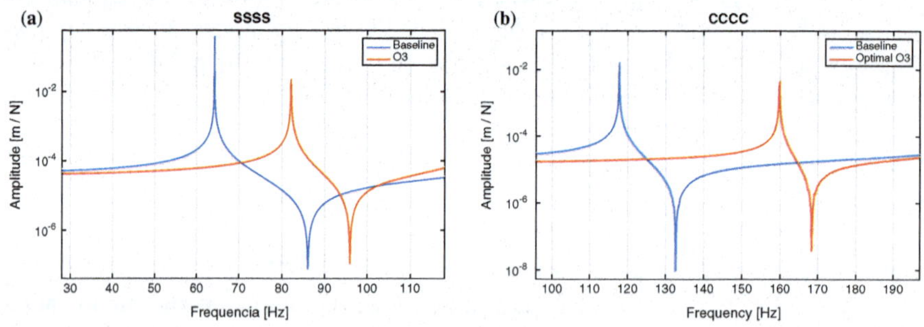

Fig. 11 FRF amplitudes for the baseline and optimized plates, for SSSS and CCCC configurations

a semi-analytical Ritz-type modeling combined with Classical Lamination Theory were capable of providing a sufficiently accurate dynamic model. Due to the typical low number of degrees-of-freedom, such a model is very convenient to alleviate the computation cost involved in optimization.

It has been found out that, at least for the particular optimization goal adopted, the achievable improvement of dynamic behavior depend both on the degree of the polynomials used to represent the fiber trajectories, which is related to the number of design variables, and on the boundary conditions. Regarding the first dependence, it has been noticed that the use of polynomials of higher degrees tends to lead to more complex fiber trajectories, which certainly raises concerns about manufacturing limitations.

Despite the existence of a number of technological challenges to be faced, it is clear that the emergence of tow steered composites broadens the possibilities of achieving improved dynamic performance.

Acknowledgements The authors are thankful to Brazilian Research Agencies CNPq (Projects 310633/2013-3, 402238/2013-3, 145439/2015-1), FAPEMIG, INCT-EIE and FAPESP (Project 2015/20363-6) for the financial support to their research work. The authors are also greatful to Fundaçao de Apoio ao Instituto de Pesquisas Tecnológicas (FIPT) for providing funding support to this research work.

References

1. Parnas, L., Oral, S., Ceyhan, A.: Optimum design of composite structures with curved fiber courses. Compos. Sci. Technol. **63**(7), 1071–1082 (2003). http://www.sciencedirect.com/science/article/pii/S0266353802003123
2. Kuo, S.Y., Shiau, L.C.: Buckling and vibration of composite laminated plates with variable fiber spacing. Compos. Struct. **90**(2), 196–200 (2009). https://doi.org/10.1016/j.compstruct.2009.02.013. http://www.sciencedirect.com/science/article/pii/S0263822309000518
3. Akhavan, H., Ribeiro, P.: Geometrically non-linear periodic forced vibrations of imperfect laminates with curved fibres by the shooting method. Compos. Part B: Eng. (2016). https://doi.org/10.1016/j.compositesb.2016.10.059. http://www.sciencedirect.com/science/article/pii/S1359836816311726
4. Lopes, C., Gürdal, Z., Camanho, P.: Variable-stiffness composite panels: buckling and first-ply failure improvements over straight-fibre laminates. Comput. Struct. **86**(9), 897–907 (2008). https://doi.org/10.1016/j.compstruc.2007.04.016. http://www.sciencedirect.com/science/article/pii/S0045794907001654
5. Gürdal, Z., Tatting, B., Wu, C.: Variable stiffness composite panels: effects of stiffness variation on the in-plane and buckling response. Compos. Part A: Appl. Sci. Manuf. **39**(5), 911–922 (2008). https://doi.org/10.1016/j.compositesa.2007.11.015. http://www.sciencedirect.com/science/article/pii/S1359835X07002643
6. Weaver, P.M., Wu, Z.M., Raju, G.: Optimisation of variable stiffness plates. In: Composite Materials and Structures in Aerospace Engineering. Applied Mechanics and Materials, vol. 828, pp. 27–48. Trans Tech Publications (2016). https://doi.org/10.4028/www.scientific.net/AMM.828.27
7. Akhavan, H., Ribeiro, P.: Natural modes of vibration of variable stiffness composite laminates with curvilinear fibers. Comput. Struct. **93**(11), 3040–3047 (2011). https://doi.org/10.1016/j.compstruct.2011.04.027. http://www.sciencedirect.com/science/article/pii/S0263822311001516
8. Akbarzadeh, A.H., Arian Nik, M., Pasini, D.: Vibration responses and suppression of variable stiffness laminates with optimally steered fibers and magnetostrictive layers. Compos. Part B: Eng. **91**, 315–326 (2016). https://doi.org/10.1016/j.compositesb.2016.02.003. http://www.sciencedirect.com/science/article/pii/S1359836816001037
9. Stodieck, O., Cooper, J.E., Weaver, P.M., Kealy, P.: Improved aeroelastic tailoring using tow-steered composites. Comput. Struct. **106**, 703–715 (2013). http://www.sciencedirect.com/science/article/pii/S0263822313003462
10. Stanford, B.K., Jutte, C.V., Chauncey Wu, K.: Aeroelastic benefits of tow steering for composite plates. Comput. Struct. **118**, 416–422 (2014). http://www.sciencedirect.com/science/article/pii/S0263822314003973

11. Ghiasi, H., Fayazbakhsh, K., Pasini, D., Lessard, L.: Optimum stacking sequence design of composite materials part II: Variable stiffness design. Comput. Struct. **93**(1), 1–13 (2010). https://doi.org/10.1016/j.compstruct.2010.06.001. http://www.sciencedirect.com/science/article/pii/S0263822310001947
12. Almeida, F., Awruch, A.: Design optimization of composite laminated structures using genetic algorithms and finite element analysis. Comput. Struct. **88**(3), 443–454 (2009). https://doi.org/10.1016/j.compstruct.2008.05.004. http://www.sciencedirect.com/science/article/pii/S0263822308001578
13. Pelletier, J.L., Vel, S.S.: Multi-objective optimization of fiber reinforced composite laminates for strength, stiffness and minimal mass. Comput. Struct. **84**(29), 2065–2080 (2006). https://doi.org/10.1016/j.compstruc.2006.06.001. http://www.sciencedirect.com/science/article/pii/S0045794906001854
14. Wu, Z., Weaver, P.M., Raju, G., Chul Kim, B.: Buckling analysis and optimisation of variable angle tow composite plates. Thin-Walled Struct. **60**, 163–172 (2012). https://doi.org/10.1016/j.tws.2012.07.008. http://www.sciencedirect.com/science/article/pii/S0263823112001930
15. Kollár, L., Springer, G.S.: Mechanics of Composite Structures. Cambridge University Press, Cambridge (2003). https://doi.org/10.1017/CBO9780511547140. https://www.cambridge.org/core/books/mechanics-of-composite-structures/804A0A5EE67784D7172E142559979445
16. Jones, R.: Mechanics of Composite Materials. CRC (1998)
17. Love, A.E.H.: The small free vibrations and deformation of a thin elastic shell. Philos. Trans. R. Soc. Lond. A: Math. Phys. Eng. Sci. **179**, 491–546 (1888). https://doi.org/10.1098/rsta.1888.0016. http://rsta.royalsocietypublishing.org/content/179/491, http://rsta.royalsocietypublishing.org/content/179/491.full.pdf
18. Guimaraes, T.A., Castro, S.G., Rade, D.A., Cesnik, C.E.: Panel flutter analysis and optimization of composite tow steered plates. In: 58th AIAAASCEAHSASC Structures, Structural Dynamics, and Materials Conference, p. 1118 (2017)

Evaluation of the Dynamic Response of Buildings with TMDs Under Earthquakes

Rúbia Mara Bosse and André Teófilo Beck

Abstract There is ample evidence that TMDs attenuate the vibration response of buildings subject to low-frequency wind loads. The same is not true for broad-banded earthquake loading. This paper investigates the effectiveness of TMDs in attenuation of vibration in buildings submitted to earthquake ground motion. A positional finite element model is developed, and employed to evaluate linear and geometrically nonlinear dynamical responses of building structures with TMDs under earthquakes. It was noted that TMDs tuned to higher frequencies work better at minimizing displacements and oscillation frequency, in contrast to what is popularly believed, that devices tuned to the building's fundamental frequencies present ideal performance. The linear regimen showed to be sufficient to estimates the displacements of the building and the imposition of earthquake loads as equivalent lateral forces was not representative and does not describe accurately the behavior of the structure. The incorporation of TMDs showed to be very effective in reducing vibrations when the structure is subjected to earthquake loads. However, benefits are only achieved when TMDs are properly designed, and when structural responses are correctly evaluated.

Keywords TMD · Dynamic responses · Geometrically nonlinear
Tuning frequency · Earthquakes

1 Introduction

Techniques and devices to suppress structural vibrations have been developed to enable safe solutions to tall building design. In this subject, the passive vibration control with absorbers as Tuned Mass Dampers (TMDs) has the advantage of being

---

R. M. Bosse (✉) · A. T. Beck
University of São Paulo, São Carlos, SP, Brazil
e-mail: rubiabosse@usp.br

A. T. Beck
e-mail: atbeck@sc.usp.br

A. de T. Fleury et al. (eds.), *Proceedings of DINAME 2017*, Lecture Notes in Mechanical Engineering, https://doi.org/10.1007/978-3-319-91217-2_13

simple and efficient in reducing the dynamic motion of structures, without demanding external power sources [1, 2]. The classic TMD is a vibration absorber composed by an auxiliary mass, connected to a linear spring, placed in parallel with a viscous damping device, commonly attached to the top of the building. Conventionally, tuning the frequency of the device to the first natural frequency of the structure is believed to lead to optimal performance.

TMDs were first studied in 1909 by Frahm [3]. Few decades later, the devices began to be employed in buildings. Hartog [4] performed first studies on optimal TMD design, developing analytical expressions for one degree of freedom systems (1DOF). Later, Warburton and Ayorinde [5] further advanced the subject by considering damping in the main system. Warburton [6] also considered random loads acting on the structure, and proposed a formulation to treat MDOF systems as an equivalent 1DOF system. Surely, when dealing with multiple degrees-of-freedom (MDOF) structures, and with uncertainties in loads, the absorber showed to be less effective in mitigating vibrations and it was found that the performance of absorbers is sensible to tuning parameters and damping coefficient, thus many strategies have been developed to enhance the performance of TMDs [7–12].

In general, absorbers have been successfully employed to reduce vibrations induced by wind forces; this is due to the fact that winds usually present a limited range of excitation frequencies. In this aspect, the TMDs are designed to control the first mode of the structure and tuned to the fundamental natural frequency of the building, which is more susceptible to be excited by winds. However, earthquakes can include a wide spectrum of excitation frequencies; hence there is no general agreement about the performance of TMDs to mitigate seismic-induced oscillations. It was found that the reduction of displacements in the main structure is dependent on the ground motion frequency [1, 7, 13–15].

Most researches in vibration control model building structures as discrete mass-spring-damper systems [14, 16, 17] and evaluate their dynamical responses in linear regimen, usually by modal superposition. In such approaches, the TMD is represented as an additional DOF attached to the main structure. Such lumped spring-mass models can be considered limited, because they assume structural members as non-deformable bodies, hence they cannot properly account for large deflections. Tall buildings are very sensitive to structural vibrations, and often present large displacements, especially under earthquake-induced base motion.

This paper presents a methodology to obtain accurate and realistic dynamical responses of building structures equipped with TMDs, and subject to earthquake ground motions. A positional finite element (FE) formulation is employed to evaluate linear and geometrically nonlinear dynamical responses. Optimal TMD design is considered, for building structures subject to *El Centro* earthquake loading.

2 Positional Finite Elements Method for Frame Structures

Geometrical nonlinear analysis is used to handle large deflections: the structure's equilibrium position in sought on its displaced state. In the so-called positional FE approach, a non-dimensional space is created and the relative curvature of beam elements is calculated for the initial and for the deformed configurations [18]. The equilibrium position is the main unknown variable, and it is obtained from the principle of stationary total potential energy. A total Lagrangian formulation is employed, using an unique reference configuration, the initial position; in this context, the mass matrix is constant and a frame element with four nodes and cubic approximation is employed. The system of nonlinear equations is solved combining the Newmark time integration with the Newton-Raphson procedure, following [19].

The linear analysis differs from the nonlinear because equilibrium is calculated in the initial position, presenting a constant stiffness matrix. In this way, the linear equilibrium equations are solved applying the Newmark time integration considering constant average acceleration, according to [20].

2.1 Frame Positional FE Model

To describe the frame element used in the article, it is necessary to map the initial and current configurations of the finite element from the reference line. The tangent (t_{ik}) and normal vectors $(\upsilon_{1k}, \upsilon_{2k})$ of the four-node elements are presented in Eq. (1), according to Fig. 1:

$$t_{ik} = \frac{d\varphi(\xi)}{d\xi}\bigg|_{\xi_k} X_{il}^{m}, \quad \nu_{1k} = \frac{-t_{2k}}{\sqrt{t_{ik}t_{ik}}}, \quad \nu_{2k} = \frac{-t_{1k}}{\sqrt{t_{ik}t_{ik}}} \tag{1}$$

where i is the coordinate direction, m represents the reference line, l the element node (shape function), ξ_k are the non-dimensional coordinates of the nodes, φ are

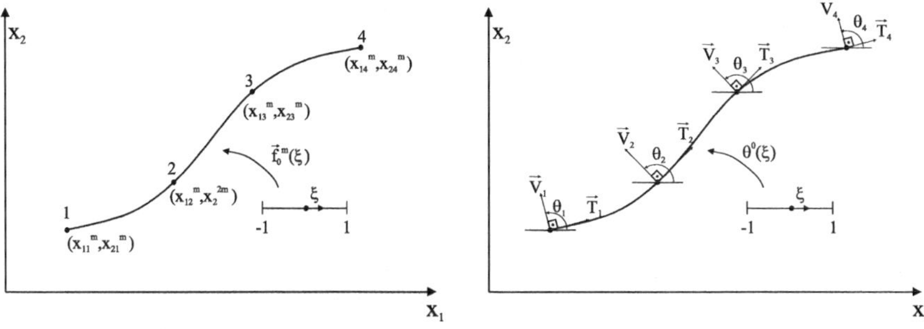

Fig. 1 Nodal vectors and reference line of the beam finite element, cubic approximation [19]

the shape functions. Figure 1 shows the initial angle between the normal vector and the horizontal direction x_1. Using the shape functions (Lagrange polynomials) to approximate $\theta^0(\xi)$, one obtains:

$$\theta_k^0 = \text{arctg}\left(\frac{\boldsymbol{v}_{2k}}{\boldsymbol{v}_{1k}}\right), \quad \theta^0(\xi) = \varphi_\ell(\xi)\theta_\ell^0. \tag{2}$$

It is possible to define the position of any point inside the element, by the vector $g_i^0(\xi,\eta)$, according to Eq. (4) and (Fig. 2).

$$\boldsymbol{x}_i(\xi,\eta) = \boldsymbol{x}_i^m(\xi) + \boldsymbol{g}_i^0(\xi,\eta), \tag{3}$$

$$\boldsymbol{g}_1^0(\xi,\eta) = \frac{\boldsymbol{h}_0}{2}\eta \cos\left[\varphi_\ell(\xi)\theta_\ell^0\right], \quad \boldsymbol{g}_2^0(\xi,\eta) = \frac{\boldsymbol{h}_0}{2}\eta \boldsymbol{sen}\left[\varphi_\ell(\xi)\theta_\ell^0\right], \tag{4}$$

where, η are the non-dimensional variables along the height h_0.

In this way, the complete mapping is obtained for both directions as:

$$\begin{aligned} \boldsymbol{f}_1^0(\xi,\eta) &= \varphi_\ell(\xi)\boldsymbol{X}_1^\ell + \frac{\boldsymbol{h}_0}{2}\eta \cos\left[\varphi_\ell(\xi)\theta_\ell^0\right], \\ \boldsymbol{f}_2^0(\xi,\eta) &= \varphi_\ell(\xi)\boldsymbol{X}_2^\ell + \frac{\boldsymbol{h}_0}{2}\eta \boldsymbol{sen}\left[\varphi_\ell(\xi)\theta_\ell^0\right]. \end{aligned} \tag{5}$$

The current configuration is the unknown parameter of the problem: it is obtained by an iterative process. However, in the first step it is assumed that the current configuration is equal to the initial one. The current configuration is defined by:

$$\boldsymbol{f}_1^1(\xi,\eta) = \varphi_\ell(\xi)\boldsymbol{Y}_1^\ell + \frac{\boldsymbol{h}_0}{2}\eta \cos[\varphi_\ell(\xi)\theta_\ell], \quad \boldsymbol{f}_2^2(\xi,\eta) = \varphi_\ell(\xi)\boldsymbol{Y}_2^\ell + \frac{\boldsymbol{h}_0}{2}\eta \boldsymbol{sen}[\varphi_\ell(\xi)\theta_\ell], \tag{6}$$

where, $\boldsymbol{Y}_i^\ell$ are the current coordinates, φ_ℓ current angles of the cross section.

With the mappings from the non-dimensional space to initial and current configurations defined, the deformation of the element can be described by the change of configuration function as $\vec{f}$ (Fig. 3). The gradient of this function is obtained from the mapping gradients.

$$\vec{f} = \vec{f}_1 \circ \left(\vec{f}_0\right)^{-1}, \quad \mathbf{A} = \mathbf{A}^1.(\mathbf{A}^0)^{-1}, \quad \mathbf{A}^0 = \begin{bmatrix} \frac{\partial f_1^0}{\partial \xi} & \frac{\partial f_1^0}{\partial \eta} \\ \frac{\partial f_2^0}{\partial \xi} & \frac{\partial f_2^0}{\partial \eta} \end{bmatrix}, \quad \mathbf{A}^1 = \begin{bmatrix} \frac{\partial f_1^1}{\partial \xi} & \frac{\partial f_1^1}{\partial \eta} \\ \frac{\partial f_2^1}{\partial \xi} & \frac{\partial f_2^1}{\partial \eta} \end{bmatrix}. \tag{7}$$

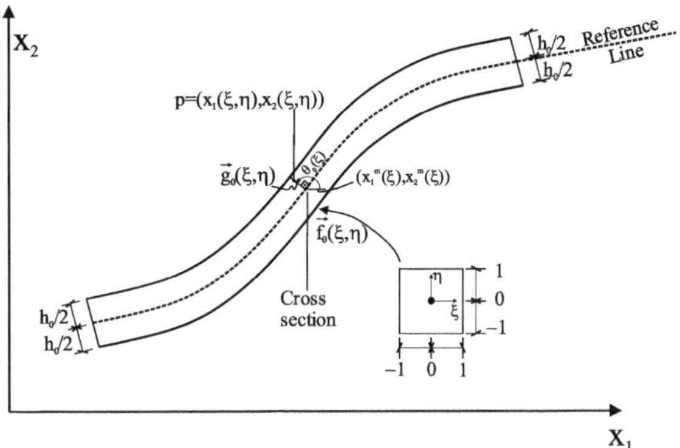

Fig. 2 Cross section point [18]

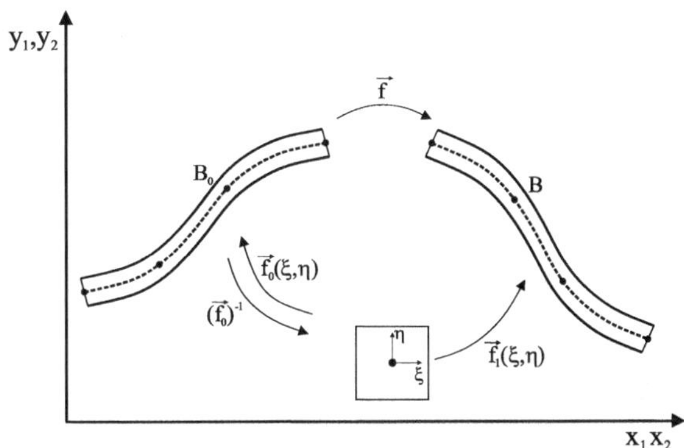

Fig. 3 Change of configuration [19]

The objective Green-Lagrange strain measure **E** is used to calculate the geometrically exact position:

$$\mathbf{E} = \frac{1}{2}(\mathbf{C} - \mathbf{I}) = \frac{1}{2}(\mathbf{A}^t \cdot \mathbf{A} - \mathbf{I}), \tag{8}$$

where **I** is the (2 × 2) identity tensor and **C** is the right Cauchy stretch.

2.2 Geometrically Nonlinear Elasto-dynamics

The Saint-Venant-Kirchhoff constitutive relation is employed to relate the Green strain (**E**) and the second Piola-Kirchhoff stress and the specific strain energy is given by:

$$u_e = \frac{E}{2}\{(E_{11}^2 + E_{22}^2) + G(E_{12}^2 + E_{21}^2)\}, \tag{9}$$

where E is the longitudinal elasticity modulus for small deformations, E_{ij} is the Green strain tensor, and $G = E/[2(1+\upsilon)]$. By considering the Poisson ratio υ to be zero, volumetric locking is avoided. The strain energy stored in the structure can be written as:

$$U_e(\vec{Y}) = \int_{V_0} u_e(\vec{Y})\, dV_0, \tag{10}$$

where V_0 is the initial volume of the structure.

The total energy of the system is obtained by adding the parcels due to the potential energy of external loads, kinetic energy, potential of external distributed forces, viscous damping and strain energy of the frame elements:

$$\prod(\vec{Y}) = \int_{V_0} u_e(\vec{Y})\, dV_0 - \vec{F} \cdot \vec{Y} - \int_{S_0} q \cdot \vec{y}^m dS_0 + \frac{1}{2} \int_{V_0} \rho_0 \dot{\vec{y}} \cdot \dot{\vec{y}} dV_0 + Q, \tag{11}$$

where $\vec{F}$ is the external nodal force vector, q is the general distributed force vector, $\vec{y}^m$ is the current position, $\dot{y}$ is the velocity, dS_0 the infinitesimal element's length, Q is the viscous damping.

The principle of stationary potential energy is applied to the total energy of the system, to impose the equilibrium of the structure:

$$\partial \prod(\vec{Y}) = \vec{F}^{\text{int}} \cdot \partial\vec{Y} - \vec{F} \cdot \partial\vec{Y} - \mathbf{L} \cdot \vec{Q} \cdot \partial\vec{Y} + \mathbf{M} \cdot \ddot{\vec{Y}} \cdot \partial\vec{Y} + \mathbf{D} \cdot \dot{\vec{Y}} \cdot \partial\vec{Y} = \vec{0}, \tag{12}$$

where $\vec{F}^{int}$ refers to the internal force vector, L is the matrix used to transform distributed loads into equivalent nodal ones, M is the constant mass matrix, D is the mass-proportional damping matrix. Equation (12) represents the geometrical non-linear dynamic equilibrium equation because of the arbitrariness of vector $\partial\vec{Y}$.

Assembled the mass and damping matrixes of the structure, and established the initial and current configuration of the elements, a temporal integration combined with the resolution of the non-linear equations is made. The resolution of the system implies the iterative calculus of the internal loads vector, the hessian matrix and update of the positions in each time step.

Vector $\vec{g}$ is the vector of unbalanced mechanical forces, it is null if $\boldsymbol{Y}$ is the correct trial position, used to calculate internal forces. The dynamic equilibrium is obtained by solving, for any time:

$$\vec{g} = \vec{F}^{int} - \vec{F} + M \cdot \ddot{\vec{Y}} + D \cdot \dot{\vec{Y}} = \vec{0}. \tag{13}$$

The resulting nonlinear system is time-integrated by the Newmark method and linearized by the Newton-Raphson algorithm; the complete technique can be consulted in [19]. The iterative process ends when a stopping tolerance (***Tol***) is satisfied, representing the correct equilibrium position.

$$\frac{\|\vec{g}(\vec{Y}_{S+1})\|}{\|\vec{F}\|} \leq Tol, \quad \frac{\|\Delta\vec{Y}_{S+1}\|}{\|\vec{X}\|} \leq Tol, \tag{14}$$

where $\Delta\vec{Y}_{s+1}$ is the current incremental trial position and $\vec{X}$ is the initial position.

3 Results and Discussion

3.1 Building Subject of the Study

In the results session to follow, a 20-storey building was modelled, to evaluate its dynamical response under the El Centro Earthquake. Modal analyses were performed to obtain the natural frequencies and vibration modes of the structure. The TMD was attached to the last floor of the building and the dynamic responses were compared for three regimens of analyses: linear with equivalent forces applied to nodal degrees-of-freedom, linear with base motion and geometrically nonlinear with ground motion.

The building under study has 20 stores, total 72 m of height, and is composed by columns and beams. The structural elements have the dimensions indicated in Fig. 4a. One sample of the El Centro accelerogram is shown in Fig. 4b.

The structure was discretized into 200 4-node frame elements; these elements can describe translations in horizontal and vertical direction, and the angle between the tangent and normal vectors of the nodes. A distributed vertical load of 4 KN/m was considered representing the live loads and a solid concrete slab of 15 cm contributes to the mass of the building. The connection between columns and beams is rigid, and the natural damping of the building was disregarded. The adopted material is reinforced concrete, with longitudinal elastic modulus of 40 GPa, transversal elastic modulus of 20 GPa, density of 2500 kg/m^3. The time of simulation was 32 s, the time step adopted was $\Delta t = 0.02$ s.

3.2 Natural Frequencies

To determine the natural frequencies of the building, modal analysis was employed. The firsts 10 natural frequencies of the building were: 0.55, 1.57, 2.73, 3.91, 5.13,

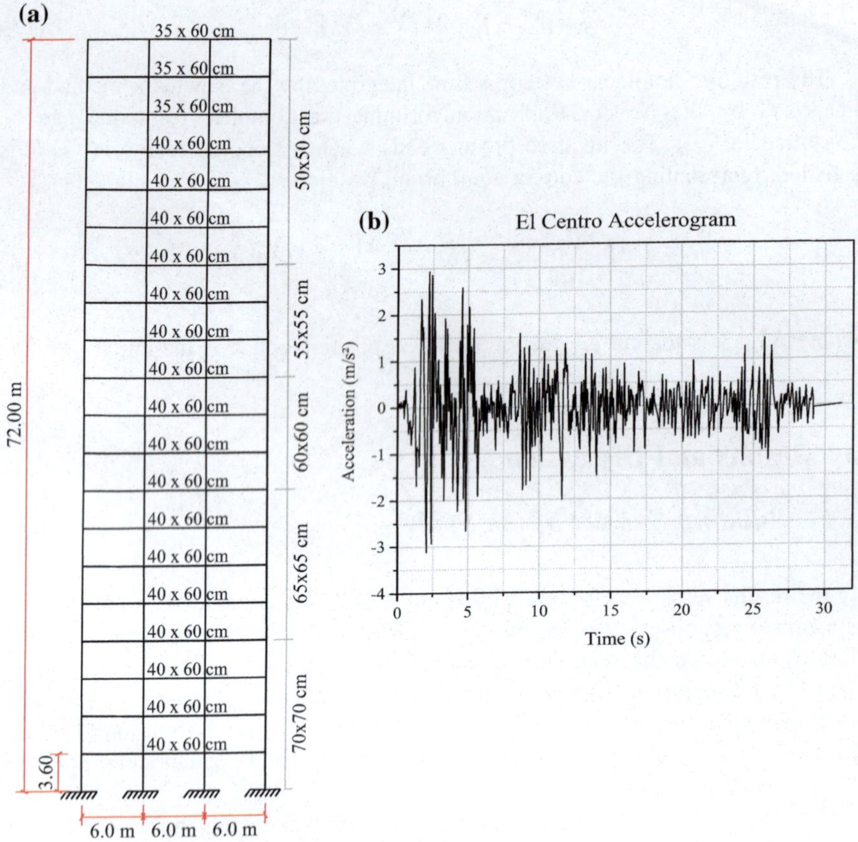

Fig. 4 **a** Dimensions of the building and its elements; **b** El Centro accelerogram

6.41, 7.15, 7.86, 8.96 and 9.29 Hz. The modes of vibration were determined by the eigenvectors of the system and are presented in Fig. 5 the firsts four natural frequencies are flexural.

3.3 TMD Tuning Frequencies

It is well established in the literature that TMDs perform well when their frequency is tuned to the first natural frequency of the building. When the first mode of vibration is excited by wind loads, for instance, the vibration energy is transferred from the building to the TMD.

Wind loads have a narrow spectrum of frequencies, which can be close to the first natural frequencies of the building. Earthquake loading, however, poses a different challenge, as excitation frequencies are broad-banded; hence, in principle,

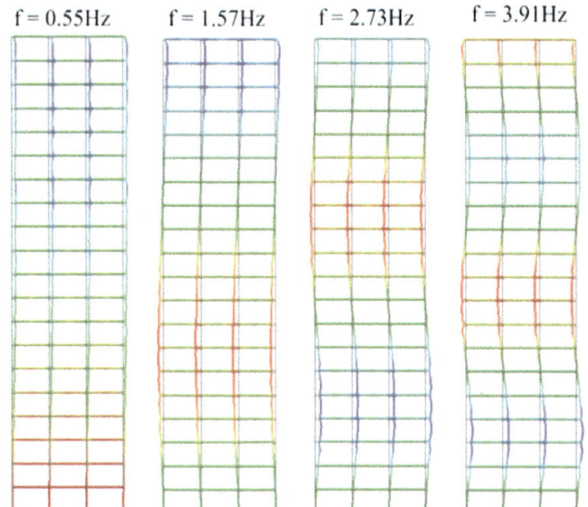

Fig. 5 Vibration modes FE model

many vibration modes of a structure can be excited. For this reason, there is no general agreement about optimal TMD design, or about their efficiency, in reducing vibrations induced by seismic loading. If the TMD is tuned to the structure's fundamental frequency, the absorber will substantially reduce just the response of the first mode, without affecting the vibration in higher modes [13].

In this paper, a study about the tuning of TMD frequency is developed. The TMD is employed in the 20 story building described above, and subject to El Centro earthquake records. Three finite element models were studied, to evaluate appropriateness of modelling assumptions: (a) Linear analysis with equivalent horizontal nodal loads applied along the buildings height; (b) Geometrical nonlinear analysis (GNL); (c) linear analysis with base-imposed earthquake displacements;

The TMD was modelled as an additional frame rigid element placed at the last storey; the element is 60 cm long and is connected to the building by two linear frame elements, whose axial stiffness is calibrated to yield the required natural frequency for the TMD. TMDs tuned to the first 10 natural frequencies of the building were tested.

For the present tuning frequency study, the mass ratio was fixed at $\bar{m} = 0.1$; and the absorber was modelled as a MCK system designed according to the simplified equations shown in [7], where the modal generalized mass and stiffness properties are used to tune the absorber to each mode of vibration of the structure. In the neutralizers was considered viscous damping acting through Rayleigh damping proportional to the mass matrix.

Table 1 presents parameters of the absorbers designed to reduce the response of the firsts 10 modes of vibration of the building. In Table 1, f_{struc} are the natural frequencies of the building obtained by modal analyses in Hz, ω_{struc} are the natural angular frequencies of the structure in each vibration motion of interest; ω_{TMD} is the

Table 1 TMDs designs

Mode	f_{struc} (Hz)	ω_{stuc} (rad/s)	ω_{TMD} (rad/s)	K_{TMD} (10^7) (N/m)	C_{TMD} (10^7)
1	0.55	3.46	3.07	0.0095	0.21
2	1.57	9.86	8.74	0.0771	0.59
3	2.73	17.18	15.22	2.34	1.02
4	3.91	24.56	21.76	4.78	1.46
5	5.13	32.21	28.54	8.22	1.91
6	6.41	40.29	35.70	12.86	2.39
7	7.15	44.91	39.80	15.98	2.67
8	7.86	49.40	43.77	19.33	2.93
9	8.96	56.28	49.86	25.08	3.34
10	9.29	58.40	51.74	27.01	3.47

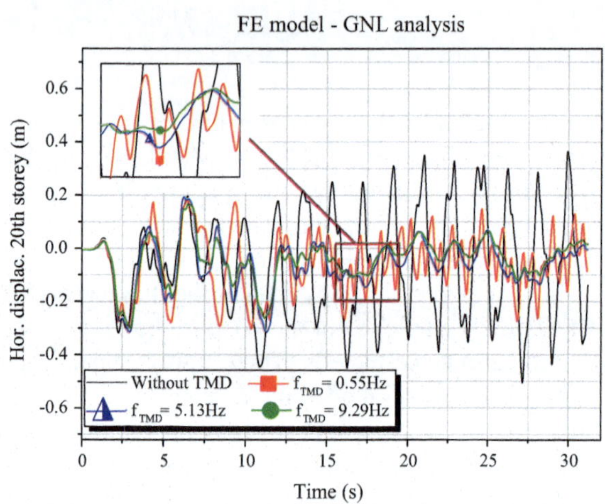

Fig. 6 Response of GNL model, at 20th storey, without and with TMDs tuned to different frequencies "f"

tuning frequency calculated for the absorber [4, 7], K_{TMD} is the spring stiffness of the absorber and C_{TMD} the damping coefficient of the TMD.

Figures 6, 7 and 8 present the horizontal displacements at the top of the building for the different absorbers tuned to the ten firsts natural frequencies. In all graphs, lines named as f_{TMD} refers to the natural frequency of the building which TMD was designed to.

Figure 6 shows that TMDs are efficient at suppressing structural vibrations. It can also be observed that TMDs tuned to higher frequencies (i.e. 9.29 Hz) have better performance.

Figures 7 and 8 compare the responses of the four models studied in this section, for the structure without TMD, and with TMDs tuned to different frequencies. It can be observed that, for the structure without TMD, responses of the three FE analysis models (linear with equivalent loads, linear with base displacements and geometrically nonlinear with base displacements) are equivalent, presenting the same

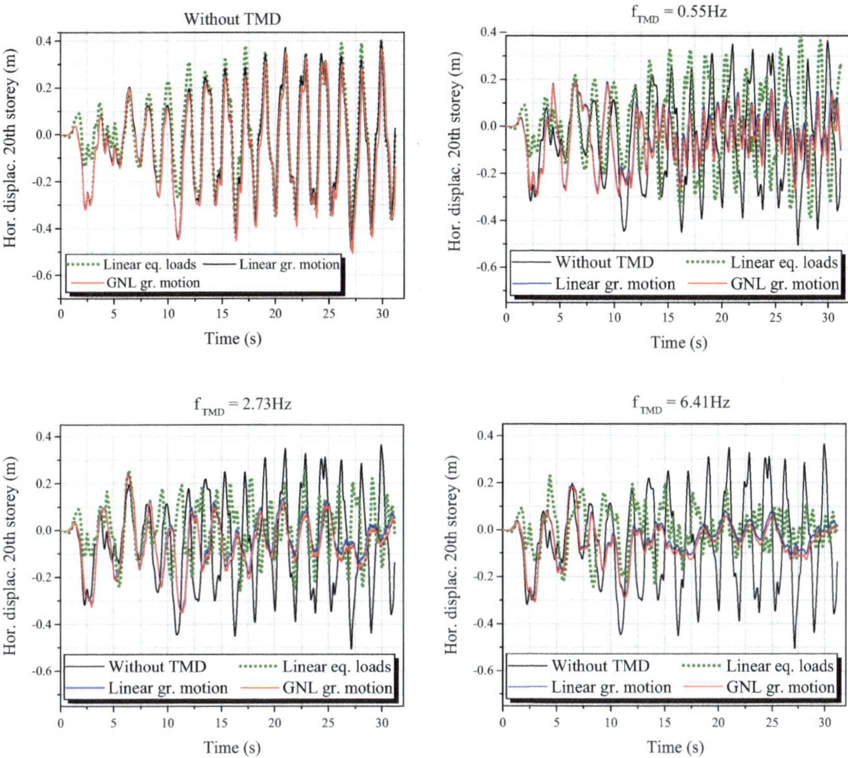

Fig. 7 Horizontal displacements at 20th storey, without and with TMDs tuned to first, third and sixth frequencies

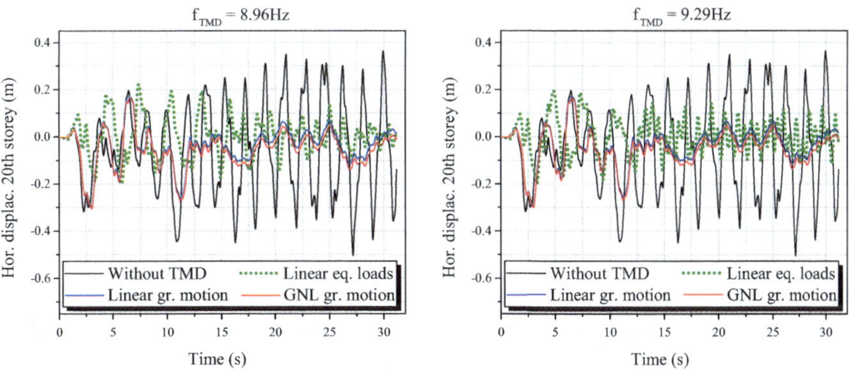

Fig. 8 Horizontal displacements at 20th storey, without and with TMDs tuned to ninth and tenth frequencies

frequency of oscillation and closely matching displacements. This shows that displacements for the structure without TMD are not too large, and can be accurately described by a linear FE analysis, and event for a simplified load model

In Figs. 7 and 8, it also becomes clear that the equivalent load model significantly overrate structural displacements; turning this model inadequate to represent structural responses under seismic loadings. It can be also observed that for higher tuning frequencies, the TMD-building system becomes stiffer, displacements and oscillation frequency are significantly reduced, and for all cases the linear and non-linear models tend to agree.

4 Concluding Remarks

This paper pursued the objective of investigating the performance of buildings equipped with absorbers as Tuned Mass Damper (TMD), when subject to earthquake loadings.

With respect to modelling assumptions, three methodologies were compared, a linear FE model with equivalent nodal loads, a linear FE model with ground movement representing *El Centro* earthquake and a Geometrically nonlinear FE model with base movement. It was found that the equivalent load model significantly underestimates building displacements, leading to oscillations around the initial position. In the comparison of the linear and nonlinear FE model with base-imposed displacements, both models presented similar results, showing that the structure subject of this study behaves in linear regime, with small to moderate displacements. Hence, for this particular structure, the more computationally intensive NL solution is not required. This study will be extended in the future to address buildings with large displacements.

Since earthquake loading excites different vibration modes of the structure, the tuning of TMD frequency is not obvious. In this paper, it was shown that TMDs present better efficiency when tuned to higher natural frequencies of the structure. For the 20-storey building studied herein, the TMDs were capable of reducing displacement amplitudes and oscillatory frequency. This is in contrast to what is commonly believed in the literature, i.e., that tuning the TMD to the first natural frequency is optimal. In general, results show that TMDs tuned to higher natural frequencies of the structure reduce displacements and oscillation frequencies guarantying more comfort to the building occupants.

The methodology proposed in this paper demonstrates the effectiveness of TMD devices in controlling structural vibrations of structures subject to earthquake loadings. However, benefits are only achieved when TMDs are properly tuned, and when structural responses are correctly evaluated.

References

1. Housner, W., Bergman, A., Caughey, K., et al.: Structural control: past, present, and future. J. Eng. Mech. **9**, 897–971 (1997)
2. Spencer, B., Nagarajaiah, S.: State of the art of structural control. J. Struct. Eng. **7**, 845–856 (2003)
3. Frahm, H.: Device for damping vibration of bodies. US Patent 989958 (1909)
4. Hartog, D.: Mechanical Vibrations, 4th edn. McGraw-Hill Inc., New York (1956)
5. Warburton, G., Ayorinde, E.: Optimum absorber parameters for simples systems. Earthq. Eng. Struct. Dyn. **8**, 197–217 (1980)
6. Warburton, G.: Optimum absorber parameters for various combinations of response and excitation parameters. Earthq. Eng. Struct. Dyn. **10**, 381–401 (1982)
7. Rana, R., Soong, T.: Parametric study and simplified design of tuned mass dampers. Eng. Struct. **3**(20), 193–204 (1998)
8. Hoang, N., Warnitchai, P.: Design of multiple tuned mass dampers by using a numerical optimizer. Earthq. Eng. Struct. Dyn. **34**, 25–144 (2005)
9. Marian, L., Giaralis, A.: Optimal design of a novel tuned mass-damper-inerter (TMDI) passive vibration control configuration for stochastically support-excited structural systems. Probab. Eng. Mech. **38**, 156–164 (2015)
10. Rüdinger, F.: Optimal vibration absorber with nonlinear viscous power law damping and white noise excitation. J. Eng. Mech. **132**, 46–53 (2006)
11. Tubaldi, E., Kougiomtzoglou, I.A.: Nonstationary stochastic response of structural systems equipped with nonlinear viscous dampers under seismic excitation. Earthq. Eng. Struct. Dyn. **44**, 121–138 (2014)
12. Kaynia, J.M., Veneziano, A.M., Biggs, D.: Seismic effectiveness of tuned mass dampers. J. Struct. Div. **107**, 1465–1483 (1981)
13. Parulekar, Y.M., Reddy, G.R.: For seismic response reduction. Int. J. Struct. Stab. Dyn. **9**, 151–177 (2009)
14. Hoang, N., Fujino, Y., Warnitchai, P.: Optimal tuned mass damper for seismic applications and practical design formulas. Eng. Struct. **30**, 707–715 (2008)
15. Soto-brito, R., Ruiz, S.E.: Influence of ground motion intensity on the effectiveness of tuned mass dampers. Earthq. Eng. Struct. Dyn. **1271**, 1255–1271 (1999)
16. Beck, A.T., Kougioumtzoglou, I.A., Santos, K.R.M.: Optimal performance-based design of non-linear stochastic dynamical RC structures subject to stationary wind excitation. Eng. Struct. **78**, 145–153 (2014)
17. Moutinho, C.: An alternative methodology for designing tuned mass dampers to reduce seismic vibrations in buildings structures. Earthq. Eng. Struct. Dyn. **41**, 2059–2073 (2012)
18. Coda, H.B., Greco, M.: A simple FEM formulation for large deflection 2D frame analysis based on position description. Comput. Methods Appl. Mech. Eng. **193**, 3541–3557 (2004)
19. Coda, H.B., Paccola, R.R.: A total-Lagrangian position-based FEM applied to physical and geometrical nonlinear dynamics of plane frames including semi-rigid connections and progressive collapse. Finite Elem. Anal. Des. **91**, 1–15 (2014)
20. Paultre, P.: Dynamics of Structures, 1st ed. Wiley (2010)

Model Based System Testing Approach for Efficient Testing of EPS Systems

Cassio T. Faria, Fabio Luis M. dos Santos, Theo Geluk, Steven Dom and Herman Van der Auweraer

Abstract Model Based System Testing (MBST) can be defined as the discipline combining physical testing and simulation models with the aim to study, identify, validate and improve the behavior of multiphysical and mechatronic systems. One of its benefits is related to the use of simulation models to improve or accelerate the testing process, using well known procedures, such as optimal sensor and excitation placement, but also more recent methodologies, such as virtual testing or human-in-the-loop interactions. In this context, these MBST methodologies can be used for Electrical power steering (EPS) system testing, to allow for better characterization of the overall system and subsystems, and to better identify and model nonlinearities. This paper presents a testing and simulation combined approach used to optimally define test conditions, such as sensor placement, test boundary conditions, excitation inputs and how they affect parameter identification.

Keywords Model based system testing · Virtual testing · Electric power steering · Nonlinear analysis · Multiphysics

---

C. T. Faria · F. L. M. dos Santos (✉) · T. Geluk · S. Dom · H. Van der Auweraer
Siemens Industry Software N.V., Interleuvenlaan 68, 3001 Louvain, Belgium
e-mail: fabio.m.santos@siemens.com

C. T. Faria
e-mail: cassio.faria@siemens.com

T. Geluk
e-mail: theo.geluk@siemens.com

S. Dom
e-mail: steven.dom@siemens.com

H. Van der Auweraer
e-mail: herman.vanderauweraer@siemens.com

H. Van der Auweraer
K.U. Leuven, Kasteelpark Arenberg 40, 3001 Louvain, Belgium

A. de T. Fleury et al. (eds.), *Proceedings of DINAME 2017*, Lecture Notes in Mechanical Engineering, https://doi.org/10.1007/978-3-319-91217-2_14

1 Introduction

The increasing challenges in product development, originating from the needs to decrease product development costs, while increasing overall performance and efficiency, have more and more led to the use of simulation methodologies in combination with testing. Physical prototypes are not only expensive and time-consuming to produce, but they are also only available late in the development cycle, when timing is even more critical. Improving testing conditions at this stage can lead to much more efficient test procedures, that not only spend less time and resources, but that can also provide data which is more useful for the validation and correlation of models, resulting in more well-performing systems, with shortened development times. However, such a combined virtual+physical approach can be less trivial when very complex, nonlinear multiphysical systems are taken in consideration. Such is the case in the automotive industry, where the increased use of mechatronics and controls also leads to higher complexity and higher needs to accurately identify system parameters in order to properly model and understand their behavior [1]. Moreover, most system components cannot be tested individually, as their behavior changes when they are integrated in the system and under the real boundary conditions. Therefore, parameters must be identified using in situ system level tests, which can be very challenging, complex and costly. In this case, simulation and virtual testing can be used to improve test reliability, and lead to the right and efficient way to instrument and carry out the test campaign.

Experimental methodologies nowadays have shifted from purely test-only routines to simulation-aided methods. As test campaigns and their objectives grow in complexity, they make use more and more of virtual models to increase the knowledge of system behavior. Moreover, testing is no longer solely related to troubleshooting analysis, and the identified parameters and models can also be used for other purposes, such as force and load estimation, in the so-called virtual sensing applications [2]. Testing has also gone beyond the boundaries of purely mechanical systems, and its applications have expanded to other fields, such as electrical motor testing [3, 4], electromechanical systems [5], multiphysical analyses [6] and mechatronic applications [7].

With the purpose of supporting the new paradigm that combines test and simulation, the model based system testing (MBST) framework was created [8]. The main purpose of MBST and its underlying methodologies is to support and improve testing and validation techniques, by using and/or combining test and simulation, with the aim to study, identify and validate multiphysical and mechatronic systems. MBST combines test and simulation into 3 categories: “Testing for Simulation”, “Simulation for Testing” and “Testing with Simulation”. The first two are cases in which both test and simulation are decoupled, and are carried out in different steps, while the last case involves test and simulation simultaneously. Figure 1 shows the MBST application tree and how the different domains are divided.

This work will focus on the “Simulation for Testing” branch of the MBST application tree. This category has two already well known procedures, optimal

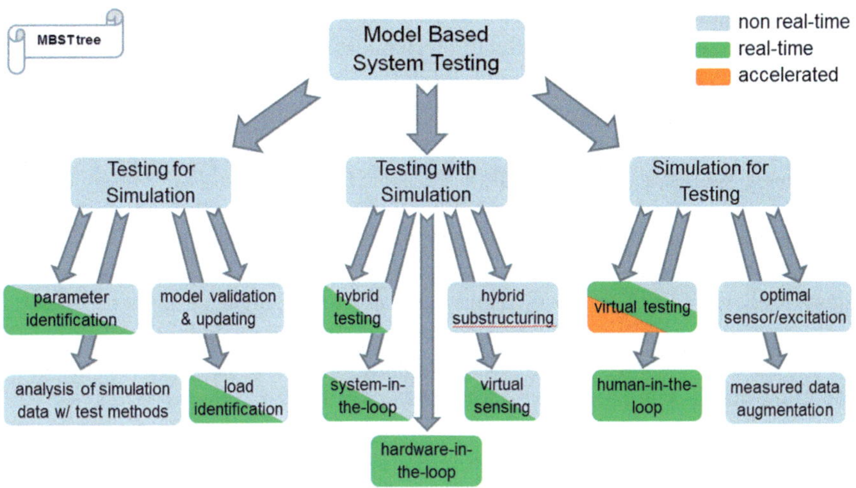

Fig. 1 MBST tree

sensor and excitation location, but also makes use of more recent methodologies, such as virtual testing and measured data augmentation. To properly illustrate and contextualize the MBST framework this paper develops the complete application for a Electric Power Steering mechatronic system.

Electrical power steering has several advantages when compared to its concurrent technology in power steering, hydraulic actuation. It is more compact, has higher efficiency and reliability, and is easier to maintain [9]. However, there can be noise and vibration issues arising from this type of system, depending on the control strategy used to alleviate the required torque from the driver [10, 11]. The EPS system can also undesirably isolate the feedback to the driver, reducing the driving feeling, which is an important sensorial feedback to the driver, giving them the right notion of the safe vehicle handling, especially with respect to unsafe conditions, such as vehicle side slip.

Model-based control strategies often prove to be an appropriate solution for EPS systems but they rely heavily on high fidelity models of the steering system which is by nature very complex, nonlinear and multiphysical. Nonlinearities such as friction and backlash are a common characteristic of the rack-pinion gear pair and also for the worm gear, some basic elements of the EPS system. Good models must include these physical properties to improve their prediction capabilities and allow for the model-based controller to properly drive the system [12].

A major portion of the mismatch between model and system behaviors arises of the poor characterization of the nonlinear components parameters. Experience from engineering projects within Siemens Industry Software dictates that the nonlinear behavior of the individual components is not the same as the behavior observed when they are interconnected and in situ. Therefore the parameters needed for the model need to be identified in situ (i.e. installed in the vehicle) at a system testing and not by a series of test bench measurements on each component. This measurement campaign of an entire electromechanical system poses several challenges, in particular to isolate and determine the influence of individual parameter on the overall system behavior.

As a main contribution, this paper implements the MBST framework on the EPS system in order to properly design a test campaign and to fully characterize the nonlinear dynamic system. The modelling and testing proposed in this paper differ from literature as the model developed and identified takes into account several nonlinearities (hysteresis, friction, backlash), in contrast to [9, 10] which consider linear models and [11, 12] which only account for a single instance of coulomb friction. Moreover characterization is carried out in situ, that is, with the EPS mounted in the vehicle, to fully capture several of the connections and boundary condition that it will experience during operation.

2 Electric Power Steering (EPS) and Its Functional Model

Electric power steering (EPS) assistance systems provide auxiliary power to the steering mechanism of the vehicle to aid the driver in performing a desired maneuver, where part of the effort is frontloaded by an electric drive combined to the steering column of the mechanism. A common schematic used to generalize this type of device [12] is provided on Fig. 2 where the main components are shown: a steering wheel (where drivers apply their inputs); an electric motor and drive to provide auxiliary power; a reduction mechanism to transmit the electric motor power into the steering column; an electronic control unit (ECU) that reads the driver's torque and commands the electric motor to assist the driver (reduce the torque required to perform a maneuver); a steering column that transmits torque to the rack and pinion; the rack and pinion that converts the rotary movement into a translation one; the tie-rods (left and right) which carry the translation motion and loads from the rack and pinion to the wheels; and finally the vehicle wheels that the driver desires to rotate around their axis to create a steering motion of the complete vehicle.

Several issues might arise from the actual construction of the EPS mechanism, such as the use of joints in the steering column to redirect the direction of the rotational movement, the safe use and adjustment mechanism in the column, and more. The simplest way to model an EPS system is to relate the assist torque with the driver torque and steering rack displacement. Figure 3 shows the diagram representing such a model, which is described by Eqs. (1) and (2).

$$J_s \ddot{\theta}_s + b\,\dot{\theta}_s + K_s\theta_s = T_d + \frac{K_s}{R_s}x_r + T_{nl} \tag{1}$$

$$M_{eq}\ddot{x}_r + b_{eq}\dot{x}_r + K_{eq}x_r = \frac{T_mG}{R_m} + \frac{K_s}{R_s}\theta_s + F_d + F_{nl} \tag{2}$$

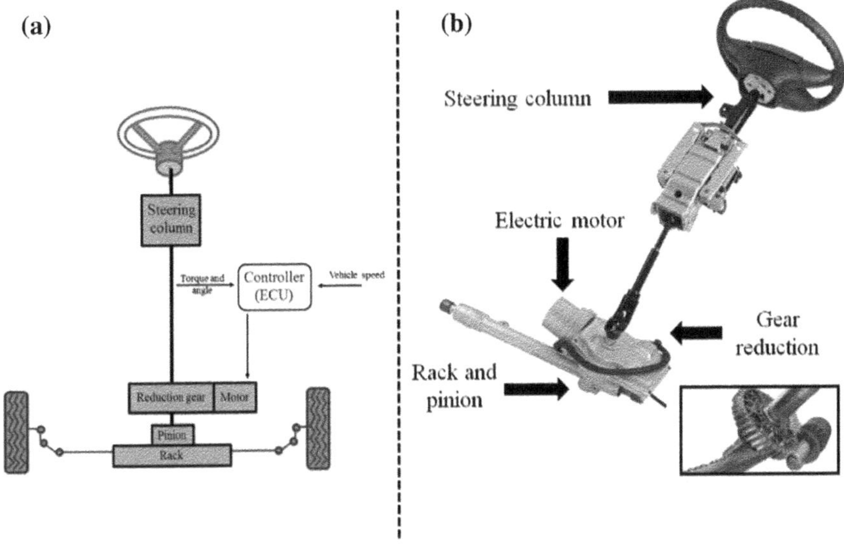

Fig. 2 Sketch of the working principle of and EPS system and component names

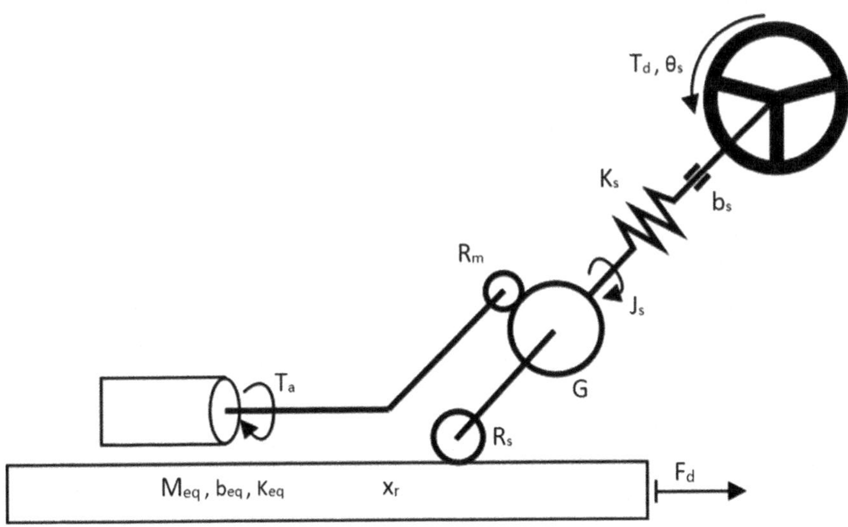

Fig. 3 Schematics for simple EPS model

where J_s, b_s and K_s are the steering column inertia, friction and stiffness, θ_s is the steering column angle, T_d is the driver torque, T_{NL} is the torque induced by all nonlinearities in the torsional system, R_s is the column pinion radius, x_r is the rack displacement, M_{eq}, b_{eq} and K_{eq} are the equivalent rack mass, friction and stiffness, T_a is the assistance torque, R_m is the assistance pinion radius, G is the gear ratio, F_d

is the disturbance from the road and F_{nl} is the sum of nonlinear forces acting on the rack. The realization of the EPS mechanism will lead to a system that has a more complex dynamic behavior than originally desired, with a series of nonlinearities, such as dry and viscous friction, backlash, hysteresis, etc. However, these phenomena are very hard to describe accurately by making use of analytical equations, so the challenge from a modeling perspective is to idealize the system and define fundamental dynamics for each of its components that can adequately represent the behavior of the real system. For this purpose, functional models can be used instead, to aid in the modelling of important and complex dynamics, while still making it simple to describe the system [13]. Functional models allocate physical systems to their (multiphysical) functionalities. In this sense, one can start from the simple electro-mechanical representation of the system by a combination of masses, springs, dampers, motors and converters (e.g. gears) of the system to reach a first functional iteration of the model.

For this paper a multi-physical functional model is constructed using Simcenter Amesim software, the sketch of the model is presented below by Fig. 4 together with a general classification used for the different subsystems considered. For each of the grouping presented a friction component is assigned (this is not the case for the control strategy—ECU) to lump all the nonlinear effects associated with each sub-system. This grouping of nonlinear effects into a single component is a particular choice of the authors of this paper, to limit the number of parameters to be identified, these components are highlighted in Fig. 4 by an orange circle.

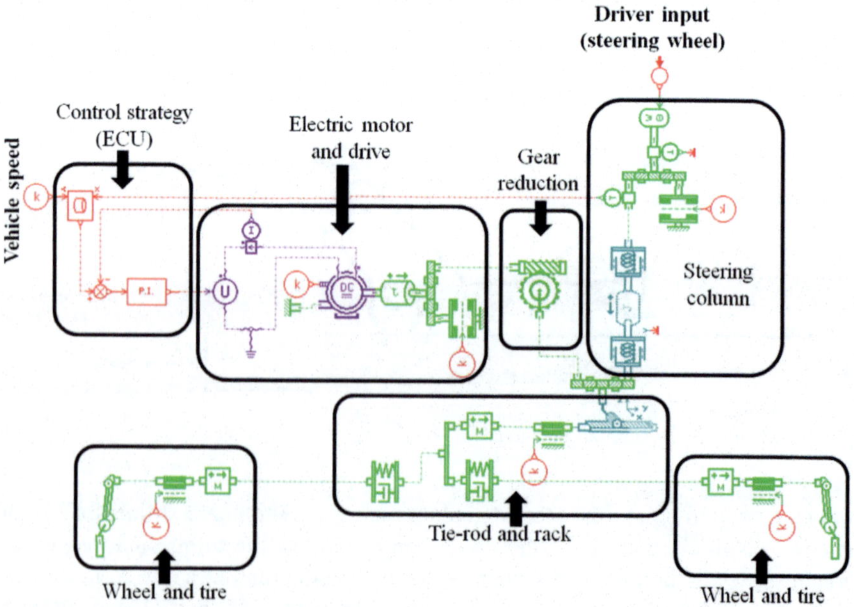

Fig. 4 EPS model realization in AMESIM

All nonlinear elements of the functional model includes a contact gap, Coulomb and stiction friction physics in their equations, more info on their definition can be found on the software technical manual [14], reference to MECFR1R0A (rotary element), MECFR1T0B (translation element) and WORMGR1A (worm gear element).

The model takes as input the angular input provided by the driver at the steering wheel and calculates for a certain vehicle speed the torque perceived by the driver. If the control strategy is turned off (zero PI gains for example), the electrical drive reacts as a passive component with certain inertia associated to it. The driver's action is cascaded down to the tie-rods which push against the tire friction to create an angular motion at the spindle. The tire friction can also be written as a function of vehicle speed to take into account the rolling velocity in the side-slip of the tire. From a practical testing perspective it is interesting to characterize the EPS dynamics when the electric drive and control strategy is not activated, to define the response of the passive mechanical system, and also considering zero vehicle speed so the tire friction can be taken at its maximum state.

Several model parameters can be identified a priory to reduce the number of unknowns in the system model. Geometrical and mechanical characteristics (mass and stiffness) can be determined a priory by the execution of simple tests and components inspection. Electrical quantities can also be directly measured at the electric motor. The largest uncertainty must be assigned to the nonlinear parameters, but a certain level of parameter freedom can also be assigned to these previously identified parameters for model fine tuning.

3 MBST Approach Applied to EPS

By applying the concepts of Model-Based System Testing (MBST), more specifically the subcategory of Simulation for Testing, one can generate a lot of insight on the testing activity by performing a series of simulations to explore the system response and help define testing configurations and methodologies that expose critical and key parameters of system of interest. To illustrate this concept the methodology is applied to the electric power steering (EPS) system in order to develop a testing procedure that allows for the composition of a high fidelity model that could later be used for control purposes.

The basic structure of the EPS model under consideration has been already introduced in the previous section of this paper where functional graphic modeling software (Simcenter Amesim) was used to sketch the structure taken for this physical system. This model can be used in a variety of ways to explore different aspects of the tests that will be carried out and to evaluate mechanism, configurations and conditions that would favor the later identification of the model parameters. A key remark is that this model already has reasonable (or even accurate) values for some of the linear dynamics components, such as masses, dimensions and stiffness. The challenge on developing an accurate model for this particular physical system lies on the characterization of the nonlinearities and viscous damping terms used.

The first exercise carried out here is to explore the behavior of smaller (called here cut-outs) of the system and subject those to different boundary condition to allow for the testing engineer to isolate the nonlinearities on a given component and precisely characterize them. It is important to recognize that not all cut-outs of the model presented in Fig. 4 can be executed in practice and also that imposing certain in situ boundary condition might be unfeasible or too costly to be executed, therefore a subset of the possibilities are explored as an intersection of three different factors: utility of such a cut-out, the effort to implement it and the cost to execute it. As an example, Fig. 5 illustrates three EPS cut-outs exploring the behavior of a section of the system under a particular boundary condition. The cut-out from the left isolates the behavior of the column, where the configuration blocks the movement of the shaft at its base and allows for the evaluation of friction given that enough flexibility is present. The use of simulation of such a subset of the original developed model can help the test engineer evaluate if this cut-out can be useful or not for their parameter identification.

The central cut-out shown in Fig. 5 consists of the disconnection of the tie rods from the wheel spindle which would allow the rack to move freely with no resistance. In this condition one can also disconnect the electric motor and have it as a free moving inertia in the system. Similar to the previous cut-out, in the one on the right side of Fig. 5, the tie-rods are blocked to create a zero displacement condition at the rack. The model realization for each cut-out can be used to evaluate and

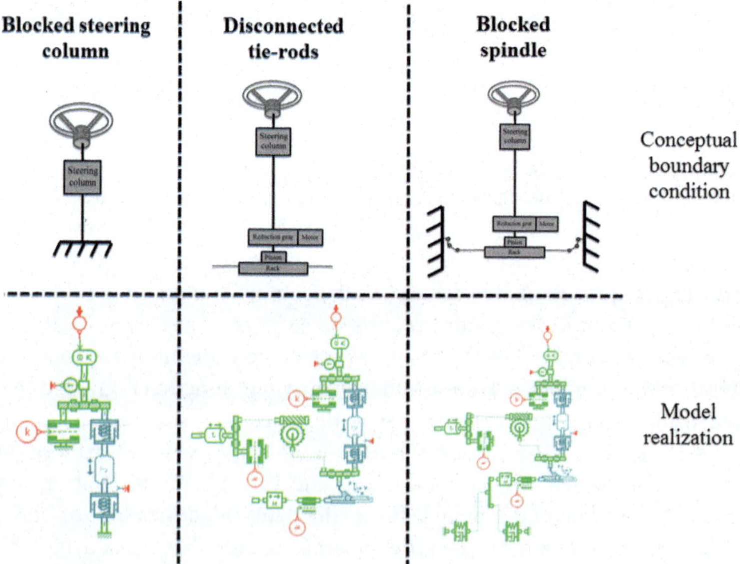

Fig. 5 Different possible boundary condition for the EPS mechanism and its model realization

investigate in detail if it is a good candidate to excite and isolate a particular nonlinear behavior in each of the components.

Besides evaluating different model configurations (cut-outs) and boundary conditions, the virtual representation of the EPS system can be used to evaluate the usage of different inputs to the system and how they affect the response. Typical input profiles to be evaluated are: harmonic inputs (sine waves or sine sweep), used to evaluate cyclic behaviors; impact inputs (sharp or wide impacts), used to evaluate linear dynamic behavior; and also ramp inputs, used to identify transition points in the system dynamics.

Moreover, a combination of these inputs or different levels of the same input can be used to exploit the dominance of a given parameter in the response of the system. By developing this knowledge over several simulations, a set of unique input profiles can be derived such that the parameters are identified more accurately. One example is shown on Fig. 6, where the same cut-out model of the disconnected tie-rods is excited using a harmonic angular input (sine wave of 30 degrees of amplitude) using two different frequencies, one low (0.5 Hz) and another higher frequency (2 Hz). A significant response difference can be noticed between the two inputs applied to the system, where on the higher frequency sine input more inertial forces have to be overcome in order to start the motion and to reverse its direction, also the higher velocities make the viscous losses in the system more prominent.

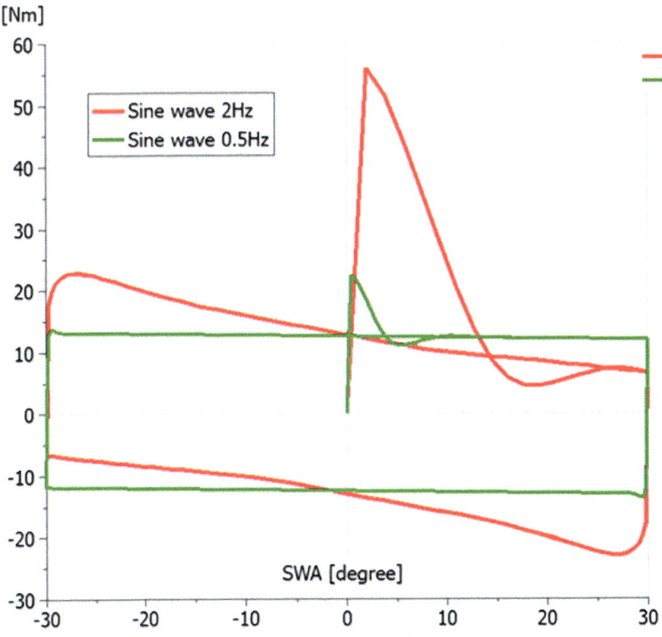

Fig. 6 Steering wheel angle by steering wheel torque plot for two different excitation harmonic inputs

As to be expected, the initial motion of the system with a higher frequency input will require a larger torque to move the rotational inertia, since the acceleration at this case is higher when compared to the 0.5 Hz input excitation. This allows the torque signal to go well beyond the noise floor level on the experimental setup and will make for a much more accurate identification of the system inertia. A similar logic can be applied to the viscous components since the system with a 2 Hz angular sinusoidal input will experience higher velocities and more viscous losses associated with it.

Another important use for the model, in order to enhance the testing of such a multi-physical system, is to evaluate how candidate sensor positions can help (or not) to detect the effect of a certain parameter on the measured system response at that location. In this situation, it is not feasible to carry out an exhaustive search of all the sensor locations and use observability criteria to determine if that measurement (or a set of measurements) can be used to identify all the parameters of interest of the system, since many of these locations and quantities cannot be instrumented due to effort, feasibility and cost related issues. Therefore, the use of experience in combination with the exploratory capabilities of the model can directly support the decision making process of the instrumentation to be installed in the EPS system.

4 Experimental Setup and Testing Results

To validate the proposed MBST approach applied to an EPS, an experimental campaign was devised starting from the principles previously described in this paper and will be detailed in this section. In order to maximize the correlation between the experimental setup and an in-vehicle mechanism in operation, the measurements were taken from an EPS installed in a vehicle (to guarantee the correct boundary conditions). Moreover, the mechanism of choice was from a world-wide commercially available vehicle of a large automotive company, which will not be named here for confidentiality issues. No previous data or knowledge on the system had been provided prior to the beginning of the MBST campaign and all the parameters of the system model (Fig. 4) were unknown.

A first inspection of the EPS under test shows that all the main components previously described in this paper are present and indeed the generic model can be used to represent the system. At this stage the same model parameters can be retrieved by direct inspection of the device, such as the dimensions of the gears, leverage arm of the tie-rod to tire, motor resistance and inductance, conversion mechanisms ratio and the mass of components (wheel, rack, etc.). Some first estimates can also be calculated for the stiffness of the steering column and the tie-rod considering the geometry of the components and the constitutive material. No prior knowledge or estimates have been assigned to any friction parameter (needed for the adequate description of the highlighted nonlinear elements of the functional model described by Fig. 4) or even to the viscous losses present in the system, but they have been chosen to be within reasonable ranges (to allow for motion of the entire system).

By setting the functional model of the system within the Simcenter Amesim environment, simulations were carried with ten candidate position of sensor locations, chosen based on the accessibility of the location and cost/time to instrument. The unknown model parameters were swept across a reasonable range of possible values and the responses of the sensors were evaluated whether they could capture a response change or not. From those candidate positions, three were selected as the minimal set of sensors needed to fully capture the system behavior. The driver steering angle was selected as the input to the system (to also be used in the simulations) and a sensor was also placed in that location.

Figure 7 is a schematic representation of the sensor positions selected for the MBST campaign, where a potentiometer is used to capture the angular input from the driver, a full-bridge of strain-gauges (062AK_350 from micro measurements) is used to measure the driver input torque and also to measure the force transmitted by the tie-rod to the wheel. A wire-draw displacement sensor (WDS-250-MPM-C-P_HG from micro-epsilon) records the displacement of the rack/tie-rod. A second force measurement of the force in the tie-rod is also done to verify the system symmetry.

Data was collected by a SCADAS mobile hardware using VB8 family modules and the acquisition and post-processing using LMS Test.Lab software (Signature

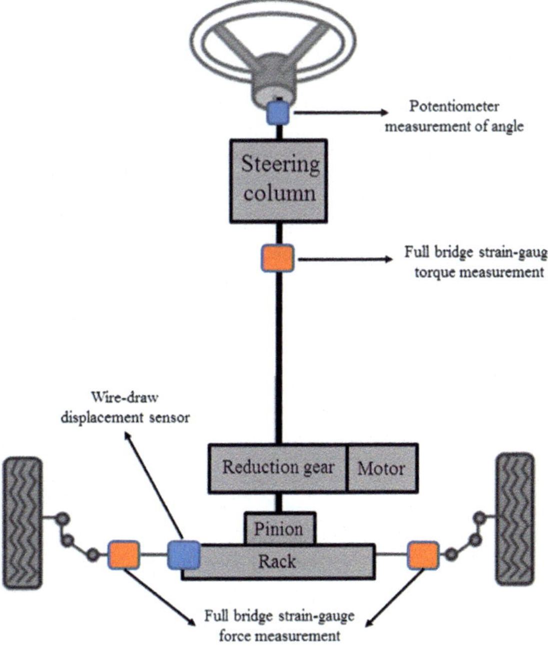

Fig. 7 Schematic representation of the test setup and instrumentation used

Acquisition module). Exponential filtering was used on the collected data to eliminate high frequency noise content of the data and since the same procedure is applied to all data phase delays should be even across the different data signals.

The models are used to evaluate possible test scenario of the EPS system considering different cut-outs and boundary conditions as explained in the previous section of this paper. In total, a combination of six different scenarios were investigated, five where the e-motor was a passive element in the system (blocked steering column, free tie-rods, blocked spindle, full system without tire friction—suspended vehicle—and full system with tires on the ground) and one scenario where the motor was actuated with a constant supply (free tie-rod). Based on the simulation results, three of these scenarios were chosen: (1) blocked steering column; (2) free tie-rods and; (3) full system with tires on the ground. These would, according to simulation results, enable the identification of all the unknown model parameters and were a good compromise between quality of the results and time/cost of the campaign.

The following step is to study how different inputs could be selected for each of the three test scenarios. For scenario (1), two sinusoidal inputs were selected, one fast (0.8 Hz) and one slow (0.2 Hz). Meanwhile, for (2) and (3), besides the two harmonic inputs, a slow ramp was also selected to allow for a clear definition of the transition points between different friction regimes in different components. At this stage, the simulations carried out provided full support to the test engineer to define the best practices for the EPS system and could then initiate the test procedure with high confidence.

The MBST procedure allowed for the execution of a fast and conscious instrumentation and testing of the EPS system given the insight provided by the simulation. The subsequent parameter identification procedure was carried out first at a model cut-out representing the test scenario (1), followed by (2) and (3) respectively. The parameters identified in the previous scenario were cascaded to the subsequent identification step and they were taken as constant to allow the algorithm to focus on the fitting based only on a subset of the model parameters. The identification procedure referred to, consists of an optimization routine, where the parameters to be identified are modified until a cost function (that is the sum of the differences between experimental and simulation responses) is minimized. A tool box on design exploration and optimization is available within Simcenter Amesim, and a Nonlinear Programing by Quadratic Lagrangian (NLPQL) was used to identify the desired parameters [14].

One of the results of the identification procedure is shown by Fig. 8, where the simulated result for a control test on scenario (2) is compared against the simulated result. As observed, the correlation between test and simulation obtained was quite good, nevertheless it is possible to see that some backlash on the experimental rack displacement (RD) measured is not captured by the model (flat top of the peak of experimental sinusoidal format).

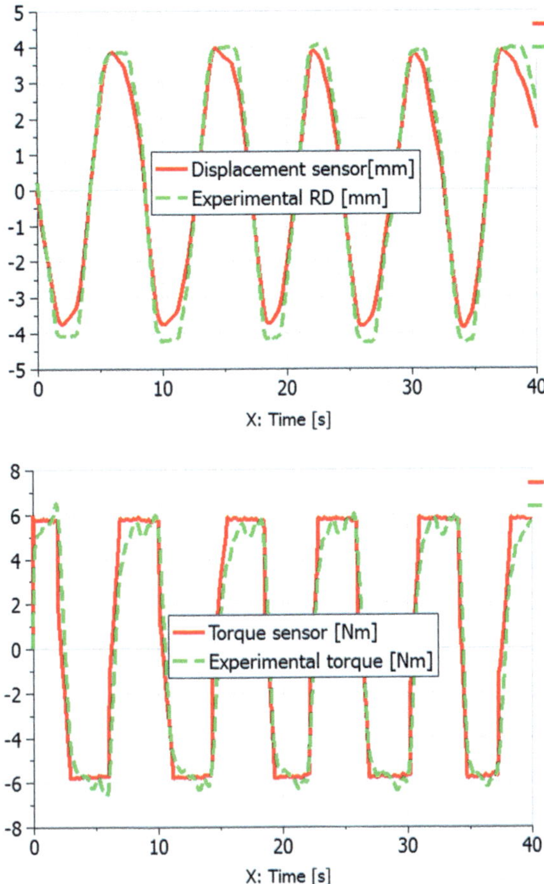

Fig. 8 Disconnected tie-rods results for the experimentally measured and simulated results after parameter identification is carried out

5 Discussion and Conclusions

This article presented the use of model based system testing applied to an electric power steering system. For that purpose, the concept of MBST was introduced, paying particular attention to the simulation for testing branch, where simulation and virtual models are used to aid the testing procedure. Then, it was shown how it is possible to efficiently test an EPS system by using a base model, to simulate different scenarios and selected the most suitable ones to be executed. The procedure proposed here makes use of a generic functional EPS model (Fig. 4) with reasonable parameters to carry out prior-to-test evaluations, in order to define optimal sensor position (in terms of feasibility, effort, observability, and cost) and also define in situ testing of sub-system denominated cut-outs. Moreover, the model was used to determine the system input profiles that could be used to identify different parameters.

Subsequently, a test campaign was carried out to validate the proposed methodology. By collecting simple geometrical and material properties, several model parameters were identified (inertia, stiffness, etc.), while the test dynamic data (using the optimum sensor location, the designed inputs and the different relevant test scenarios) was used to identify the nonlinear and viscous parameters, by means of a NLPQL optimization algorithm. The results showed good correlation between model and test, especially taking into account the complexity of the system.

In conclusion, it was observed that the combination of simulation and test (the MBST framework) can be very beneficial to aid in testing and parameter identification of complex systems. By using a model of the EPS system, it was possible to speed-up the testing procedure, reduce the number of used sensors and improve accuracy of the results.

Responsibility Notice
The authors are the only responsible for the material included in this paper.

Acknowledgements The research presented in this paper was partly performed in the context of the ITEA2 project 11004 MODRIO. The authors gratefully acknowledge the support of VLAIO, The Flemish agency for Innovation & Entrepreneurship.

References

1. Van der Auweraer, H., Anthonis, J., De Bruyne, S., Leuridan, J.: Virtual engineering at work: the challenges for designing mechatronic products. J. Simul. Based Eng.: Eng. Comput. **29**(3), 389–408 (2012)
2. Lourens, E., Reynders, E., De Roeck, G., Degrande, G., Lombaert, G.: An augmented Kalman filter for force identification in structural dynamics. J. Mech. Syst. Signal Process. **27**, 446–460 (2012)
3. Cai, W., Pillay, P., Tang, Z., Omekanda, A.M.: Low-vibration design of switched reluctance motors for automotive applications using modal analysis. IEEE Trans. Ind. Appl. **39**(4), 971–977 (2003)
4. Chauvicourt, F., Faria, C.T., Dziechciarz, A., Martis, C.: Influence of rotor geometry on NVH behavior of synchronous reluctance machine. In: Proceedings of the Tenth International Ecological Vehicles and Renewable Energies (EVER) Conference, Monaco, pp. 1–6 (2015)
5. Ozdoganlar, B., Hansche, B.D., Carne, T.G.: Experimental modal analysis for microelectromechanical systems. J. Exp. Mech. **45**(6), 498–506 (2005)
6. dos Santos, F.L.M., Anthonis, J., Naclerio, F., Gyselinck, J.J.C., Van der Auweraer, H., Goes, L.: Multiphysics NVH modeling: simulation of a switched reluctance motor for an electric vehicle. IEEE Trans. Ind. Electron. **61**(1), 469–476 (2014)
7. Samin, J.-C., Bruls, O., Collard, J.-F., Sass, L., Fisette, P.: Multiphysics modeling and optimization of mechatronic multibody systems. J. Multibody Syst. Dyn. **18**(3), 345–373 (2007)
8. dos Santos, F.L.M., Pastorino, R., Peeters, B., Faria, C.T., Desmet, W., Góes, L.C.S., Van Der Auweraer, H.: Model based system testing: bringing testing and simulation close together. In: Damage Detection and Mechatronics: In Structural Health Monitoring, vol. 7, pp. 91–97 (2016)

9. Badawy, A., Zuraski, J., Bolourchi, F., Chandy, A.: Modeling and analysis of an electric power steering system. In: Proceedings of the Symposium on Steering and Suspension Technology, No. 1991-01-0399 (1999)
10. Zaremba, A.T., Liubakka, M.K., Stuntz, R.M.: Control and steering feel issues in the design of an electric power steering systems. In: Proceedings of the American Control Conference, vol. 1, pp. 36–40 (1998)
11. Li, X., Zhao, X.-P., Chen, J.: Controller design for electric power steering system using T-S fuzzy model approach. Int. J. Autom. Comput. **6**(2), 198–203 (2009)
12. Data, S., Pesce, M., Reccia, L.: Identification of steering system parameters by experimental measurements processing. J. Automob. Eng. **218**, 783–792 (2004)
13. Van Beek, T.J., Erden, M.S., Tomiyama, T.: Modular design of mechatronic systems with function modeling. Mechatronics **20**(8), 850–863 (2010)
14. Simcenter Amesim user manual, contents of the mechanical library. Simcenter Amesim, Siemens Software NV (2017)

Experimental Assessments of the Added Mass of Flexible Cylinders in Water: The Role of Modal Shape Representation

Rafael Salles and Celso Pupo Pesce

Abstract A flexible vertical cylinder model, fixed at both ends, is tested experimentally immersed in water and then in air. Galerkin's decomposition is applied to obtain a Reduced Order Model (ROM) from a continuum one. Two closed-form trial modal shapes are chosen for the modal decomposition process. Then, modal added mass is assessed using classical Fourier and Hilbert transform (HT) signal analyses, comparing the model *eigenvalues* with the frequency evaluated from the experimental signals. The choice of modal shape is shown to alter significantly added mass experimental assessment. Similarity to classic results with rigid cylinders is achieved by taking a sufficiently proper modal representation. Moreover, the first mode added mass coefficient attains the same value of that previously determined for a cantilevered flexible circular cylinder, by Pesce and Fujarra in (Pesce and Fujarra, Int J Offshore Polar Eng 10:26–33, 2000) [1].

Keywords Modal decomposition · Added mass · Flow induced vibration Flexible cylinder · Hilbert transform

1 Introduction

Oil and gas exploitation has been a major world economic activity, over the years. In the offshore activities scenario, risers—long tubular structures connecting the floating unity to the sea bed—play important roles in drilling, prospection and transport of those commodities. Flexible vertical risers are commonly used in offshore operations and they are subjected to hydrodynamic loads due to current—what causes VIV (Vortex Induced Vibrations), as well as parametric and internal resonance caused by movements imposed at the top by the floating unit vessel; see Fig. 1a. The dynamics of such structures is usually assessed through nonlinear

R. Salles (✉) · C. P. Pesce
Offshore Mechanics Laboratory, Escola Politécnica, University of São Paulo, São Paulo, Brazil
e-mail: rafael.salles@usp.br

A. de T. Fleury et al. (eds.), *Proceedings of DINAME 2017*, Lecture Notes in Mechanical Engineering, https://doi.org/10.1007/978-3-319-91217-2_15

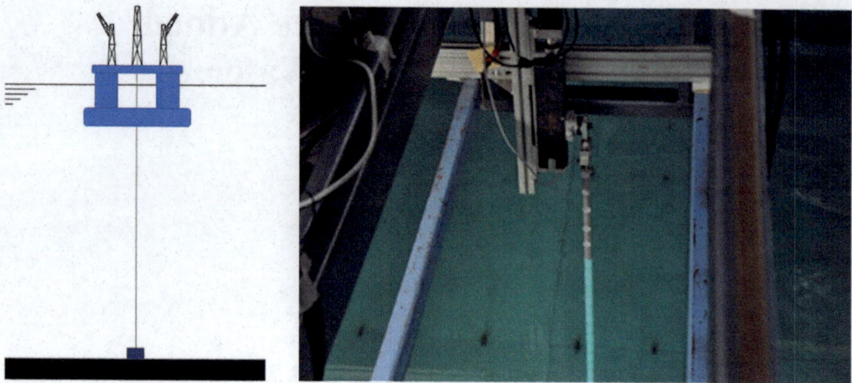

Fig. 1 Vertical riser schematic representation and vertical flexible cylinder scaled model. Left: Vertical riser schematic installation in a semi-submersible drilling platform. Right: Experimental model of flexible vertical cylinder in a towing tank. Extracted from Salles [2]

analytical and numerical models, which are commonly validated with laboratorial experiments, carried out on small scale physical models.

On this road, by using an optical tracking system (Qualisys®), composed by underwater and aerial cameras, experimental tests conducted with a flexible vertical cylinder in water were carried out; Pesce [3]. Through a Galerkin's modal decomposition scheme, assuming a simple sinusoidal modal shape representation, parametric resonances excited by the periodic variation of tension were addressed; see Franzini et al. [4]. Such a simple choice regarding modal shapes was also followed in Fu et al. [5] and in Franzini et al. [6, 7], analyzing the behavior of a long flexible cylinder subjected to hydrodynamic loads. Thorsen et al. [8] proposed a semi-empirical numerical VIV model, verifying their calibrated coefficients by comparing the numerical results with the experimental data obtained in Fu et al. [5]. However, neither Franzini et al. [4], Fu et al. [5] or Thorsen et al. [8] assessed the modal added mass parameter, restricting themselves to adopting the potential flow asymptotic limit, $\widehat{C}_a = 1$, for all modes. As a matter of fact, a first assessment on the influence of drag and added mass coefficients on the response of a parametrically excited vertical flexible cylinder was made by Franzini et al. [9], through a parametric study with a nonlinear reduced order model. Nonetheless, fundamental studies on modal added mass of flexible cylinders are not commonly found in the technical literature. A single experimental assessment is reported in Pesce and Fujarra [1], concerning the first vibration mode of a cantilevered flexible cylinder, where $C_{a,1} = 1.17$ is obtained. This is not the case for rigid cylinders. Early in the '70s, Sarpkaya [10] provided an extensive experimental study on rigid cylinders subjected to oscillatory flows, aiming at a better evaluation of hydrodynamics coefficients, as drag and inertia parameters, for a large range of Reynolds (*Re*) and Keulegan-Carpenter (*KC*) numbers. Sarpkaya [10] showed experimentally that, for low *KC* values, the inertia coefficient tends to the expected asymptotic value

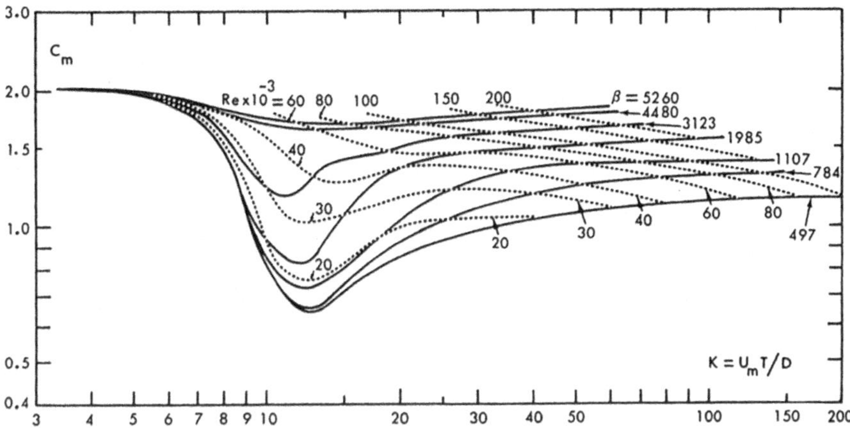

Fig. 2 Inertia coefficients, $C_m = C_a + 1$, for a rigid circular cylinder as function of Keulegan-Carpenter number, K. Extracted from Sarpkaya [10]

$C_m = 1 + C_a \approx 2$, in which $\widehat{C}_a \approx 1$ is the classic potential flow added mass coefficient of a rigid cylinder; see Fig. 2.

Recently, Salles [2] assessed the modal added mass coefficient of a vertical flexible cylinder using the same experimental data base used in Franzini et al. [4]. Salles used sinusoidal modes and obtained values of added mass coefficients larger than $C_a = 1$.

The present study aims at identifying the role modal shape representation could play on the assessment of the corresponding added mass coefficient. Taking sinusoidal or 'quasi-Bessel' modes in Galerkin's projection, modal added mass coefficients are evaluated and compared. The results are interpreted on the light of Sarpkaya [10] rigid cylinders experimental results, contributing to the discussion on the ad hoc assumption, $C_a = 1$, usually taken by many authors when dealing with vibrations of flexible cylinders in water.

1.1 *Sarpkaya's (1977) Experiments on Added Mass of a Rigid Cylinder Under Oscillatory Flow*

One of the main points of riser dynamics modeling is the correct evaluation of hydrodynamic forces. In particular, inertial and viscous forces. The well-known concepts of 'added mass' and 'damping' play then roles of paramount importance.

The classic work by Sarpkaya [10], with a rigid cylinder under oscillatory flow, brings quite comprehensive results about the inertia coefficient, $C_m = 1 + C_a$—where $C_a = m_s/m_d$, being m_a the added mass and m_d the displaced mass, both per unit length—, related to two important parameters: the Keulegan–Carpenter number $KC = K = UT/D \approx 2\pi A/D$ and the Reynolds number, $Re = UD/\nu$ or the

β-parameter, $\beta = Re/K$. For the non specialized reader, one of the seminal results from Sarpkaya's extensive experimental study is summarized in Fig. 2, where the small amplitude asymptotic limit for the inertia coefficient is recovered as $C_m \rightarrow 2^+$. For flexible circular cylinders, however, no extensive and comprehensive parallel study may be found in the technical literature, at least to the authors' knowledge.

1.2 The Vertical Flexible Cylinder Experimental Set Up

In the core of an extensive research project on nonlinear dynamics of risers, a small scale model was designed and built as a flexible cylinder, under proper similarity laws. The Froude scale was used as the leading similarity law, as offshore structures are subjected to loads arising from the floating unit vessel motions, mainly caused by the action of free surface waves.

Pereira et al. [11] presented a study of how the similitude parameters were chosen, leading to the construction of a small-scale model made from a silicon tube filled with stainless steel micro-spheres; see Fig. 1b. In the small scaling methodology, within a large set of dimensionless parameters, geometric rigidity (tensioning), axial stiffness, bending stiffness, immersed weight and added mass play dominant roles. See also Pereira [12] and Salles [2].

An optical tracking system composed by aerial and submerged cameras was used to measure cartesian coordinates of reflective targets placed all along the model length. This measurement technique is noninvasive, cleaner and easier to be implemented, compared to traditional ones that use strain gages and/or accelerometers; see, e.g., Pesce and Fujarra [1]; Morooka and Tsukada [13]. The small model *elastica* may then be reconstructed at any instant of time from the tracked targets. A load cell was installed at the top of the model to register tension. The bottom and top extremities are fixed in the supporting structure.

The experimental set up was designed for a comprehensive series of tests, carried out at IPT towing tank; see Fig. 3a. Such tests involved three kinds of excitation loadings Pesce [3]: (i) sinusoidal vertical displacement imposed at the top; (ii) relative constant current profile, by the towing carriage; (iii) combining (i) and (ii). For an extensive report on the main experimental results, see Franzini et al. [4], Franzini et al. [6, 7], Pereira et al. [14]. All those experiments were preceded by decaying tests in water, in order to assess natural frequencies, damping and, focus of the present analysis, added mass coefficients.

For such an assessment, another experimental campaign was performed outside the towing tank. The experimental tests in air, see Fig. 3b, were carried out for structural characterization, disregarding the effects of added mass and drag included in the experimental campaign in water. By comparing decaying tests in air and in water, added mass and hydrodynamic damping can be assessed, as shown below, essentially following a methodology used in Pesce and Fujarra [1].

The experiments in air were done with the same small scale riser model used in water. A vertical configuration in air was established, making sure that a first modal

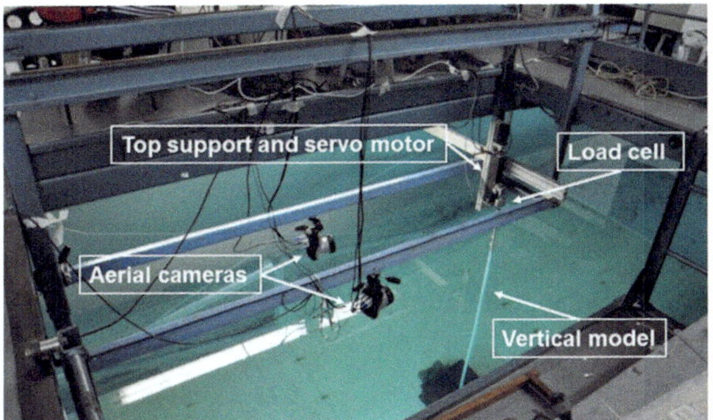

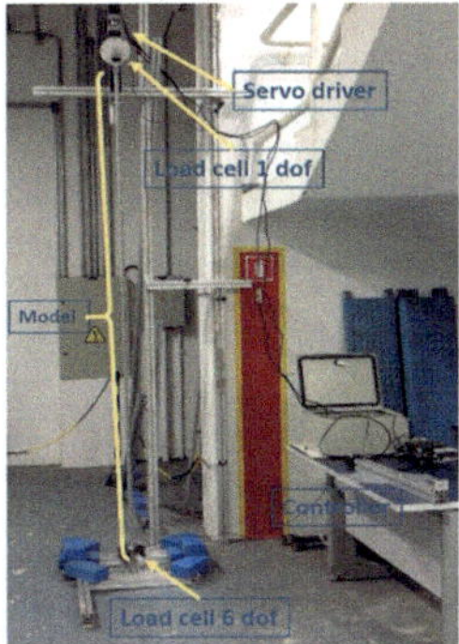

Fig. 3 Vertical flexible cylinder scaled model: experimental campaigns in water and in air. Left: Experimental set up of the flexible vertical cylinder in water. Extracted from Salles [2]. Right: Experimental set up of the flexible cylinder in air. Extracted from Salles [2]

Table 1 Model physical properties for both free-decay experimental tests in air and water, as built

Properties		Value	
		Air	Water
Unstretched length, L_0	(m)	2.613	2.552
Stretched length, L	(m)	2.665	2.602
Immersed length, L_i	(m)	–	2.257
Diameter, D	(mm)	22.2	
Linear mass, m_s	(kg/m)	1.14	1.19
Linear weight, γ	(N/m)	11.13	–
Immersed linear weight, γ_i	(N/m)	–	7.87
Static tension at the top, T_t	(N)	47.56	40
Axial stiffness, EA	(N)	1200	
Bending stiffness, EI	(Nm^2)	0.056	

rigidity estimation—using a vertical bi-supported heavy ideal string together with trigonometric *Eigenfunctions*, see Salles [2]—is dynamically equivalent to the value measured in water, where buoyance forces take place. Table 1 presents the physical properties of the aforementioned experimental models, as built.

1.3 Modal Decomposition: Galerkin's Method

Considering an Euler-Beam model as in, Eq. (1), it is possible to use standard Galerkin's decomposition techniques in order to represent the structure dynamics as a sum of a finite mode numbers. The called modes, or *Eigenfunctions*—ψ_k, are smooth functions that do not violate any kinematic system constraint. Notice that hydrodynamic drag forces are essentially nonlinear, though.

$$m_s\frac{\partial^2 u}{\partial t^2}+EI\frac{\partial^4 u}{\partial z^4}+c_s\frac{\partial u}{\partial t}=\frac{\partial}{\partial z}\left(T(z,t)\frac{\partial u}{\partial z}\right)-m_a\frac{\partial^2 u}{\partial t^2}-\frac{1}{2}\rho_w DC_D\left\|\frac{\partial u}{\partial t}\right\|\frac{\partial u}{\partial t} \tag{1}$$

For the ideal vertical heavy string, the kinematic system constraints are simply written $u(0)=u(L)=\mathbf{0}$, since bending effects are non existent. In such idealized case, where transversal rigidity is due only to tension, Bessel functions of first and second kinds of zeroth order are shown to form the *eigenfunctions* set; see Pesce et al. [15]. Usually, for long beams, geometrical rigidity dominates bending effects, which are relevant just in the neighborhood of the constraints or for higher vibration modes.

A wise integral averaging technique applied by Mazzilli et al. [16] takes extensibility and bending effects into account through the definition of an additional equivalent tension. Such a technique gives rise to analytical representations for the *eigenmodes* which are called by the authors '*quasi*-Bessel' modes. Those *eigenmodes* are used in the present paper. They resemble a closed form WKB solution given in Pesce et al. [15] where extensibility and bending stiffness were disregarded for a catenary shape.

Sinusoidal functions can be used as the simplest 'trial functions' instead, what is also done in the present paper, as a first approximation. See also Franzini et al. [4], Franzini et al. [9] or Salles [2] for additional considerations.

Equation (1) is rewritten as Eq. (2), in which an alternative form for the added mass coefficient, $a=m_a/m_s$, is introduced. The non-linear model presented in Eqs. (1)–(2) considers the effect of flexural rigidity (*EI*), structural damping (assumed to be viscous linear, c_s), the variation of geometric stiffness due to the traction along the model spanwise ($T(z,t)=T(L)-\gamma(L-z)$), added inertia (m_a) and drag force (considered to be quadratic in the relative velocity with respect to the flow, as in Morison's formula, being C_D the drag coefficient).

$$m_s(1+a)\frac{\partial^2 u}{\partial t^2}+EI\frac{\partial^4 u}{\partial z^4}+c_s\frac{\partial u}{\partial t}+\frac{1}{2}\rho_w DC_D\left\|\frac{\partial u}{\partial t}\right\|\frac{\partial u}{\partial t}-\frac{\partial T}{\partial z}\frac{\partial u}{\partial z}-T\frac{\partial^2 u}{\partial z^2}=0 \tag{2}$$

The next step is to obtain a Reduced-Order Model (ROM) from Eq. (2), using Galerkin's method, through a classic separation of variable procedure,

$$u(z,t)\approx A^k(t)\psi_k(z) \tag{3}$$

where summation is implied. Proceeding with Galerkin's projection, the dynamic equations can be written:

$$M_{kj}\ddot{A}^k+C^s_{kj}\dot{A}^k+C^h_{kj}\left\|\dot{A}^k\right\|\dot{A}^k+\eta_{kj}A^k=0 \tag{4}$$

in which

$$M_{kj} = m_s(1 + a_k) \int_0^L \psi_k \psi_j \, dz$$

$$C^s_{kj} = c_s \int_0^L \psi_k \psi_j \, dz$$

$$C^h_{kj} = \frac{1}{2} \rho_w D C_{Ds} \int_0^L \psi_k \|\psi_k\| \psi_j dz$$

$$\eta_{kj} = - \int_0^L \left(T' \psi'_k \psi_j + T \psi''_k \psi_j - EI \psi^{IV}_k \psi_j \right) dz.$$

Assuming orthogonality conditions to hold,[1] non-damped and damped natural frequencies would then be given by,

$$\omega_{n,k} = 2\pi f_{n,k} = \sqrt{\frac{\eta_{kk}}{M_{kk}}} \tag{5}$$

and

$$\omega_{d,k} = 2\pi f_{d,k} = \omega_{n,k} \sqrt{1 - \zeta^2_{kk}} = \frac{C^s_{kk}}{2 M_{kk} \omega_{n,k}}. \tag{6}$$

1.4 Modal Basis Representation

As aforementioned, the main criterion to define a function as an approximation for an *eigenfunction* is to respect all kinematic constraints. For the flexible cylinder, the system constraints are simply $u(0) = u(L) = \mathbf{0}$, which could be written as $\psi_k(0) = \psi_k(L) = 0$ for every $k \in \mathbb{N}^\star$. Mainly, the present work aims to assess how the choice of the projection basis would affect the assessment of the added mass coefficient.

The first basis considered is the exact solution of the classical tensioned string problem, in the absence of a gravitational force field. Known as the Pythagorean harmonics of a string, the *eigenfunctions* are trigonometric as defined in Eq. (7). Hereinafter, the trigonometric modes will form a basis called *sinusoidal* in the following figures and tables.

$$\psi_k(z) = \sin\left(k\pi \frac{z}{L}\right) \tag{7}$$

[1] *Quasi*-Bessel, or Bessel-like, modes are non-orthogonal to each other. An orthogonalization procedure would then be needed.

The second basis is formed by *quasi*-Bessel modes presented in Mazzilli et al. [16], determined from a Timoshenko's beam with non-linearities. After an integral averaging process, extensibility and flexural stiffness effects are transformed into an equivalent traction, thus reducing the order of the partial differential equation in space, from fourth to second. Such a procedure arrives at an equivalent vertical heavy string problem, which has Bessel's functions as *eigenfunctions*. By means of an asymptotic approximation, Mazzilli et al. reached a closed-form solution, Eq. (8),

$$\psi_k(z,\xi_k) = \frac{1}{\sqrt[4]{1+\underline{\alpha}z}}\sin\left[\underline{\beta}\left(\sqrt{1+\underline{\alpha}z}-1\right)\right] \tag{8}$$

in which:

$$\underline{\alpha} = \frac{\gamma}{N_{bk}}$$

$$N_{bk} = N_{bt(0)} + \left(\frac{k\pi}{L}\right)^2 EI\left(1+\frac{3}{16}\xi_k^2\right)$$

$$N_{b(0)} = \frac{EA(L-L_0)}{L_0}$$

$$\underline{\beta} = \frac{k\pi}{\sqrt{1+\underline{\alpha}L}-1}.$$

It should be noted that Eq. (8) has the same mathematical structure of a WKB closed form solution obtained by Pesce et al. [15] in the case of a catenary-like heavy string.

Henceforward, the *quasi*-Bessel modes will be also called Bessel-like modes. In the present work, linear Bessel-like modes, i.e., $\xi_k = 0$, were used for the analysis, as the vibration amplitudes, ξ_k, are very small and would not affect the modal representation. Figure 4 shows the first three modes for the Sinusoidal and Bessel-like basis normalized to have its maximum equal to 1. The solid lines represent the Bessel-like modes and the dashed lines the sinusoidal ones. The Bessel-like modes are more representative of the flexible vertical heavy string. In fact, their maxima occur inside the half lower part, as it would be expected from the analytical solution of the ideal bi-articulated vertical heavy string, the classic Bessel *eigenfunctions*.

Hereafter, only the first mode will be addressed, so that an orthogonalization procedure for the Bessel-like modes may be abandoned in the Galerkin's projection.

2 Analysis

The modal decomposition is performed using both trigonometric and *quasi*-Bessel fundamental modes. The first (fundamental) mode amplitude series for both decompositions are presented in Figs. 5, 6, 7 and 8. Notice that time scales are quite different, as decaying in air is much slower than in water.

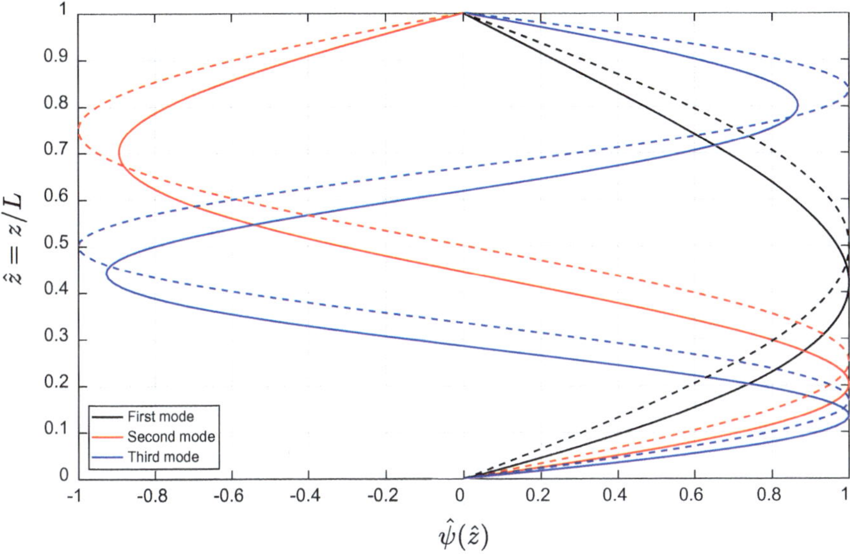

Fig. 4 Graphical representation of the first three sinusoidal (dashed lines) and orthogonalized *quasi*-Bessel model (solid lines)

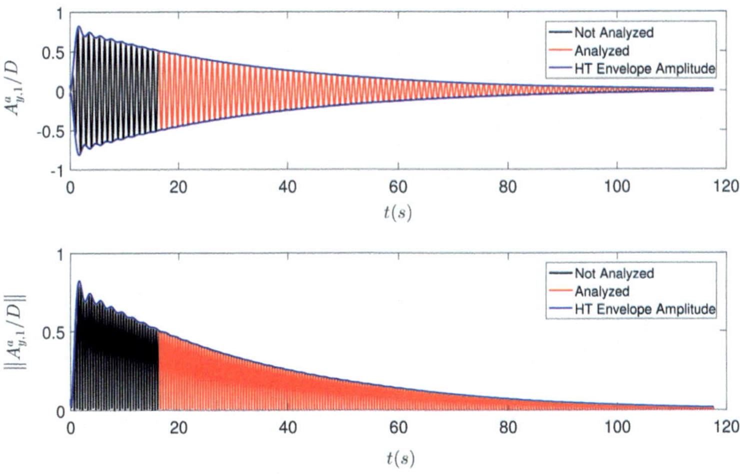

Fig. 5 Fundamental modal amplitude from sinusoidal decomposition: free-decay in air

It is interesting to give the reader some details on the methodology behind the modal analysis adopted. Figure 5 shows the full free-decay amplitude time series upon which the Hilbert Transform (HT) was used in order to determine the signal envelope amplitude. The HT envelope amplitude is then applied to determine the linear viscous equivalent structural damping and to study the instant damped

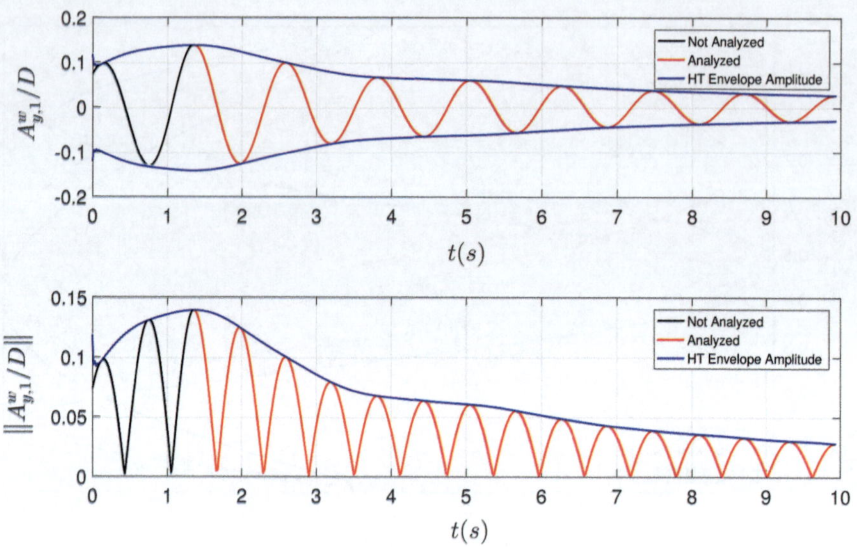

Fig. 6 Fundamental modal amplitude from sinusoidal decomposition: free-decay in water

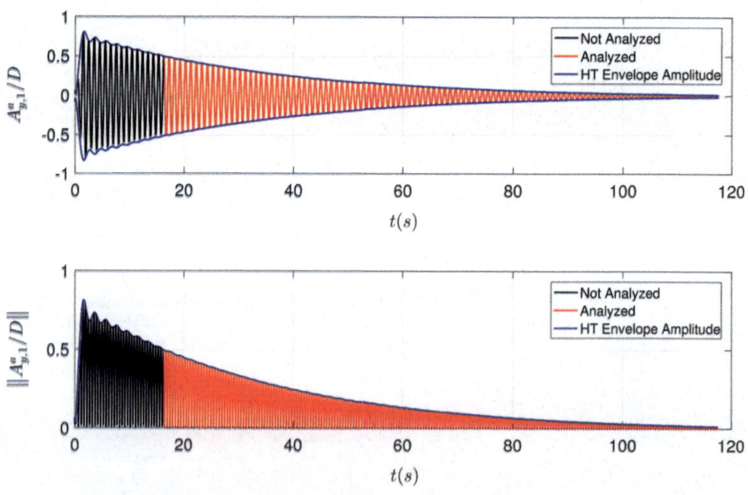

Fig. 7 Fundamental modal amplitude from Bessel-like decomposition: free-decay in air

frequency. Figure 5 also shows that there is a time interval chosen for the analysis, marked in red color.

Besides the HT procedure, a standard Fourier (FFT) analysis was carried out, in order to directly assess the damped natural frequency, assuming its invariance with respect to the vibration amplitude.

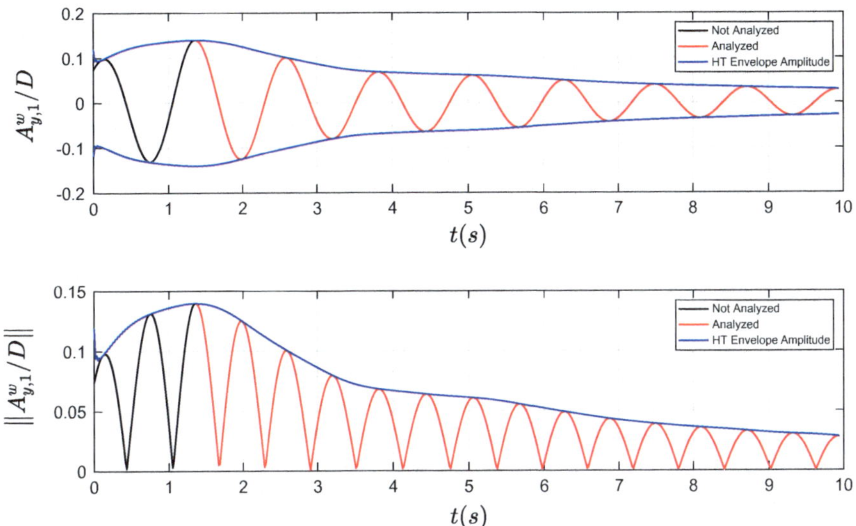

Fig. 8 Fundamental modal amplitude from Bessel-like decomposition: free-decay in water

Table 2 Linear viscous damping for the first modal temporal free-decay series

	$\zeta_1(\%)$	
	Sinusoidal	Bessel-like
Air	0.49	0.49
Water	4.17	4.00

2.1 *Modal Structural Damping*

From the HT envelopes in air, a simple exponential fitting is used to assess the linear structural damping coefficient, as presented in Table 2.

On the other hand, besides structural damping, the free decay in water is characterized by a nonlinear hydrodynamic dissipation, usually modeled quadratically by Morison's drag formula, see Eq. (2). It would then be much more complicated to isolate both dissipation mechanisms. For the sake of simplicity, an equivalent linear damping coefficient for the free-decay in water is considered, as shown in Table 2, after an exponential fitting on the signal amplitude envelope is done.

2.2 *Instant Frequency*

The phase signal obtained from the Hilbert transform is used to determine the instant damped frequency of the free-decay experimental tests via numerical time differentiation. Figures 9 and 10 present the instant frequency determined for each

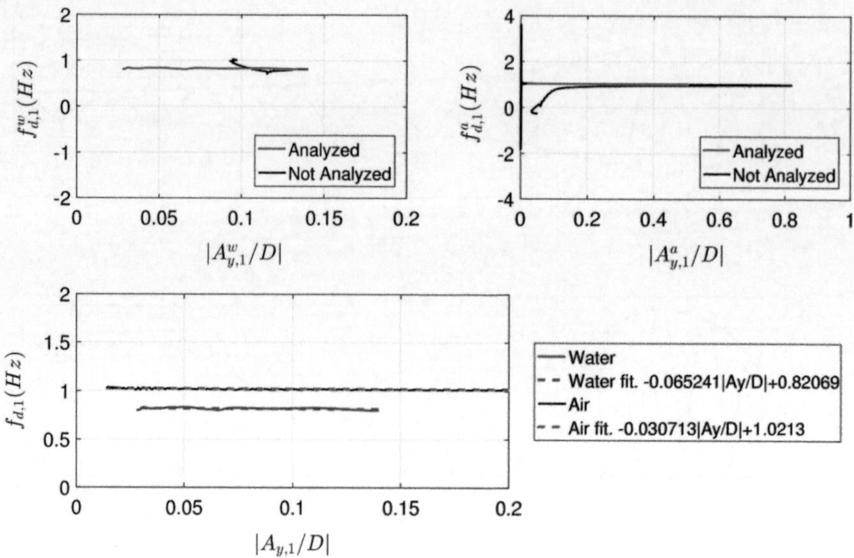

Fig. 9 Flexible vertical cylinder first mode instant damped frequency from free-decay tests—sinusoidal basis. Above: in water (left), in air (right); below: in air and water with respective fitting

free-decay test for both decompositions, sinusoidal and Bessel-like. The instant frequency is plotted as a function of the HT envelope normalized with respect to the diameter, in order to address its dependence on the vibration amplitude.

Figures 9 and 10 also provide a linear fitting for the instant damped frequency calculated within the analysed vibration amplitude range. As it can be seen, the instant damped frequencies do not vary significantly with the vibration amplitude, being almost constant, either in air or water, irrespective the projection basis chosen.

Regarding both modal decompositions—Sinusoidal and Bessel-like—for the first mode amplitude, Table 3 shows the fundamental natural damped frequency measured using a standard Fourier analysis (FFT) and the results obtained with the HT instant frequency linear fittings presented in Figs. 9 and 10. Deviations between FFT and HT frequencies are also given, showing that both techniques meet results closely.

2.3 Added Mass Assessment

For the first mode, using modal mass and modal rigidity terms given in Eq. (4) and presented in Table 4, non damped frequencies, $\hat{f}_{n,1}$ as function of the first mode added mass coefficient, a_1, were calculated from Eqs. (5)–(6) and are given in Table 5 for both conditions in water and in air. The damped natural frequencies,

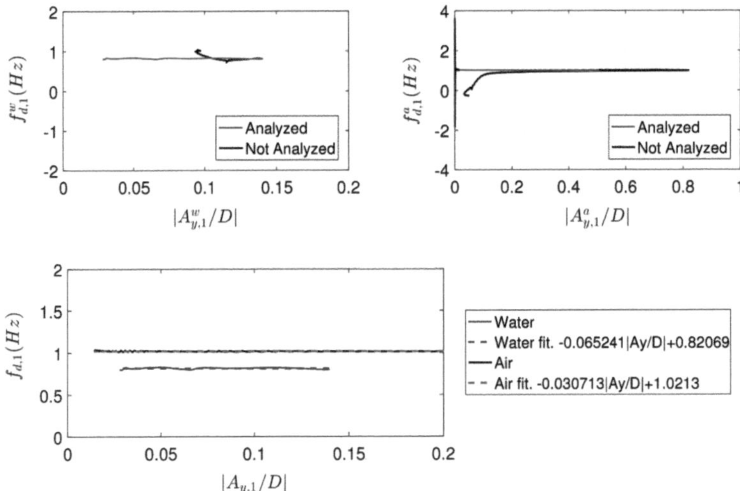

Fig. 10 Flexible vertical cylinder first mode instant damped frequency from free-decay tests—Bessel-like basis. Above: in water (left), in air (right); below: in air and water with respective fitting

Table 3 First model natural damped frequencies, $f_{d,1}$ (Hz), measured with FFT and HT for both modal representations

	Sinusoidal			Bessel-like		
	FFT	HT	(FFT-HT)/FFT (%)	FFT	HT	(FFT-HT)/FFT (%)
Air	1.014	1.0213	–0.72	1.014	1.0213	–0.72
Water	0.8211	0.8207	0.05	0.8211	0.8207	0.05

$f_{d,1}$, experimentally obtained from the FFT analysis, already presented in Table 3, are reproduced in Table 5 for reference sake.

Assume that the calculated natural damped frequency, $\hat{f}_{n,1}\sqrt{1-\zeta_1^2}$ (using the modal damping coefficient presented in Table 2 and the natural frequencies in Table 5), is a good estimate for the measured damped frequency, $f_{d,1}$. Then, by taking the ratio between the damped frequency, $f_{d,1}$, of both free-decay tests in air and in water, and recalling that the added mass in air is practically null, see Eq. (9),

$$\frac{f^a_{d,1}}{f^w_{n,1}} \approx \frac{\hat{f}^a_{n,1}}{\hat{f}^w_{n,1}}\sqrt{1+a_1}, \quad \frac{\sqrt{1-\left(\zeta^a_1\right)^2}}{\sqrt{1-\left(\zeta^w_1\right)^2}} \approx 1, \tag{9}$$

it is possible to assess the modal added mass coefficient in water, as given in Eq. (10),

Table 4 Modal mass and modal rigidity. First mode only. Sinusoidal and Bessel-like projections

	Sinusoidal		Bessel-like	
	M_1 (kg)	η_1 (N/m)	M_1 (kg)	η_1 (N/m)
Air	1.5698	57.6515	1.2142	54.7226
Water	1.5482	56.8190	1.3083	52.8053

Table 5 Damped natural frequencies from decay tests, using standard Fourier analysis (FFT), and non damped ones, having mass coefficient as parameter

	$f_{d,1}$ (Hz)[a]	$\hat{f}_{n,1}\sqrt{1+a_1}$ (Hz)[b]	
		Sinusoidal	Bessel-like
Air	1.0140	0.9645	1.0685
Water	0.8211	0.9610	1.0112

[a]Measured (FFT)
[b]Calculated

$$a_1 = \left(\frac{\hat{f}^{w}_{n,1} f^{a}_{d,1}}{\hat{f}^{a}_{n,1} f^{w}_{d,1}}\right)^2 - 1. \tag{10}$$

Defining a modal reduced mass parameter as the quotient between the modal structural mass and the modal displaced water mass,

$$a_1 = \frac{M_{a,1}}{M_{s,1}} = \frac{M_{a,1}}{M_{d.1}}\frac{M_{d,1}}{M_{s.1}} = \frac{C_{a,1}}{m_1^\star} \tag{11a}$$

where

$$m_1^\star = \frac{M_{s,1}}{M_{d.1}} = \frac{4m_s\int_0^L \psi_1^2(z)\,dz}{\pi\rho_w D^2\int_0^{L_i}\psi_1^2(z)\,dz}, \tag{11b}$$

the added mass coefficient related to the modal displaced mass, $C_{a,1}$, can be evaluated from the modal added mass coefficient related to the modal structural mass, a_1, in the form

$$C_{a,1} = m_1^\star a_1. \tag{12}$$

Using the modal integrals given in Table 6, two modal reduced mass parameters related to the model immersed length may be defined, as follows:

1. $m^\star_{I,1}$: model structure (as built) with length L and immersed length L_i;
2. $m^\star_{II,1}$: model structure hypothetically completely immersed, $L = L_i$.

The hypothetical case in which the structure would be completely immersed in the water, $m^\star_{II,1}$, is proposed in order to enforce some sort of similarity with the rigid circular cylinder studied in Sarpkaya [10]. The fact the model is or not completely

Table 6 Free-decay in water: the first mode integral mass term

	Based on	
	Total length (L)	Immersed length (L_i)
	$\int_0^L \psi_1^2(z)\,dz\,(m)$	$\int_0^{L_i} \psi_1^2(z)\,dz\,(m)$
Sinusoidal	1.3010	1.1285
Bessel-like	1.0994	1.0883

Table 7 First mode added-mass coefficients—a_1 and $C_{a,1}$—and corresponding mass-ratios. $a_1 = m_{a,1}/m_{s,1}$ determined from Eq. (10), with damped natural frequencies obtained from Fourier analysis

		$L_i < L$		$L_i = L$	
	a_1	$m_I^\star$	$C_{a,1}^I$	$m_{II}^\star$	$C_{a,1}^{II}$
Sinusoidal	0.514	3.455	1.176	2.997	1.541
Bessel-like	0.387	3.028	1.172	2.997	1.160

immersed alters the modal reduced mass, inasmuch as the quantity of modal displaced mass also changes.

Table 7 shows the values of the modal reduced mass for both cases defined before and both modal shapes considered. Along with the modal reduced mass, the modal added mass related to the displaced mass is also calculated. The Bessel-like modal shape is physically closer to the actual modal shape and, as Table 7 shows, choosing the modal representation affects the modal added mass coefficient substantially.

Although the modal added mass result obtained using the Bessel-like mode is around 17% larger than the 'potential-flow' one, $\widehat{C}_{a,1} \approx 1$, the same asymptotic behavior as in the rigid cylinder case studied by Sarpkaya [10] is observed, i.e., $C_a \to 1^+$ for low KC, in the present experiment of order 1; see Fig. 2. Notice the logarithm scale used. It should be also noted that, by using essentially the same methodology, with a proper *eigenmode* for a cantilevered flexible circular cylinder, Pesce and Fujarra [1] had experimentally obtained the first modal added mass coefficient $C_{a,1} = 1.17$, in a remarkable agreement with the present assessment.

Figures 11 and 12 present the modal added mass parameters, a_1 and $C_{a,1}$, as function of the HT envelope amplitude obtained using the instant damped frequency linear fittings determined in Figs. 9 and 10. The values found using the HT methodology at zero amplitude meet the outcomes presented in Table 7 with a standard Fourier analysis. Notice that the quadratic form for a_1, as per Eq. (10), increases the differences found in the damped natural frequencies, if HT or FFT analysis are used (compare Figs. 9 and 10 and Table 3).

Finally, Fig. 12 shows the first mode added mass coefficient, determined with respect to the modal displaced mass, $m_{I,1}^\star$, which is based on the as built immersed length. On the other hand, Fig. 13 uses the hypothetical case in which the structure

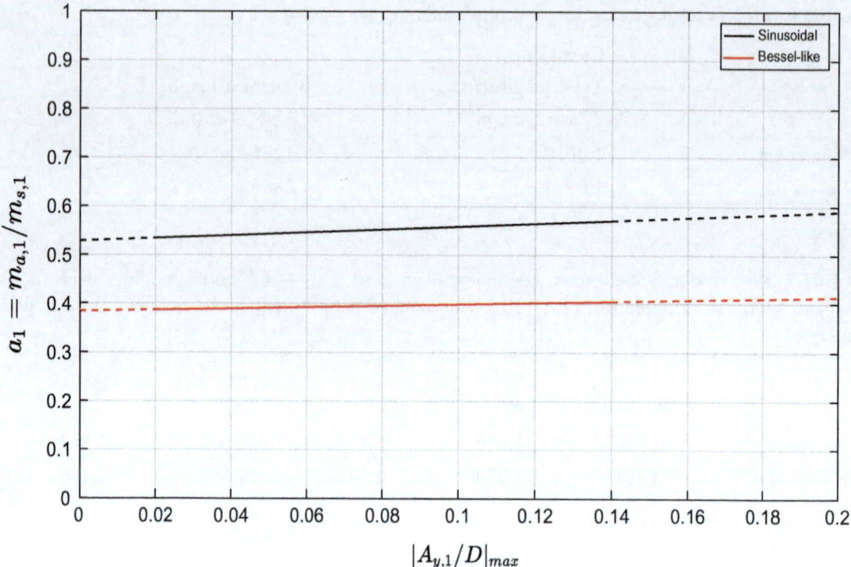

Fig. 11 First mode added mass coefficient, a_1, defined with respect to the structural mass, as function of the HT envelope amplitude for sinusoidal and Bessel-like modal projections

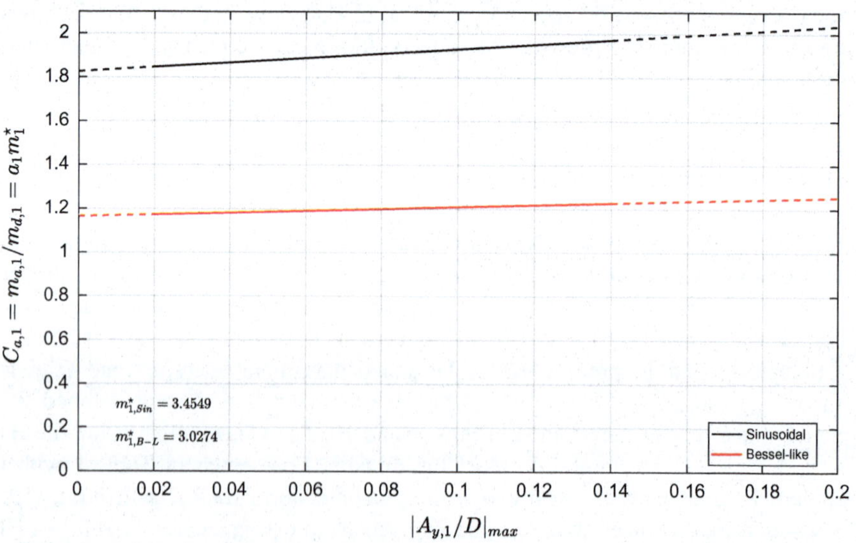

Fig. 12 Added mass coefficient, $C_{a,1}$, defined with respect to the as built displaced mass, for the first mode, as function of the HT envelope amplitude for sinusoidal and Bessel-like modal projections

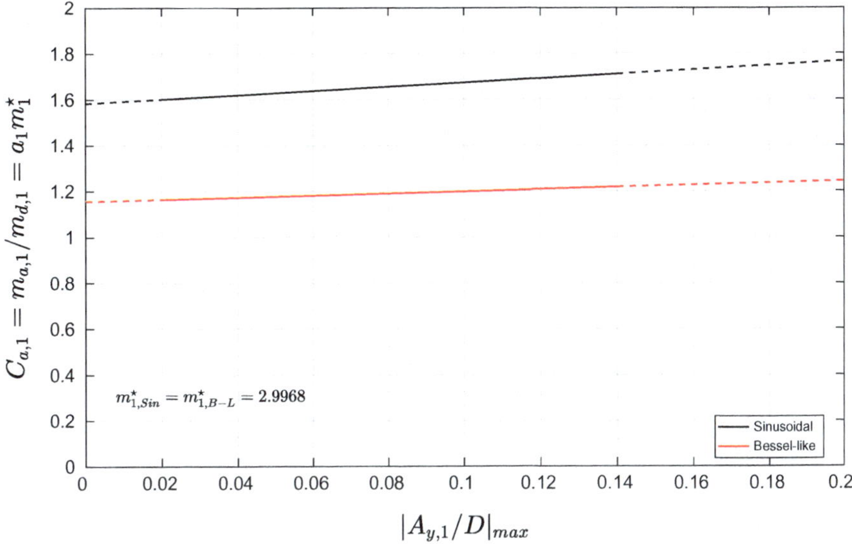

Fig. 13 Added mass coefficient, $C_{a,1}$, defined with respect to an hypothetical full-length displaced mass, for the first mode, as function of the HT envelope amplitude for sinusoidal and Bessel-like modal projections

is considered completely immersed in the water. Comparing Figs. 12 and 13, it can be readily noticed that the way the modal reduced mass is defined does not affect significantly the results. However, the way the decomposition basis is taken, has indeed a significant effect. Moreover, if the Bessel-like projection is taken, the simple and standard Fourier analysis procedure leads to results which are in remarkable agreement with that reported by Pesce and Fujarra [1] for a cantilevered flexible cylinder, $C_{a,1} = 1.17$. As a matter of fact, such a value is less than 20% larger than $C_{a,1} = 1$, the small amplitude added mass value that should be expected from a presumed rigid cylinder potential flow similarity.

3 Conclusions

A modal added mass coefficient assessment using two free-decay experimental tests for the same model, in immersed and non-immersed conditions, was successfully carried out using standard (Fourier) and non-standard (Hilbert Transform) analysis methodologies. Applying usual Galerkin's projection schemes, the first mode added mass coefficient was assessed, by using two distinct trial functions basis: the simplest one—sinusoidal; and a Bessel-like *eigenfunction* set, coming from an accurate asymptotic modelling regarding the vertical tensioned beam dynamic problem. The choice of a sufficiently representative modal shape has shown to be essential, altering significantly the experimental assessment of the first mode added mass coefficient, for a vertical flexible cylinder configuration. Along with the choice of a representation basis, the modal added mass, due to its quadratic dependence, presents a significant numerical sensitivity for small deviations in the frequency values obtained using FFT or HT methodologies.

The present experimental assessment for the first mode added mass of a vertical flexible circular cylinder fixed at both extremities showed a remarkable agreement with that reported by Pesce and Fujarra [1] for a cantilevered one, $C_{a,1} = 1.17$. This value is just 17% larger than the potential flow added mass coefficient, $\widehat{C}_a = 1$, commonly assumed in riser engineering, in an ad hoc manner. It should be also noticed that the well-known experimental chart by Sarpkaya [10], for rigid cylinders in oscillatory flow, shows—in log scale—an asymptotic limit for the inertia coefficient, $C_m = 1 + C_a$, slightly larger than 2.

The next step for the modal added mass coefficient assessment would be expanding the analysis for higher modes, looking for a wider scenario involving the evaluation of added mass parameters for Reduced Order Models. The Hilbert Transform methodology is also an asset of the present work, being possible to use it in other experimental tests to determine the dependency of the system natural frequency on the vibration amplitude. A subsidiary and practical result, emerged during the HT methodology application, is the possibility of an a posteriori assessment of the optical tracking system measurement, being it in air or water, as shown in the Appendix.

Acknowledgements The results presented in the paper were obtained from a complementary analysis of data collected during a comprehensive research project on non-linear dynamics of risers sponsored by Petrobras and carried out in 2011–2013. The Coordination for the Improvement of Higher Education Personnel, CAPES, is acknowledged by the first author for the PhD grant 33002010049-P9. The National Council for Scientific and Technological Development, CNPq, is acknowledged by the second author for the research grant 308990/2014-5. Special thanks to Drs. Guilherme Franzini, Rodolfo Gonçalves and to the IPT towing tank technical staff.

Appendix 1—The HT Methodology as a Tool to Assess Optical Tracking Accuracy

During the experimental set up in air and water, the calibration of the optical tracking system revealed a measurement accuracy about a decimal of millimetre (0.1 mm). On the other hand, by using the HT procedure to evaluate the instant

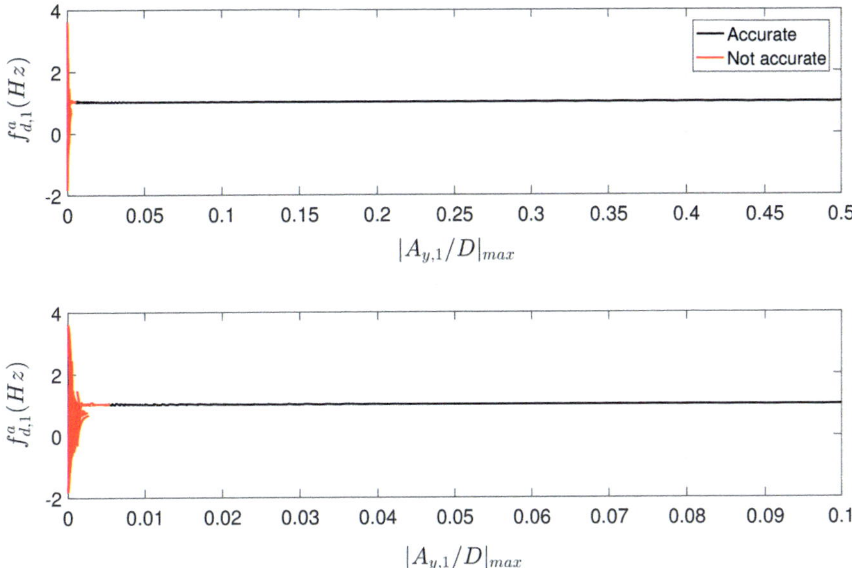

Fig. 14 Instant damped frequency from free-decay in air showing a clear bound for the optical tracking system accuracy

damped frequency as function of vibration amplitude, resolution is clearly obtained, revealing a figure better than $0.005 \times 22.2\,\text{mm} \approx 0.1\,\text{mm}$, as shows Fig. 14, for the decay test in air.

Appendix 2—Nomenclature

Latin Symbols

a: added mass coefficient, $a = m_a/m_s$	A: modal amplitude	C_a: added mass coefficient, $C_a = m_a/m_d$
$\widehat{C}_a$: 'potential flow' added mass coefficient, $\widehat{C}_a \approx 1$	C_D: drag force coefficient	C^h: modal drag force coefficient
C_m: inertial coefficient, $C_m = 1 + C_a$	C^s: modal linear viscous damping coefficient	c_s: linear viscous damping coefficient
D: diameter	EA: axial stiffness	EI: bending stiffness
f_d: damped natural frequency (measured)	$\hat{f}_n$: natural frequency (calculated)	k: mode number
K, KC: Keulegan-Carpenter number	L: stretched length	L_0: unstretched length
L_i: immersed length	L_t: total length	M: modal mass
m_a: added mass per unit length	m_d: displaced mass per unit length	m_s: structural mass per unit length
M_a: modal added mass	M_d: modal displaced mass	M_s: modal structural mass
$m^\star$: reduced mass parameter	$m_1^\star$: first mode reduced mass	N_b: equivalent normal traction
$N_{b(0)}$: traction at the cylinder bottom	Re: Reynolds number	t: time
$T(z, t)$: tension	$u(z, t)$: displacement vector	U: mean velocity
x: cartesian coordinate	y: cartesian coordinate	z: cartesian coordinate

Greek Symbols

$\underline{\alpha}$: *quasi*-Bessel mode parameter	β: Sarpkaya's β-parameter	$\underline{\beta}$: *quasi*-Bessel mode wave number
γ: linear weight	γ_i: immersed linear weight	η: modal rigidity
ζ: linear viscous damping coefficient	ν: kinematic viscosity	ξ: dimensionless modal amplitude
ρ_w: water specific mass	ψ: modal shape	ω: angular frequency

References

1. Pesce, C.P., Fujarra, A.L.C.: Vortex induced vibrations and jump phenomenon: experiments with a clamped flexible cylinder in water. Int. J. Offshore Polar Eng. **10**, 26–33 (2000)
2. Salles, R.: Experimental analysis of fluid-structure interaction phenomena on a vertical flexible cylinder: modal coefficients and parametric resonance. Master's thesis, Escola Politécnica da Universidade de São Paulo, São Paulo, Brazil (2016)
3. Pesce, C.P.: Riser dynamics: experiments with small scale models. In: LabOceano—Ten-Years Anniversary Celebration Workshop, 29–30 April 2013

4. Franzini, G.R., Pesce, C.P., Salles, R., Gonçalves, R.T., Fujarra, A.L.C., Mendes, P.: Experimental analysis of a vertical and flexible cylinder in water: response to top motion excitation and parametric resonance. J. Vib. Acoust. **137** (2015). https://doi.org/10.1115/1.4025759
5. Fu, S., Wang, J., Baarholm, R., Wu, J., Larsen, C.M.: Features of vortex-induced vibration in oscillatory flow. ASME J. Offshore Mech. Arct. Eng. **136**(1), 011801 (2014)
6. Franzini, G.R., Pesce, C.P., Gonçalves, R.T., Fujarra, A.L.C., Mendes, P.: Experimental Investigations on Vortex-Induced Vibrations with a Long Flexible Cylinder. Part I: Modal-Amplitude Analysis with a Vertical Configuration. FIV (2016)
7. Franzini, G.R., Pesce, C.P., Gonçalves, R.T., Fujarra, A.L.C., Mendes, P.: Experimental Investigations on Vortex-Induced Vibrations with a Long Flexible Cylinder. Part II: Effect of Axial Motion Excitation in a Vertical Configuration. FIV (2016)
8. Thorsen, M.J., Saevik, S., Larsen, C.M.: Time domain simulation of vortex-induced vibrations in stationary and oscillating flows. J. Fluids Struct. **61**, 1–19 (2016). https://doi.org/10.1016/j.jfluidstructs.2015.11.006
9. Franzini, G.R., Santos, C.C.P., Pesce, C.P., Mazzilli, C.E.N.: Parametric excitation of an immersed, vertical and slender beam using reduced-order models: influence of hydrodynamic coefficients. Mar. Syst. Ocean Technol. https://doi.org/10.1007/s40868-016-0013-z (2016)
10. Sarpkaya, T.: In-line and transverse forces on cylinders in oscillatory flow at high Reynolds number. J. Ship Res. 200–2016 (1977). https://doi.org/10.1016/0167-2789(83)90298-1
11. Pereira, F.R., Pesce, C.P., Gonçalves, R.T., Franzini, G.R., Fujarra, A.L.C., Salles, R., Mendes, P.: Risers model test: scaling methodology and dynamic similarity. In: 22nd Proceedings on International Society of Offshore and Polar Engineers (ISOPE2012), Greece (2012)
12. Pereira, F.R.: Investigação das Vibrações Induzidas pela Emissão de Vórtices em Modelos Reduzidos de Riser Lançados em Catenária. Ph.D. thesis, Escola Politécnica, Universidade de São Paulo, São Paulo, Brasil (2014)
13. Morooka, C.K., Tsukada, R.I.: Experiments with a steel catenary riser model in a towing tank. Appl. Ocean Res. **43**, 244–255 (2013)
14. Pereira, F.R., Pesce, C.P., Gonçalves, R.T., Fujarra, A.L.C., Franzini, G.R., Mendes, P.: Experimental Investigations on Vortex-Induced Vibrations with a Long Flexible Cylinder. Part III: Modal-Amplitude Analysis with a Catenary Configuration. FIV (2016)
15. Pesce, C.P., Fujarra, A.L.C., Simos, A.N., Tanuri, E.A.: Analytical and closed-form solution for deep water riser-like eigenvalue problem. In: 9th Proceedings on International Ocean and Polar Engineering Conference (ISOPE), France (1999)
16. Mazzilli, C.E.N., Lenci, S., Demeio, L.: Nonlinear free vibrations of tensioned vertical risers. In: 8th Proceedings on European Nonlinear Dynamics Conference (ENOC2014), Austria (2014)

Part III
Robotics and Mechatronic Systems

Dynamic Modeling and Simulation of a Parallel Planar Manipulator with Linear Electric Actuators Using Power Flow Approach

A. N. Albuquerque, M. Speranza Neto and M. A. Meggiolaro

Abstract This work presents the analytical form determination of the dynamic model of a parallel planar mechanism with three degrees of freedom through the characterization of the power flow between its components. From the geometrical relations associated to the displacement of their degrees of freedom, the kinematic relations associated to their speeds are determined. Considering the power flow between the degrees of freedom, and also between these and the actuating elements (linear electric actuators) the equilibrium relations of the forces and torques are obtained. Accounting for inertial effects of system components, the stiffness and damping effects, the equations of motion or the state equations are analytically determined. Besides, the relation between the inverse kinematics and the direct dynamics is presented. The proposed methodology is generalized and applicable in any type of mechanism (open or closed, planar or spatial). Thus, this methodology (power flow) is more efficient to achieve the dynamic analytical (closed) models of parallel mechanisms. Simulations are performed to validate this approach, using the real data (geometry, inertia, damping, actuators forces, etc.) from a planar mechanism designed and built especially for the purpose to compare the simulated and experimental results. The analytical equations lead to a more efficient simulation process and real-time control of these systems.

Keywords Parallel mechanisms · Inverse kinematic · Direct dynamic
Power flow · Causal relations · Bond graphs

A. N. Albuquerque (✉) · M. Speranza Neto · M. A. Meggiolaro
Department of Mechanical Engineering, Pontifical Catholic University,
Rua Marquês de São Vicente, 225, Gávea, Rio de Janeiro, RJ 22241-900, Brazil
e-mail: allan@puc-rio.br

M. Speranza Neto
e-mail: msn@puc-rio.br

M. A. Meggiolaro
e-mail: meggi@puc-rio.br

A. de T. Fleury et al. (eds.), *Proceedings of DINAME 2017*, Lecture Notes in Mechanical Engineering, https://doi.org/10.1007/978-3-319-91217-2_16

1 Introduction

Despite of having a smaller workspace, higher inertia and a harder dynamic analysis, parallel systems have great advantages when compared to serial manipulators, as better stability and accuracy, ability to handle relatively large loads, high velocities and accelerations and low power operation [1]. The improvement in the modeling of parallel mechanisms contributes to solve problems in fields such as vehicle dynamics and robotics. In some robot tasks such as when both feet of an anthropomorphic robot find a restriction (such as the floor, for example), the kinematic chain closes and thus, to estimate the robot's hip movement in order to balance it, multi-branch mechanisms or parallel mechanisms modeling techniques are used [2].

Zhao and Gao [3] used the Bond Graphs Technique to model the kinematics and dynamics of a Stewart platform. A comparison with experimental tests proved the feasibility and efficiency of the model, whose method can be used to model other types of parallel mechanisms. In his work, [4] represented the Stewart platform dynamics using a novel spatial visualization form of the bond graphs. By using the power flow diagram, with the noted causal relationships, creating system equations becomes a relatively straightforward task [5]. Thus, this methodology (bond graphs or power flow) is more efficient to achieve the dynamic analytical (closed) models of parallel mechanisms.

The proposed methodology in this work is generalized and applicable in any type of mechanism (open or closed, planar or spatial). For a better comprehension of the methodology, a planar case will be discussed in this work. The inverse kinematic model of the closed chain mechanism, which has easy solution when compared to the direct model, can be developed by any known methodology, without the need for a systematic approach. It begins by determining the inverse geometric model and its derivation to obtain the kinematic relations, and therefore the inverse Jacobian matrix. With the inverse kinematic model, the inverse kinematics bond graph is built and, from the cause and effect relations, the direct dynamic model of the mechanism is found.

2 Inverse Kinematics of the Planar Platform

Figure 1 shows the 3-R$\underline{\text{P}}$R parallel manipulator considered in this study. Three limbs connects to the mobile platform and the fixed base by rotational joints in points B_i and A_i, i = 1, 2 and 3. To describe its geometry, a referential frame A(X, Y) fixed to the platform base is added and other frame, B(x, y), is coupled to the mobile platform. Another reference frame, C(x_i, y_i), is fixed to each rotational joint, thus having its origin at the point A_i (i = 1, 2 and 3). The y_i axis of this system points from A_i to B_i. For convenience, the origin of the frame B is located at the center of the mobile platform. The position of the mobile platform can be described

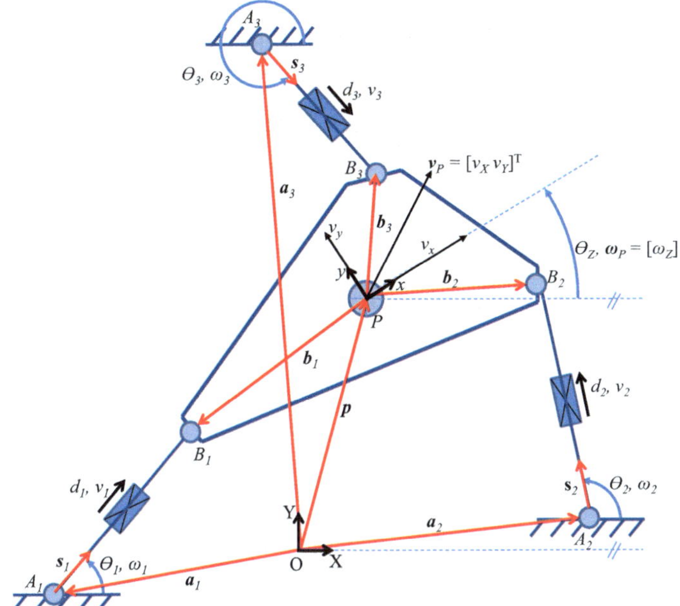

Fig. 1 Planar platform with three degrees of freedom

by the vector $\boldsymbol{p} = [p_X, p_Y]^T = [X, Y]^T$ and by the rotation matrix ${}^{A}\mathbf{R}_{B}$. Hence, the velocities state of the mobile platform is defined as a three dimensional vector with the absolute linear velocity and the angular velocity of the mobile platform (Eq. 1).

$$\dot{\mathbf{x}} = \mathbf{v} = \begin{bmatrix} \mathbf{v}_{\mathrm{P}} \\ \boldsymbol{\omega}_{\mathrm{P}} \end{bmatrix} = \begin{bmatrix} \dot{X} \\ \dot{Y} \\ \theta \end{bmatrix} = \begin{bmatrix} v_X \\ v_Y \\ \omega_z \end{bmatrix} \tag{1}$$

For this manipulator, the input vector is given by $\boldsymbol{v}_A = [v_1, v_2, v_3]^T$ and the output vector can be described by the centroid velocity P and the angular velocity of the mobile platform, $\boldsymbol{v} = [v_x, v_y, \omega_z]^T$. Using the vector loop technique and then, applying the differential with respect to time, the relationship between the variables which describe the angular and linear velocity of the mobile platform and the velocities of the links of the planar platform is found. With this relation, the inverse Jacobian of the manipulator is obtained, as shown in Eq. 2 [6].

$$\dot{\mathbf{q}} = \begin{bmatrix} v_1 \\ v_2 \\ v_3 \end{bmatrix} = \mathbf{J}^{-1}\dot{\mathbf{x}} = \begin{bmatrix} cos\,\theta_1 & sin\,\theta_1 & b_{1X} sin\,\theta_1 - b_{1Y} cos\,\theta_1 \\ cos\,\theta_2 & sin\,\theta_2 & b_{2X} sin\,\theta_2 - b_{2Y} cos\,\theta_2 \\ cos\,\theta_3 & sin\,\theta_3 & b_{3X} sin\,\theta_3 - b_{3Y} cos\,\theta_3 \end{bmatrix} \begin{bmatrix} \dot{X} \\ \dot{Y} \\ \theta \end{bmatrix} \tag{2}$$

in which θ_i are given by Eq. 3 (with i = 1, 2 and 3). Rewriting Eq. 2 in function of $tan(\theta_i)$, differentiating both sides, and manipulating the terms in order to put in evidence the absolute linear velocities and angular velocity of the platform, we obtain the inverse Jacobian that relates these velocities to the angular velocity of each of the members (Eq. 4).

$$\theta_i = tan^{-1}\left(\frac{b_{iY} - a_{iY}}{b_{iX} - a_{iX}}\right) = tan^{-1}\left(\frac{Y + b_{ix} sin\,\theta + b_{iy} cos\,\theta - a_{iY}}{X + b_{ix} cos\,\theta - b_{iy} sin\,\theta - a_{iX}}\right) \tag{3}$$

$$\boldsymbol{\omega}_A = \begin{bmatrix} \omega_1 \\ \omega_2 \\ \omega_3 \end{bmatrix} = \mathbf{J}_\theta^{-1} \begin{bmatrix} v_X \\ v_Y \\ \omega_Z \end{bmatrix} = \begin{bmatrix} \frac{(a_{1Y} - b_{1Y}) cos^2(\theta_1)}{(b_{1X} - a_{1X})^2} & \frac{cos^2(\theta_1)}{(b_{1X} - a_{1X})} & \frac{cos^2(\theta_1)}{(b_{1X} - a_{1X})^2} j_{\theta_1} \\ \frac{(a_{2Y} - b_{2Y}) cos^2(\theta_2)}{(b_{2X} - a_{2X})^2} & \frac{cos^2(\theta_2)}{(b_{2X} - a_{2X})} & \frac{cos^2(\theta_2)}{(b_{2X} - a_{2X})^2} j_{\theta_2} \\ \frac{(a_{3Y} - b_{3Y}) cos^2(\theta_3)}{(b_{3X} - a_{3X})^2} & \frac{cos^2(\theta_3)}{(b_{3X} - a_{3X})} & \frac{cos^2(\theta_3)}{(b_{3X} - a_{3X})^2} j_{\theta_3} \end{bmatrix} \begin{bmatrix} \dot{X} \\ \dot{Y} \\ \theta \end{bmatrix} \tag{4}$$

In Eq. 4, $j_{\theta i}$ are given by Eq. 5, with $i = 1, 2, 3$, $c\theta = cos(\theta)$ and $s\theta = sin(\theta)$. In order to obtain the relation between the linear and angular velocities and accelerations of the moving platform and the linear accelerations of the actuators of the mechanism, the differential of the inverse Jacobian has to be calculated, as shown in Eq. 6. The matrix of the derivatives of the inverse Jacobian is given by Eq. 7.

$$j_{\theta i} = (b_{ix}\, c\,\theta - b_{iy}\, s\,\theta)(b_{iX} - a_{iX}) + (b_{ix}\, s\,\theta + b_{iy}\, c\,\theta)(b_{iY} - a_{iY}) \tag{5}$$

$$\ddot{\mathbf{q}} = \begin{bmatrix} a_1 \\ a_2 \\ a_3 \end{bmatrix} = \dot{\mathbf{J}}^{-1} \dot{\mathbf{x}} + \dot{\mathbf{J}}^{-1} \ddot{\mathbf{x}} = \dot{\mathbf{J}}^{-1} \begin{bmatrix} v_X \\ v_Y \\ \omega_Z \end{bmatrix} + \mathbf{J}^{-1} \begin{bmatrix} a_X \\ a_Y \\ \alpha_Z \end{bmatrix} \tag{6}$$

$$\dot{\mathbf{J}}^{-1} = \begin{bmatrix} -s\,\theta_1\,\theta_1 & c\,\theta_1\,\theta_1 & \dot{b}_{1X}\, s\,\theta_1 - \dot{b}_{1Y}\, c\,\theta_1 + b_{1X}\, c\,\theta_1\,\theta_1 + b_{1Y}\, s\,\theta_1\,\theta_1 \\ -s\,\theta_2\,\theta_2 & c\,\theta_2\,\theta_2 & \dot{b}_{2X}\, s\,\theta_2 - \dot{b}_{2Y}\, c\,\theta_2 + b_{2X}\, c\,\theta_2\,\theta_2 + b_{2Y}\, s\,\theta_2\,\theta_2 \\ -s\,\theta_3\,\theta_3 & c\,\theta_3\,\theta_3 & \dot{b}_{3X}\, s\,\theta_3 - \dot{b}_{3Y}\, c\,\theta_3 + b_{3X}\, c\,\theta_3\,\theta_3 + b_{3Y}\, s\,\theta_3\,\theta_3 \end{bmatrix} \tag{7}$$

In Eq. 7, $\dot{b}_{iX}$ and $\dot{b}_{iY}$ are given by Eqs. 8 and 9, respectively, for $i = 1, 2$ and 3. The same method is applied with the relation between the velocities and accelerations of the moving platform and the angular accelerations of the actuators (Eq. 10). The matrix of the derivatives of $\mathbf{J}_\theta^{-1}$ was calculated by the same methodology used for the derivatives of $\mathbf{J}^{-1}$ (Eq. 7).

$$\dot{b}_{iX} = v_X - \omega_Z (b_{ix}\, s\,\theta + b_{iy}\, c\,\theta) \tag{8}$$

$$\dot{b}_{iY} = v_Y + \omega_Z (b_{ix}\, c\,\theta - b_{iy}\, s\,\theta) \tag{9}$$

$$\boldsymbol{\theta} = \begin{bmatrix} \dot{\omega}_1 \\ \dot{\omega}_2 \\ \dot{\omega}_3 \end{bmatrix} = \begin{bmatrix} \alpha_1 \\ \alpha_2 \\ \alpha_3 \end{bmatrix} = \dot{\mathbf{J}}_\theta^{-1} \dot{\mathbf{x}} + \dot{\mathbf{J}}_\theta^{-1} \ddot{\mathbf{x}} = \dot{\mathbf{J}}_\theta^{-1} \begin{bmatrix} v_X \\ v_Y \\ \omega_Z \end{bmatrix} + \mathbf{J}_\theta^{-1} \begin{bmatrix} a_X \\ a_Y \\ \alpha_Z \end{bmatrix} \tag{10}$$

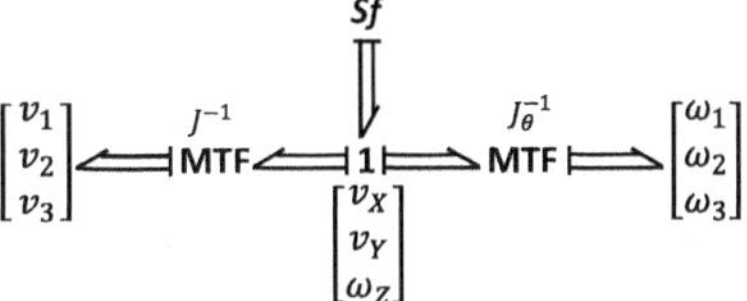

Fig. 2 Multibond graphs representation of the planar platform inverse kinematics

In a graph that correctly describes the kinematics (1 and 0 junctions, transformers and gyrators), the dynamics (capacitors, inertias and resistors) can be imposed without the risk of creating models where the main constraints of mechanical systems are violated: geometric or kinematic ties [7].

In this model, speed conditions are imposed by ideal velocity sources, that is, a source of velocity for v_X, v_Y and ω_Z. Besides these velocities, the other 1 junctions (of common velocities) indicate the linear (v_1, v_2 and v_3) and angular velocities (ω_1, ω_2 and ω_3) of the actuators. Thus, the inverse kinematics of the planar platform via multibond graphs is represented as shown in Fig. 2, whereby the modulated transformer type represents the matrices $\mathbf{J}^{-1}$ (Eq. 2) and $\mathbf{J}_\theta^{-1}$ (Eq. 4).

3 Dynamic Model of the Parallel Mechanism Using Power Flow Approach

According to [8], when possible, both completely match the power variables on the inputs and outputs of the subsystems (same type and direction of power flow) and a consistent cause and effect relation (which variables enter and which come out of the models to be coupled), the resulting model is fully equivalent to that which would be obtained analytically using other methodologies, allowing its simulation from the simple connection of the modules. Considering this, the diagram (Fig. 3) that illustrates the relationships of cause and effect of the planar platform with three degrees of freedom is mounted.

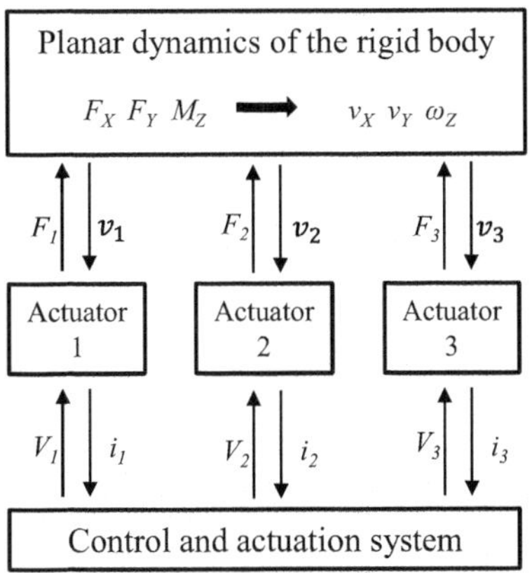

Fig. 3 Cause and effect relations of the planar platform

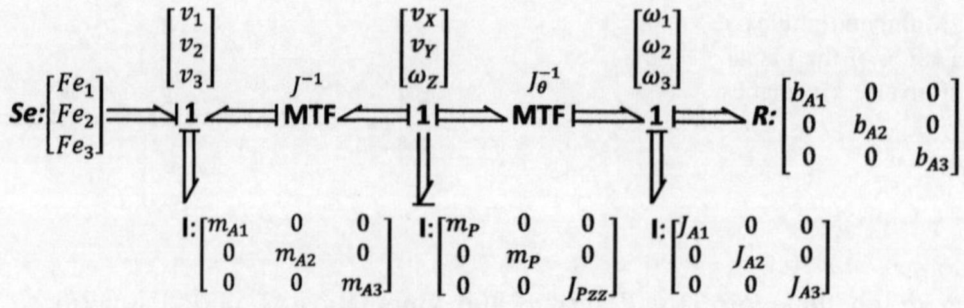

Fig. 4 Multibond graphs representation of the planar platform dynamics

With the kinematic relationships of this parallel mechanism comes the relation of consequence of the power conservation on the actuators coupling with the rigid body (Eqs. 11 and 12). Considering the inertia effects of the moving platform, with mass m_P and mass moment of inertia J_{Pzz}, the multibond graphs structure of the direct dynamics model of the planar platform with three degrees of freedom is shown in Fig. 4. Using the concepts, elements and the graphical representation of the Bond Graph Technique, was further added the inertial effects of the bodies that compound the actuators, introducing the terms m_{Ai} and J_{Ai}, which correspond to the mass and moments of inertia of the actuators, with $i = 1, 2$ and 3. This is made by applying the correct direction of power flow (indicated by the half arrow) and the correct direction in which the effort signal is directed (indicated by the causal stroke). It was also included in this model the equivalent viscous friction in the rotation joints.

$$[F_X \quad F_Y \quad M_Z] . \begin{bmatrix} v_X \\ v_Y \\ \omega_Z \end{bmatrix} = [F_1 \quad F_2 \quad F_3] \cdot \mathbf{J}^{-1} \begin{bmatrix} v_X \\ v_Y \\ \omega_Z \end{bmatrix} \Rightarrow \begin{bmatrix} F_X \\ F_Y \\ M_Z \end{bmatrix} = (\mathbf{J}^{-1})^{\mathrm{T}} \begin{bmatrix} F_1 \\ F_2 \\ F_3 \end{bmatrix} \tag{11}$$

$$[F_X \quad F_Y \quad M_Z] = [M_1 \quad M_2 \quad M_3] \cdot \mathbf{J}_\theta^{-1} \Rightarrow \begin{bmatrix} F_X \\ F_Y \\ M_Z \end{bmatrix} = (\mathbf{J}_\theta^{-1})^{\mathrm{T}} \begin{bmatrix} M_1 \\ M_2 \\ M_3 \end{bmatrix} \tag{12}$$

From the model in the Fig. 4, the constitutive equations of the inertia elements (*I*) with integral (or natural) causality are written in their differential form. Thus, making explicit the efforts, inserting this equation into the junction structures equations and replacing the constitutive equations of the inertial elements with differential (forced) causality, the resistors elements (*R*) and the modulated transformers (*MTF*), the Eq. 13 is obtained.

$$\mathbf{M}_P \dot{\mathbf{v}} = \mathbf{J}^{-\mathrm{T}} \mathbf{f}_e - \mathbf{J}^{-\mathrm{T}} \mathbf{M}_A \, \dot{\mathbf{v}}_A - \mathbf{J}_\theta^{-\mathrm{T}} \mathbf{J}_A \, \dot{\omega}_A - \mathbf{J}_\theta^{-\mathrm{T}} \mathbf{B}_A \boldsymbol{\omega}_A \tag{13}$$

Substituting the equations from the derivatives of the Jacobian matrices (Eqs. 6 and 10) and Eq. 4 into Eq. 13 and solving the algebraic loops associated to the storage elements with differential causality, the state-space equations are obtained (Eq. 14), with $\mathbf{M}_1$ and $\mathbf{M}_2$ given by Eqs. 15 and 16, respectively.

$$\dot{\mathbf{v}} = \left(\mathbf{M}_1^{-1}\mathbf{M}_2\right)\mathbf{v} + \left(\mathbf{M}_1^{-1}\mathbf{J}^{-T}\right)\mathbf{f}_e \tag{14}$$

$$\mathbf{M}_1 = \mathbf{M}_P + \mathbf{J}^{-T}\mathbf{M}_A\mathbf{J}^{-1} + \mathbf{J}_\theta^{-T}\mathbf{J}_A\mathbf{J}_\theta^{-1} \tag{15}$$

$$\mathbf{M}_2 = -\mathbf{J}^{-T}\mathbf{M}_A\,\dot{\mathbf{J}}^{-1} - \mathbf{J}_\theta^{-T}\mathbf{J}_A\,\dot{\mathbf{J}}_\theta^{-1} - \mathbf{J}_\theta^{-T}\mathbf{B}_A\mathbf{J}_\theta^{-1} \tag{16}$$

3.1 Dynamic Model of the Actuation System

The property of modularity, one of the major advantages of the technique, enables the development of complex systems models from simple subsystems, since these are created predicting the manner in which they will engage each other [6]. Figure 5 presents the electric actuator scheme used in this modeling. An electric motor provides power to the actuation system through a torque T_m and an angular velocity ω_m. This power is then transmitted to a leadscrew by a gear set. In bond graphs modeling, motors can, in general, be considered, as effort sources.

In the dynamic model of the actuation system were considered the inertia of the motor (J_m), of the gear train (J_C), of the actuator rod (m_A) and also the viscous friction coefficients b_m, b_C and b_A associated with these elements. Figure 6 presents the bond graph structure of the actuation system, where n_e is the transmission ratio between the gears A and C. The leadscrew D has the same velocity of C, ω_C. Through the leadscrew nut, which is coupled to the actuator rod, this movement becomes linear with velocity $\dot{d}$. This relation is given by $n_P = 0.5 \cdot \pi^{-1} \cdot p \cdot Ne$, where p is the leadscrew pitch and Ne refers to type of thread. In the electrical circuit model, R, L and K_e are the resistance, the inductance and the electromagnet constant of the motor, respectively.

The elimination of the electric dynamics, which has time constants of smaller orders of magnitude than the mechanical dynamics, is made by considering the values of L_i approximately equal to zero (for i = 1, 2 and 3). The actuator system

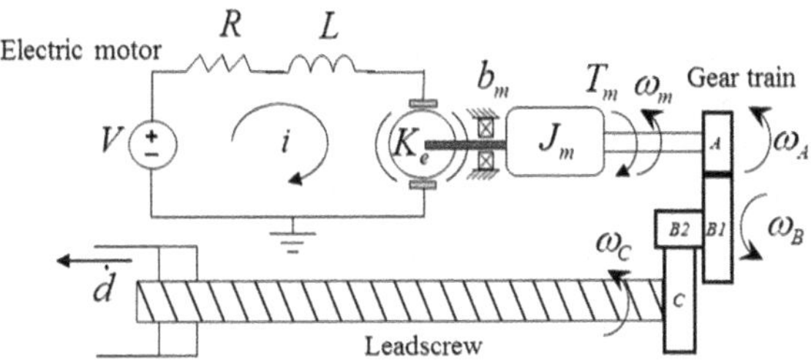

Fig. 5 Electric actuator scheme

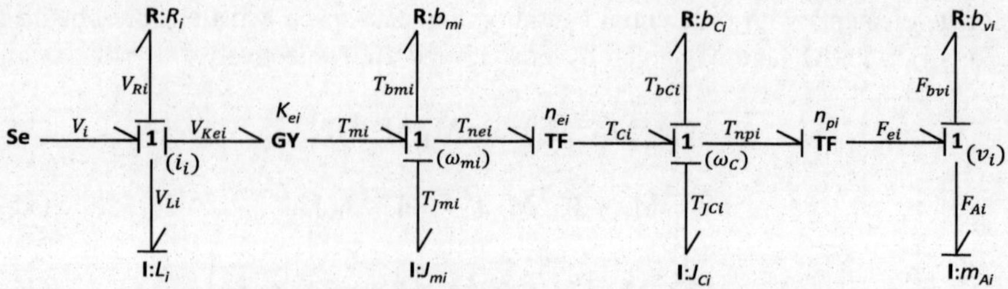

Fig. 6 Bond graphs for the electric linear actuator

model parameters were obtained by a set of experiments using a small scale electric motor dynamometer.

3.2 Coupled Dynamic Model

Figure 7 shows the coupled dynamic model represented using multibond graphs. The actuator models are coupled to the planar platform model through the 1 junctions that represents the actuator output speed, v_i, with $i = 1, 2$ and 3. Using the Bond Graph Technique formulation, the state-space equation, where $\boldsymbol{v}$ is the state vector and $\boldsymbol{s}_e$ is the input vector (Eq. 19) is obtained, with $\mathbf{M}_3$ and $\mathbf{M}_4$ given by Eqs. 17 and 18, respectively.

$$\mathbf{M}_3 = \mathbf{M}_\mathrm{P} + \mathbf{J}^{-\mathrm{T}}\mathbf{M}_{\mathrm{Aa}}\mathbf{J}^{-1} + \mathbf{J}_\theta^{-\mathrm{T}}\mathbf{J}_\mathrm{A}\mathbf{J}_\theta^{-1} \tag{17}$$

$$\mathbf{M}_4 = -\mathbf{J}^{-\mathrm{T}}\mathbf{B}_{\mathrm{Av}}\mathbf{J}^{-1} - \mathbf{J}^{-\mathrm{T}}\mathbf{M}_{\mathrm{Aa}}\dot{\mathbf{J}}^{-1} - \mathbf{J}_\theta^{-\mathrm{T}}\mathbf{J}_\mathrm{A}\dot{\mathbf{J}}_\theta^{-1} - \mathbf{J}_\theta^{-\mathrm{T}}\mathbf{B}_\mathrm{A}\mathbf{J}_\theta^{-1} \tag{18}$$

$$\dot{\mathbf{v}} = \left(\mathbf{M}_3^{-1}\mathbf{M}_4\right)\mathbf{v} + \left(\mathbf{M}_3^{-1}\mathbf{J}^{-\mathrm{T}}\right)\mathbf{s}_\mathrm{e} \tag{19}$$

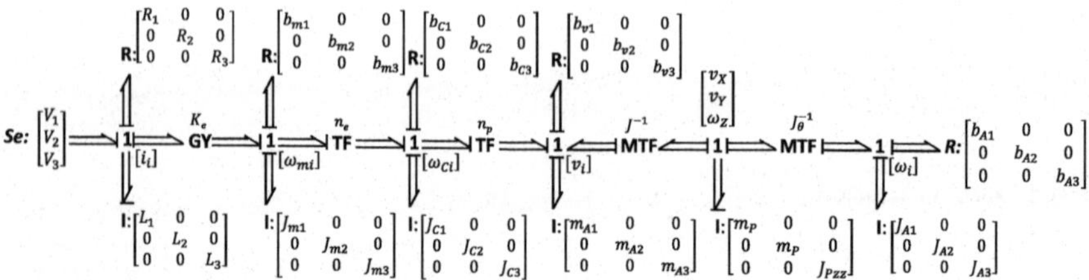

Fig. 7 Complete multibond graph representation for the 3-RPR parallel mechanism

4 Results

For the purpose of providing real data (geometry, inertia, damping, etc.), a planar mechanism was designed and built (Fig. 8a). The actuator system model parameters were obtained by a set of experiments using a small scale electric motor dynamometer (Fig. 8b).

4.1 Inverse Kinematic Model Validation

Simulations were made to validate the inverse geometric model (vector loop equation) and the inverse kinematic model (using the matrices $\mathbf{J}^{-1}$ and $\mathbf{J}_{\theta}^{-1}$). Table 1 presents the geometric parameters of the mechanism.

Using the derivatives of the Jacobian matrices from Eqs. 2 and 4, the time response of the limbs was obtained for the input functions shown in Eq. 20. Figure 9b and 9d show the linear and angular accelerations of the actuators and Fig. 9a and 9c show the linear and angular velocities of the actuators, respectively, by integrating (with the corresponding initial conditions) the accelerations of the actuators.

$$\begin{cases} \ddot{X} = 5.00\sin(\pi t)\ \mathrm{mm/s^2} \\ \ddot{Y} = -5.00\sin(\pi t)\ \mathrm{mm/s^2} \\ \ddot{\theta} = 0.0873\ \mathrm{rad}/s^2 \end{cases} \quad (20)$$

4.2 Direct Dynamic Model Validation

In the simulation of the direct dynamic model were considered the mass and the mass moment of inertia of the moving platform, m_P and J_{Pzz}, the mass and the mass

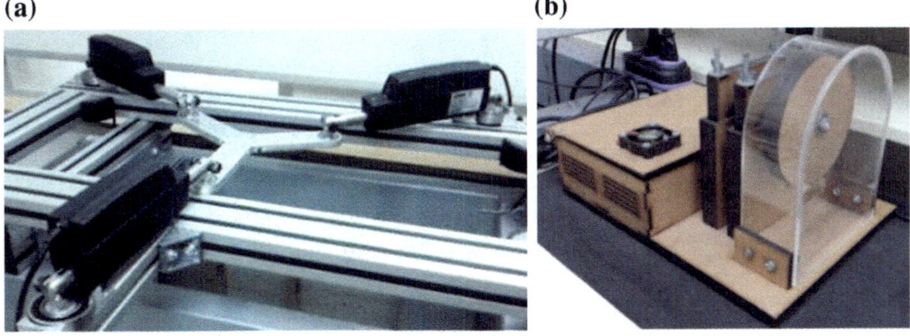

Fig. 8 Built planar mechanism (**a**) and the small scale dynamometer (**b**)

Table 1 Geometric parameters

Identification	Symbol	Value
A1 joint coordinates in reference frame A (mm)	a_1	[−389.14 −224.67]
A2 joint coordinates in reference frame A (mm)	a_2	[389.14 −224.67]
A3 joint coordinates in reference frame A (mm)	a_3	[0.00 449.34]
B1 joint coordinates in reference frame B (mm)	b_1	[−125.00 −72.17]
B2 joint coordinates in reference frame B (mm)	b_2	[125.00 −72.17]
B3 joint coordinates in reference frame B (mm)	b_3	[0.00 144.34]
Linear actuator fixed length (mm)	L_{min}	255.00
Stroke of the linear actuator (mm)	S	100.00

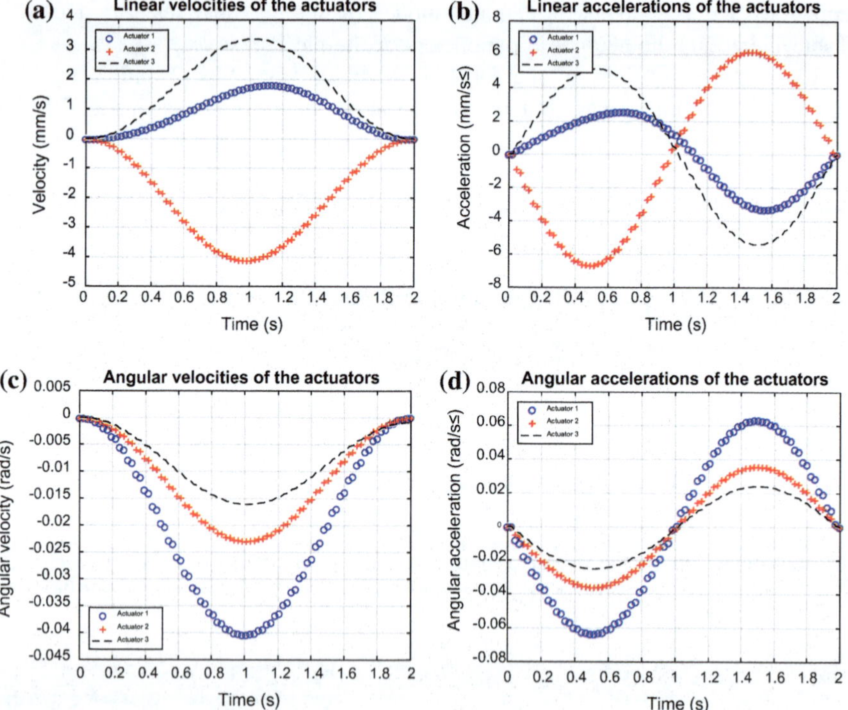

Fig. 9 Linear and angular velocities and accelerations of the actuators

moment of inertia of the actuators, m_{A1}, m_{A2}, m_{A3} and J_{A1}, J_{A2}, J_{A3}, and the viscous friction coefficients from the actuators joints, b_{A1}, b_{A2} and b_{A3}. Table 2 presents the parameters used in this simulation. Two pulses with amplitudes 5 and –5 N, widths of 0.1 s and interval of 0.1 s between them were given by the actuator 1. Figures 10a show the linear accelerations of the moving platform and Fig. 10b show the linear velocities of the moving platform. These velocities and

Table 2 Planar mechanism simulation parameters

Identification	Symbol	Value
Mass of the platform (kg)	m_P	0.578
Mass moment of inertia of the platform (kg m^2)	J_{Pzz}	4.50×10^{-3}
Mass of the actuator rod (kg)	m_{A1}, m_{A2}, m_{A3}	0.175
Mass moment of inertia of the actuator (kg m^2)	J_{A1}, J_{A2}, J_{A3}	7.28×10^{-3}
Viscous friction coefficient of the joints (N s m^{-1})	b_{A1}, b_{A2}, b_{A3}	0.006

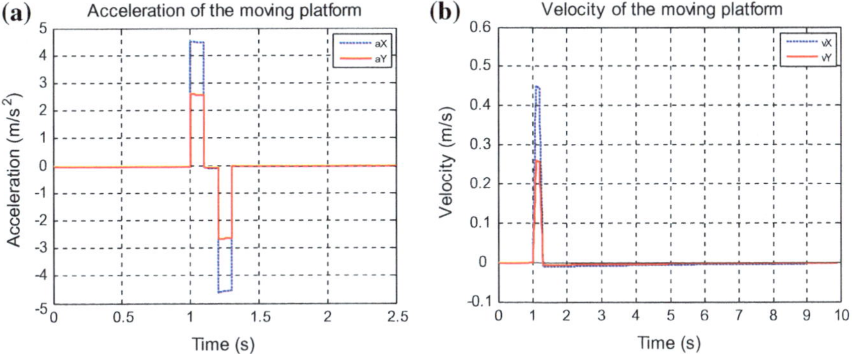

Fig. 10 Linear accelerations and velocities of the moving platform

accelerations occur on the expected directions and with amplitudes compatible with the type of system studied. Further analysis, comparing the results with real data, will be performed on the coupled system model.

4.3 Coupled Dynamic Model Validation

In the simulation of the coupled dynamic model, several parameters were considered besides the ones already discussed: the total transmission ratio of the actuators gear train and leadscrew, n_{ep}, the equivalent mass moment of inertia of the actuators, J_{mC}, and the equivalent viscous friction coefficients from motor and gear train of the actuators, b_{mC}. Table 3 presents the parameters used in this simulation.

Figure 12 shows the time response of the actuators for different values of proportional gain for a given input ([$X = 0.0$ mm, $Y = 20.0$ mm, $\theta = 0.00$ rad]) using the control strategy shown in Fig. 11, where G^{-1} represents the inverse geometric model an J^{-1} represents the inverse Jacobian model of the mechanism.

Table 3 Actuating system parameters

Identification	Symbol	Value
Total transmission ratio of the actuator i (–)	n_{epi}	0.1156
Friction coefficient of the motor and gear train i (N m s rad^{-1})	b_{mCi}	0.003
Mass moment of inertia of the motor and gear train i (kg m^2)	J_{mCi}	5.62×10^{-5}

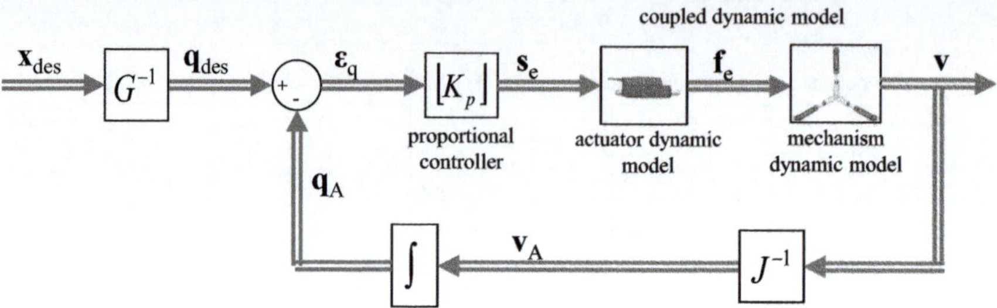

Fig. 11 Position control strategy

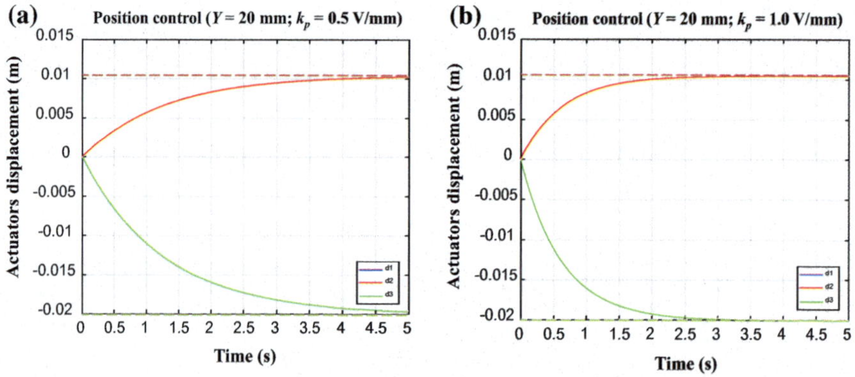

Fig. 12 Time response for $Y(t) = 20$ mm. **a** $k_p = 0.5$; **b** $k_p = 1.0$

The reference values for the steady state are [$d_1 = 10.5$ mm, $d_2 = 10.5$ mm, $d_3 = -20.0$ mm]. The system showed sensitivity to gain with time constants compatible with the physical characteristics of the mechanism. The actuators displacement tends to the values estimated by the inverse geometry model, which is compatible with the one measured by the built platform when it is on the proper configurations (same as the simulation inputs).

5 Conclusions and Future Work

In this work a procedure for the determination of the analytical form of dynamic models of a 3-RPR parallel manipulator with linear electric actuation through the characterization of the power flow between its components was presented. From the geometrical relations associated to the displacement of their degrees of freedom, the kinematic relations associated to their velocities were determined. Considering the power flow between the degrees of freedom and between these and the actuating elements, the equilibrium relations of the forces and torques were obtained. Also, inertial effects of system components, stiffness and damping effects were taken into account and the equations of motion for the direct dynamics of a parallel manipulator were analytically determined. This approach adopted the same fundamentals, concepts and elements of the Bond Graph Technique.

Simulations were performed to validate this approach, using the real data (geometry, inertia, damping, actuators forces, etc.) from a small scale electric motor dynamometer and a planar mechanism designed and built especially for the purpose to compare the simulated and experimental results. The ongoing work focuses in implement these models in the built platform in order to verify these responses on real environment.

References

1. Wang, Y.: Symbolic Kinematics and Dynamics Analysis and Control of a General Stewart Parallel Manipulator. State University of New York at Buffalo, Department of Mechanical and Aerospace Engineering, Master's Thesis, Buffalo, USA (2008)
2. Khandelwal, S., Chevallereau, C.: Estimate of the trunk attitude of a humanoid by data fusion of inertial sensors and joint encoders. In: Proceedings of the Sixteenth International Conference on Climbing and Walking Robots, Sydney, Australia (2013)
3. Zhao, Q., Gao, F.: Bond graph modelling of hydraulic six-degree-of-freedom motion simulator. Proc. Inst. Mech. Eng. Part C: J. Mech. Eng. Sci. USA (2012)
4. Yildiz, I., Ömürlü, V.E., Sagirli, A.: Dynamic modeling of a generalized Stewart platform by bond graph method utilizing a novel spatial visualization technique. Int. Rev. Mech. Eng. (Turkey) (2008)
5. McBride, R.T.: System Analysis Through Bond Graph Modeling. University of Arizona, Department of Electrical and Computer Engineering, Doctoral dissertation, Tucson, USA (2005)
6. Albuquerque, A.N., Speranza Neto, M., Meggiolaro, M.A.: Dynamic models of parallel mechanisms using power flow approach. In: CONEM 2016, IX Congresso Nacional de Engenharia Mecânica, Fortaleza, Brazil (2016)
7. Speranza Neto, M., Silva, F.R.: Modelagem e análise de sistemas dinâmicos (in Portuguese). Pontifical Catholic University, Department of Mechanical Engineering, Lecture notes, Rio de Janeiro, Brazil (2005)
8. Karnopp, D.C., Margolis, D.L., Rosenberg, R.C.: System Dynamics: A Unified Approach, 2nd ed. Wiley (1990)

Numerical and Experimental Analysis of a Parallel 2-DOF Manipulator

William S. Cardozo and Hans I. Weber

Abstract A parallel two degrees of freedom manipulator designed for a variable orientation of a body in space is analyzed. The manipulator consists of one universal joint with one axis fixed to the base and the other axis fixed to a moving platform. A similar device is used in spacecraft to orient the rocket nozzle. In this work, two parallel linear hydraulic actuators move the platform. A novel proportional digital hydraulic valve is used to control the actuators. Each fork of the universal joint has an angular position sensor mounted to measure the relative motion of the cross and an inertial measurement unit (IMU) is fixed to the moving platform. Load cells and pressure transducers are mounted on the actuators to measure force and chambers pressure. Numerical simulations are presented using a desired trajectory as input for a proportional controller (P-controller). An experimental apparatus is used to validate the numerical results.

Keywords Kinematics · Electrohydraulic servosystem · Servoactuator
Universal joint

1 Introduction

In this work, a platform is designed with two linear hydraulic actuators mounted around a universal joint, thus forming a parallel manipulator with two degrees of freedom (2-DOF). Figure 1 shows the concept of this device with a load over the moving platform. The hydraulic actuators are connected to the base through smaller universal joints and to the moving platform through ball joints.

W. S. Cardozo (✉) · H. I. Weber
PUC-Rio, R. Marquês de São Vicente, 225, Rio de Janeiro, RJ, Brazil
e-mail: billi83@globo.com

H. I. Weber
e-mail: hans@puc-rio.br

A. de T. Fleury et al. (eds.), *Proceedings of DINAME 2017*, Lecture Notes in Mechanical Engineering, https://doi.org/10.1007/978-3-319-91217-2_17

Fig. 1 Kinematic scheme

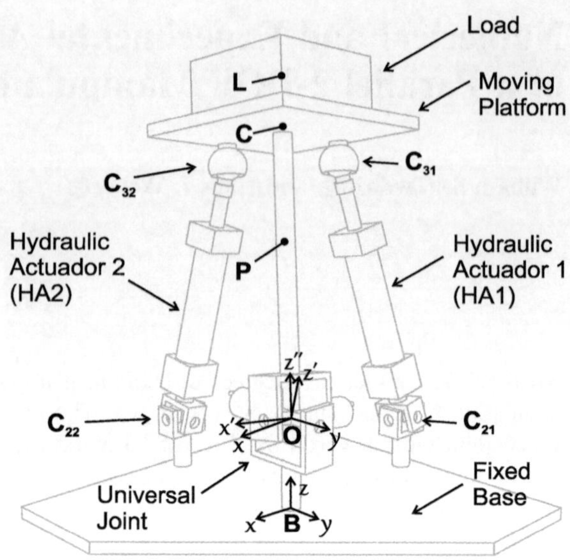

The thrust vector control (TVC) scheme is similar to this platform and orients the rocket nozzle in spacecraft. The TVC produces torque around the center of mass of the spacecraft in order to steer and keep it on the desired flight path [1].

Lazic and Ristanovic [2] proposed a test bench for a TVC system based on a fixed base robotic system. They presented an analysis of the kinematics and a control strategy for a gimbaled 2-DOF parallel platform with two electrohydraulic actuators (EHAs).

Wekerle et al. [3] presented the requirements of an actuation system for TVC systems. The working envelope of several TVC actuators from the literature is analyzed. A simplified and linearized model of a spacecraft as a rigid body is presented considering wind disturbances. During the development of this work, the actuator requirements presented in [3] were used as a reference. In [1] Wekerle et al. presented a 2-dof mockup of a rocket motor nozzle with two EHAs. The actuators performance is identified in a mass-spring test bench. The TVC mockup presented in [1] uses two commercial servovalves; however, in this work low-cost valves were manufactured. The custom-built valves reduce the cost and the complexity of the system, because the commercial servo valves need a super clean fluid and a sophisticated peripheral control system. Indeed, the EHA system presented in [1] is closer to the most widespread control system in the industry. Here, a low-cost hydraulic system is proposed, but it needs a lot of development to achieve the robustness of the TVC presented in [1].

Ghosh et al. [4] developed a 2-dof parallel hydraulic actuated system for a heave-and-pitch motion simulator. Two low-cost commercial solenoid proportional valves control the hydraulic actuators. These types of valves have a dead-band and non-linear behavior [5]. The authors have considered different types of controllers, and the self-tuning fuzzy proportional–integral–derivative (PID) with bias control showed the best performance. The PID controller had the worst response [4]. The commercial proportional valves used in [4] have a dead-band of 15% (usually, 5–20% is common [6]). In this work, the dead-band of the proposed valve is less than 0.5%, hence a better performance of a linear controller is expected.

On the way to construct this platform, Cardozo and Weber [7] have introduced a 2-DOF mechanism similar to the platform of this work. In addition, Cardozo and Weber [8] presented a 2-DOF mechanism the same platform presented in Fig. 1, but with a homokinetic joint instead of a gimbal. Numerical simulations have shown the efficacy of the homokinetic platform and using an experimental apparatus the numerical results are validated.

In this work, the same proportional digital hydraulic valve (PDHV) used in [8] controls the oil flow in the hydraulic actuators chambers. The proposed PDHV has much lower price and less contamination sensitivity, when compared with a commercial sliding spool servovalve. However, the PDHV is low-bandwidth modulating. Numerical and experimental trials validate the proposed system. The test bench has an angular position sensor on each fork of the universal joint to measure the relative motion of the cross. An inertial measurement unit (IMU) measures the embarked angular velocity of the moving platform and the linear acceleration of a point. Moreover, load cells and pressure transducers mounted on the actuators measure the force and the chambers pressure.

2 Hydraulic Control System

In this work, the platform is actuated by two independent EHA systems. Each EHA system has a four-way valve-controlled single-rod linear actuator, hence this is one of the most basic kind of hydraulic controlled system. However, even though this is a simple system, the cost is very high if a servo-valve is necessary. Aiming a low-cost system alternative, a novel control valve is proposed specially developed for this manipulator, called PDHV. This valve is a proportional four-way valve with an underlapped symmetric closed-center rotary spool. And an open-loop stepper motor drives the valve spool through an elastic-coupling. Each actuator of the platform is controlled by one PDHV, as shown in Fig. 2. By changing the valve rotor angle θ_V, the pressure drops and the flow rate through the valve change, which changes the pressure in the pipelines and in the actuator chambers, thereby generating the force that moves the piston.

In Fig. 2, p_S and Q_S are the supply pressure and flow rate, respectively. The arrows in the pipelines indicate the positive sense of the flow rates Q_1, Q_2, Q_3 and Q_4. p_1, p_2, p_3 and p_4 are the pressures in the valve connections. p_A and p_B are the pressures in chambers A and B, respectively; A_e and A_c are the areas of the embolus and the crow, respectively; F_a is the force acting on the rod; and $\dot{l}_a$ is the piston speed. For a simpler kinematic analysis of the platform, l_a is the distance between the centers of the mounting joints of the actuators.

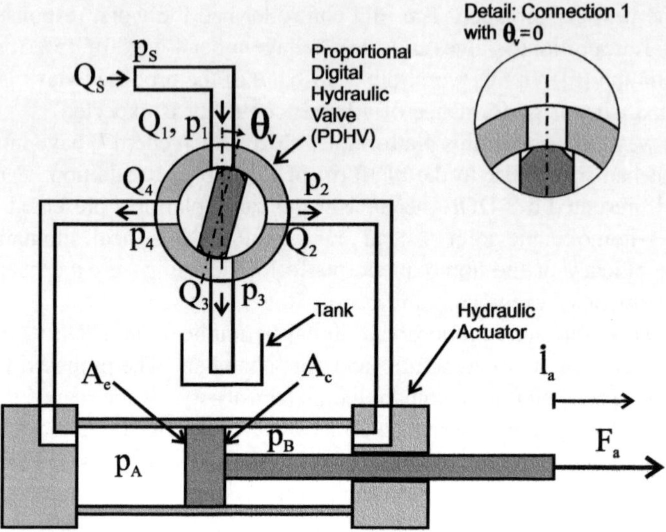

Fig. 2 Hydraulic scheme

Using the kinematic model and the measured angular platform position, the actual actuators position l_a is calculated. The controller compares the actual positions with the desired ones and changes the valve rotor angle θ_V proportionally to the error.

3 Modeling

In the study of the platform kinematics three reference frames are defined, F fixed to the base, R attached to the main crosshead and S attached to the moving platform. The rotation between the frames are summarized by the following indication,

$$\underset{\substack{\text{Fixed}\\ \text{Base}\\ (x,y,z)}}{F} \xrightarrow{\beta(y)} \underset{\substack{\text{Main}\\ \text{Crosshead}\\ (x',y,z')}}{R} \xrightarrow{\alpha(x')} \underset{\substack{\text{Moving}\\ \text{Platform}\\ (x',y'',z'')}}{S}$$

where β is the angle of rotation around y-axis and α is around x′-axis. Both angles are measured directly using angular transducers.

The vector of angular momentum of the moving platform and the load using *S* frame about its center of gravity is introduced by,

$$ {}^{S}\mathbf{H}_{P} = {}^{S}\mathbf{I}_{P}\,{}^{S}_{F}\boldsymbol{\omega}_{S},\quad {}^{S}\mathbf{H}_{L} = {}^{S}\mathbf{I}_{L}\,{}^{S}_{F}\boldsymbol{\omega}_{S} \tag{1,2} $$

where ${}^{S}_{F}\boldsymbol{\omega}_{S}$ is the angular velocity vector of the platform in body coordinates. The linear momentum vectors of the platform and the load are given by Eqs. (3) and (4) using S frame,

$$ {}^{S}\mathbf{G}_{P} = m_{P}\,{}^{S}\mathbf{v}_{P},\quad {}^{S}\mathbf{G}_{L} = m_{L}\,{}^{S}\mathbf{v}_{L}. \tag{3,4} $$

Equation (5) shows the angular momentum of the system composed by the moving platform and the embarked load, about the center of the main crosshead, which is a fixed point, using S frame [8].

$$ {}^{S}\mathbf{H}_{O} = {}^{S}\mathbf{H}_{P} + {}^{S}_{O}\tilde{\mathbf{r}}_{P}\,{}^{S}\mathbf{G}_{P} + {}^{S}\mathbf{H}_{L} + {}^{S}_{O}\tilde{\mathbf{r}}_{L}\,{}^{S}\mathbf{G}_{L} \tag{5} $$

where ${}^{S}_{O}\tilde{\mathbf{r}}_{P}$ and ${}^{S}_{O}\tilde{\mathbf{r}}_{L}$ are the tilde matrices of the position vector of the platform and load CM using S frame. Using Euler's Law, Eq. (6) gives the torque ${}^{S}\mathbf{M}_{O}$ acting on the moving platform [9].

$$ {}^{S}\mathbf{M}_{O} = {}^{S}\dot{\mathbf{H}}_{O} + {}^{S}_{F}\tilde{\boldsymbol{\omega}}_{S}\,{}^{S}\mathbf{H}_{O} \tag{6} $$

The torque acting on the moving platform is generated by the force vector of the actuators, by the gravity force and by the universal joint constrains, as shown in Eq. (7).

$$ {}^{S}\mathbf{M}_{O} = {}^{S}\mathbf{T}^{F}\left(\left(m_{P}\,{}^{F}_{O}\tilde{\mathbf{r}}_{P} + m_{L}\,{}^{F}_{O}\tilde{\mathbf{r}}_{L}\right){}^{F}\mathbf{g} + {}^{F}_{O}\tilde{\mathbf{r}}_{C31}\,{}^{F}\mathbf{F}_{a1} + {}^{F}_{O}\tilde{\mathbf{r}}_{C32}\,{}^{F}\mathbf{F}_{a2}\right) + {}^{S}\mathbf{M}_{J} \tag{7} $$

where ${}^{S}\mathbf{T}^{F}$ is the rotation matrix that leads from F to S frame, m_P and m_L are the platform and the load masses, ${}^{F}\mathbf{g}$ is the gravity vector, ${}^{F}_{O}\tilde{\mathbf{r}}_{C31}$ and ${}^{F}_{O}\tilde{\mathbf{r}}_{C32}$ are the tilde matrices of the position vector of the points C_{31} and C_{32}, and ${}^{S}\underline{\mathbf{M}}_{J}$ is the torque due to the universal joint constraint.

Looking at the crosshead of the main universal joint, it consists of two orthogonal revolute joints: one in the y-axis and one in the x′-axis. Hence, all the torque transmitted by the joint is orthogonal to the y-axis and to the x′-axis. Thus, the torque constraint of main universal joint is on the z′-axis $\left({}^{R}\mathbf{M}_{J} = [0 \quad 0 \quad M_{J}]^{T}\right)$. Thereby, the torque due to the universal joint constraint using the S frame is given by,

$$ {}^{S}\mathbf{M}_{J} = {}^{S}\mathbf{T}^{R}\,{}^{R}\mathbf{M}_{J} = {}^{S}\mathbf{T}^{R}[0 \quad 0 \quad M_{J}]^{T} \tag{8} $$

where ${}^{S}\mathbf{T}^{R}$ is the rotation matrix that leads from R to S frame.

Using Eqs. (6) and (7), the angular accelerations are calculated as

$$\dot{\beta} = \frac{\tau_2 - \tau_3 \tan\alpha - c_4 \dot{\alpha}\beta}{c_3}, \ \ddot{\alpha} = \frac{\tau_1 - c_2 \beta^2}{c_1} \tag{9, 10}$$

where the vector components $[\tau_1 \quad \tau_2 \quad \tau_3]$ and the constants are introduced as:

$$[\tau_1 \quad \tau_2 \quad \tau_3]^T = {}^S\mathbf{T}^F\left(\left(m_P\, {}^F_O\tilde{\mathbf{r}}_P + m_L\, {}^F_O\tilde{\mathbf{r}}_L\right){}^F\mathbf{g} + {}^F_O\tilde{\mathbf{r}}_{C31}\, {}^F\mathbf{F}_{a1} + {}^F_O\tilde{\mathbf{r}}_{C32}\, {}^F\mathbf{F}_{a2}\right), \tag{11}$$

$$c_1 = I_{1P} + m_p l_p^2 + I_{1L} + m_L l_L^2, \tag{12}$$

$$c_2 = (c_1 - I_{3P} - I_{3L}) \sin\alpha \cos\alpha, \tag{13}$$

$$c_3 = c_1 \cos\alpha + (I_{3P} + I_{3L}) \tan\alpha \sin\alpha, \tag{14}$$

$$c_4 = -(2c_1 - I_{3P} - I_{3L}) \sin\alpha + (I_{3P} + I_{3L}) \tan\alpha \cos\alpha. \tag{15}$$

Neglecting the actuators inertia, the forces of the HA1 and HA2 are given by

$${}^F\mathbf{F}_{a1} = \frac{{}^F_O\mathbf{r}_{C31}}{\left|{}^F_O\mathbf{r}_{C31}\right|} F_{a1}, \ {}^F\mathbf{F}_{a2} = \frac{{}^F_O\mathbf{r}_{C32}}{\left|{}^F_O\mathbf{r}_{C32}\right|} F_{a2}, \tag{16, 17}$$

where $\frac{{}^F_O\mathbf{r}_{C31}}{\left|{}^F_O\mathbf{r}_{C31}\right|}$ and $\frac{{}^F_O\mathbf{r}_{C32}}{\left|{}^F_O\mathbf{r}_{C32}\right|}$ are unit vectors in the HA1 and HA2 directions, respectively. F_{a1} and F_{a2} are the forces magnitudes of the HA1 and HA2, respectively, and are calculated using Eq. (18) [6].

$$F_{a1} = p_{A1} A_e - p_{B1} A_c - F_f - m\ddot{l}_{a1} \tag{18}$$

where the index "1" denotes actuator 1, m is the assembly mass which moves with the piston, and F_f is the friction force. Changing the index from 1 to 2 on Eq. (18) the force on the second actuator is obtained.

In the present work, the friction force is obtained experimentally at constant actuator speed, without a load, the fitted function is depicted in [8]. For this kind of numerical simulation, the friction force is well represented as a function of the actuator speed using a Stribeck curve. The good result is due to the fact that in the analyzed conditions slip velocity on the actuator is significant and a relative small amount of static/kinetic friction appears in the transition [10]. A more precise model, like LuGre model, shows the time dependence of the friction and could be used in a more general condition or for real time friction compensation [11].

The pressure drop through a PDHV is calculated using a variable-area orifice model [6],

$$Q_{ij} = \alpha_{Dij} A_{ij} \sqrt{\frac{2}{\rho} \left|p_i - p_j\right|} \operatorname{sign}\left(p_i - p_j\right). \tag{19}$$

Q_{ij} is the flow rate between connections i and j, where these indexes vary from 1 to 4 accordingly to the valve connection depicted in Fig. 2. α_{Dij} is the discharge coefficient. A_{ij} is the orifice area, which is a function of the valve rotor angle θ_V. In this work, the pipeline pressure loss due to friction is negligible, and it is not considered. But it is considered that the pump is capable of maintaining constant supply pressure during the experiments. This hypothesis was validated.

The continuity equation applied on the hydraulic scheme, considering an incompressible flow, gives the oil flow in the actuators chambers:

$$Q_a = Q_{14} - Q_{43}, \; Q_b = Q_{12} - Q_{23}. \tag{20, 21}$$

Equations (22) and (23) show the relationships between the flow rate in the chambers and the speed of the actuators:

$$Q_a = A_e \dot{l}_a, \; Q_b = A_c \dot{l}_a. \tag{22, 23}$$

During the experiment that identifies the friction force on the actuator, a discrete number of valve opening positions are demanded. These data are also used to identify the valve discharge coefficients.

Figure 3 gives the flow rate characteristic versus the rotary spool shift for several pressures drops.

The non-zero flow for a null spool shift is characteristic of an underlapped valve. In addition, the single rod actuator used in this work gives a positive velocity in a null spool shift condition. The numerical and experimental trials identify that when the spool is deflected form the central position of $-0.08°$, the actuators have null velocity in the absence of external forces. Hence, the null valve position ($\theta_V = 0$) is

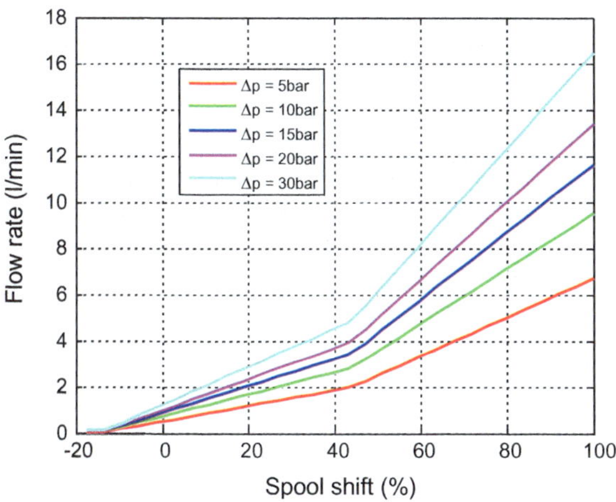

Fig. 3 Flow rate versus spool shift

deflected −0.08° from the actual central position. This procedure helps the controllability of the single rod cylinder.

About the bandwidth, until frequency of 11.5 Hz, the valve shift is limited by its maximum stroke of 14.4°. Between 11.5 and 15.4 Hz, the limitation is caused by the inversion time (6.5 ms), maximum angular velocity (21.0 rad/s) and acceleration (1,613 rad/s^2) of the open-loop stepper control. Between 15.4 and 76.9 Hz, the limitation is caused by the inversion time and the maximum angular acceleration. After 76.9 Hz, the valve spool does not move, because the minimum inversion time is not reached.

4 Controller

The platform controller is a decentralized proportional controller; hence the position control of each actuator is independent. Figure 4 depicts the control loop.

The control loop receives a desired angular position for the platform. Using the kinematic model, the desired actuator positions are calculated and compared with the actual positions. The errors are multiplied by a gain and sent to the PDHV controller. The PDHV controller has an open loop to control a stepper motor attached to the valve spool. Accordingly to the spool position, the oil flow to the actuators changes, thus its position changes and the orientation of the platform changes. The proportional gain is chosen trying several values in the simulations, Fig. 5 shows the step response of HA1 for several gains on the gimbaled platform.

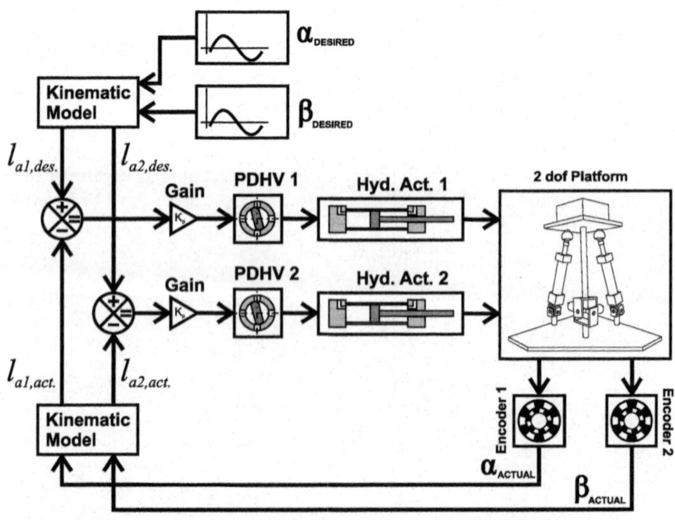

Fig. 4 2-DOF platform controller

Fig. 5 Step response of HA1 with different gains

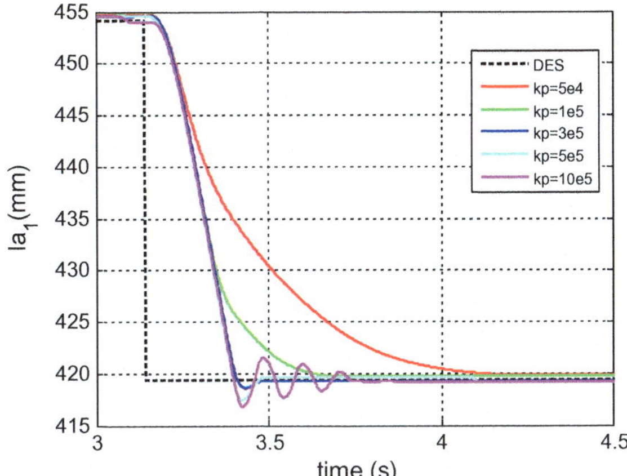

Using Fig. 5, the $k_P = 1 \times 10^5$ is chosen because there are no oscillations and no overshoots. Due to the symmetry of the actuating system, the HA2 has a similar response, thus the same gain is used. The data acquisition system (DAQ) consist in two customized microcontrolled circuits with an acquisition frequency of 50 Hz. One, receives the sensors data, sends these data to a computer with a MATLAB running script. The other one receives the desired valves positions from MATLAB script and sends it to the valves controllers. The valve controllers are a variable frequency open loop microcontrolled circuit specially built for the PDHV with 25 kHz control frequency.

Fig. 6 Simulated and experimental platform orientation

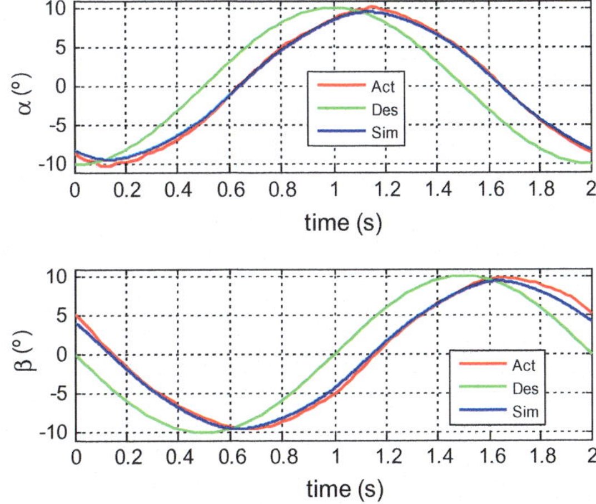

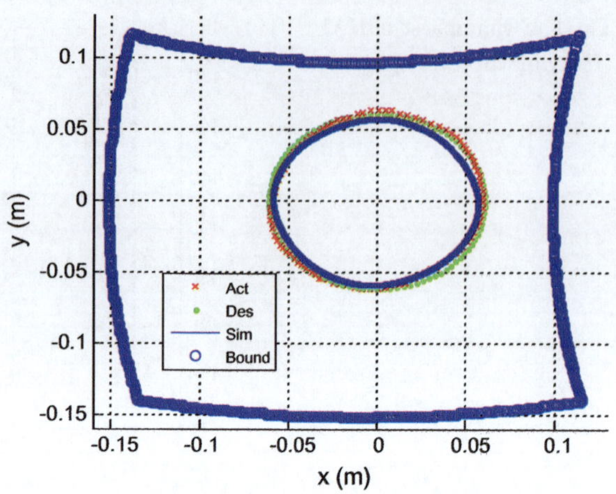

Fig. 7 Top view of the platform table center trace

5 Simulation and Experiments

The experimental data are compared with the numerical simulation and the desired position. The simulations and the experimental parameters can be found in [8]. Figure 6 shows α and β the experimental (Act) data compared with the numerical simulation (Sim) and the desired position (Des).

The deviation between α simulated and the experimental result is 2.2° in maximum and 1.2° in average. The β deviation is 1.8° in maximum and 0.9° in average. The average error between simulated and the experimental actuators length result is 2 mm, and this error reaches 4 mm for actuator one and 5 mm for actuator two in the maximum. The top view of the table center trace can be seen in Fig. 7 with the platform limits (Boundaries).

The deviations between simulated and the experimental results are mainly due to model simplifications and platform assembly problems, like backlash and joint friction. The main model simplifications are no friction losses in the pipelines, no actuators inertia and an actuators friction dependent of speed only. The difference between actual, or the simulated and the desired position is due the extremely simple control algorithm. This proportional control with bigger gain leads to stability problems, mainly chattering.

6 Conclusions

In this work, two configurations of a 2-DOF electrohydraulic actuated platform were developed for a low-cost TVC system. The most widespread TVC solution for liquid-propellant rockets is investigated using a gimbaled platform.

The kinematic and the dynamic modeling were presented. The models of the electrohydraulic system and the controller were also shown. These models were used to perform numerical simulations. The comparison between the numerical and the experimental results reveal that the presented model provides a good insight into the behavior of the system.

This work proposed a novel hydraulic control valve built specially for the control of the platform. It provided a feasible electrohydraulic control system. The operation of the novel control valve was validated to control the actuators position of the platform. Finally, it is concluded that, in spite of the low bandwidth, the proposed hydraulic control system is able to control a low-cost TVC system. This approach is interesting for upper stage TVCs in vacuum, where high precision and not velocity is the challenge.

References

1. Wekerle, T., Barbosa, E.G., Batagini, C.M., Costa, L.E.V.L., Trabasso, L.G.: Closed-loop actuator identification for Brazilian Thrust Vector Control development. In: Proceedings of the 20th IFAC Symposium on Automatic Control in Aerospace, Sherbrooke, Quebec, Canada (2016)
2. Lazic, D.V., Ristanovic, M.R.: Electrohydraulic thrust vector control of twin rocket engines with position feedback via angular transducers. Control Eng. Pract. **15**, 583–594 (2007)
3. Wekerle, T., Brito, A., Trabasso, L.G.: Requirements for an aerospace actuation system derived from the control design point of view. In: 23rd ABCM International Congress of Mechanical Engineering, Rio de Janeiro, RJ, Brazil, 6–11 December 2015
4. Ghosh, B.B., Sarkar, B.K., Saha, R.: Realtime performance analysis of different combinations of fuzzy–PID and bias controllers for a two degree of freedom electrohydraulic parallel manipulator. Robot. Comput. Integr. Manuf. **34**, 62–69 (2015)
5. Manring, N.: Hydraulic Control Systems. Wiley, New Jersey, USA (2005)
6. Jelali, M., Kroll, A.: Hydraulic servo-systems modelling, identification and control, 2nd edn. Springer, London (2003)
7. Cardozo, W., Weber, H.: Analysis and development of a parallel 2-dof manipulator. In: 23rd ABCM International Congress of Mechanical Engineering, Rio de Janeiro (2015)
8. Cardozo, W., Weber, H.: Dynamic modeling of a 2-dof parallel electrohydraulic-actuated homokinetic platform. Mech. Mach. Theory **118** (2017)
9. Weber, H.I.: Raciocinando Dinâmica de Rotação. Book under development, not published yet (2017)
10. Piatkowski, T.: Dahl and LuGre dynamic friction models—the analysis of selected properties. Mech. Mach. Theory **73**, 91–100 (2014)
11. Yao, J., Deng, W., Jiao, Z.: Adaptive control of hydraulic actuators with LuGre model-based friction compensation. IEEE Trans. Ind. Electron. **62**(10) (2015)

Parameter Optimization and Active Control of Electromechanical Suspension Systems

Willian Minnemann Kuhnert, Marcos Silveira and Paulo José Paupitz Gonçalves

Abstract This paper has as object of study a simplified model for the the automobile suspension system, which can become a regenerative system by coupling a RLC electric circuit to the mechanical system. The main objectives of this paper are to study and optimize a simplified electromechanical suspension model that, when in passive mode, harvests energy, while maintaining the handling stability and passenger comfort, and when in active mode, uses energy to improve comfort for passengers and handling stability with least effort. A multi-objective optimization procedure was carried out and Pareto frontier was obtained for the objective functions when considering the passive mode. When considering active control, changes were proposed to the optimal control in order to reduce control effort for feedforward strategy, while for feedback strategies, the stability gain range was obtained by Routh-Hurwitz criterion. The proposed control sets have particular advantages regarding isolation, energy harvested and control effort.

Keywords Optimization · Electromechanical suspension · Control · Energy harvesting · Regenerative suspension

1 Introduction

The study of regenerative systems to use wasted energy has become an important subject in the development of new mechanical systems. This is the case of automobile suspension systems which have two main important objectives: handling stability and comfort. These are usually competing objectives, where improving one degrades the other. In a typical vehicular damping system, viscous dampers dissi-

W. M. Kuhnert (✉) · M. Silveira · P. J. P. Gonçalves
School of Engineering, São Paulo State University (UNESP), Bauru, Brazil
e-mail: willian.kuhnert@feb.unesp.br

M. Silveira
e-mail: m.silveira@feb.unesp.br

P. J. P. Gonçalves
e-mail: paulo.jpg@feb.unesp.br

A. de T. Fleury et al. (eds.), *Proceedings of DINAME 2017*, Lecture Notes in Mechanical Engineering, https://doi.org/10.1007/978-3-319-91217-2_18

pate energy as heat, i.e., energy is lost to the environment. The implementation of an energy recovery system allows part of this energy to be reused. These devices have advantages compared to conventional viscous damper, such as ease of application of an active control and parameter setting, broader band of vibration attenuation, and the possibility of system energy recovery [1, 16, 19].

The electromechanical system may also be used in a semi-active suspension, with the control being realized by the electrical components of the device, facilitating its implementation with good to fair improvement, while systems with active control without look-ahead capabilities have proven to be little better or even worse when compared to them [2, 16].

Experimental electromechanical and electromagnetic suspension systems have already been built, such as systems using rotary motors and linear to angular motion conversion, and showed good relationship of recovered energy and weight [1, 19]. In some systems the type of configuration considerably reduces the total weight compared to other systems already developed. Final weight was only about 20% higher than a hydraulic suspension system, which can be considered a relatively low weight. This shows that the development of this type of suspension system has advanced significantly towards the feasibility of its application.

In Ref. [20], the authors developed a damper which can harvest energy and demonstrated the balance between the amount of energy recovered from the system and passenger comfort. Available energy in the system varies between 100 and 400 W in midsize cars running in good and medium condition roads, and depends on vehicle speed, road roughness and tire stiffness. Therefore, the compromise between recovered energy, passenger comfort and stability of the vehicle depends on the type of road on which the vehicle travels, and a control algorithm must be developed to optimize the dynamic behavior in every situation [6].

Nagode [13] designed prototypes of electromechanical suspension-based energy harvesting systems and tested them for railroad applications in order to convert vibration energy into useful electrical energy for the vehicle, and Kjellqvist [9] evaluated an electromechanical actuator for railroad applications experimentally, claiming that it can have the capability to improve the dynamics of a rail vehicle and showing that a bandwidth of 30 Hz is sufficient, although it can be improved.

On Ref. [8] an active electromechanical wheel suspension system was designed and evaluated, allowed to work on fully active mode. The authors claim that the preferred compromise between passenger comfort, handling stability and energy dissipation can be controlled by adapting the dimensioning method concerning the actuator and the control parameters during the development process. The proposed suspension could improve significantly body isolation, however for the control law adopted, the transmissibility close to the wheel-hop frequencies worsened.

The conventional elements used in suspension systems are usually springs and dampers. In some cases, the spring and damper elements are developed in a way that results in nonlinear behavior. In other situations, the springs and dampers can be modeled as linear elements, but the kinematics of the suspension system results in geometric nonlinear behavior. In [14], the authors explored the dynamical behavior of an electromagnetic suspension system with nonlinear stiffness. It was observed that the electrical parameters are highly responsible for the behavior of the resulting mechanical damping effect, and indicates that the capacitance can be used as a control parameter.

The aim of this paper is to expand the results on [10], which studied a simplified suspension model which takes into account a secondary complete RLC electrical circuit that converts mechanical into electrical energy, thus acting as a damper. The vertical dynamics is analyzed, with a view to transfer functions of displacement transmissibility and charge. As charge produced by the passive isolator and isolation provided are competing objectives, a multi-objective optimization procedure with genetic algorithm is used in order to optimize the passive system in Sect. 3. Feedforward and feedback control strategies are employed to reduce vibration and a view to control effort is given. Feedforward strategy could be used as another possible solution to the lack of look-ahead capability described by [2] in active systems. Modifications to the optimal feedforward control strategy are suggested in order to decrease control effort in Sect. 4.1, and a feedback controller is proposed in Sect. 4.2. In short, the paper studies the passive and active electromechanical suspension systems and proposes: a multi-objective optimization procedure for the passive system, two modified feedforward optimum control in order to decrease control effort, and a displacement feedback controller.

2 Problem Formulation

The simplified electromechanical suspension model considered in this paper is depicted in Fig. 1 which represents a mass supported by mechanical components (spring and damper) and coupled electrical circuit including a resistor, an inductor and a capacitor. The system is subjected to harmonic base excitation x_0, which induces system displacement x and electrical charge q. Thus the system has two generalized coordinates. The equations of motion of this system can be achieved by Lagrangian approach for the mechanical and electrical subsystems, as described in the literature [4, 14, 15]. The set of two differential equations describing the dynamics of this system can be written as:

$$m\ddot{x} + c\dot{x} + kx - B\dot{q} = c\dot{x}_0 + kx_0 \tag{1}$$

$$L\ddot{q} + R\dot{q} + \frac{q}{C_0} + B\dot{x} = B\dot{x}_0 + e \tag{2}$$

where the mechanical system parameters m, k, and c are, respectively, the mass, stiffness and damping coefficients. The electrical system parameters L, R and C_0 are, respectively, the inductance, resistance and capacitance coefficients, and e is the external electric voltage (when present). The term B, which couples the two equations, is defined as the *Transducer Constant* (equal to the product of the length of the exposed coil and the magnetic flux). This term has SI unit Volt × second/meter

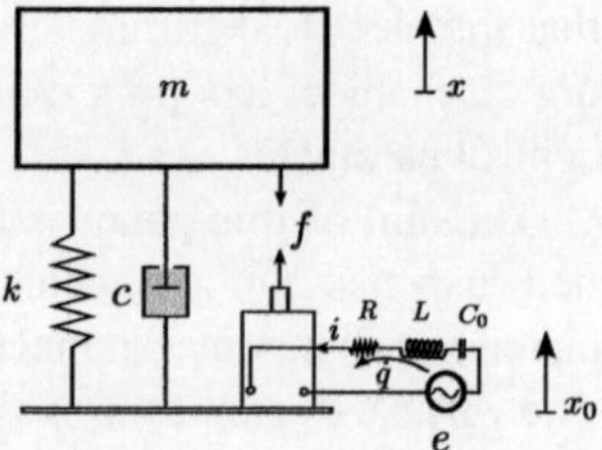

Fig. 1 Electromechanical suspension system subjected to base excitation

or Newton/Ampère. The dependent variables are the displacement of the mass (x) and the charge (q), while x_0 is the base displacement.

Frequency Response Analysis

Considering harmonic motion of the system described by set of Eqs. 1 and 2 in the form of complex exponential, such that $x(t) = Xe^{j\omega t}$, $x_0(t) = X_0e^{j\omega t}$, $q(t) = Qe^{j\omega t}$ and $e(t) = Ee^{j\omega t}$, the equations of motion are written in the frequency domain as:

$$\left(-\omega^2 m + j\omega c + k\right) X - j\omega BQ = (k + j\omega c)\, X_0 \tag{3}$$

$$\left(-\omega^2 L + j\omega R + \frac{1}{C_0}\right) Q + j\omega BX = j\omega BX_0 + E \tag{4}$$

A procedure to obtain the transmissibility frequency response $|X/X_0|$ consists of eliminating the charge amplitude Q. The motion of electromechanical system is better described in terms of the dynamic stiffness (or impedance), since it allows the modeling and analysis of numerical models that can be easily measured by experimental tests. Defining the impedance terms Z_m, Z_e and base impedance Z_s as:

$$Z_m = -\omega^2 m + j\omega c + k,\ Z_e = -\omega^2 L + j\omega R + \frac{1}{C_0},\ Z_s = j\omega c + k, \tag{5}$$

it is possible to write the displacement (X) in terms of the impedance as:

$$X = \left(Z_e Z_m - \omega^2 B^2\right)^{-1} \left(\left(Z_e Z_m - \omega^2 B^2\right) X_0 + j\omega BE\right) \tag{6}$$

For the case of the electrical charge Q, it is also possible to obtain the transfer function relating the base motion with the electrical charge by eliminating the displacement variable X such that:

$$Q = \left(Z_e Z_m - \omega^2 B^2\right)^{-1} \left(j\omega B\left(Z_m - Z_s\right) X_0 + Z_m E\right) \tag{7}$$

The transfer functions (X/X_0 and Q/X_0) for base excitation are shown in Fig. 2. The displacement transfer function (X/X_0) indicates the gain the sprung mass has when it is harmonically excited in its base under each excitation frequency. The charge transfer function (Q/X_0) indicates how much energy is obtained in the electrical subsystem when the mass is excited as aforementioned. Neglecting the parameters

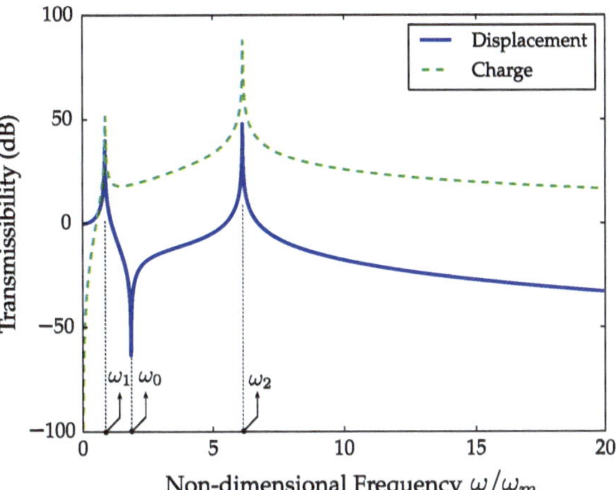

Fig. 2 Displacement and charge transfer functions as functions of the excitation frequency (a small amount of damping was considered to plot this figure)

related to the energy dissipation by making $c = R = 0$ which physically means no mechanical damping nor electrical resistance, it is possible to obtain Eq. 8 defining the system poles:

$$\omega_{1,2}^2 = \frac{\omega_m^2 + \omega_e^2 + \alpha\tau^2 \pm \sqrt{\left(\omega_m^2 - \omega_e^2\right)^2 + \alpha\tau^2\left(\alpha\tau^2 + 2\left(\omega_m^2 + \omega_e^2\right)\right)}}{2} \tag{8}$$

These roots are shown in Fig. 2 marked as ω_1 and ω_2. The normalization terms are $\omega_m = \sqrt{k/m}$, $\omega_e = \sqrt{1/LC_0}$, which are the natural frequencies of the uncoupled mechanical and electrical systems, $\tau = B/m$ and $\alpha = m/L$. The system also presents anti-resonance frequency (or zero) defined by $\omega_0 = \omega_m\omega_e/\sqrt{\omega_m^2 + \alpha\tau^2}$.

3 Parameters Optimization for Passive Vibration Control

In this section the system is optimized by using Genetic Algorithm based on the NSGA II algorithm [5] for multi-objective optimization, using the implementation in C-language of this algorithm. However, other approaches could be used as the one described in [12] or an analytical approach to analyze the power flow as in [18]. The objectives are to minimize the area under the curve of function $|X/X_0|$ in frequency domain and to maximize the area under the curve $|Q/X_0|$ in frequency domain. The optimization is done by adjusting coefficients c and R, according to the following constrained problem:

$$\underset{c,R}{minimize} \quad J_1, J_2$$
$$\text{subjected to} \quad 1 \le c \le 2000$$
$$1 \le R \le 200$$

where the objective functions J_1 and J_2 are defined as:

$$J_1 = \frac{1}{n}\sum_{i=1}^{n}\left|\frac{X(\omega_i)}{X_0(\omega_i)}\right| \quad \text{and} \quad J_2 = \frac{1}{\frac{1}{n}\sum_{i=1}^{n}\left|\frac{Q(\omega_i)}{X_0(\omega_i)}\right|} \tag{9}$$

For the optimization, the adopted parameters were: $m = 250$ (kg), $k = 40,000$ (N/m), $L = 0.76$ (H), $C_0 = 0.0014$ (F) and $B = 500$ (N/A) as used in Ref. [14].

The adopted population size for the optimization was 100, running for 80 generations, with probability for crossover of 90% and mutation probability of 50%.

The results of the optimization are summarized in a Pareto frontier, shown in Fig. 3a. As expected, the Pareto frontier shows an inverse relationship between the improvement of isolation and improvement of charge flowing in the electrical circuit. Figure 3b shows the effect of varying the mass on the Pareto frontier. It can be seen that changing the mass value shifts the Pareto frontier and does not change its shape significantly. Using the parameters obtained by the optimization procedure in the Pareto Frontier, defined by the points (1), (2) and (3) in Fig. 3a, it was possible to obtain the transfer functions ($|X/X_0|$ and $|Q/X_0|$) illustrated in Fig. 4. The point (1) corresponds to a situation where the objective function J_1 is a minimum, while objective function J_2 is a maximum, thus the area under curve $|Q/X_0|$ is minimum (least charge produced) and the area under curve $|X/X_0|$ is also minimum (best isolation). The point (3) corresponds to a situation where objective function J_2 is a minimum, thus maximum area under $|Q/X_0|$ curve, while objective function J_1 is a maximum, thus maximum area under $|X/X_0|$ curve. Point (2) corresponds to a trade-off relationship between the obtained response and the generated charge. Figure 4 reinforces aforementioned conclusions. The red lines correspond to the point (3) condition in Fig. 3 and it is unquestionable the best condition from the Pareto frontier to generate charge and the worst to obtain isolation. The dashed green lines correspond to point (2) and they show a situation in which there is better isolation than the previous condition, but generates lesser charge. The blue lines correspond to point (1) and it is unquestionable the best condition from the Pareto frontier for isolation and the worst for charge generation.

4 Active Control Optimization

In this section the strategies of feedforward and feedback control are investigated. Since active control requires energy, it is important to reduce the control effort necessary to act on the system in order to have efficient energy usage.

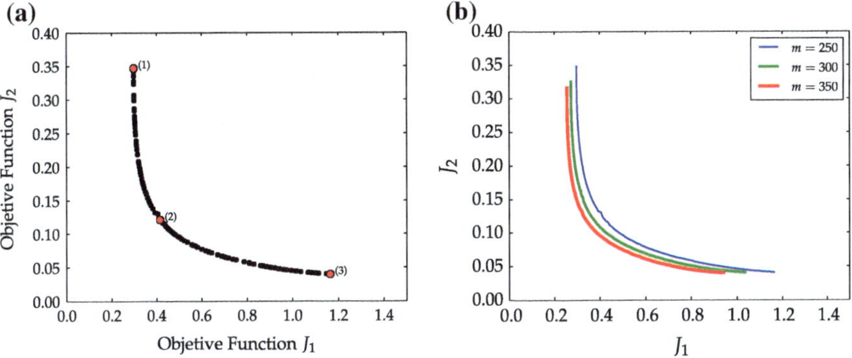

Fig. 3 Pareto frontier obtained for values of viscous damping c and electrical resistance R (**a**) and influence of values of the system mass m (**b**)

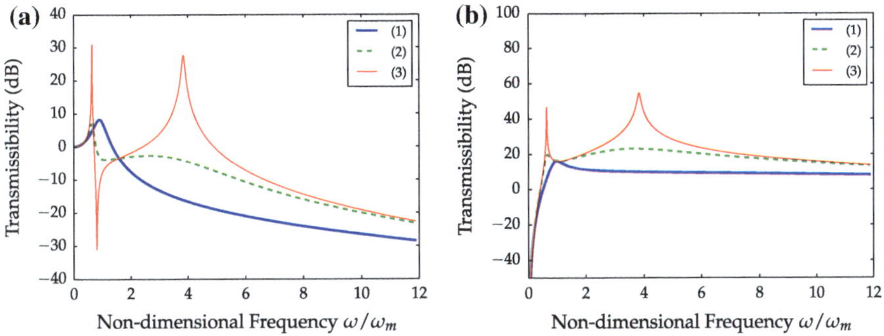

Fig. 4 Frequency response for the three values extracted from Fig. 3. Displacement transmissibility (**a**) and charge transmissibility (**b**). (1) $c = 3.92$ (N s/m) and $R = 177.48$ (Ω), (2) $c = 1.03$ (N s/m) and $R = 38.99$ (Ω), (3) $c = 1.00$ (N s/m) and $R = 1.00$ (Ω)

4.1 Feedforward Control Strategy

Feedforward is considered here as a solution to reduce the vibration of the system. Feedforward control was extensively studied in many applications and its principles are discussed in the book by [7] and also in references such as [3, 17].

The diagrams shown in Fig. 5 illustrate the idea of feedforward control, which makes use of a sensor to measure the disturbance to a control input. In this case, it converts the signal to an electrical voltage.

In simple terms, the idea of feedforward control is to provide a secondary input (*control input*) that cancels out (or minimizes) the effect of the disturbance. The control input is linear related to the disturbance input.

Consider the block diagram shown in Fig. 5b, where the block H_{ff} describes the feedforward control filter, and the blocks H_d and H_c describe the transfer functions

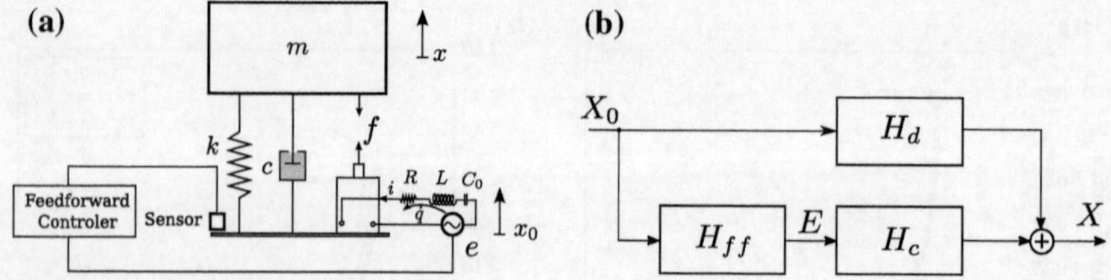

Fig. 5 The feedforward strategy for active control. Schematic idea of feedforward control (**a**) and block diagram (**b**)

relating the contributions of the disturbance X_0 to the response X and the contributions of the applied voltage E to the response X, respectively.

If the interest is to control the displacement of the mass, it is necessary to consider the expression relating X, X_0 and E, which is Eq. 6.

The transfer function H_d is obtained by making $E = 0$ and H_c is obtained by making $X_0 = 0$ in Eq. 6. The total response X can be written as:

$$X = H_d X_0 + H_c E = (H_d + H_c H_{ff}) X_0 \tag{10}$$

In this case, it is possible to find optimal feedforward control by making $X = 0$, which gives the following:

$$E_{opt} = j\left(\frac{Z_e}{\omega B}(j\omega c + k) - \omega B\right) X_0 \tag{11}$$

If this value of voltage (Eq. 11) is applied to the Eq. 6, the displacement amplitude is canceled ($X = 0$). This occurs because the impedance match between the mechanical and electrical inputs.

However, the control effort in terms of the total disturbance X_0, shown in Fig. 6a, is very high at low frequencies, which can make the controller implementation impracticable. If the transfer function of the optimal control E_{opt} in Eq. 11 is expanded, there is a term defined by $jk/\omega C_0 B$ which has a frequency dependence in the denominator. This term is responsible for large values of control voltage in the low frequency range of Fig. 6. There are some possibilities of reducing the control effort in the low frequency region. To do this, it is necessary to change the term $jk/\omega C_0 B$.

Two possible changes in the controller were investigated and the magnitude of the control voltage is shown in Fig. 6b. The first proposed solution is to remove the term $jk/\omega C_0 B$ from the feedforward controller, which produces a control voltage E_1. In the second case E_2, the term is changed to $jk/\omega_a C_0 B$, where $\omega_a = \sqrt{k/C_0\left(B^2 + kL\right)}$ which was set to match the control effort for ω_0 between control E_2 and E_{opt} (see Fig. 6b). By changing the feedforward optimal controller, there is a reduction in the performance of the controller. The performances are shown in the results of Fig. 7a (and in the detail for the first resonance frequency Fig. 7b). The performance with-

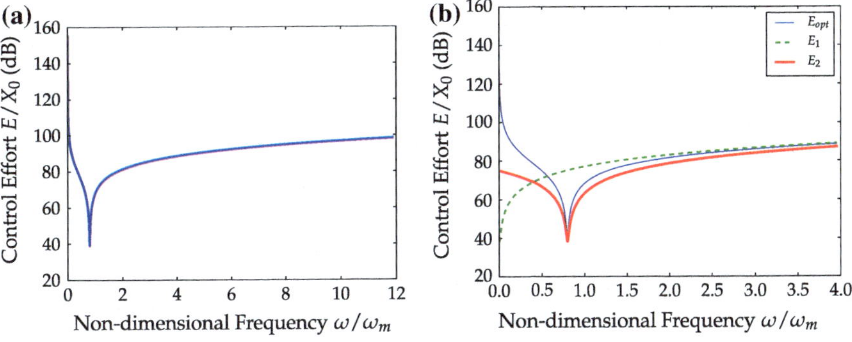

Fig. 6 Optimal feedforward control. Control effort (**a**) and the control effort for feedforward control (**b**)

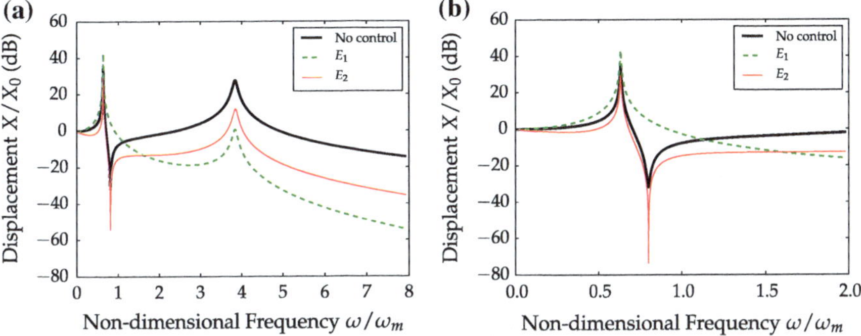

Fig. 7 Comparison of the transmissibility for the system without control and with modified feedforward control (**a**) and the detail of the first resonance (**b**)

out control is the third situation shown in Fig. 4, while the performances E_1 and E_2 are both from the aforementioned adapted optimal controllers. By observing Fig. 7, control E_1 is the best choice applied for the system at higher excitation frequencies, while control E_2 is the best choice applied for the system at low frequencies.

4.2 Feedback Control Strategy

Feedback control is now considered as a solution to control the vibration of the system. The case investigated is the displacement feedback, such that $E = g_d X$, where g_d is the displacement gain and $s = j\omega$. Figure 8 shows the system with feedback control, which consists of a sensor positioned in a point of interest (mass which vibrates due to the base excitation) that feeds back the electrical circuit in order to input a voltage to control the system vibration. For the case with displacement feedback,

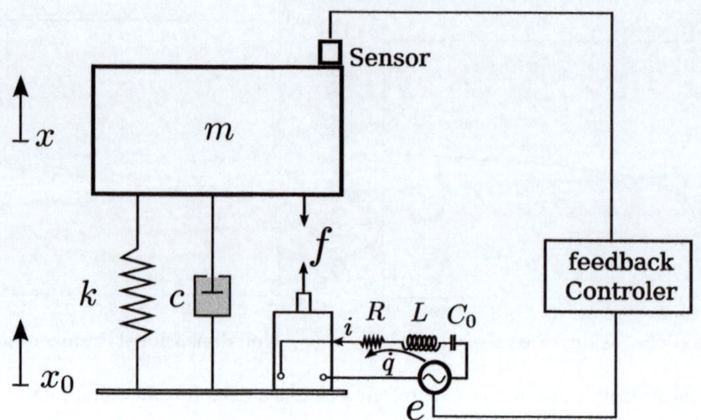

Fig. 8 Scheme of the system with feedback control

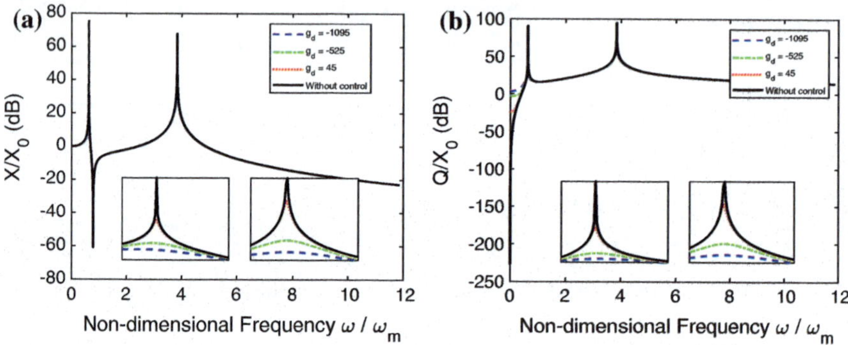

Fig. 9 Frequency responses for displacement feedback control and zoomed first and second resonance regions. Transmissibility (**a**) and control effort (**b**)

the denominator of the new transfer function becomes Eq. 12,

$$(mL)s^4 + (mR + cL)s^3 + \left(B^2 + cR + kL + \frac{m}{C_0}\right)s^2 + \left(kR + \frac{c}{C_0} - g_d B\right)s + \left(\frac{k}{C_0}\right) = 0 \quad (12)$$

According to [11], the Routh-Hurwitz stability criterion provides stability for the solution of Eq. 13 in gain g_d, for each feedback condition:

$$a_3 a_2 a_1 - a_3^2 a_0 - a_4 a_1^2 = 0 \quad (13)$$

where a_4, a_3, a_2, a_1 and a_0 are the terms multiplying, respectively, s^4, s^3, s^2, s^1 and s^0 in Eq. 12. Using the system parameters described in the previous sections, the conditions for stability states that $g_a > -1098$ and $g_b < 49.4$. Figure 9 show for both transmissibility and control effort (frequency responses) that as the absolute value of the gain is the highest as possible, the isolation is better. There is minimal or no difference between a system without active control and the system with active control for stable gains $g_a < g_d < g_b$ in regions far from the resonance regions.

5 Conclusions

Characterization of the frequency response of a linear electromechanical model was carried out, pointing out the equations of its poles and zeros. Optimization of the passive system was performed by using Genetic Algorithm and a Pareto frontier was obtained in order to group pairs of values of mechanical damping coefficient c and electrical resistance R which lead to better isolation, better energy harvesting or intermediate situations, for the proposed system. The influence of the sprung mass on the Pareto frontier was also investigated and it was concluded it does not change the shape of the frontier, although as the mass increases, the values of both objective functions decrease, i.e., increasing the sprung mass can improve both comfort and energy harvested. Active control with feedforward strategy was proposed. Since the optimal control given by Eq. 11 has a high control effort in low frequencies, two modifications were also proposed leading to the modified control laws E_1 and E_2 from that E_1 is the best for higher frequencies and E_2 is best for lower frequencies, considering both isolation (or resonance limitation) and control effort. E_2 is also a viable choice when maintaining the zero in transmissility frequency since it requires the same power as the system with optimal control, while E_1 has a significant effort and cannot maintain this zero. Finally, active control with displacement feedback was proposed, with its stability gain range obtained by Routh-Hurwitz criterion. It presented better isolation when compared to the passive system near the resonance regions.

Acknowledgements The authors would like to acknowledge FAPESP (process 2016/17083-4) for financial support providing the studentship of the first author.

References

1. Amati, N., Festini, A., Tonoli, A.: Design of electromagnetic shock absorbers for automotive suspensions. Veh. Syst. Dyn. **49**(12), 1913–1928 (2011)
2. Beno, J.H., Hoogterp, F.B., Bresie, D.A., Weeks, D.A., Ingram, S.K., Weldon, W.F.: Electromechanical suspension for combat vehicles. In: Proceedings of Society of Automotive Engineers, Technical Paper 950775 (1995)
3. Brennan, M.C., McGowan, A.M.R.: Piezoelectric power requirements for active vibration control. In: Proceedings of the 4th Smart Structures and Materials, pp. 660–669 (1997)
4. Crandall, S.H.: Dynamics of mechanical and electromechanical systems, McGraw-Hill (1968)

5. Deb, K., Pratap, A., Agarwal, S., Meyarivan, T.A.M.T.: A fast and elitist multiobjective genetic algorithm: NSGA-II. IEEE Trans. Evol. Comput. **6**(2), 182–197 (2002)
6. Ebrahimi, B., Bolandhemmat, H., Khamesee, M.B., Golnaraghi, F.: A hybrid electromagnetic shock absorber for active vehicle suspension systems. Veh. Syst. Dyn. **49**(1–2), 311–332 (2011)
7. Fuller, C.C., Elliott, S., Nelson, P.A.: Active control of vibration, Academic Press (1996)
8. Jonasson, M., Roos, F.: Design and evaluation of an active electromechanical wheel suspension system. Mechatronics **18**(4), 218–230 (2008)
9. Kjellqvist, P.: Experimental evaluation of an electromechanical suspension actuator for rail vehicle applications. In: Proceedings of the International Conference on Power Electronics, Machines and Drives (IET) (Conf. Publ. No. 487), pp. 165–170 (2002)
10. Kuhnert, W.M., Silveira, M., Gonçalves, P.J.P.: Optimization of passive and active electromechanical suspension systems. In: Pre Proceedings of the International Symposium on Dynamic Problems of Mechanics (DINAME) (Conf. Publ. No. 166), 8 pp. (2017)
11. Meirovitch, L.: Methods of analytical dynamics, Courier Corporation (2010)
12. Messine, F., Nogarede, B., Lagouanelle, J.L.: Optimal design of electromechanical actuators: a new method based on global optimization. IEEE Trans. Magn. **34**(1), 299–308 (1998)
13. Nagode, C.M.J.: Electromechanical suspension-based energy harvesting systems for railroad applications (2013)
14. Pontes, B.R. Jr., Silveira, M., Mazotti, A.C., Gonçalves, P.J., Balthazar, J.M.: Contribution of electrical parameters on the dynamical behaviour of a nonlinear electromagnetic damper. Nonlinear Dyn. **79**(3), 1957–1969 (2015)
15. Preumont, A.: Mechatronics: dynamics of electromechanical and piezoelectric systems, vol. 136. Springer Science & Business Media (2006)
16. Silveira, M., Pontes, B.R., Balthazar, J.M.: Use of nonlinear asymmetrical shock absorber to improve comfort on passenger vehicles. J. Sound Vib. **333**(7), 2114–2129 (2014)
17. Snyder, S.D., Tanaka, N.: On feedforward active control of sound and vibration using vibration error signals. J. Acoust. Soc. Am. **94**(4), 2181–2193 (1993)
18. Stephen, N.G.: On energy harvesting from ambient vibration. J. Sound Vib. **293**(1), 409–425 (2006)
19. Tonoli, A., Amati, N., Detoni, J.G., Galluzzi, R., Gasparin, E.: Modelling and validation of electromechanical shock absorbers. Veh. Syst. Dyn. **51**(8), 1186–1199 (2013)
20. Zuo, L., Scully, B., Shestani, J., Zhou, Y.: Design and characterization of an electromagnetic energy harvester for vehicle suspensions. Smart Mat. Struct. **19**(4), 045003 (2010)

A Neural Network Observer for Injection Rate Estimation in Common Rail Injectors with Nozzle Wear

Oliver Hofmann, Manuel Kiener and Daniel Rixen

Abstract The objective of this study is to present a neural observer that estimates changing injection behavior due to wear and aging effects within the nozzle of a common rail diesel injector. Using a dynamic identification system in combination with a modified learning rule, the neural observer is applicable to a wide range of problem sets. A multilayer perceptron (MLP) network with three layers and few neurons in the hidden layer ensures fast computing and high efficiency; network learning is based on quasi-Newton optimization and an additional line search algorithm. Modeling the bottom part of the injector introduces a simulation model, which is validated with experimental data from a solenoid common rail diesel injector. Estimation results conform well with the altered plant and therefore demonstrate the significant benefit of using the proposed neural network observer concept.

Keywords Neural network observer · Diesel injector · Injection rate estimation Nozzle wear · Injector aging

1 Introduction

The injection process significantly influences the performance of internal combustion engines. Aging effects such as coking or wear, which develop in the injector nozzle over its lifetime, deteriorate the injection behavior and result in increasing soot and nitrogen oxide (NO_x) emissions. These effects cause large uncertainties in injection rate estimation and require novel modeling, identification and control strategies. Krogerus et al. [7] presented a survey of analysis, modeling, and diagnostics of diesel fuel injection systems, which showed that fault effects such as nozzle wear influence

O. Hofmann (✉) · M. Kiener · D. Rixen
Department of Mechanical Engineering, Chair of Applied Mechanics,
Technical University of Munich, 85748 Garching, Germany
e-mail: oliver.hofmann@tum.de

A. de T. Fleury et al. (eds.), *Proceedings of DINAME 2017*, Lecture Notes in Mechanical Engineering, https://doi.org/10.1007/978-3-319-91217-2_19

the injection behavior over an engine's lifetime. Furthermore, the negative impact of injection faults on the combustion behavior, such as reduced combustion efficiency and higher emissions, was demonstrated [2, 6].

The estimation of the injection behavior in direct injection engines was subject of various studies. In [8], an injection rate observer for a solenoid injector was presented that was based on a sliding-mode approach. Another estimator, which used the signal of an in-line pressure sensor in a piezoelectric injector was shown in [13]. Since nozzle wear and aging effects in the injectors were not considered in the mentioned studies, previous work by the authors dealt with model-based estimation methods and state observers that consider these effects [4, 5], with the drawback that detailed pre-knowledge about the aging effects was required. Artificial neural networks and map-based methods are often used for the modeling and control of the injection behavior and its impacts on the combustion performance. Concepts that combine state observers and neural networks to identify nonlinearities can be found in various publications. Abdollahi et al. [1] and Talebi et al. [14] derived a stable neural network observer for coping with unknown system faults. Using multilayer perceptron (MLP) networks, the observer scheme can be applied to nonlinear MIMO systems by assuming observability. The learning rule for the network's weight adaption is based on a stable back-propagation training algorithm using Lyapunov's direct method. Hintz [3] introduced another state observer concept employing radial basis function (RBF) networks. Learning rules based on the failure models of Narendra and Annaswamy [9] made an identification of static and dynamic nonlinearities possible. Stringent requirements imposed on the system structure, such as the occurrence of isolated nonlinearities, limit the approach's applicability.

In this paper, a neural network observer scheme is proposed, which robustly estimates the injection rate of a common rail diesel injector. The observer's layout uses a neural network approximation to take additive fault into account. In contrast to state observers without adaptive models, the proposed neural network observer can cope with large uncertainties in the estimated states. It therefore ensures a robust estimation behavior considering wear and aging effects within the injector nozzle. The design of the neural network observer is discussed in Sect. 2. System identification provides information to the observer about the existing network output error. We determine a multilayer perceptron (MLP) feed-forward neural network structure with three layers of neurons. The network is trained using a modified quasi-Newton optimization approach to minimize a predefined cost function. A line search algorithm helps to optimize the observer performance by adapting the learning step size. Section 3 deals with the application of the proposed observer to a common rail diesel injector. The model focuses on the nozzle area, as aging effects are expected to be most evident there. We derive a nonlinear state-space model using measurement signals of the rail pressure and needle lift as model inputs and the pressure within the lower feed line as model output. Simulation results of the state-space model proposed in this study are compared to available measurement data in Sect. 4. Furthermore, we analyze the performance of the injection rate estimation, comparing the results to simulation data with modified nozzles.

2 Neural Observer Design

We consider the general nonlinear plant model, which is assumed to be observable

$$\begin{aligned} \dot{\boldsymbol{x}} &= \boldsymbol{f}(\boldsymbol{x},\boldsymbol{u}) + \boldsymbol{\Phi}(\boldsymbol{x},\boldsymbol{u}) \\ \boldsymbol{y} &= \boldsymbol{C}\boldsymbol{x} \end{aligned} \tag{1}$$

where $\boldsymbol{u}$ is the input, $\boldsymbol{y}$ is the output, and $\boldsymbol{x}$ is the state vector of the system. We suppose the model $\boldsymbol{f}(\boldsymbol{x},\boldsymbol{u})$ with additive fault $\boldsymbol{\Phi}(\boldsymbol{x},\boldsymbol{u})$ that we want to approximate using a neural network model. The nonlinear neural network observer is given by

$$\begin{aligned} \dot{\hat{\boldsymbol{x}}} &= \boldsymbol{f}(\hat{\boldsymbol{x}},\boldsymbol{u}) + \boldsymbol{\phi}(\hat{\boldsymbol{x}},\boldsymbol{u}) \\ \hat{\boldsymbol{y}} &= \boldsymbol{C}\hat{\boldsymbol{x}} \end{aligned} \tag{2}$$

where $\hat{\boldsymbol{x}}$ is the observed state. The neural network approximation $\boldsymbol{\phi}(\hat{\boldsymbol{x}},\boldsymbol{u})$ aims to identify the additive fault occurring within the plant. Consequently, the output error between plant and observer is defined as

$$\boldsymbol{e} = \boldsymbol{y} - \hat{\boldsymbol{y}} \tag{3}$$

The introduced observer system is depicted in Fig. 1. Additive fault influences the original plant model Eq. (1), which is denoted as $\boldsymbol{\Phi}(\boldsymbol{x},\boldsymbol{u})$. The neural observer estimates the derived state vector $\dot{\hat{\boldsymbol{x}}}$ by summing up the results of the nonlinear vector valued function $\boldsymbol{f}(\hat{\boldsymbol{x}},\boldsymbol{u})$ and the neural network output $\boldsymbol{\phi}(\hat{\boldsymbol{x}},\boldsymbol{u})$. The three-layer MLP feed-forward neural network is illustrated next to the observer structure. Weight adjustments within the neural observer are based on the network output error $\boldsymbol{\epsilon}$. The transfer function $\boldsymbol{E}(\boldsymbol{u},\hat{\boldsymbol{x}},\tilde{\boldsymbol{x}},\boldsymbol{e})$ describes the transition of the system error $\boldsymbol{e}$ to the network error $\boldsymbol{\epsilon}$ and is examined below.

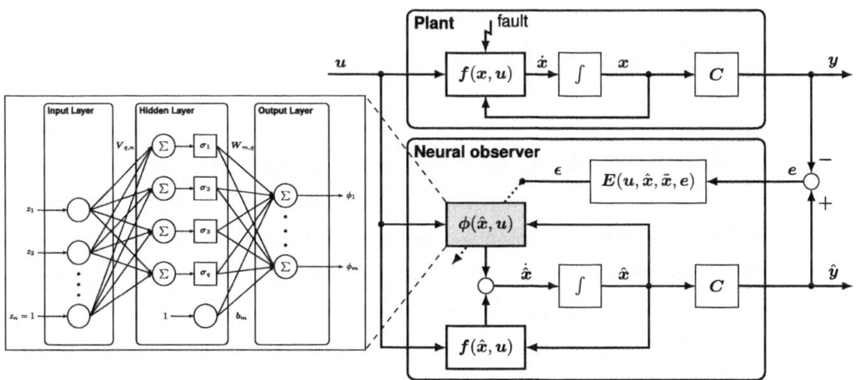

Fig. 1 Neural observer design with three-layer MLP feed-forward network

The observer's objective is to minimize the network output error defined as

$$\boldsymbol{\epsilon} = \boldsymbol{\Phi}(\boldsymbol{x}, \boldsymbol{u}) - \boldsymbol{\phi}(\hat{\boldsymbol{x}}, \boldsymbol{u}) \tag{4}$$

The neural network thereby relies on the information of this specific error to apply a learning algorithm. Further information needs to be supplied because the fault on the plant $\boldsymbol{\Phi}(\boldsymbol{x}, \boldsymbol{u})$ is an unknown quantity. We introduce a separate dynamic system to identify the system fault:

$$\dot{\tilde{\boldsymbol{x}}} = \boldsymbol{f}(\tilde{\boldsymbol{x}}, \boldsymbol{u}) - \boldsymbol{C}^T(\boldsymbol{y} - \boldsymbol{C}\tilde{\boldsymbol{x}}) \tag{5}$$

where $\tilde{\boldsymbol{x}}$ denotes the state vector of the novel identification system. We obtain approximated information about the fault behavior within the plant, which we use for further investigation. Subsequently, comparing the network output to the identified fault provides the required error. After applying mathematical transformations, the network output error is given by

$$\boldsymbol{\epsilon} = \boldsymbol{L}\left[\dot{\boldsymbol{e}} + \boldsymbol{C}\boldsymbol{f}(\hat{\boldsymbol{x}}, \boldsymbol{u}) - \boldsymbol{C}\boldsymbol{f}(\tilde{\boldsymbol{x}}, \boldsymbol{u})\right] \tag{6}$$

where $\boldsymbol{L}$ describes an observer design matrix, to assign $\boldsymbol{\epsilon}$ to the neural network output $\boldsymbol{\phi}$ by applying a gain value. To deal with faults in the physical plant, we model the identified error using a neural network approximation. The considered neural network is a multilayer perceptron (MLP) feed-forward network with three layers of neurons, which we define as

$$\begin{aligned} \boldsymbol{\phi}(z) &= \boldsymbol{W}\sigma(\boldsymbol{V}z) + \boldsymbol{b} \\ z &= \left[\boldsymbol{x}\ \boldsymbol{u}\ 1\right]^T \end{aligned} \tag{7}$$

The configurable weights $\boldsymbol{V}$ and $\boldsymbol{W}$, as well as the bias of the output layer $\boldsymbol{b}$, are given by

$$\begin{aligned} \boldsymbol{V} &= \left[\boldsymbol{v}_1{}^T \ \dots \ \boldsymbol{v}_q{}^T\right]^T \in \mathbb{R}^{q \times n} \\ \boldsymbol{W} &= \left[\boldsymbol{w}_1 \ \dots \ \boldsymbol{w}_q\right] \in \mathbb{R}^{m \times q} \\ \boldsymbol{b} &\in \mathbb{R}^{m \times 1} \end{aligned} \tag{8}$$

where n is the number of network inputs, m is the number of network outputs, and q is the number of hidden layer neurons. Note that the bias of the hidden layer is included in the weight matrix $\boldsymbol{V}$. The weight matrices consist of parameters that the learning algorithm determines, and we use a tangent hyperbolic as transfer function

$$\sigma_i = \frac{2}{1 + \exp\left(-2\boldsymbol{v}_i{}^T z\right)} - 1 \tag{9}$$

The training of the network weights and biases aims to minimize the network output's remaining approximation error $\boldsymbol{\epsilon}$. In order to measure the goodness of the fit between identified fault data and network output, we use a real time recurrent learning algorithm [15] for the observer application. Unlike during conventional backpropagation, the network parameters are not updated while sweeping through the training data set from t_0 to t_{end}, but use the information of all time steps instead. Therefore, the cost function J characterizes the total sum-squared error of all time samples in the network output [10]

$$J = J(t_0) + J(t_1) + \cdots + J(t_{end}) = \sum_{t=0}^{t_{end}} \frac{1}{2} \boldsymbol{\epsilon}^T \boldsymbol{\epsilon} \tag{10}$$

The objective of the network training is to minimize the cost function without constraints, $\min_{\boldsymbol{p}} J(\boldsymbol{p})$, which results in the following optimality conditions of the problem

$$\begin{aligned} \nabla J(\boldsymbol{p}^*) &= \mathbf{0} \\ \nabla^2 J(\boldsymbol{p}^*) &\geq \mathbf{0} \end{aligned} \tag{11}$$

where $\bullet^*$ denotes the parameter set with optimal conditions and all network parameters to be optimized are arranged within the parameter set

$$\boldsymbol{p} = \left[\boldsymbol{v}_1{}^T \ldots \boldsymbol{v}_q{}^T \ \boldsymbol{w}_1{}^T \ldots \boldsymbol{w}_q{}^T \ \boldsymbol{b}^T \right]^T \tag{12}$$

The cost function gradient, $\nabla J(\boldsymbol{p})$, of the given optimization problem is derived by applying the chain rule and results in

$$\nabla J(\boldsymbol{p}) = \sum_{t=0}^{t_{end}} \frac{\partial J}{\partial \boldsymbol{p}} = \sum_{t=0}^{t_{end}} \left(\frac{\partial J}{\partial \boldsymbol{\epsilon}} \cdot \frac{\partial \boldsymbol{\epsilon}}{\partial \boldsymbol{\phi}} \cdot \frac{\partial \boldsymbol{\phi}}{\partial \boldsymbol{p}} \right) = \sum_{t=0}^{t_{end}} \begin{bmatrix} -\frac{\boldsymbol{\epsilon}^T \boldsymbol{w}_1 (\sigma_1 + 1)^2}{\exp(2 \boldsymbol{v}_1 \boldsymbol{z})} \boldsymbol{z} \\ \vdots \\ -\frac{\boldsymbol{\epsilon}^T \boldsymbol{w}_q (\sigma_q + 1)^2}{\exp(2 \boldsymbol{v}_q \boldsymbol{z})} \boldsymbol{z} \\ -\epsilon_1 \boldsymbol{\sigma} \\ \vdots \\ -\epsilon_m \boldsymbol{\sigma} \\ -\boldsymbol{\epsilon} \end{bmatrix} \tag{13}$$

To solve the optimization problem, we iteratively update the parameters using a line search approach with search direction $\boldsymbol{\Delta p}^k$ and step size α^k

$$\boldsymbol{p}^{k+1} = \boldsymbol{p}^k + \alpha^k \boldsymbol{\Delta p}^k \tag{14}$$

The search direction is obtained by approximating the cost function using a quadratic model from the second-order Taylor series [11]

$$J(\boldsymbol{p} + \boldsymbol{\Delta p}) \approx J(\boldsymbol{p}) + \boldsymbol{\Delta p}^T \nabla J(\boldsymbol{p}) + \frac{1}{2} \boldsymbol{\Delta p}^T \nabla^2 J(\boldsymbol{p}) \boldsymbol{\Delta p} \tag{15}$$

which results in the so-called "Newton search direction", $\boldsymbol{\Delta p}^k$, assuming that the Hessian $\nabla^2 J(\boldsymbol{p}^k)$ is positive definite

$$\boldsymbol{\Delta p}^k = - \left[\nabla^2 J(\boldsymbol{p}^k)\right]^{-1} \nabla J(\boldsymbol{p}^k) \tag{16}$$

As direct calculation of the Hessian is numerically expensive, an approximation of the Hessian, $\boldsymbol{H}^k \approx \nabla^2 J(\boldsymbol{p}^k)$, is often used. The very common BFGS updating method proposed by Broyden, Fletcher, Goldfarb, and Shanno is given by [11]

$$\boldsymbol{H}^{k+1} = \boldsymbol{H}^k - \frac{\boldsymbol{H}^k \boldsymbol{d}^k (\boldsymbol{d}^k)^T \boldsymbol{H}^k}{(\boldsymbol{d}^k)^T \boldsymbol{H}^k \boldsymbol{d}^k} + \frac{\boldsymbol{v}\boldsymbol{v}^T}{\boldsymbol{d}^T \boldsymbol{v}} \tag{17}$$

where

$$\begin{aligned} \boldsymbol{v} &= \boldsymbol{p}^{k+1} - \boldsymbol{p}^k \\ \boldsymbol{d} &= \nabla J(\boldsymbol{p}^{k+1}) - \nabla J(\boldsymbol{p}^k) \end{aligned}$$

The step size is obtained using the Amijio backstepping algorithm. Starting with an initial value $\alpha^k = \alpha_0^k$, the step size is reduced to satisfy the inexact line search condition for sufficient decrease with the constant value $\rho \in (0, 1)$

$$J(\boldsymbol{p}^k + \alpha^k \boldsymbol{\Delta p}^k) < J(\boldsymbol{p}^k) + \rho \alpha^k \nabla J(\boldsymbol{p}^k) \boldsymbol{\Delta p}^k \tag{18}$$

The complete algorithm used for the neural network training is summarized in Algorithm 1. We initialize the network weights as well as the Hessian approximation at $k = 0$ using the scaled identities $\beta \boldsymbol{I}$ and $\gamma \boldsymbol{I}$, respectively. Furthermore, the convergence tolerance is set to $\mu > 0$, which is used as the optimization abort criterion. In each iteration, a search direction $\boldsymbol{\Delta p}^k$ and a step size α^k are computed according to the Newton direction with Hessian approximation, Eq. (16), and the backstepping method to satisfy the Amijio condition, Eq. (18). The parameter set, $\boldsymbol{p}^{k+1}$, is updated using line search according to Eq. (14) and the Hessian, $\boldsymbol{H}^{k+1}$, of the next iteration is approximated using the BFGS algorithm, Eq. (17). When reaching the convergence tolerance, μ, the optimal network weights are derived from the optimal parameter set, $\boldsymbol{p}^*$.

3 Application: Injection Rate Observer

Aging effects within the nozzles of common rail diesel injectors result in deteriorating injection behavior. We use the neural network observer to estimate the injection rate of a diesel injector, taking the changed dynamics due to nozzle wear into con-

Algorithm 1: Neural network learning for fault identification.

Input : $\boldsymbol{p}^0$	// Initial network weights
	// $\boldsymbol{V}^0 = \beta\boldsymbol{I}, \boldsymbol{W}^0 = \beta\boldsymbol{I}$, and $\boldsymbol{b}^0 = \beta\boldsymbol{I}$
$\boldsymbol{H}^0 = \gamma\boldsymbol{I}$	// Initial Hessian approximation
$\mu > 0$	// Convergence tolerance
$k = 0$	
while $\|\|\nabla J^k\|\| > \mu$ **do**	
compute search direction $\boldsymbol{\Delta p}^k$	// Newton direction with Hessian approximation, Eq. (16)
compute step size α^k	// Backstepping to satisfy Amijio condition, Eq. (18)
update parameter set $\boldsymbol{p}^{k+1}$	// Line search update, Eq. (14)
compute $\boldsymbol{H}^{k+1}$	// BFGS update, Eq. (17)
$k = k + 1$	
end	
Output: $\boldsymbol{p}^* = \boldsymbol{p}^k$	// Optimal solution of the weights $\boldsymbol{V}^*, \boldsymbol{W}^*$, and $\boldsymbol{b}^*$

sideration. The injector of interest is a solenoid CRIN3.18 injector manufactured by Bosch, which is depicted is Fig. 2. Modeling the bottom part of the injector is sufficient, since this study investigates the effects of nozzle wear. Measurement data at different operating conditions is available for the marked signals rail pressure, p_R, lower feed line pressure, p_L, needle lift x_N, and the injection rate, Q_H, which is also used to analyze the observer's performance.

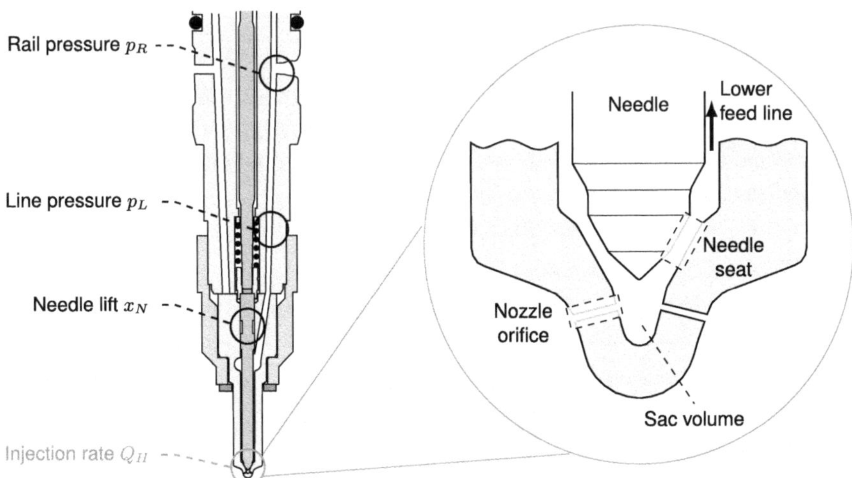

Fig. 2 Bottom part of the considered common rail diesel injector CRIN3.18 with available measurement signals and detailed nozzle section

A simplified model describing the injection dynamics is used for observer design. We model the one-dimensional flow of incompressible diesel fluid in the lower feed line using the physical principles of mass conservation, Newton's second law, and energy conservations. To obtain an analytical solution of these equations, several simplifications are applied. Neglecting the convective terms as well as gravity effects and applying the Galerkin method [12] with steady friction in the pipe leads to the following equation

$$\dot{Q}_L = \frac{A_L}{\rho_L L_L}\left(p_R - p_L\right) - \frac{\xi}{2A_L L_L} Q_L |Q_L| \tag{19}$$

The storage capability of the lower feed line is considered using a compressible model of the cavity

$$\dot{p}_L = \frac{K_L}{V_L}\left(Q_L - Q_H\right) \tag{20}$$

The volumetric flow rate at the needle seat Q_S and the injected flow rate through the nozzle orifices are obtained using Bernoulli's equations. The sac volume is considered by applying the continuity equation to the cavity

$$Q_S = \operatorname{sgn}(p_L - p_S)\alpha_S A_S \sqrt{\frac{2\left|p_L - p_S\right|}{\rho_C}} \tag{21}$$

$$\dot{p}_S = \frac{K_C}{V_S}\left(Q_S - Q_H\right) \tag{22}$$

$$Q_H = \operatorname{sgn}(p_S - p_C)\alpha_H A_H \sqrt{\frac{2\left|p_S - p_C\right|}{\rho_C}} \tag{23}$$

where p_C is the pressure in the combustion chamber, which is assumed as constant.

We obtain a normalized state-space representation of the equations using the rail pressure and the area at the needle seat as system inputs, and line pressure, line flow rate, and sac pressure as states.

$$\begin{aligned} \boldsymbol{u} &= \begin{bmatrix} \frac{p_R - p_C}{p_0} & \frac{A_S}{A_0} \end{bmatrix}^T \\ \boldsymbol{x} &= \begin{bmatrix} \frac{p_L - p_C}{p_0} & \frac{Q_L}{Q_0} & \frac{p_S - p_C}{p_0} \end{bmatrix}^T \end{aligned} \tag{24}$$

Equations (19)–(23) can then be rewritten for $A_S > 0$ as

$$\begin{aligned} \dot{\boldsymbol{x}} &= \boldsymbol{f}(\boldsymbol{x},\boldsymbol{u}) = \begin{bmatrix} a_1 x_2 - a_2\sqrt{x_3} \\ a_3\left(u_1 - x_1\right) - a_4 x_2 |x_2| \\ a_5 u_2\sqrt{x_1 - x_3} - a_6\sqrt{x_3} \end{bmatrix} \\ \boldsymbol{y} &= \boldsymbol{C}\boldsymbol{x} \end{aligned} \tag{25}$$

The measurement of the line pressure is used to identify nozzle wear, which results in $y = x_1$.

Analyzing the observer's performance, we designed a simulated experiment, which was based on the state-space model of the injector and the neural network observer

$$\begin{aligned} \dot{\hat{\boldsymbol{x}}} &= \boldsymbol{f}(\hat{\boldsymbol{x}}, \boldsymbol{u}) + \boldsymbol{\phi}(\hat{\boldsymbol{x}}, \boldsymbol{u}) \\ \hat{\boldsymbol{y}} &= \boldsymbol{C}\hat{\boldsymbol{x}} \end{aligned} \tag{26}$$

where we modeled the system plant by adding a fault model, representing nozzle wear, as follows

$$\boldsymbol{\Phi}(\boldsymbol{x}, \boldsymbol{u}) = \begin{bmatrix} -v_H a_2 \sqrt{x_3} \\ 0 \\ v_S a_5 u_2 \sqrt{x_1 - x_3} - v_H a_6 \sqrt{x_3} \end{bmatrix} \tag{27}$$

The fault model describes parameter variation within the needle seat and the nozzle orifices with $v_S, v_H \in (-\frac{1}{2}, \frac{1}{2})$ and results in additional nonlinear system dynamics to be identified.

4 Results

The validation results of the simplified state-space model are presented in Fig. 3. We used measured signals of the rail pressure and the needle lift as input signals for the model and compared the resulting feed line pressure and the mass injection rate to experimental data. The figure shows the results at an energizing time of 1.9 ms and set rail pressures of 60, 100, and 140 MPa. It can be seen that the simulation results, despite the simplification of the model, agree well with the measured data

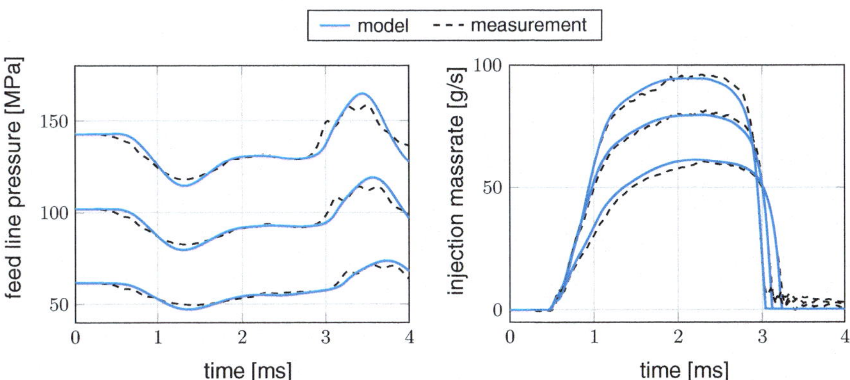

Fig. 3 Validation of the simplified state-space injector model with experimental data at energizing time 1.9 ms and set rail pressures of 60, 100, and 140 MPa

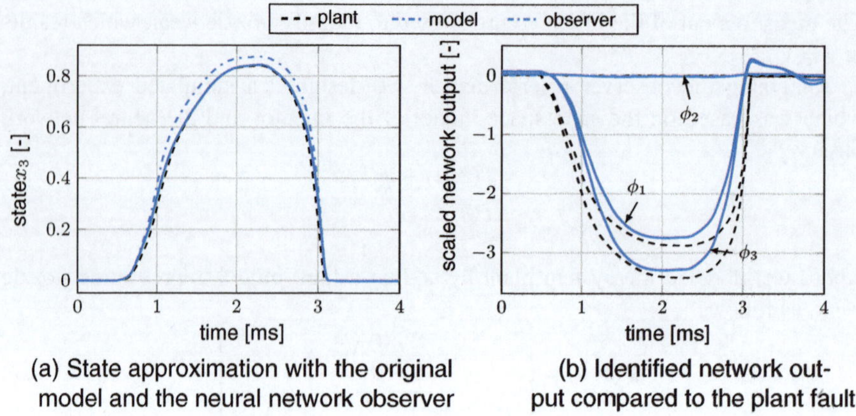

Fig. 4 Identification of the nonlinearity due to nozzle wear at energizing time 1.9 ms and set rail pressure 100 MPa. Artificial nozzle wear of $v_H = 0.3$ is identified using the injection rate observer algorithm after 30 training epochs

during the event of an injection ($Q_H > 0$) at the different working conditions. The accuracy deteriorates after the end of injection, which is irrelevant for the observer application since the nozzle is separated from the rail supply and aging effects cannot be identified. In summary, it can be said that the state-space model is capable of predicting significant signals during the injection process and that it has therefore proved to be suitable to estimate the injection rate.

Applying the observer algorithm for injection rate estimation, we used a neural network with $q = 5$ hidden layer neurons. The observer design vector was experimentally determined as $\boldsymbol{L} = \begin{bmatrix} 1 & 0 & 12 \end{bmatrix}^T$. We used a scaled identity instead of random numbers to initialize the network weights in order to improve comparability. Furthermore, network training was stopped after 30 training epochs for the same reason. As an example, Fig. 4 shows a state approximation result at energizing time 1.9 ms and set rail pressure 100 MPa using the network observer. The fault model, Eq. (27), was added to the initial plant model with the aging coefficients of $v_S = 0$ and $v_H = 0.3$. In Fig. 4a, the estimation of state x_3, which is relevant for estimating the injection rate, is compared to the output of the model without observer. It can be seen that the neural network observer improves the prediction compared to the initial model. Figure 4b shows the states, ϕ_1 to ϕ_3, of the identified fault model. The observer tracks the artificial plant fault quite well by using the considered design vector $\boldsymbol{L}$. Note that the identification of faults with more complex dynamical behavior is challenging, since a constant gain $\boldsymbol{L} = \text{const}$ was used in this study to distribute the measured error to all states.

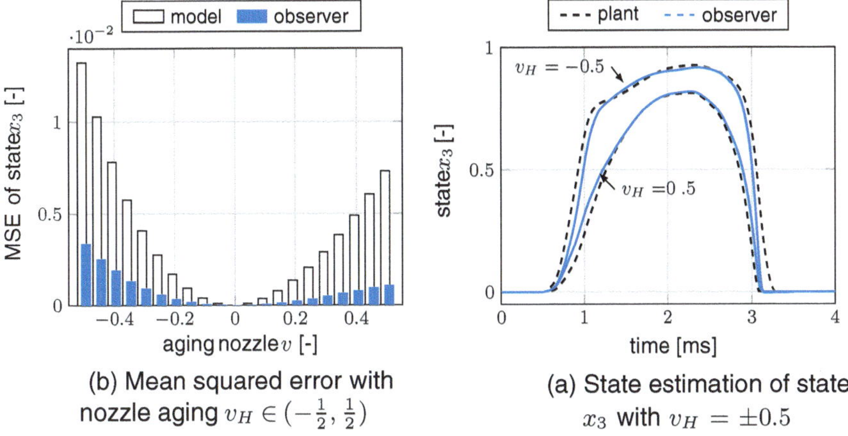

Fig. 5 Performance of the network observer with simulated aging parameter in the nozzle $v_H \in (-\frac{1}{2}, \frac{1}{2})$

We analyzed the performance of the network observer by using the mean squared error (MSE) of the state x_3, which is defined as

$$\text{MSE} = \frac{1}{N} \sum_{i=0}^{N} \left[x_3(t_i) - \hat{x}_3(t_i) \right]^2 \tag{28}$$

The observer was tested by varying the nozzle parameter within the range of $v_H \in (-\frac{1}{2}, \frac{1}{2})$. The mean squared error was evaluated for each configuration, which is depicted in Fig. 5. Compared to the initial model, the estimation error improves at each analyzed fault condition, using the neural network observer. Additionally, the resulting state of interest is shown for the extreme nozzle aging parameters, $v_H = \pm 0.5$. It can be seen that the observer predicts the shape of the faulty state in both cases, but slightly better for $v_H = 0.5$ as the observer is unable to estimate the modified end of injection for $v_H = -0.5$. In terms of nozzle wear, which can be described as a linear parameter change of the nozzle orifice discharge coefficient, the neural network observer exhibited very good overall performance.

5 Conclusion

In this paper, a neural network state observer scheme was proposed for application to a common rail diesel injector exhibiting aging phenomena. The observer design included a three-layer MLP feed-forward network for fast, flexible adoptions of the observer's characteristics. Applying a Newton optimization procedure combined with a backstepping method yields good learning process efficiency. Separate fault identification guarantees correct network output error for the weight adoption process. As the presented concept combines system identification with a classical observer scheme, a dynamical model of the fault in the plant could be obtained. Simulation results confirm the applicability of the proposed observer to the injector model under wear and aging effects. Due to the reliable performance and the additional advantage of an identified fault model, the neural network observer is beneficial in terms of its use for control methods in future investigations.

Acknowledgements This research is supported by the German Research Foundation (DFG, grant number RI2451/1). The authors would like to thank Sebastian Schuckert and Georg Wachtmeister of the Chair of Internal Combustion Engines, Technical University of Munich, for providing the experimental data used for validating the injector model.

References

1. Abdollahi, F., Talebi, H., Patel, R.: A stable neural network-based observer with application to flexible-joint manipulators. IEEE Trans. Neural Netw. **17**, 118–129 (2006). https://doi.org/10.1109/TNN.2005.863458
2. dAmbrosio, S., Ferrari, A.: Diesel injector coking: optical-chemical analysis of deposits and influence on injected flow-rate. Fuel spray and engine performance. J. Eng. Gas Turbines Power **134** (2012). https://doi.org/10.1115/1.4005991
3. Hintz, C.: Identifikation nichtlinearer mechatronischer Systeme mit strukturierten rekurrenten Netzen. PhD thesis, Technical University of Munich (2003)
4. Hofmann, O., Strau, P., Schuckert, S., Huber, B. et al.: Identification of aging effects in common rail diesel injectors using geometric classifiers and neural networks. SAE Technical Paper 2016-01-0813, SAE International (2016). https://doi.org/10.4271/2016-01-0813
5. Hofmann, O., Han, S., and Rixen, D.: Common rail diesel injectors with nozzle wear: modeling and state estimation. SAE Technical Paper 2017-01-0543, SAE International (2017). https://doi.org/10.4271/2017-01-0543
6. Ikemoto, M., Omae, K., Nakai, K., Ueda, R., Kakehashi, N., Sunami, K.: Injection nozzle coking mechanism in common-rail diesel engine. SAE Int. J. Fuels Lubric. **5**(2011-01-1818), 78–87 (2011). https://doi.org/10.4271/2011-01-1818
7. Krogerus, T.R., Hyvnen M.P., Huhtala K.J.: A survey of analysis, modeling, and diagnostics of diesel fuel injection systems. J. Eng. Gas Turbines Power **138**, 081501-1–081501-11 (2016). https://doi.org/10.1115/1.4032417
8. Nam, K., Yoon, M., Park, S., Sunwoo, M.: Development of a sensorless estimation algorithm of the injection timing and rate for an HSDI common-rail injector. JSME Int. J. Series C **47**, 882–888 (2004). https://doi.org/10.1299/jsmec.47.882
9. Narendra, K., Annaswamy, A.: Stable adaptive systems. Int. J. Adapt. Control Signal Process. **4**, 185–186 (1990). https://doi.org/10.1002/acs.4480040212
10. Nelles, O.: Nonlinear system identification: from classical approaches to neural networks and fuzzy models. Springer, Berlin Heidelberg, Germany (2001). https://doi.org/10.1007/978-3-662-04323-3
11. Nocedal, J., Wright, S.: Numerical Optimization. Springer Science & Business Media, New York, USA (2006). https://doi.org/10.1007/978-0-387-40065-5
12. Pfeiffer, F., Borchsenius, F.: New hydraulic system modelling. J. Vib. Control **10**, 1493–1515 (2004). https://doi.org/10.1177/1077546304042063

13. Satkoski, C.A., Ruikar, N.S., Biggs, S.D., Shaver, G.M.: Piezoelectric fuel injection: cycle-to-cycle control of tightly spaced injections. Control Eng. Practice **20**, 1175–1182 (2012). https://doi.org/10.1016/j.conengprac.2012.06.002
14. Talebi, H., Abdollahi, F., Patel, R., Khorasani, K.: Neural network-based state estimation of nonlinear systems. Lect. Notes Control Inf. Sci. **395**, 15–35 (2010). https://doi.org/10.1007/978-1-4419-1438-5
15. Williams, R., Zipser D.: A learning algorithm for continually running fully recurrent neural networks. Neural Comput. **1**, 270–280 (1989). https://doi.org/10.1162/neco.1989.1.2.270

Lagrange's, Maggi's and Kane's Equations Applied to the Dynamic Modelling of Serial Manipulator

Fernando Malvezzi, Renato M. M. Orsino and Tarcisio Antonio Hess Coelho

Abstract Robot Manipulators have been employed in many types of industries, such as pharmaceutical, chemical, automotive, aerospace, etc. A manipulator is a mechanism used to move an object along a given trajectory. Topologically, the mechanism can be constituted by parallel or serial chains. The serial kinematic chain is constituted by links connected sequentially by joints. The aim of this work is to obtain a qualitative comparison among three approaches typically applied to the modelling of multibody mechanical systems. The chosen system is a 5-DOF serial robot manipulator and the three approaches are based on the use of Lagrange's, Maggi's and Kane's equations. The purpose of the modelling is to obtain the equations of motion for this serial robotic manipulator. Some numerical simulations are performed to illustrate how the obtained models can be used to predict the dynamic behavior of the chosen system.

Keywords Multibody dynamics · Analytical mechanics · Lagrangian formulation · Kane's formalism · Maggi's equations · Robotic manipulators

F. Malvezzi (✉)
Department of Mechanical Engineering, Maua Institute of Technology,
São Caetano do Sul, Brazil
e-mail: fernando.malvezzi@maua.br

R. M. M. Orsino
Department of Mechanical Engineering, Polytechnic School,
University of São Paulo, São Paulo, Brazil
e-mail: renato.orsino@gmail.com

T. A. H. Coelho
Department of Mechatronics and Mechanical Systems Engineering,
Polytechnic School, University of São Paulo, São Paulo, Brazil
e-mail: tarchess@usp.br

A. de T. Fleury et al. (eds.), *Proceedings of DINAME 2017*, Lecture Notes in Mechanical Engineering, https://doi.org/10.1007/978-3-319-91217-2_20

1 Introduction

The equations of motion for multibody systems can be obtained from different methodologies such as Newton-Euler equations, Lagrange equations, Kane's Method and Maggi's equations.

The main advantage of Newton-Euler Method [1] is that the equations of motion will always have the same fundamental form independently of the geometry, inertia or constraints of motion of a rigid body. On the other hand, the constraint forces or torques must be determined, which may lead to difficulties when the system is composed by many bodies. Regarding to Lagrangian formalism [2, 3], it allows obtain constraint-free ordinary differential equations, which is an important advantage when compared to Newton-Euler Method. Many commercial softwares for Multibody Dynamic Systems, such as ADAMS, DADS and DYMAC, apply Lagrangian formulation [4]. Another method used in commercial softwares for Multibody Dynamic Systems, such as SD-EXACT, NBOD2 and SD/FAST [4], is Kane's.

Developed at the time of the first applications of computational tools to the study of multibody systems, at the 1960s, Kane's method [5] is claimed by some authors not to be original as a theoretical formalism [6–8], once it is based on the developments of Appell and Maggi dating from the early twentieth century. Kane's method, however, can be considered as a specialization of previously developed methodologies for optimized computational analysis of multibody systems. As well as in Lagrangian formalism, Kane's approach allows obtain constraint-free ordinary differential equations. Maggi's equations inspired not only Kane's developments but also are associated to several methodologies based on the use of orthogonal complement projections [9–11]. Such an approach allows the use of redundant variables in a given formulation without having to deal with the inconvenience of introducing undetermined multipliers. In its original form, Maggi's equations are an extension of the Lagrangian formalism, in which the application of a projection operator (orthogonal complement matrix) eliminates the terms containing the multipliers.

This work deals with a qualitative comparison of three approaches typically applied to the modeling of multibody mechanical systems. The chosen system is a serial robot manipulator and the three approaches are based on the use of Lagrange's, Maggi's and Kane's equations. The aim of modeling is to obtain the equations of motion for this serial robot manipulator. Moreover, the numerical simulations are performed with these models in order to analyze some typical motions performed by the system. Finally, a qualitative assessment of each approach is performed considering four features associated with these approaches.

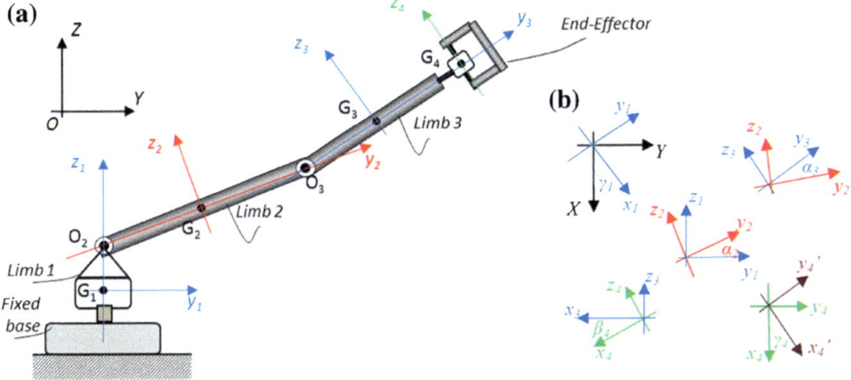

Fig. 1 **a** Serial manipulator model; **b** generalized coordinates

Table 1 Modies movements of the manipulator robot

Body number	Robot component	Generalized coordinates
1	Limb 1	γ_1
2	Limb 2	α_2
3	Limb 3	α_3
4	End-effector	β_4 and γ_4

2 Dynamic Modeling

The serial robot manipulator shown in Fig. 1a has 5-DOF. Table 1 and Fig. 1b shows the robot movements and the generalized coordinates chosen in the Kane's, Lagrange's and Maggi's approaches. In order to simplify the dynamic modelling of this multibody mechanical system, the following hypotheses are adopted:

- All limbs of the robot are considered rigid bodies.
- The frictional torque at all joints is negligible.
- The following axes are ones of symmetry: x_1, z_1; x_2, y_2, z_2; x_3, y_3, z_3. Consequently, the product of inertia of the bodies 1, 2 and 3 are nulls.
- The moments of inertia of body 4 (end-effector) are negligible. The mass of the object moved by the robot was added to the mass of end-effector.
- The G_4 point of the frame $G_4x_4y_4z_4$ represents the centroid of the end-effector.

2.1 Formulation Based on Kane's Method

The equations of motion of this serial manipulator robotic exposed in Fig. 1 can be obtained by the Kane's equations [5]:

$$F_r + F_r^* = 0 \quad (r = 1, 2, 3) \tag{1}$$

where F_r is the generalized active forces associated to r-th coordinate and F_r^* is the generalized inertia forces associated to r-th coordinate. The generalized active forces F_r are obtained by the following equation:

$$F_r = \sum \mathbf{F_k} \cdot \frac{\partial \mathbf{v_{G_k}}}{\partial u_r} + \sum \mathbf{M_k} \cdot \frac{\partial \boldsymbol{\omega}_\mathbf{k}}{\partial u_r} \tag{2}$$

where $\mathbf{F_k}$ and $\mathbf{M_k}$ are the forces and moments actuating on body k, respectively; $\mathbf{v_{G_k}}$ and $\boldsymbol{\omega}_\mathbf{k}$ are the velocity of the centre of mass and the angular velocity of body k, respectively; $\partial \mathbf{v_{G_k}}/\partial u_r$ and $\partial \boldsymbol{\omega}_\mathbf{k}/\partial u_r$ are the partial velocities and partial angular velocities of body k, respectively. The variables u_r are called "generalized speeds", and correspond either the to time derivatives of generalized coordinates or to any other set of variables that could replace them in the description of the motion of the system. The generalized inertia forces F_r^* are calculated by Eq. (3). Notice that F_r* are obtained by dot-multiplying the inertia terms of Newton-Euler equations [6] by the partial velocities and partial angular velocities.

$$F_r^* = -\sum \left[m_k \mathbf{a_{G_k}} \cdot \frac{\partial \mathbf{v_{G_k}}}{\partial u_r} + (J_k \dot{\omega}_\mathbf{k} + \boldsymbol{\omega}_\mathbf{k} \times J_k \boldsymbol{\omega}_\mathbf{k}) \cdot \frac{\partial \boldsymbol{\omega}_k}{\partial u_r} \right] \tag{3}$$

where $\mathbf{a_{G_k}}$ is the acceleration of the centre of mass of body k. By applying Eqs. (1–3), we obtain the equations of motion for the serial mechanism exposed in Fig. 1:

$$\mathbf{F} = \mathbf{M}(\mathbf{q})\,\dot{\mathbf{u}} + \mathbf{V}(\mathbf{q}, \mathbf{u}) \tag{4}$$

In Eqs. (4–7), $\mathbf{q}$ is the set of generalized coordinates; m_k is the mass of body k ($k = 1, 2, 3, 4$); J_{x_k}, J_{y_k}, J_{z_k} are the central inertia moments of body k ($k = 1, 2, 3$); ℓ_2, ℓ_3 are the length of bodies 2 and 3, respectively; T_1, T_2 and T_3 are the actuators torque applied to the bodies 1, 2 and 3, respectively.

$$\mathbf{F} = \begin{bmatrix} T_1 \\ T_2 \\ T_3 \end{bmatrix} \quad \mathbf{M} = \begin{bmatrix} M_{11} & M_{12} & M_{13} \\ M_{21} & M_{22} & M_{23} \\ M_{31} & M_{32} & M_{33} \end{bmatrix} \quad \dot{\mathbf{u}} = \begin{bmatrix} \ddot{\gamma}_1 \\ \ddot{\alpha}_2 \\ \ddot{\alpha}_3 \end{bmatrix} \quad \mathbf{V} = \begin{bmatrix} V_1 \\ V_2 \\ V_3 \end{bmatrix} \tag{5}$$

$$\begin{cases} M_{11} = J_{z_1} + m_2 \frac{\ell_2^2}{4} c_{\alpha_2}^2 + J_{y_2} s_{\alpha_2}^2 + J_{z_2} c_{\alpha_2}^2 + m_3 [\ell_2^2 c_{\alpha_2}^2 + \frac{\ell_3^2}{4} c_{(\alpha_2+\alpha_3)}^2 + \ell_2 \ell_3 c_{(\alpha_2+\alpha_3)} c_{\alpha_2}] \\ \qquad + J_{y_3} s_{(\alpha_2+\alpha_3)}^2 + J_{z_3} c_{(\alpha_2+\alpha_3)}^2 + m_4 [\ell_2^2 c_{\alpha_2}^2 + 2_2 \ell_3 c_{(\alpha_2+\alpha_3)} c_{\alpha_2} + \ell_3^2 c_{(\alpha_2+\alpha_3)}^2] \\ M_{12} = M_{13} = M_{21} = M_{31} = 0 \\ M_{22} = m_2 \frac{\ell_2^2}{4} + J_{x_2} + m_3 (\ell_2^2 + \frac{\ell_3^2}{4} + \ell_2 \ell_3 c_{\alpha_3}) + J_{x_3} + m_4 (\ell_2^2 + \ell_3^2 + 2\ell_2 \ell_3 c_{\alpha_3}) \\ M_{23} = M_{32} = m_3 (\frac{\ell_3^2}{4} + \frac{\ell_2 \ell_3}{2} c_{\alpha_3}) + J_{x_3} + m_4 (\ell_3^2 + \ell_2 \ell_3 c_{\alpha_3}) \\ M_{33} = m_3 \frac{\ell_3^2}{4} + J_{x_3} + m_4 \ell_3^2 \end{cases} \tag{6}$$

$$\begin{cases} V_1 = -m_2 \frac{\ell_2^2}{2} s_{\alpha_2} c_{\alpha_2} \dot{\gamma}_1 \dot{\alpha}_2 + 2(J_{y_2} - J_{z_2}) s_{\alpha_2} c_{\alpha_2} \dot{\gamma}_1 \dot{\alpha}_2 + 2J_{y_3} s_{(\alpha_2+\alpha_3)} c_{(\alpha_2+\alpha_3)} (\dot{\alpha}_2 + \dot{\alpha}_3) \dot{\gamma}_1 \\ \quad - 2J_{z_3} s_{(\alpha_2+\alpha_3)} c_{(\alpha_2+\alpha_3)} (\dot{\alpha}_2 + \dot{\alpha}_3) \dot{\gamma}_1 + m_3 [-2\ell_2^2 s_{\alpha_2} c_{\alpha_2} \dot{\alpha}_2 - \frac{\ell_3^2}{2} s_{(\alpha_2+\alpha_3)} c_{(\alpha_2+\alpha_3)} (\dot{\alpha}_2 + \dot{\alpha}_3) \\ \quad - \ell_2 \ell_3 s_{(\alpha_2+\alpha_3)} (\dot{\alpha}_2 + \dot{\alpha}_3) c_{\alpha_2} - \ell_2 \ell_3 c_{(\alpha_2+\alpha_3)} (\dot{\alpha}_2 + \dot{\alpha}_3) s_{\alpha_2} \dot{\alpha}_2] \dot{\gamma}_1 + m_4 [-2\ell_2^2 s_{\alpha_2} c_{\alpha_2} \dot{\alpha}_2 \\ \quad - 2\ell_3^2 s_{(\alpha_2+\alpha_3)} c_{(\alpha_2+\alpha_3)} (\dot{\alpha}_2 + \dot{\alpha}_3) - 2\ell_2 \ell_3 s_{(\alpha_2+\alpha_3)} (\dot{\alpha}_2 + \dot{\alpha}_3) c_{\alpha_2} - 2\ell_2 \ell_3 c_{(\alpha_2+\alpha_3)} s_{\alpha_2} \dot{\alpha}_2] \dot{\gamma}_1 \\ V_2 = -m_3 [\ell_2 \ell_3 s_{\alpha_3} (\dot{\alpha}_2 + \frac{\dot{\alpha}_3}{2}) \dot{\alpha}_3] - m_4 [\ell_2 \ell_3 s_{\alpha_3} (2\dot{\alpha}_2 + \dot{\alpha}_3) \dot{\alpha}_3] - \{ -m_2 \frac{\ell_2^2}{4} s_{\alpha_2} c_{\alpha_2} \dot{\gamma}_1^2 \\ \quad + (J_{y_2} - J_{z_2}) s_{\alpha_2} c_{\alpha_2} \dot{\gamma}_1^2 - m_3 [\ell_2^2 s_{\alpha_2} c_{\alpha_2} + \frac{\ell_3^2}{4} s_{(\alpha_2+\alpha_3)} c_{(\alpha_2+\alpha_3)} + \ell_2 \ell_3 s_{(2\alpha_2+\alpha_3)}] \dot{\gamma}_1^2 \\ \quad + (J_{y_3} - J_{z_3}) s_{(\alpha_2+\alpha_3)} c_{(\alpha_2+\alpha_3)} \dot{\gamma}_1^2 - m_4 [-\ell_2^2 s_{\alpha_2} c_{\alpha_2} - \ell_2 \ell_3 s_{(2\alpha_2+\alpha_3)} - \ell_3^2 s_{(\alpha_2+\alpha_3)} c_{(\alpha_2+\alpha_3)}] \dot{\gamma}_1^2 \} \\ \quad + m_2 g \frac{\ell_2}{2} c_{\alpha_2} + m_3 g [\ell_2 c_{\alpha_2} + \frac{\ell_3}{2} c_{(\alpha_2+\alpha_3)}] + m_4 g [\ell_2 c_{\alpha_2} + \ell_3 c_{(\alpha_2+\alpha_3)}] \\ V_3 = -\ell_2 \ell_3 s_{\alpha_3} \dot{\alpha}_2 \dot{\alpha}_3 (\frac{m_3}{2} + m_4) - \{ -m_3 [\frac{\ell_3^2}{4} s_{(\alpha_2+\alpha_3)} c_{(\alpha_2+\alpha_3)} \dot{\gamma}_1^2 + \frac{\ell_2 \ell_3}{2} s_{(\alpha_2+\alpha_3)} c_{\alpha_2} \dot{\gamma}_1^2 \\ \quad + \ell_2 \ell_3 s_{\alpha_3} (\dot{\alpha}_2 + \dot{\alpha}_3) \dot{\alpha}_2] + (J_{y_3} - J_{z_3}) s_{(\alpha_2+\alpha_3)} c_{(\alpha_2+\alpha_3)} \dot{\gamma}_1^2 \} - m_4 [-\ell_2 \ell_3 s_{(\alpha_2+\alpha_3)} c_{\alpha_2} \\ \quad - \ell_3^2 s_{(\alpha_2+\alpha_3)} c_{(\alpha_2+\alpha_3)}] \dot{\gamma}_1^2 - \ell_2 \ell_3 s_{\alpha_3} (\dot{\alpha}_2 + \dot{\alpha}_3) \dot{\alpha}_2] + m_3 g \frac{\ell_3}{2} c_{(\alpha_2+\alpha_3)} + m_4 g \ell_3 c_{(\alpha_2+\alpha_3)} \end{cases} \tag{7}$$

2.2 Formulation Based on the Lagrangian Formalism

The Lagrange equations [2] for the serial mechanism exposed in Fig. 1 are given by:

$$\frac{d}{dt}\left(\frac{\partial L}{\partial \dot{\gamma}_1}\right) - \frac{\partial L}{\partial \gamma_1} = Q_{\gamma_1} \quad \frac{d}{dt}\left(\frac{\partial L}{\partial \dot{\alpha}_2}\right) - \frac{\partial L}{\partial \alpha_2} = Q_{\alpha_2} \quad \frac{d}{dt}\left(\frac{\partial L}{\partial \dot{\alpha}_3}\right) - \frac{\partial L}{\partial \alpha_3} = Q_{\alpha_3} \tag{8}$$

where L is called Lagrangian, which is equal to the kinetic energy of system minus the potential energy. By developing Eq. (8), we obtain the equations of motion for the serial mechanism shown in Fig. 1:

$$\mathbf{Q} = \mathbf{M}(\mathbf{q})\ddot{\mathbf{q}} + \mathbf{V}(\mathbf{q}, \dot{\mathbf{q}}) \tag{9}$$

where

$$\mathbf{Q} = \begin{bmatrix} T_1 \\ T_2 \\ T_3 \end{bmatrix} \quad \mathbf{M} = \begin{bmatrix} M_{11} & M_{12} & M_{13} \\ M_{21} & M_{22} & M_{23} \\ M_{31} & M_{32} & M_{33} \end{bmatrix} \quad \ddot{\mathbf{q}} = \begin{bmatrix} \ddot{\gamma}_1 \\ \ddot{\alpha}_2 \\ \ddot{\alpha}_3 \end{bmatrix} \quad \mathbf{V} = \begin{bmatrix} V_1 \\ V_2 \\ V_3 \end{bmatrix} \tag{10}$$

where T_1, T_2 and T_3 are the actuators torque applied to the bodies 1, 2 and 3, respectively. The elements of the matrices **M** and **V** in Eq. (10) were shown in Eqs. (6) and (7).

2.3 Formulation Based on Maggi's Equations

After deriving the Lagrangian equations of motion for the mechanism, it can be noticed that most of the difficulties in this procedure arise from the complexity of the expressions of the energy functions involved. The hypothesis concerning the negligible moments of inertia of body 4 is rather fortunate, otherwise the algebraic complexity of the terms in these equations would be even greater. These difficulties could be overcome if the positions, orientations, velocities and angular velocities of the rigid-bodies could be described in terms of simpler expressions. The most straightforward alternative for doing so is to define redundant generalized coordinates and quasi-velocities so that the energy functions become as simple as possible. Such an approach has proven to be successful for the mathematical modeling of parallel mechanisms [12, 13]. In case of redundancy in the definition of generalized coordinates, the application of the Lagrangian formalism would require the use of undetermined multipliers. Maggi's equations, however, provide an alternative formalism in which one can still take advantage of a modeling based on redundant variables, without needing to use Lagrangian multipliers.

Let (q_i) denote a generic set of redundant generalized coordinates adopted for the description of the motion of a given multibody system. In order to this set of variables be able to represent configurations and states compatible with the constraints of the system, it must be required for them to satisfy some constraint equations. These equations might be written in one of the following forms:

$$h_k(t, q_i) = 0 \tag{11}$$

$$n_k(t, q_i, \dot{q}_i) = 0 \tag{12}$$

Holonomic constraints can be represented by expressions in both forms (11) and (12). Constraints which can only be represented by an expression in the form of (12) but not by any expression in the form of (11) are nonholonomic. Maggi's equations are applicable both to holonomic and nonholonomic systems whose constraints can at least be expressed in the form of (12). Both the second time derivative of (11) and the first time derivative of (12) can be expressed as an affine equation in terms of $\ddot{q}_r$:

$$\sum_r A_{kr}(t, q_i, \dot{q}_i)\ddot{q}_r + b_k(t, q_i, \dot{q}_i) = 0 \tag{13}$$

In a given time instant t^*, consider that the state $(t^*; \overset{*}{q_i}; \overset{*}{\dot{q}_i})$ is known, so that the variations of the coordinates and of its time derivatives can be assumed to be zero. In order for an infinitesimal variation of $\ddot{q}_r$ not to violate any constraint of the system, the following condition must be satisfied:

$$\sum_r A_{kr}\left(t, \overset{*}{q_i}, \overset{*}{\dot{q}_i}\right)\delta\ddot{q}_r = 0 \tag{14}$$

Therefore, in a time interval defined in a neighborhood of t^*, any virtual displacement must satisfy the condition [14]:

$$\sum_r A_{kr}\delta q_r = 0 \tag{15}$$

Let $A = [A_{kr}]$ denote the matrix constituted by the coefficients A_{kr}. A general solution for (15) involves finding a matrix $\mathbf{C} = [C_{rs}]$ which is an orthogonal complement of $\mathbf{A}$, i.e. a maximal rank matrix satisfying the condition $\mathbf{AC} = 0$ [14]. Thus, the variations δq_r associated to the generalized coordinates q_r can be expressed as a linear combination of as much arbitrary variations (denoted by $\delta\theta_s$) as the number ν of degrees of freedom of the system, i.e.:

$$\delta q_r = \sum_s C_{rs}\delta\theta_s \tag{16}$$

Therefore, applying this result to the extended Hamilton's Principle, it can be stated that the Maggi's equations for this system are given by the following expression:

$$\sum_r C_{rs}\left[\frac{d}{dt}\left(\frac{\partial L}{\partial \dot{q}_r}\right) - \frac{\partial L}{\partial q_r} - Q_{q_r}\right] = \sum_r C_{rs}\Phi_{q_r} = 0 \quad s = 1, \ldots, \nu \tag{17}$$

Analysing the foregoing derivations, it is convenient to define an extra angular generalized coordinate $\chi_3 = \alpha_2 + \alpha_3$, and ($x_j$; y_j; z_j) representing the Cartesian coordinates of the centres of mass G_j of the limbs j = 2 and j = 3, and of the end-effector $j = 4$:

$$\begin{aligned} x_j &= -e_j\ell_2 s_{\gamma_1} c_{\alpha_2} - f_j\ell_3 s_{\gamma_1} c_{(\alpha_2+\alpha_3)} \\ y_j &= e_j\ell_2 c_{\gamma_1} c_{\alpha_2} + f_j\ell_3 c_{\gamma_1} c_{(\alpha_2+\alpha_3)} \\ z_j &= e_j\ell_2 s_{\alpha_2} + f_j\ell_3 s_{(\alpha_2+\alpha_3)} \end{aligned} \tag{18}$$

With $e_2 = ½$, $f_2 = 0$, $e_3 = 1$, $f_3 = ½$, $e_4 = 1$ and $f_4 = 1$. In this case, taking $\delta q = (\delta\gamma_1, \delta\alpha_2, \delta\alpha_3, \delta\chi_3, \delta x_j, \delta y_j, \delta z_j)$ and $\delta\theta = (\delta\gamma_1, \delta\alpha_2, \delta\alpha_3)$ it can be stated that $\delta q = \mathbf{C}\delta\theta$, with:

$$\mathbf{C} = \begin{bmatrix} 1 & 0 & 0 \\ 0 & 1 & 0 \\ 0 & 0 & 1 \\ 0 & 1 & 1 \\ -c_{\gamma_1}\left(e_j\ell_2 c_{\alpha_2} + f_j\ell_3 c_{\chi_3}\right) & -s_{\gamma_1}\left(e_j\ell_2 s_{\alpha_2} + f_j\ell_3 s_{\chi_3}\right) & -f_j\ell_3 s_{\gamma_1} s_{\chi_3} \\ -s_{\gamma_1}\left(e_j\ell_2 c_{\alpha_2} + f_j\ell_3 c_{\chi_3}\right) & -c_{\gamma_1}\left(e_j\ell_2 s_{\alpha_2} + f_j\ell_3 s_{\chi_3}\right) & -f_j\ell_3 c_{\gamma_1} s_{\chi_3} \\ 0 & e_j\ell_2 c_{\alpha_2} + f_j\ell_2 c_{\chi_3} & f_j\ell_3 c_{\chi_3} \end{bmatrix} \tag{19}$$

Also:

$$\begin{aligned} L = {} & \frac{1}{2}J_{z_1}\dot{\gamma}_1^2 + \frac{1}{2}m_2\left(\dot{x}_2^2 + \dot{y}_2^2 + \dot{z}_2^2\right) + \frac{1}{2}J_{x_2}\dot{\alpha}_2^2 + \frac{1}{2}J_{y_2}\left(s_{\alpha_2}\dot{\gamma}_1\right)^2 \\ & + \frac{1}{2}J_{z_2}\left(c_{\alpha_2}\dot{\gamma}_1\right)^2 + \frac{1}{2}m_3\left(\dot{x}_3^2 + \dot{y}_3^2 + \dot{z}_3^2\right) + \frac{1}{2}J_{x_3}\dot{\chi}_3^2 + \frac{1}{2}J_{y_3}\left(s_{\chi_3}\dot{\gamma}_1\right)^2 \\ & + \frac{1}{2}J_{z_3}\left(c_{\chi_3}\dot{\gamma}_1\right)^2 + \frac{1}{2}m_4\left(\dot{x}_4^2 + \dot{y}_4^2 + \dot{z}_4^2\right) + m_2 g z_2 + m_3 g z_3 + m_4 g z_4 \end{aligned} \tag{20}$$

Thus:

$$\begin{cases} \Phi_{\gamma_1} = \left[\left(J_{y_2} + J_{z_2}\right)\left(s_{\alpha_2}\right)^2 + \left(J_{y_3} + J_{z_3}\right)\left(s_{\chi_3}\right)^2 + J_{z_1}\right]\ddot{\gamma}_1 \\ \qquad + \left(J_{y_2} + J_{z_2}\right)\left(s_{2\alpha_2}\right)\dot{\gamma}_1\dot{\alpha}_2 + \left(J_{y_3} + J_{z_3}\right)\left(s_{2\chi_3}\right)\dot{\gamma}_1\dot{\chi}_3 - T_1 \\ \Phi_{\alpha_2} = J_{x_2}\ddot{\alpha}_2 - \left(J_{y_2} + J_{z_2}\right)s_{\alpha_2}c_{\alpha_2}\dot{\gamma}_1^2 - T \\ \Phi_{\alpha_3} = -T_3 \\ \Phi_{\chi_3} = J_{x_3}\ddot{\chi}_3 - \left(J_{y_3} + J_{z_3}\right)\dot{\gamma}_1^2 s_{\chi_3} c_{\chi_3} \\ \Phi_{\chi_j} = m_j\ddot{x}_j \\ \Phi_{y_j} = m_j\ddot{y}_j \\ \Phi_{z_j} = m_j\left(\ddot{z}_j - g\right) \end{cases} \tag{21}$$

Table 2 Robot parameters employed in the simulations

Robot parameters	Symbol	Value
Moment of inertia of the body 1 with respect to the z_1 axis (kg m^2)	J_{z_1}	3
Mass of body 2 (kg)	m_2	3
Length of body 2 (m)	ℓ_2	1
Moment of inertia of the body 2 with respect to the x_2 axis (kg m^2)	J_{x_2}	0.36
Moment of inertia of the body 2 with respect to the y_2 axis (kg m^2)	J_{y_2}	0.07
Moment of inertia of the body 2 with respect to the z_2 axis (kg m^2)	J_{z_2}	0.36
Mass of body 3 (kg)	m_3	3
Length of body 3 (m)	ℓ_3	1
Moment of inertia of the body 3 with respect to the x_3 axis (kg m^2)	J_{x_3}	0.36
Moment of inertia of the body 3 with respect to the y_3 axis (kg m^2)	J_{y_3}	0.07
Moment of inertia of the body 3 with respect to the z_3 axis (kg m^2)	J_{z_3}	0.36
Mass of body 4 (kg)	m_4	1

The system of Maggi's equations for the manipulator can be readily obtained by Eq. (17), using the expressions of C_{rs} from (19) and the expressions for Φ_{qr} from (21).

3 Numerical Simulations

The equations of motion obtained by Kane's, Lagrange's and Maggi's approaches are the same. In this section, some numerical simulations are performed to illustrate how the obtained models can be used to predict the dynamic behavior of the manipulator. Table 2 shows the robot parameters employed in the simulations.

In order to perform numerical simulations, the expressions of the redundant variables χ_3, x_j, y_j and z_j ($j = 2; 3; 4$) and their time derivatives in terms of γ_1, α_2 and α_3 can be either replaced in Maggi's equations, leading to a system of ordinary differential equations (ODEs) with as much equations as the number of degrees of freedom of the manipulator (3, in this case), or can simply be taken along with the already obtained Maggi's equations, leading to an extended system of equations which are typically referred as differential-algebraic equations (DAEs). The former option, in this case, for the particular matrix **C** in Eq. (20), would lead to a system of equations identical to the foregoing Lagrangian equations of motion. Therefore, in the numerical simulations performed, the latter option is chosen. Two inverse simulations and one forward simulation are performed. In the inverse ones, Maggi's equations are used to calculate the time histories of the torques provided by the actuators in order to perform a given prescribed motion. In the forward one, slight variations of the static values of the torques (the values that ensure the equilibrium of the system in the reference configuration in which the values of γ_1, α_2 and α_3 and their time derivatives are zero) are considered, and the output obtained are the time

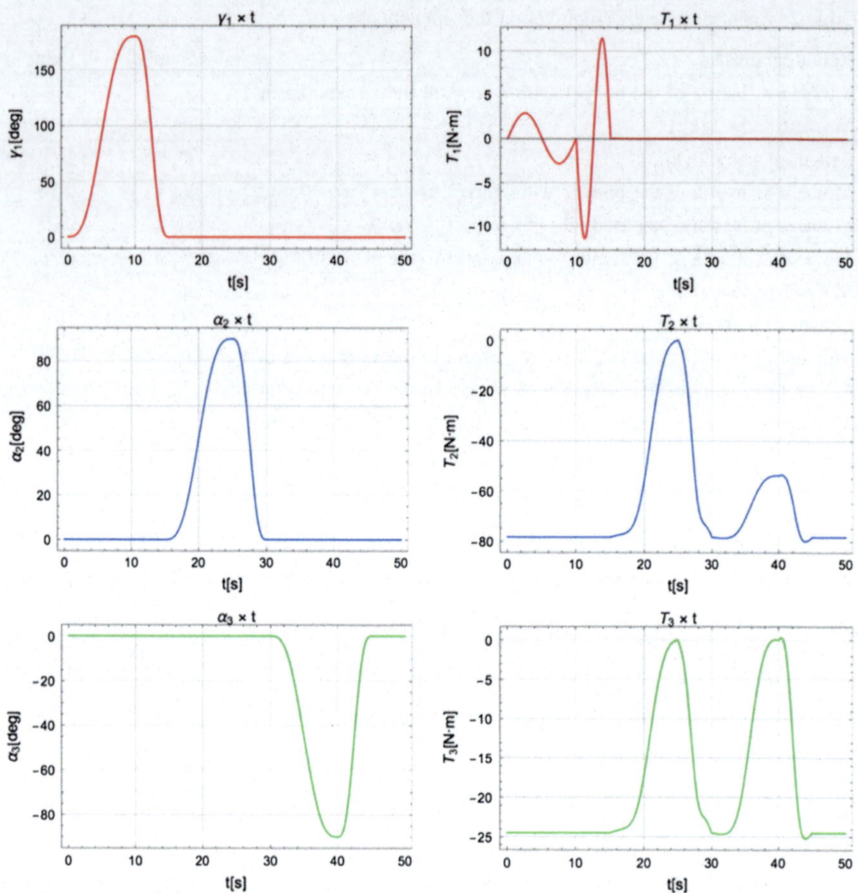

Fig. 2 First inverse simulation: slow prescribed motion

histories of the associated motion, provided by the integration of the corresponding equations of motion.

Both in the inverse and forward simulations, the interpolating curves use to define the time histories of the inputs respect two properties: their second time derivatives are sinusoidal and the extremum points are also inflection points (i.e. first and second time derivatives are simultaneously zero). In the case of the inverse simulations, two similar scenarios: in the second, the rates are 2.5 times faster than in the first. This allows making an assessment of the influence of the inertial effects in the motion of this mechanism (once they must be much more influent in this second scenario). The results of these numerical simulations are shown in Figs. 2, 3 and 4.

The results seem to be consistent with the existing cylindrical symmetry of the mechanism (the motions with respect to the vertical plane passing through the

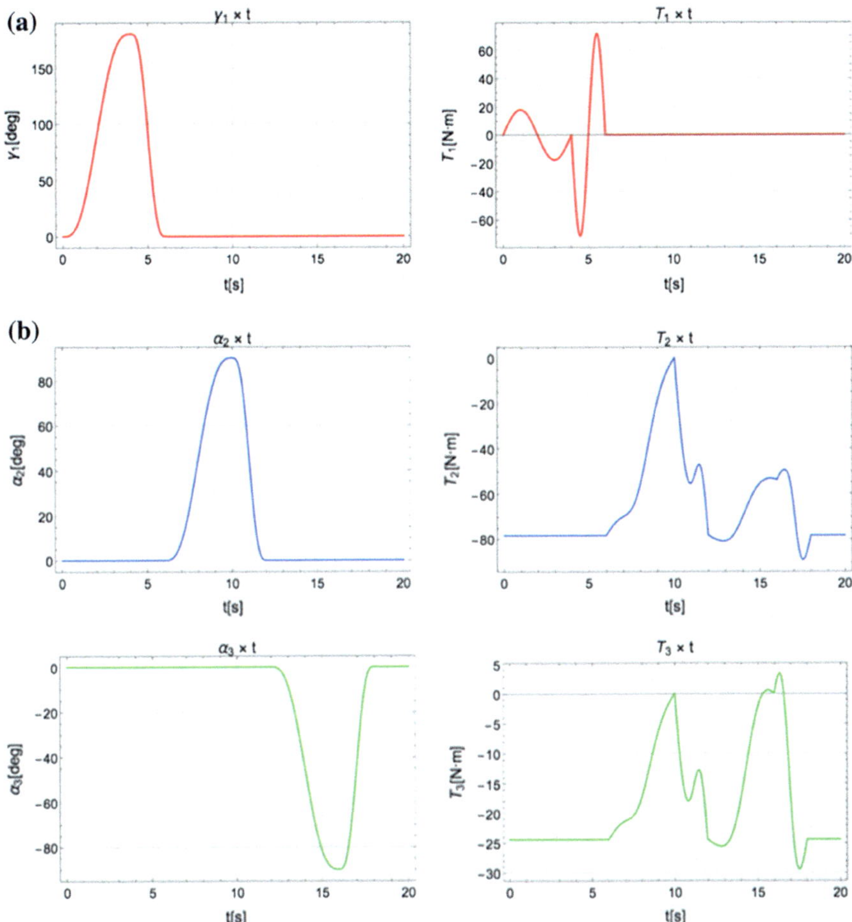

Fig. 3 **a** Second inverse simulation: faster prescribed motion: γ_1 versus t; T_1 versus t. **b**. Second inverse simulation: faster prescribed motion: α_2 versus t; T_2 versus t; α_3 versus t; T_3 versus t

centres of mass of 2, 3 and 4 are not influenced by the torque T_1) and reveal, as expected for a serial mechanism, a significant influence of the inertial effects in the torque imposed (being greater the closer the actuator is to the base of the mechanism).

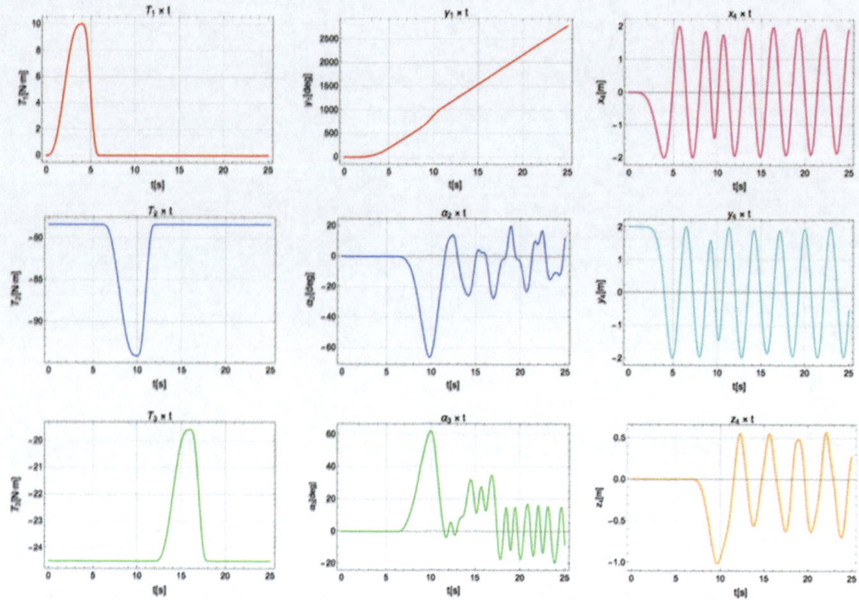

Fig. 4 Forward simulation

Table 3 Qualitative assessment of the models: Kane's, Lagrange's and Maggi's approach

Model	Derivation	Interpretability	Modularity	Constraint forces elimination
Kane	2	5	2	5
Lagrange	4	3	2	5
Maggi	5	2	5	5

4 Qualitative Assessment of Models

In order to conclude the discussion on the suitability of each analytical mechanics approach to model the chosen system, a qualitative assessment is performed based on the results presented in the previous sections. This assessment consists of comparing the following four characteristics associated with these approaches:

1. Derivation: the effort to obtain dynamic equations of the model in a form that is suitable to numerical simulations.
2. Interpretability: the ability to understand the meaning of each term of dynamic equations.
3. Modularity: how simple is the procedure of adding components in a model.
4. Constraint forces elimination: how simple is the procedure of obtaining equations of motion without terms related to constraint forces.

For each of these characteristics, Kane's, Lagrange's and Maggi's approach were ranked in order to guide the selection of an appropriate method to model other kinds of mechanical systems similar to the serial mechanism analyzed in this work. The three approaches receive a grade from 1 to 5, being awarded with an index equal to 5 the method that stands out in a given characteristic in relation to others (Table 3).

In terms of derivation, Maggi's approach stand out due to the possibility of obtaining the generalized forces from the partial derivatives of scalar energy functions which, differently from the conventional Lagrangian formulation, do not need to be expressed in terms of a minimal set of variables and thus, can be further simplified by an adequate choice of redundant variables.

Kane's equations, on the other hand, require the vector expressions of accelerations and angular accelerations to be obtained, representing the most complex derivation procedure.

Concerning interpretability, however, Kane's equations are the most physically insightful, once the interpretation of each term within the equations of motion follows automatically from their derivation. Such a natural interpretability is not immediate in Lagrangian formulation and is even more difficult when Maggi's equations are used, due to the use of redundant variables and of a projection operator (matrix $\mathbf{C}$).

In terms of modularity, the fact that Maggi's equations allow the use of redundant variables makes it easier to include extra components in a model, once the dynamic description of them does not require the use of the same minimal set of variables adopted for the system itself, which characterizes Maggi's approach as an effective modular one. Finally, none of the methods applied presents any further complexity in terms of constraint force elimination.

5 Conclusions

In this work, a 5-DOF serial manipulator robot was modeled using Kane's, Lagrange's Equations, and Maggi's equations. The equations of motion obtained by the three methods are the same. There was a certain difficulty related to the dynamic modelling due to the relative movements between the bodies of the chosen system.

Once we apply the Kane's approach it seems easy to work out a computational procedure, which allows to model a system methodically. However, to apply the Kane's approach it is necessary to calculate the acceleration of the centre of mass of each body, which is a disadvantage when compared to the Lagrangian formulation, in which only the velocities of centres of mass and angular velocity expressions of each body are required.

Maggi's formalism also requires the computation of partial derivatives of the Lagrangian of the system. However, it is possible to take advantage of the use of redundant coordinates to simplify the energy terms as much as possible, which can make the modeling procedure much simpler when compared to the conventional

Lagrangian formalism. Finally, the simulations results are consistent and reveal a significant influence of the inertial effects in the torque imposed as expected for a serial mechanism.

Acknowledgements Renato Maia Matarazzo Orsino acknowledges grant #2016/09730-0, São Paulo Research Foundation (FAPESP).

References

1. Tenenbaum, R.A.: Fundamentals of Applied Dynamics, Ed. Springer Science & Business Media (2006)
2. Gantmacher, F.: Lectures in Analytical Mechanics, Ed. Mir Publishers, Moscow, p. 264 (1970)
3. Meirovitch, L.: Methods of Analytical Dynamics, Ed. McGraw-Hill, New York, p. 806 (2001)
4. Tsai, L.-W.: Robot Analysis: The Mechanics of Serial and Parallel Manipulators, Ed. John Wiley & Sons, New York, p. 505 (1999)
5. Kane, T.R., Levinson, D.A.: Dynamics: Theory and Applications, Ed. McGraw-Hill, New York, p. 400 (1985)
6. Baruh, H.: Analytical dynamics, Ed. McGraw-Hill, Boston, p. 718 (1999)
7. Desloge, E.A.: Relationship between Kane's equations and the Gibbs-Appell equations. J. Guid. Control, Dyn. Am. Inst. Aero. Astro. **10**(1), 120–122 (1987)
8. Papastavridis, J.G.: On the nonlinear Appell's equations and the determination of generalized reaction forces. Int. J. Eng. Sci. **26**(6), 609–625 (1988)
9. Khan, W.A., Krovi, V.N., Saha, S.K., Angeles, J.: Modular and recursive kinematics and dynamics for parallel manipulators. Multibody System Dynamics, Kluwer Academic Publishers, vol. 14, no. 3–4, pp. 419–455 (2005)
10. Laulusa, A., Bauchau, O.A.: Review of contemporary approaches for constraint enforcement in multibody systems. J. Comput. Nonlinear Dyn. **3**(1), 011005 (2008)
11. Saha, S.K., Angeles, J.: Dynamics of nonholonomic mechanical systems using a natural orthogonal complement. J. Appl. Mech. **58**(1), 238–243 (1991)
12. Orsino, R.M.M., Hess-Coelho, T.A., Pesce, C.P.: Analytical mechanics approaches in the dynamic modelling of delta mechanism. Robotica **33**(04), 953–973 (2015)
13. Orsino, R.M.M., Hess-Coelho, T.A.: A contribution for developing more efficient dynamic modelling algorithms of parallel robots. Int. J. Mech. Robot. Syst. **1**(1), 15–34 (2013)
14. Orsino, R.M.M.: A contribution on modeling methodologies for multibody systems. PhD thesis, University of Sao Paulo (2016)

Modelling of the Manipulator of a Mini Hydraulic Excavator

Éverton Lins de Oliveira and Décio Crisol Donha

Abstract Small hydraulic excavators are versatile machines used in a wide range of operations such as digging, removal of debris, transportation of cargo and earthmoving. Operating a hydraulic excavator in certain environments is a difficult and dangerous task, especially in hazardous environments subject to natural disturbances or inadequate health conditions for human work. Because of these conditions, the automation of excavators or its components has been the subject of many studies in recent years. In this paper, to enable the development of a control system for the manipulator of a mini excavator, a complete mathematical model of the manipulator dynamics is developed. The work includes the validation of the model of the manipulator by means of simulation of a computational model and by comparison with the results obtained by a commercial software of dynamic analysis. Results are discussed and evaluated and suggestions for future work are enclosed.

Keywords Hydraulic excavator · Hydraulic manipulator · Excavator model
Manipulator model · Dynamic model · Computational model

1 Introduction

Hydraulic excavators are versatile machines used in various types of operations, such as digging, removing debris, cargo, ground, and earthworks in general. Operating a hydraulic excavator in certain environments is a difficult task, especially when it comes to dangerous environments subject to natural disturbances or inadequate health conditions for human work. In recent years, the automation of an excavator has been the subject of many studies, seeking for high efficiency and

É. L. de Oliveira (✉) · D. C. Donha
University of São Paulo, São Paulo, SP 2231, Brazil
e-mail: ev_lins@usp.br

D. C. Donha
e-mail: decdonha@usp.br

A. de T. Fleury et al. (eds.), *Proceedings of DINAME 2017*, Lecture Notes in Mechanical Engineering, https://doi.org/10.1007/978-3-319-91217-2_21

improved safety [1]. To develop a suitable control system for the manipulator of a hydraulic excavator, within the control strategy in mind, it is necessary to develop at first a complete mathematical model that accounts for mechanical and hydraulic dynamic of the manipulator. Many papers in the literature focus on the modeling of manipulator's mechanical dynamics [2–5], however, few of them are dedicated to the manipulator's complete modeling [6, 7]. In this works, to enable the control and automation of the manipulator, a complete mathematical model is developed considering the manipulator's mechanical and hydraulic parts. The manipulator is then divided into two subsystems, a mechanical and a hydraulic one. The mechanical subsystem is modeled by Kane's method and Newton's second law. The hydraulic subsystem is modeled through the application of fundamental principles of fluids mechanics. In the sequence, these models are coupled, simulated and analyzed. After the validation of the coupled model presented here, it is possible to synthesize and test a control system.

2 Manipulator Modeling

Here will be presented the modeling of the manipulator's mechanical and hydraulic subsystems and its coupling.

2.1 Mechanical Subsystem Modeling

In this section, the mechanical subsystem dynamic model will be derived. To this end, the following simplifying assumptions are adopted:

- The manipulator's links are assumed as perfect rigid bodies;
- Only the main bodies are considered relevant in links dynamics;
- The friction in the revolute joints is negligible.

Forward Kinematics. To describe the kinematics of the bodies, it is used an inertial frame called I and a system of rectangular coordinates (O_0, x_0, y_0, z_0), fixed in the center of the joint of the manipulator base, and also a local frame 1 solidary to the base, with origin coincident with the origin of the inertial frame and coordinates (O_1, x_1, y_1, z_1). This frame describes the position of the base relative to the frame I using the angular displacement θ_1, which is defined positive when moving in the counterclockwise sense, according to the right-hand rule. To describe the boom position relative to base frame and to describe the bucket position relative to stick frame, the local frames 2, 3 and 4 are required, which are solidary to boom, stick and the bucket, respectively. The coordinate systems (O_2, x_2, y_2, z_2), (O_3, x_3, y_3, z_3) and (O_4, x_4, y_4, z_4) with the angular displacements θ_2, θ_3 and θ_4 are

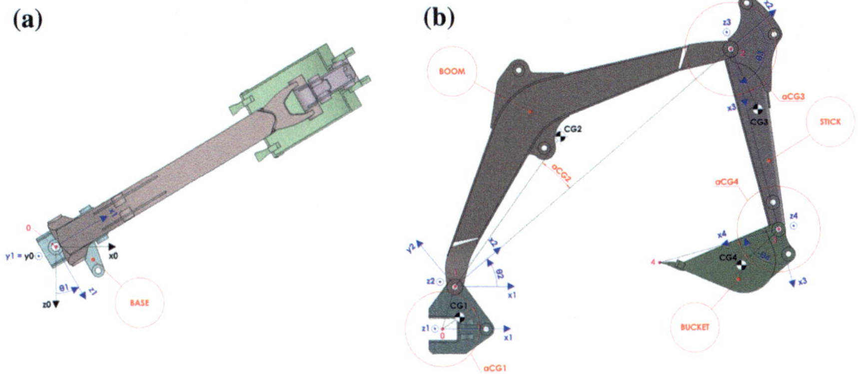

Fig. 1 Frames, generalized coordinates and points. **a** *xz* plane; **b** *xy* plane

relate to the frames 2, 3 and 4, respectively. The set of angular displacements ($\theta_1, \ldots, \theta_4$) are the generalized coordinates. For convenience, a number is assigned to the center of the joints and links of the manipulator. The base joint center is the point 0, the boom joint center is the point 1, the stick joint center is the point 2, the bucket joint center is the point 3 and the tip of bucket teeth is the point 4. The base, boom, stick and bucket are the links 1, 2, 3 and 4 respectively. Figure 1 shows all frames, links, generalized coordinates and points defined above.

Next, will be formalized the description of the orientation and the position of any link and point of the manipulator, relative to the frame *I*. Starting with the description of the orientation, considering initially a frame *j* and a frame $j-1$, with $j=1, \ldots, m$ frames, where origin O_j coincide with origin O_{j-1}. The rotation of *j* to $j-1$ is given by the matrix ${}^{j-1}\mathbf{R}_j$, now as direction cosines matrix [8]. This matrix has the property of being orthogonal, so, ${}^{j-1}\mathbf{R}_j^{-1} = {}^{j-1}\mathbf{R}_j^T = {}^{j}\mathbf{R}_{j-1}$.

Let ${}^{j-1}\boldsymbol{\omega}_{j-1\_j}$ be the angular velocity vector of *j* relative and expressed in $j-1$. To write de absolute angular velocity vector of *j* in *I*, it is necessary to consider the absolute angular velocity vector ${}^{I}\boldsymbol{\omega}_{j-1}$ of $j-1$, more the rotation of ${}^{j-1}\boldsymbol{\omega}_{j-1\_j}$ from $j-1$ to *I*, as shown below:

$$ {}^{I}\boldsymbol{\omega}_j = {}^{I}\boldsymbol{\omega}_{j-1} + {}^{I}\mathrm{R}_{j-1}\,{}^{j-1}\boldsymbol{\omega}_{j-1\_j} = {}^{I}\boldsymbol{\omega}_{j-1} + {}^{I}\boldsymbol{\omega}_{j-1\_j} \tag{1} $$

Now, considering that $j=i$, with $i=1, \ldots, N$ links, thus, Eq. (1) can be rewritten as:

$$ {}^{I}\boldsymbol{\omega}_i = {}^{I}\boldsymbol{\omega}_{i-1} + {}^{I}\mathbf{R}_{i-1}\,{}^{i-1}\boldsymbol{\omega}_{i-1\_i} = {}^{I}\boldsymbol{\omega}_{i-1} + {}^{I}\boldsymbol{\omega}_{i-1\_i} \tag{2} $$

Associating ${}^{I}\boldsymbol{\omega}_i$ with the angular velocity vector of link *i* of the manipulator, it is possible to get the angular acceleration vectors of that link by making: ${}^{I}\boldsymbol{\alpha}_i = {}^{I}\dot{\boldsymbol{\omega}}_i$.

With respect to the position, considering any point CGi belonging to the manipulator, so, its position vector on the origin of frame I can write as the absolute position vector of the point O_i, predecessor of CGi, more the rotation of vector ${}^{i}\mathbf{r}_{O_i\_CGi}$, between the points O_i and CGi, from $j=i$ to I:

$$ {}^{I}\mathbf{r}_{CGi} = {}^{I}\mathbf{r}_{O_i} + {}^{I}\mathbf{R}_i {}^{i}\mathbf{r}_{O_i\_CGi} = {}^{I}\mathbf{r}_{O_i} + {}^{I}\mathbf{r}_{O_i\_CGi} \tag{3} $$

Taking ${}^{I}\mathbf{r}_{CGi}$ as the position vector of the center of mass of link i, it is possible to get the velocity and acceleration vectors of that center of mass by making: ${}^{I}\mathbf{v}_{CGi} = {}^{I}\dot{\mathbf{r}}_{CGi}$ and ${}^{I}\mathbf{a}_{CGi} = {}^{I}\dot{\mathbf{v}}_{CGi}$, respectively.

To simplify the kinematic expressions, the absolute vectors will be written in local frame. To do so, the rotation $\boldsymbol{\eta}_i = {}^{I}\mathbf{R}_i^{T}\,{}^{I}\boldsymbol{\eta}_i$, with ${}^{I}\boldsymbol{\eta}_i$ being an arbitrary absolute vector, will be performed. From here on, the notation without superscript on the left will be used to indicate an absolute vector written in local frame's base ($\boldsymbol{\omega}_i$, $\boldsymbol{\alpha}_i$, $\mathbf{v}_{CGi}$, $\mathbf{a}_{CGi}$), i.e., in frame $j=i$, with the exception of inertia tensor $\mathbf{I}_i$ of the link i, which in fact is expressed in the local frame.

Links Dynamics. By the Kane's method the links dynamics model is derived based on the dynamic equilibrium between the generalized inertial forces, F_k^*, and generalized active forces, F_k, according to Eq. (4) adapted from Baruh [8]:

$$ \underbrace{\sum_{i=1}^{N}\left(m_i \mathbf{a}_{CGi}^{T} \frac{\partial \mathbf{v}_{CGi}}{\partial \dot{q}_k} + \dot{\mathbf{H}}_i^{T} \frac{\partial \boldsymbol{\omega}_i}{\partial \dot{q}_k}\right)}_{F_k^*} = \underbrace{\sum_{i=1}^{N}\left(\mathbf{F}_i^{T} \frac{\partial \mathbf{v}_{CGi}}{\partial \dot{q}_k} + \mathbf{M}_i^{T} \frac{\partial \boldsymbol{\omega}_i}{\partial \dot{q}_k}\right)}_{F_k} \tag{4} $$

where m_i is the mass of link i, with $i=1, \ldots, N$ links, q_k is the generalized coordinate, thus, $q_k = \theta_k$, with $k=1, \ldots, n$ generalized coordinates, and $\dot{\mathbf{H}}_i$ is the vector of angular momentum variation of link i, that is given by:

$$ \dot{\mathbf{H}}_i = \mathbf{I}_i \boldsymbol{\alpha}_i + \boldsymbol{\omega}_i \wedge (\mathbf{I}_i \boldsymbol{\omega}_i) \tag{5} $$

The active forces vector $\mathbf{F}_i$ and the sum of generalized active moments vectors $\mathbf{M}_i^{T}(\partial \boldsymbol{\omega}_i / \partial \dot{q}_k)$ are given, respectively, by:

$$ \mathbf{F}_i = \mathbf{G}_{CGi} = {}^{I}\mathbf{R}_i^{T}\,{}^{I}\mathbf{G}_{CGi} \tag{6} $$

$$ \sum_{i=1}^{N} \mathbf{M}_i^{T} \frac{\partial \boldsymbol{\omega}_i}{\partial \dot{q}_k} = \tau_k \tag{7} $$

where ${}^{\mathrm{I}}\mathbf{G}_{CGi} = [0 \quad m_i g \quad 0]^{T}$ is the weight force vector of link i, with g as the gravity acceleration in vertical direction, and τ_k is the motor torque.

Hydraulic Force. The motor torque τ_k can be related to the hydraulic force F_{hk} by the principle of the virtual works [5], resulting in: $\tau_k = F_{hi}(\partial l_{CiHi} / \partial q_k)$, where l_{CiHi} is the length between the points Ci and Hi, with $i=1, \ldots, N$ cylinders and $i=k$.

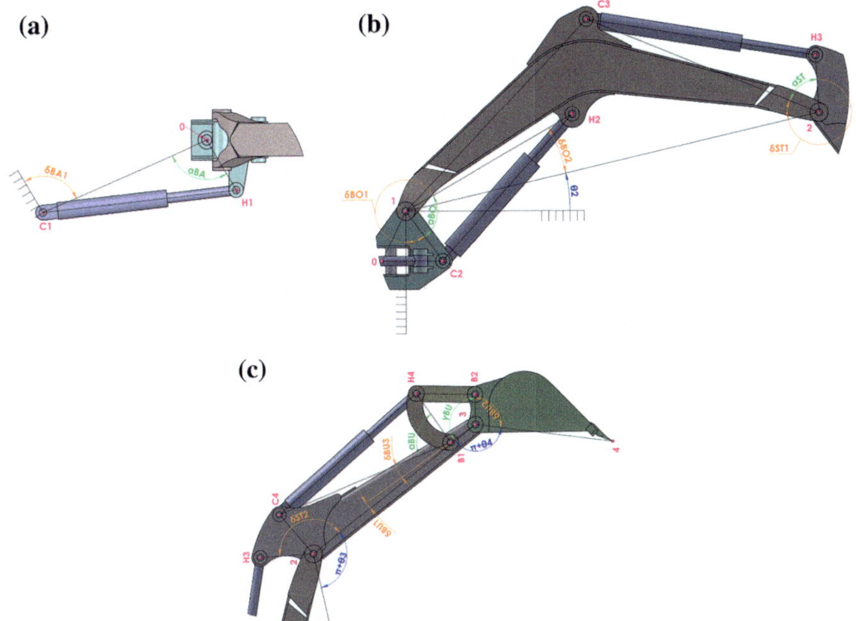

Fig. 2 Hydraulic cylinders' inverse kinematics. **a** Inverse kinematic of cylinder 1; **b** inverse kinematic of cylinders 2 and 3; **c** inverse kinematic of cylinder 4

To determine the cylinders lengths as functions of the generalized coordinates, the cylinders inverse kinematics analysis is performed [5], taking the Fig. 2 as reference.

Based on Fig. 2a one obtains for cylinder 1:

$$\alpha_{BA} = \theta_1 + \delta_{BA1} \tag{8}$$

$$l_{C1\_H1} = \sqrt{l_{0\_C1}^2 + l_{0\_H1}^2 - 2\, l_{0\_C1}\, l_{0\_H1} cos\alpha_{BA}} \tag{9}$$

From Fig. 2b, c one obtains for cylinders 2 and 3, respectively:

$$\alpha_{BO} = \frac{\pi}{2} + \theta_2 - (\delta_{BO1} - \delta_{BO2}) \tag{10}$$

$$l_{C2\_H2} = \sqrt{l_{1\_H2}^2 + l_{1\_C2}^2 - 2\, l_{1\_H2}\, l_{1\_C2} cos\alpha_{BO}} \tag{11}$$

$$\alpha_{ST} = \pi - \theta_3 - (\delta_{ST1} + \delta_{ST2}) \tag{12}$$

$$l_{C3\_H3} = \sqrt{l_{2\_C3}^2 + l_{2\_H3}^2 - 2 l_{2\_C3} l_{2\_H3} cos\alpha_{ST}} \tag{13}$$

For the cylinder 4 in Fig. 2c follows the relationships:

$$\gamma_{BU} = \pi - \theta_4 - (\delta_{BU1} + \delta_{BU2}) \tag{14}$$

$$l_{B1\_B2} = \sqrt{l_{3\_B2}^2 + l_{3\_B1}^2 - 2l_{3\_B2}l_{3\_B1}cos\gamma_{BU}} \tag{15}$$

$$\alpha_{BU} = acos\left(\frac{l_{B2\_H4}^2 - l_{B1\_B2}^2 - l_{B1\_H4}^2}{2l_{B1\_B2}l_{B1\_H4}}\right) - asin\left(\frac{l_{3\_B2}sin\gamma_{BU}}{l_{B1\_B2}}\right) - \delta_{BU3} \tag{16}$$

$$l_{C4\_H4} = \sqrt{l_{C4\_B1}^2 + l_{B1\_H4}^2 - 2l_{C4\_B1}l_{B1\_H4}cos\alpha_{BU}} \tag{17}$$

The analysis of the kinematics of the fourth cylinder is more intricate because of the presence of the six bars mechanism, formed by the cylinder 4 with the two-bar linkage, stick and bucket. With the lengths of the cylinders in function of the generalized coordinates, it is possible to determine the hydraulic forces.

Matrix Form of Links Dynamics. The links dynamics, obtained by Kane's method, can be written in the following form [3]:

$$\mathbf{J}^T(\mathbf{q})\mathbf{F} = \mathbf{M}(\mathbf{q})\,\ddot{\mathbf{q}} + \mathbf{C}(\mathbf{q}, \dot{\mathbf{q}})\,\dot{\mathbf{q}} + \mathbf{G}(\mathbf{q}) \tag{18}$$

where $\mathbf{q} \in \Re^n$ is the generalized coordinates vector with $\dot{\mathbf{q}} \in \Re^n$ and $\ddot{\mathbf{q}} \in \Re^n$ as its first and second time derivative, respectively, $\mathbf{M}(\mathbf{q}) \in \Re^{n \times n}$ is the inertia matrix, $\mathbf{C}(\mathbf{q}, \dot{\mathbf{q}}) \in \Re^{n \times n}$ is the Coriolis and centripetal efforts matrix, $\mathbf{G}(\mathbf{q}) \in \Re^n$ is the gravitational torque vector, $\mathbf{F} \in \Re^n$ is the force vector resulting from cylinders dynamics, and $\mathbf{J}^T(\mathbf{q})$ is the Jacobian matrix, which is given by: $\mathbf{J}(\mathbf{q}) = [J_{i,k}]$, where $J_{i,k} = \partial l_{CiHi}/\partial q_k$.

Inclusion of Cylinders Mechanical Parts. Applying Newton's second law to describe the dynamics of cylinders mechanical parts, and realizing the conversion from force to torque that was presented, thus, ones obtains Eq. (19).

$$\mathbf{J}^T\mathbf{F}(\mathbf{q}) = \mathbf{J}^T\mathbf{A}_a\mathbf{p}_l - \mathbf{J}^T\mathbf{M}_c\ddot{\mathbf{y}}_c - \mathbf{J}^T\mathbf{G}_c(\mathbf{q}) - \mathbf{J}^T\mathbf{F}_f(\mathbf{q}, \dot{\mathbf{q}}) \tag{19}$$

where $\mathbf{A}_a \in \Re^{n \times n}$ is the diagonal matrix of cylinders bore transversal section area, $\mathbf{p}_l \in \Re^n$ is the load pressure vector, $\mathbf{M}_c \in \Re^{n \times n}$ is the diagonal matrix of cylinders mass, $\ddot{\mathbf{y}}_c \in \Re^n$ is the cylinders linear acceleration vector, $\mathbf{G}_c(\mathbf{q}) \in \Re^n$ is the cylinders gravitational force vector, where the calculations details can be seen in Santos [9], and $\mathbf{F}_f(\mathbf{q}, \dot{\mathbf{q}}) \in \Re^n$ is the cylinders friction vector, which is given, in scalar form, by [10]:

$$F_f = F_c[1 + (K_{st} - 1)exp(-c_v|\dot{y}_c|)]sign(\dot{y}_c) + B_v\dot{y}_c \tag{20}$$

with F_c being the Coulomb friction, K_{st} is the static friction coefficient, c_v is the velocity transition coefficient, associated with Stribeck friction, and B_v is the viscous friction coefficient. If there are discontinuity problems, the term $sign(\dot{y}_c)$ can be approximated by a sigmoid, or a boundary layer can add in Eq. (20).

The cylinders linear velocity and acceleration vectors, $\dot{\mathbf{y}}_c$ and $\ddot{\mathbf{y}}_c$, can be written as a function of generalized velocity and acceleration vectors, $\dot{\mathbf{q}}$ and $\ddot{\mathbf{q}}$, with the kinematic relations that transform the linear motion into the rotational one [9], i.e., $\dot{\mathbf{y}}_c = \mathbf{J}^T\dot{\mathbf{q}}$ and $\ddot{\mathbf{y}}_c = \mathbf{J}^T\ddot{\mathbf{q}} + \dot{\mathbf{J}}^T\dot{\mathbf{q}}$. Replacing these relations in Eq. (19) one obtains:

$$\mathbf{J}^T\mathbf{F} = \mathbf{J}^T\mathbf{A}_a\mathbf{p}_l - (\mathbf{J}^T\mathbf{M}_c\mathbf{J}^T)\ddot{\mathbf{q}} - \left(\mathbf{J}^T\mathbf{M}_c\dot{\mathbf{J}}^T\right)\dot{\mathbf{q}} - \mathbf{J}^T\mathbf{G}_c - \mathbf{J}^T\mathbf{F}_f \tag{21}$$

Putting Eq. (21) in (18) results in the coupled model of mechanical subsystem:

$$\mathbf{J}^T\mathbf{F}_h = \mathbf{M}_m\,\ddot{\mathbf{q}} + C_m\dot{\mathbf{q}} + \mathbf{G}_m + \mathbf{D}_m \tag{22}$$

where $\mathbf{M}_m = \mathbf{M} + \mathbf{J}^T\mathbf{M}_c\mathbf{J}^T$, $\mathbf{C}_m = \mathbf{C} + \mathbf{J}^T\mathbf{M}_c\dot{\mathbf{J}}^T$, $\mathbf{G}_m = \mathbf{G} + \mathbf{J}^T\mathbf{G}_c$, $\mathbf{D}_m = \mathbf{J}^T\mathbf{F}_f$ and $\mathbf{F}_h = \mathbf{A}_a\,\mathbf{p}_l$.

2.2 Hydraulic Subsystem Modeling

The modeling presented here is based on the study of the hydraulic servo system shown in Fig. 3. This system consists of a double action differential cylinder controlled by a 4/3-way proportional directional control valve. For the hydraulic subsystem modeling, the following simplifying assumptions are adopted:

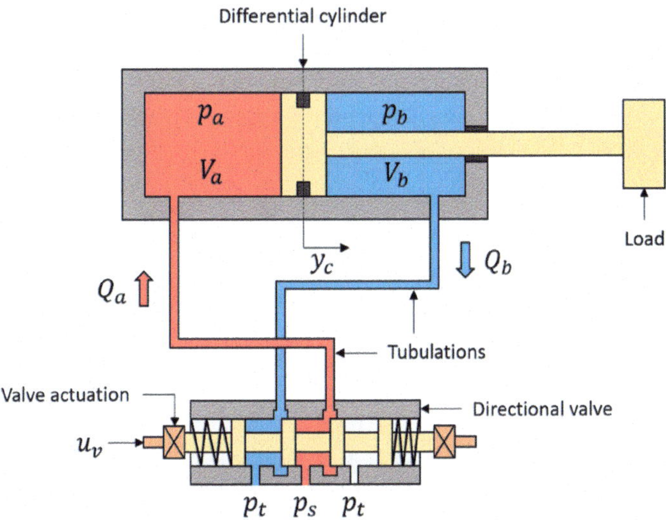

Fig. 3 Hydraulic servo system (*Source* Adapted from Valdiero [12], p. 19)

- The supply unit provides constant pressure and flow to the system;
- The valve dynamics is fast enough so that the ratio between the applied tension to the valve and spool displacement is given by a constant;
- The valve internal leakage is not significant;
- The valve dead zone is considered negligible;
- The flow regime through the valve orifices is assumed as turbulent;
- The dynamics of tabulations are insignificant, therefore, can be modeled as inefficient volumes;
- The leakage between cylinders and environment is not considered;
- The leakage between cylinders chambers is not considered;
- The fluid bulk modulus is admitted as constant, even though it depends on the effective pressure, temperature and the amount of air mixed with the fluid [11].

Load Flow Equation. In Fig. 2, the flow rates Q_a and Q_b through valve ports a and b, respectively, are given by the flow rate equation through orifices [13], as shown in the sequence:

$$Q_a = u_v K_a g_a(p_a, u_v), \text{ with } g_a(p_a, u_v) = \begin{cases} \sqrt{p_s - p_a} & \text{if} \quad u_v \geq 0 \\ \sqrt{p_a - p_t} & \text{if} \quad u_v < 0 \end{cases} \tag{23}$$

$$Q_b = -u_v K_b g_b(p_b, u_v), \text{ with } g_b(p_b, u_v) = \begin{cases} \sqrt{p_b - p_t} & \text{if} \quad u_v \geq 0 \\ \sqrt{p_s - p_b} & \text{if} \quad u_v < 0 \end{cases} \tag{24}$$

where u_v is the voltage applied to the valve, K_a and K_b are the flow rate coefficients of ports a and b, respectively, p_a and p_b are, correspondingly, the effective pressures in chambers a and b, and the terms p_s and p_t are the supply and tank effective pressures, respectively.

In steady state, it is assumed that $Q_a = -\alpha_c^{-1} Q_b$, where $\alpha_c = A_b/A_a$ is the actuator cross section area ratio between chambers a and b. Therefore, with Eqs. (23), (24) and the definitions of load flow rate, $Q_l \triangleq Q_a = -\alpha_c^{-1} Q_b$, and load pressure, $p_l \triangleq p_a - \alpha_c p_b$, for an asymmetric cylinder [11], and after some algebraic manipulation, one obtains:

$$Q_l = u_v K_l g_l(p_l, u_v), \text{ with } g_l(p_l, u_v) = \begin{cases} \sqrt{p_s - p_l - \alpha_c p_t} & \text{if} \quad u_v \geq 0 \\ \sqrt{\alpha_c p_s + p_l - p_t} & \text{if} \quad u_v < 0 \end{cases} \tag{25}$$

where $K_l = (\sigma_v K_a)/\sqrt{\sigma_v^2 + \alpha_c^3}$ is the load flow rate coefficient, with $\sigma_v = K_b/K_a$ as the valve flow rate coefficients ratio.

Load Pressure Dynamic. Applying the continuity equation in the cylinder chambers results in [11]:

$$\dot{p}_a = \frac{\beta_e}{V_a}(Q_a - A_a \dot{y}_c) \tag{26}$$

$$\dot{p}_b = \frac{\beta_e}{V_b}(Q_b + A_b \dot{y}_c) \tag{27}$$

where β_e is the effective fluid bulk modulus, $V_a = A_a(\tilde{l}_a + y_c) + V_{tub}$ is the volume of chamber a, $V_b = A_b(\tilde{l}_b - y_c) + V_{tub}$ is the volume of chamber b, $\tilde{l}_{a,\,b} = l_{a,\,b}(t_0)$ is the chambers initial lengths, and V_{tub} is the fluid volume in the tabulations.

Substituting Eqs. (26) and (27) in the load pressure first time derivative, that is given by: $\dot{p}_l = \dot{p}_a - \alpha_c \dot{p}_b$, and after some algebraic manipulation, one obtains:

$$\dot{p}_l = \frac{\beta_e}{V_a}\left(\frac{r_V + \alpha_c^2}{r_V}\right)(Q_l - A_a \dot{y}_c) \tag{28}$$

where $r_V = V_b / V_a$ is the chambers volume ratio.

Matrix Form of Hydraulic Subsystem Dynamics. Based in Santos [9], and considering $i = 1, \ldots, N$ actuators, the hydraulic subsystem dynamics can be written as follows:

$$\dot{\mathbf{p}}_l = \mathbf{E}_l\,(\mathbf{q}, \mathbf{p}_l, \mathbf{u}_v)\mathbf{u}_v - \mathbf{F}_l(\mathbf{q})\,\mathbf{J}^T\,(\mathbf{q})\,\dot{\mathbf{q}} \tag{29}$$

where $\mathbf{p}_l \in \Re^n$ is the load pressure vector, and terms $\mathbf{E}_l \in \Re^{n \times n}$ and $\mathbf{F}_l \in \Re^{n \times n}$ are diagonal matrices of non-linear functions, which are given, respectively, by:

$$\mathbf{E}_l\,(\mathbf{q}, \mathbf{p}_l, \mathbf{u}_v) = diag\left[\frac{\beta_e}{A_{ai}(\tilde{l}_{ai} + y_{ci}) + V_{tubi}}\left(\frac{r_{Vi} + \alpha_{ci}^2}{r_{Vi}}\right)K_{li}g_{li}\,(p_{li}, u_{vi})\right] \tag{30}$$

$$\mathbf{F}_l(\mathbf{q}) = diag\left[\frac{\beta_e}{A_{ai}(\tilde{l}_{ai} + y_{ci}) + V_{tubi}}\left(\frac{r_{Vi} + \alpha_{ci}^2}{r_{Vi}}\right)A_{ai}\right] \tag{31}$$

with $K_{li} = (\sigma_{vi} K_{ai}) / \sqrt{\sigma_{vi}^2 + \alpha_{ci}^3}$. In Eqs. (30) and (31) the linear displacement y_{ci} is given by: $y_{ci} = l_{CiHi} - \tilde{l}_{CiHi}$, where $\tilde{l}_{CiHi} = l_{CiHi}(t_0)$ is the initial length of the cylinder i. Substituting the relations $F_{hi} = p_{li} A_{ai}$ and $p_{li} = F_{hi} / A_{ai}$ in Eq. (29), one obtains:

$$\dot{\mathbf{F}}_h = \mathbf{A}_a\,\mathbf{E}_l(\mathbf{q}, \mathbf{F}_h, \mathbf{u}_v)\,\mathbf{u}_v - \mathbf{A}_a\,\mathbf{F}_l(\mathbf{q})\,\mathbf{J}^T\,(\mathbf{q})\dot{\mathbf{q}} \tag{32}$$

The Eq. (32) expresses the hydraulic subsystem dynamics as a function of the hydraulic force.

2.3 Manipulator's Coupled Model

Gathering the equations of mechanical and hydraulic subsystems, one obtains the manipulator's coupled model as can be seen bellow:

$$\ddot{\mathbf{q}} = \mathbf{M}_m^{-1}(\mathbf{q})\left[\mathbf{J}^T(\mathbf{q})\mathbf{F}_h - \mathbf{C}_m(\mathbf{q},\dot{\mathbf{q}})\,\dot{\mathbf{q}} - \mathbf{G}_m(\mathbf{q}) - \mathbf{D}_m(\mathbf{q},\dot{\mathbf{q}})\right] \quad (33)$$

$$\dot{\mathbf{F}}_h = \mathbf{A}_a\,\mathbf{E}_l(\mathbf{q},\mathbf{F}_h,\mathbf{u}_v)\,\mathbf{u}_v - \mathbf{A}_a\,\mathbf{F}_l(\mathbf{q})\,\mathbf{J}^T(\mathbf{q})\dot{\mathbf{q}} \quad (34)$$

Without loss of generality, it can be said that Eqs. (33) and (34) represent a mechanical subsystem that is actuated by a hydraulic force, $\mathbf{F}_h$, generated by a hydraulic subsystem. This hydraulic force arises from the pressure difference between cylinder's chambers, when a tension, $\mathbf{u}_v$, is applied to the valve [9]. Analyzing the Eqs. (33) and (34) is verified that the coupled model captures the main equivalent effects of the coupling, such as the conversion of hydraulic to mechanical power, and the variation of the cylinders chambers volumes due to the moving parts displacement and due to the hydraulic fluid compressibility [7].

3 Simulation

In this section, the results of the manipulator's model simulation are presented. In Table 1 are gathered the parameters of mechanical and hydraulic subsystems used in the simulation. These parameters are from the project of a mini excavator prototype.

Other hydraulic subsystem parameters used in the simulations are: $\beta_e = 1.30$ GPa, $F_{pr} = 100$ N,$K_{st} = 1.25$ and $c_v = 100$ s/m.

The manipulator model is simulated with a step type input, referring to a time variable tension applied to the valves. The step command used is applied simultaneously in all valves, and it is given by:

$$u_v(\mathrm{V}) = \begin{cases} 0 & \text{if} \quad 0 > t \leq 2(\mathrm{s}) \\ -0.50 & \text{if} \quad 2 > t \leq 5(\mathrm{s}) \end{cases} \quad (35)$$

To verify the manipulator model, it is performed a comparison of the results obtained with the mathematical model, developed here, and those provided by a commercial software of dynamic systems analysis. The comparison model was created by the authors with the standard blocks of Simscape Multibody™ and Simscape Fluids™ in MATLAB/Simulink©. In Fig. 4 is shown a block diagram of the comparison model created with Simscape™.

The mathematical model was simulated through the numerical integration of its differential equations in MATLAB/Simulink©. The integration was performed with the ODE4 (solver based in the 4th order Runge-Kutta method) with a fixed step of

Table 1 Manipulator parameters

Symbol's vector	Values
$[\delta_{BA1}, \delta_{BA2}, \delta_{BO1}, \delta_{BO2}, \delta_{ST1}, \delta_{ST2}]$	$[1.71, 1.28, 0.64, 0.29, 0.60, 2.54]$ (rad)
$[\delta_{BU1}, \delta_{BU2}, \delta_{BU3}, \alpha_{CG1}, \alpha_{CG2}, \alpha_{CG3}, \alpha_{CG4}]$	$[0.05, 1.70, 0.22, 0.60, 0.25, 0.17, 0.49]$ (rad)
$[l_{1\_C2}, l_{1\_H2}, l_{2\_C3}, l_{2\_H3}, l_{0\_C1}, l_{0\_H1}]$	$[0.30, 0.93, 1.22, 0.28, 0.86, 0.27]$ (m)
$[l_{C4B1}, l_{3\_B1}, l_{B1\_H4}, l_{B2\_H4}, l_{3\_B2}]$	$[0.95, 0.16, 0.30, 0.30, 0.15]$ (m)
$[l_{0\_1}, l_{1\_2}, l_{2\_3}, l_{3\_4}, l_{0\_CG1}, l_{1\_CG2}, l_{2\_CG3}, l_{3\_CG4}]$	$[0.25, 2.07, 1.05, 0.71, 0.12, 1.04, 3.70, 2.92]$ (m)
$[m_1, m_2, m_3, m_4]$	$[32.49, 180.23, 67.30, 52.96]$ (kg)
$[m_{c1}, m_{c2}, m_{c3}, m_{c4}]$	$[17.27, 28.74, 21.44, 17.73]$ (kg)
$[I_{xx1}, I_{xy1}, I_{xz1}, I_{yy1}, I_{yz1}, I_{zz1}]$	$[0.67, -0.05, 0.02, 0.48, -0.04, 0.69]$ (kgm^2)
$[I_{xx2}, I_{xy2}, I_{xz2}, I_{yy2}, I_{yz2}, I_{zz2}]$	$[7.00, -1.19, 0, 79.27, 0, 84.88]$ (kgm^2)
$[I_{xx3}, I_{xy3}, I_{xz3}, I_{yy3}, I_{yz3}, I_{zz3}]$	$[0.67, -1.36, 0, 10.76, 0, 11.19]$ (kgm^2)
$[I_{xx4}, I_{xy4}, I_{xz4}, I_{yy4}, I_{yz4}, I_{zz4}]$	$[1.50, -0.04, 0, 3.12, 0, 2.49]$ (kgm^2)
$[p_s, p_t]$	$[10, 0.1]$ (MPa)
$[K_a, K_b, K_l]$	$[7.37, 7.37, 5.21] \times 10^{-8}$ $(\text{m}^4\text{V}^{-1}\text{s}^{-1}\text{N}^{-1/2})$
$[A_{a1}, A_{a2}, A_{a3}, A_{a4}, A_{b1}, A_{b2}, A_{b3}, A_{b4}]$	$[3.10, 5.03, 3.10, 3.10, 1.50, 2.60, 1.50, 1.50] \times 10^{-3}$ (m^2)
$[\tilde{l}_{a1}, \tilde{l}_{a2}, \tilde{l}_{a3}, \tilde{l}_{a4}, \tilde{l}_{b1}, \tilde{l}_{b2}, \tilde{l}_{b3}, \tilde{l}_{b4}]$	$[0.31, 0.40, 0.58, 0.43, 0.20, 0.15, 0.09, 0.08]$ (m)
$[V_{tub1}, V_{tub2}, V_{tub3}, V_{tub4}]$	$[3.93, 5.89, 7.85, 9.82] \times 10^{-5}$ (m^3)
$[B_{v1}, B_{v2}, B_{v3}, B_{v4}]$	$[55.30, 70.20, 55.30, 55.30]$ (kNs/m)
$[F_{c1}, F_{c2}, F_{c3}, F_{c4}]$	$[1.13, 1.43, 1.13, 1.13]$ (kN)

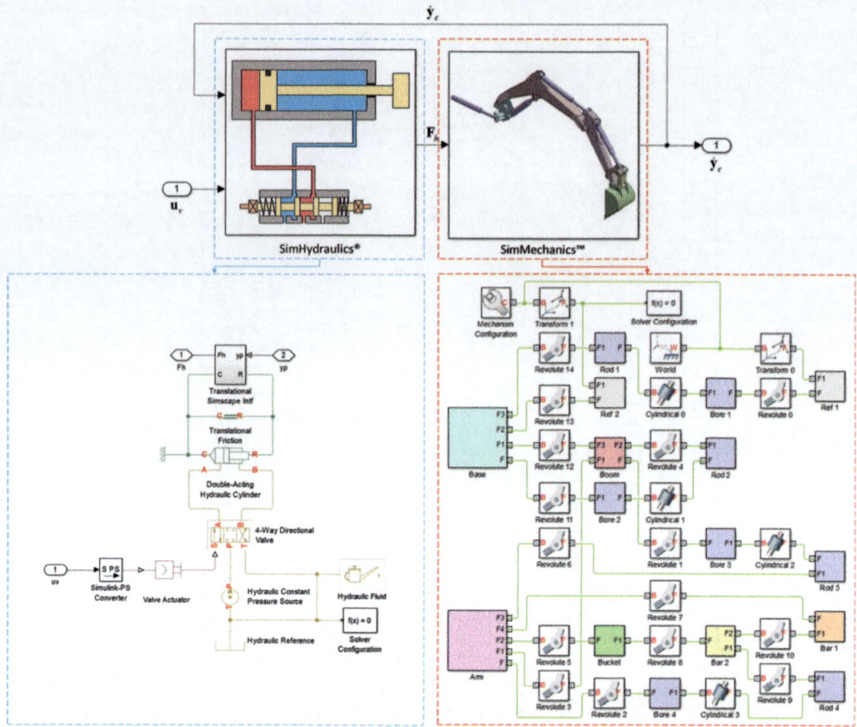

Fig. 4 Block diagram of comparison model created with Simscape™

1×10^{-3} s. In Fig. 5 are show the comparison of the angular displacements, angular velocities and hydraulic forces (vertical axis on the left), and the relative error between the models' states (vertical axis on the right).

In the angular displacements' comparison, no significant difference is noticed as can been seen in the relative errors graphs, where the maximum relative error between angular displacements is less than 1%. However, in the comparisons of the angular velocities and hydraulic forces, it is noticed that the Simscape™ model presents a slightly more damped behavior than the mathematical model in the oscillatory part. This is mainly due to the hypothesis that the fluid bulk modulus is constant and that there is no valve leakage, since in Simscape™ the simplifying hypotheses mentioned are not considered [14]. But even with these differences, both models still present similar results, with a maximum relative error in the oscillatory part less than 30% between the angular velocities, and less than 35% between the hydraulics forces. In the stationary part the relative error for those states is almost zero.

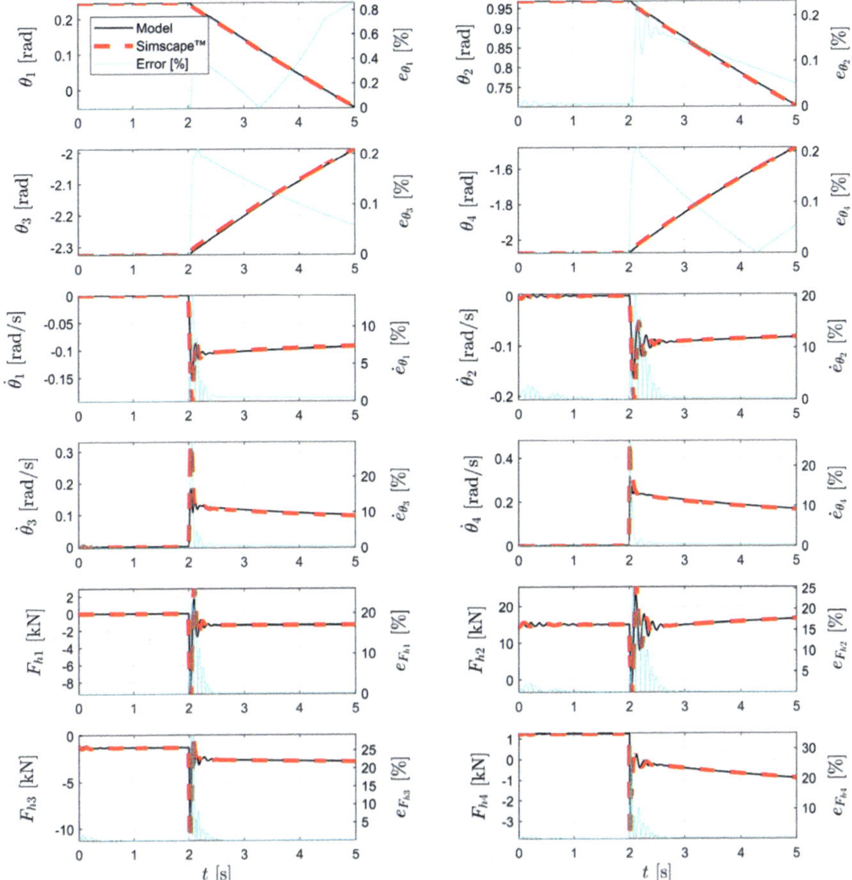

Fig. 5 States comparison and relative error

4 Conclusions

In this work, a mathematical model of the manipulator of a mini hydraulic excavator was developed. The manipulator is divided into two subsystems, one mechanical and the other hydraulic. The mechanical subsystem is modeled with Kane's method and Newton's second law, and the modeling of the hydraulic subsystem is performed with the application of fundamental principles of fluid mechanics. Subsequently, the models of the subsystems were coupled to get the manipulator's model. To conclude, a comparison was performed with the results obtained with the mathematical model and those provided by a commercial software of dynamic systems analysis. In this comparison, the mathematical model showed good results, even with the simplifications performed during modeling. For future works, the control of the manipulator will be performed with the developed model.

References

1. Hoan, L.Q., Jeong, C.S., Kim, H.S., Yang, H.L., Yang, S.Y.: Study on modeling and control of excavator. Int. J. Proj. Manage. **28**(3), 969–974 (2011)
2. Vähä, P.K., Skibniewski, M.J.: Dynamic model of excavator. J. Aerosp. Eng. **6**(2), 148–158 (1993)
3. Koivo, A.J., Thoma, M., Kocaoglan, E., Andrade-Cetto, J.: Modeling and control of excavator dynamics during digging operation. J. Aerosp. Eng. **9**(1), 10–18 (1996)
4. Patel, B.P., Prajapati, J.M.: Dynamics of mini hydraulic backhoe excavator: a Lagrange-Euler (L-E) approach. Int. J. Mech. **2**(5), 195–204 (2014)
5. Šalinić, S., Bošković, G., Nikolić, M.: Dynamic modelling of hydraulic excavator motion using Kane's equations. Autom. Constr. **44**, 56–62 (2014)
6. Qing-hua, H., Da-quing, Z., Peng, H., Hai-tao, Z.: Modeling and control of hydraulic excavator's arm. J. Central S. Univ. Technol. **13**(4), 422–427 (2006)
7. Muvengei, M., Kihiu, J.: Bond graph modeling of mechanical dynamics of an excavator for hydraulic system analysis and design. Int. J. Mech. **3**(3), 326–334 (2009)
8. Baruh, H.: Analytical Dynamics. McGraw-Hill, Singapore (1999)
9. Santos, C.H.: Modeling and cascade control of a hydraulic manipulator (In Portuguese). M. Sc. Thesis, Mechanical Engineering Department, Federal University of Santa Catarina, Brazil (2001)
10. Armstrong-Hélouvry, B., Dupont, P., de Wit, C.C.: A survey of models, analysis tools and compensation methods for the control of machines with friction. Automatica **30**(7), 1083–1138 (1984)
11. Merrit, H.: Hydraulic Control Systems. John Wiley & Sons, New York (1967)
12. Valdiero, A.C.: Hydraulic robot control with friction compensation (In Portuguese). Ph.D. Thesis, Mechanical Engineering Department, Federal University of Santa Catarina, Brazil (2005)
13. Bu, F., Yao, B.: Nonlinear adaptive robust control of hydraulic actuators regulated by proportional directional control valves with deadband and nonlinear flow gains. In: Proceedings of the American Control Conference, vol. 6, pp 4129–4133, Chicago (2000)
14. MathWorks, Inc.: Simscape User's Guide, Natick (2017)

Part IV
Control of Mechanical Systems

Hardware-in-the-Loop Optimization of an Interaction Controller for Improved Coupled Dynamics

Gustavo J. G. Lahr, Henrique B. Garcia, Thiago H. S. Silva and Glauco A. P. Caurin

Abstract This paper presents the implementation of an optimization method to find, without knowledge of the environment characteristics, the best interaction controller parameters to revamp the coupled dynamics. The objective is to improve various industrial robot applications that involves mechanical contact. An enhanced contact is accomplished by lowering the following metrics: rise time, total variation and steady state error. Hence, the impedance controller was the interaction control technique chosen to be optimized. Contact is established between a Kuka KR16 robot TCP and an aluminum platform, where the force data was acquired by a 6-axis force-torque sensor located in the robot's end-effector. Using a hardware-in-the-loop optimization approach, the force feedback is processed by a NSGA-II algorithm. Each individual of the GA represents a specific impedance controller and as the generations passes, these values get more suitable for lowering the metrics. Results show convergence in 5 generations.

Keywords Industrial robot · Interaction controller
Impedance control · Multi-objective optimization

G. J. G. Lahr (✉) · H. B. Garcia · T. H. S. Silva · G. A. P. Caurin
São Carlos School of Engineering, University of São Paulo, São Carlos, SP 13566-590, Brazil
e-mail: gjgl.lahr@gmail.com
URL: http://www.mecatronica.eesc.usp.br/mecatronica/index.php/en

H. B. Garcia
e-mail: thiagohsegreto@gmail.com

T. H. S. Silva
e-mail: henriqueborgesgarcia@gmail.com

G. A. P. Caurin
e-mail: gcaurin@sc.usp.br

A. de T. Fleury et al. (eds.), *Proceedings of DINAME 2017*, Lecture Notes in Mechanical Engineering, https://doi.org/10.1007/978-3-319-91217-2_22

1 Introduction

Mechanical contact is a weighting factor when it comes to a wide variety of robotic applications. Performing delicate surgeries [12], manufacturing in shared environments with humans [9] and stabilizing bionic orthesis [3] are some examples of contact involved in processes developed by robots. To achieve a suitable contact behavior, an estimation of the forces and torques acting on robot's TCP is required, which is implemented with an interaction controller.

One of the most successful interaction controller is the impedance control [8, 11]. It provides stability on free motion if the controller itself is stable. However, although this property is guaranteed, excess of oscillations may happen during contact. This effect, characterized by intermittent interaction, is due to the transition from unconstrained to constrained motion and is called chatter or bounce [18]. The desired scenario, where the bounce effect is minimized, is established using a coupled dynamic in which the rise time and overshoot are the smallest. These two metrics are directly related to the amount of bouncing, and are intrinsically related to the impedance controller and to the environment characteristics.

A possibility to enhance the performance during interaction is to estimate the environment parameters. An usual model for interaction in the literature is shown at Fig. 1a. The robot-controller is modeled as a linear second order system, and the environment as a linear first order. The resulting dynamics is also modeled, so during interaction, the force measured at the sensor is obtained for the coupled system. Different methods for estimation are possible: Erickson et al. [7] compared four different methods, either on-line and off-line; Countinho and Cortesao [4, 5] implemented a multiple observers method and another method which does not need the position values of robot's TCP. Multiple applications rely on environment parameters estimation for improved performance [13, 16]. Robot and environment, however, contain uncertainties. Lahr et al. [14] experimented a stiff contact task and it proved to be different from the second order model proposed, in a controlled situation. The manipulator is also often a source of not modeled dynamics, which could be neglected depending on the application [15], but for many situations it should be considered.

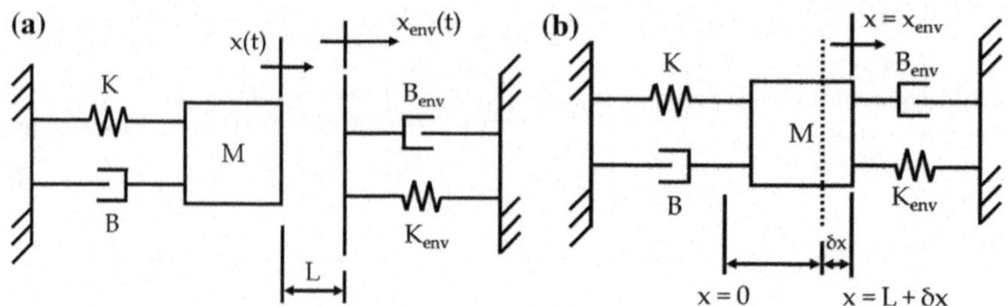

Fig. 1 Modeling for interaction situation along 1 DoF: **a** before contact; **b** during contact

We propose a method which does not need to estimate the environment's parameters, neither needs to take into account the nonlinearities. With a hardware-in-the-loop optimization, where the metrics are obtained from real experiments within the optimization loop, it is possible to optimize the impedance controller parameters without considering non-modeled dynamics on robot-tool-environment. This is possible due the use of a genetic algorithm for a simplified one degree of freedom contact task. We demonstrate that stiff contact interaction contains errors, not predicted by [17], and it is solved by a constrained optimization statement and validated through experimental results with an industrial robot.

1.1 Impedance Control

It is possible to approach the problem along one degree of freedom (DoF), in order to simplify the modeling, where the robot may adopt three parameters to emulate: mass (M), spring (K), and damper (B). If the robot is required to follow a desired position $x_0(t)$, with derivatives $\dot{x}_0(t)$ and $\ddot{x}_0(t)$, is must be compared to the actual position, $x(t)$, and errors between these represent movement from unconstrained to constrained motion. Figure 1a shows the robot modeled as a second order system, and the environment as a first order with K_{env} and B_{env} parameters.

When contact is achieved (Fig. 1b), the coupled system robot/environment now displays a displacement between desired and actual position, denoted by $\delta x(t)$, which is $\delta x(t) = x(t) - L$, where L is the distance between robot's position and the environment at time $t = 0$. The model contains a desired contact force, F_d, and the actual interaction force, F_{int}. Applying Laplace transform with initial conditions null, we have Eq. (1).

$$M[\ddot{x}(t) - \ddot{x}_0(t)] + B[\dot{x}(t) - \dot{x}_0(t)] + K[x(t) - x_0(t)] = F_d(t) - F_{int}(t) \Rightarrow$$

$$\Rightarrow (Ms^2 + Bs + K)\Delta X(s) = \Delta F(s) \tag{1}$$

Proposed by Hogan [11], the impedance controller has gained ground in research and development processes. Its wide applicability is given by the fact that it implements a controller relating the mechanical quantities flow and effort: velocities and forces, respectively. Although Hogan makes the case that an impedance does not have to be linear, it is a common implementation a controller which is a second order and linear time invariant [2], and implemented in this work for a one degree of freedom. Its implementation on discrete systems, as industrial robots, may be found at [14].

1.2 Related Work

Optimization methods for impedance controllers have already been subjects of study in related works. Particle Swarm Optimization algorithm was used for tuning impedance controller parameters [19], although in a simulated model of a planar 3 DOF manipulator. Yet, the transition between unconstrained to constrained motion isn't modeled, letting the nonlinearities of the contact overlooked. Li [17] tunes the controller parameters considering a multi-objective optimization in which the objective functions are: overshoot, settling time, and steady-state error. Using only force data feedback, the technique embodies a Pareto optimality along one DoF through genetic algorithm. However, it still in constrained movement only and do not comprehends experimental results. Our proposal embraces the nonlinearity of the changes between unconstrained and constrained motion, applying in a industrial robot with a commercial force torque sensor.

2 Experiments

2.1 Multi-objective Optimization

Many classical optimization methods use gradients, which requires knowledge of the objective function in question. Often this is not known or has nonlinearities that makes it a computationally expensive method. Some modern approaches deal with the use of values of this function without knowing its derivatives [22], such as genetic algorithms (GA). Using the concept of population, which is composed of several individuals, the GA seeks to generate compositions of the objective function values for each individual within the space of possible solutions. This allows to attenuate the local minima problem, given the variability in the results obtained. This work uses NSGA-II algorithm [6], which is capable of deal with multi-objective problems and multiple variables. The input vector for the metrics evaluation is the force vector in the Z axis, denoted by F_z.

NSGA-II being multi-objective means that the optimization can be conducted based on more than one desirable criterion to obtain the best population through evolution. It is an important feature in the case of this work, since a good contact has low rise time and low overshoot+settling time. These metrics, however, are trade-offs between themselves: will be found higher overshoot and settling time if the rise time is reduced. This motivates the search for the best values through these metrics.

As settling time and overshoot are quantities that have different numerical values, the use of weights for balancing both during optimization would be necessary. The process may be facilitated by using a quantity which relates both, called total variation (TV). This is obtained by the sum of all the differences between consecutive peaks and valleys of the objective function, F_z, per Eq. (2) [1]. Being i related to the vector index which the evaluation algorithm is analyzing, N is the size of the force vector recorded.

$$TV(f) = \sup_{0 \le t_1 \le \cdots \le t_N} \sum_{i=1}^{N-1} \left| F_z(t_i) - F_z(t_{i+1}) \right| \tag{2}$$

For the rise time rule, which is the metric that measures the time taken for the system to reach its final value for the first time, using the rule from 0 to 100%, as stated by Eq. (4) [20]. The final value, $\bar{F}_{z_{end}}$, is obtained by the mean of the 50 final values of the force vector (to decrease the influence of signal noise), F_z, with N number of points, as described by Eq. (3).

$$\bar{F}_{z_{end}} = \frac{\sum_{i=N-50}^{N} F_{z_i}}{50} \tag{3}$$

$$T_r = t_r,\ \text{where } \bar{F}_{z_{end}} - F_z(t_r) < 0 \text{ for the first time} \tag{4}$$

Also, since the impedance controller does not consider an integrative portion by itself and the contact is with stiff environment [21], meaning there is a steady state error associated. This metric was also inserted as another objective, denoted by E_{ss} and described by Eq. (5), obtained by the subtraction of the reference force value, F_{d_z}, and the final force vector value, $\bar{F}_{z_{end}}$.

$$E_{ss} = \left| F_{d_z} - \bar{F}_{z_{end}} \right| \tag{5}$$

So the problem is characterized by three variables ($\mathbf{X} = [M, B, K]^T$) with three goals defined as $f_1(\mathbf{X}) = TV, f_2(\mathbf{X}) = T_r$ and $f_3(\mathbf{X}) = E_{ss}$, and the optimization problem is presented in Eq. (6). The constraints are defined as each variable boundary as $10\,\text{kg} \le M \le 60\,\text{kg}$, $100\,\text{N s/m} \le B \le 1000\,\text{N s/m}$ e $100\,\text{N/m} \le K \le 4000\,\text{N/m}$ and, from empirical experimentation, constraint for the damping value, $\zeta = B/(2\sqrt{MK})$.

$$\begin{aligned} \min_{\mathbf{X}} \quad & (f_1(\mathbf{X}), f_2(\mathbf{X}), f_3(\mathbf{X})) \\ & \mathbf{X} = [M, B, K]^T \\ \text{s.t.} \quad & 10\ \text{kg} \le M \le 20\ \text{kg} \\ & 100\ \text{N s/m} \le B \le 1000\ \text{N s/m} \\ & 1000\ \text{N/m} \le K \le 4000\ \text{N/m} \\ & 0.5 \le \zeta \le 0.95 \end{aligned} \tag{6}$$

2.2 Experimental Implementation

To achieve the desired behavior for the study, a contact is carried out between a tool mounted at the extremity of a Kuka KR16 Robot and an aluminum profile structure. The robot movement is constrained in the robot Z world's coordinate frame (only one DoF), which is perpendicular to the environment's surface. Then the NSGA-II algorithm is used to find the improved solution during the generations' evolution: each individual had about 20 s for its task, leading to 1,766 force points acquired. A total of 20 generations were used for this study, where each one had 20 individuals, resulting in a total of 400 experimental repetitions. The implementation was only possible due the use of the Robot Sensor Interface 2.3 library, implementing an impedance controller with a force-torque sensor from ATI Industrial Automation, model Delta SI-660-60.

The environment is part of a workbench used for general experimental purposes, being a fast-assembly aluminum square profile, which had four other profiles fixed perpendicularly to it. Since the algorithm is supposed to enhance the impedance controller without knowing environment's properties, it is not necessary to calculate the equivalent stiffness of the workbench. The tool is a spindle holder with flat tip and nonlinear structure, due to the spindle hole (Fig. 2).

A block diagram of the implementation is illustrated at Fig. 3. The arrows portray the information flow as inputs or outputs, and the blocks are the representation of each device in the loop. The group *Robot and coupled dynamics* represents the real system composed by robot, sensor, tool, and environment, which returns the values of forces F_z to the computer algorithms. The computer, in its turn, represented by the group *Algorithms and threads*, receives the force input data (F_z), calculates the objective metrics ($[T_r, TV, E_{ss}]^T$) via *Data info* algorithm, and sends these data to NSGA-II algorithm's input. Therefore, the GA will produce new gains for the *Impedance controller* ($[M, B, K]^T$), which operates at a constant 12 ms control loop time and is responsible to keep the robot working properly via *Control actions* during the whole experiment.

All codes run in C# within an i7 processor Windows 10 based PC. Communication is established using UDP/IP protocol network via XML data packages exchange between PC and robot controller. These packages contain information about robot position, force-torque values, and communication status. The thread *Impedance Controller* is responsible to guarantee that the real time communication will not be lost, and it also calculates the impedance controller's control action, the one needed to move the robot and deal with the mechanical interaction.

At the moment the interaction controller is started, after all systems have been initialized, the robot's TCP is placed 5 mm distant from the environment. Since it will take some time to the tool establish contact, during this period the forces are basically zero, and this elapsed time is taken into account for the rise time metric. This way, one may notice that the values of T_r are higher in the order of a few seconds instead of

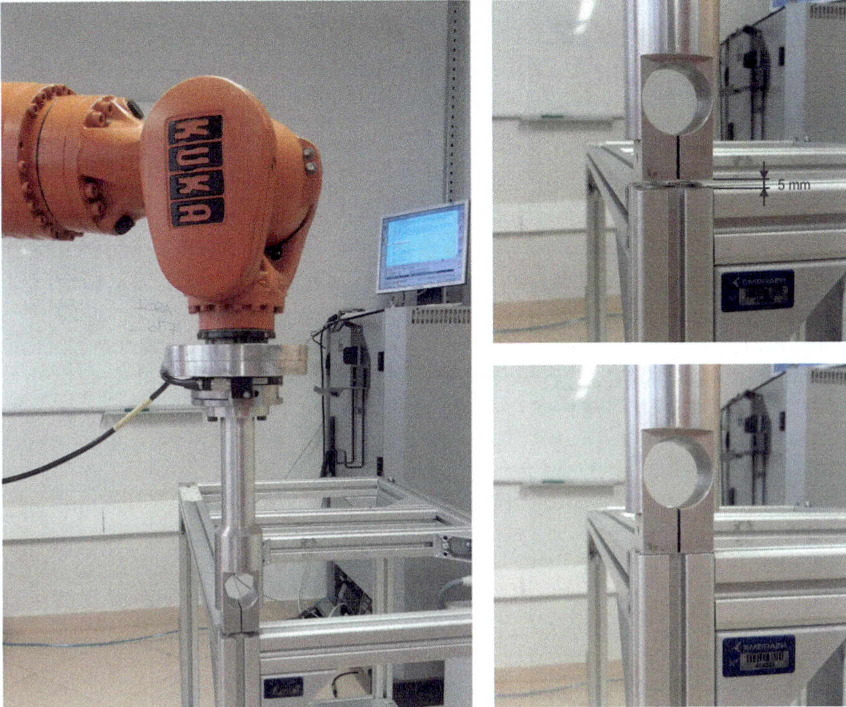

Fig. 2 Experimental setup. Left—Kuka Robot KR16 with a 6 axis force/torque sensor attached to the wrist and the corresponding tool in contact with the aluminum frame; top right—initial position; bottom right—final position

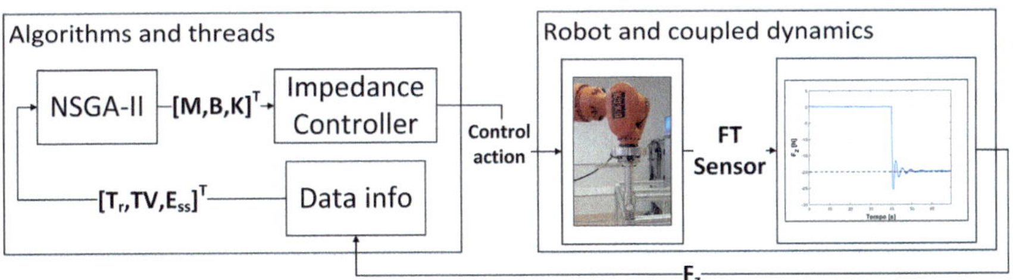

Fig. 3 Information flow diagram. Left—running algorithms and real time threads for communication and GA functioning; right—real and environment setup with force acquisition

milliseconds, which is usual for this metric. The reference trajectory desired is null, $x_0(t) = 0$. This simplifies the contact task, where the desired force is $F_d(t) = 60$ N for all individuals. The whole experiment took around 2.5 h to complete all individuals from all generations.

3 Results

Results show that the objective functions were properly diminished. This decrease is manifested in the series of graphics shown in Fig. 4 in which each color represents an individual. It can be noted that the rise time suffers a draw back through the generations and the amplitudes of oscillation, which is described by the total variation, are tightened. Also, the steady state error is decreased, since the last values, after 20 s, are closer to 60 N. It is possible to note the amount of time elapsed before the contact, which decreased from the 1st generation (starting almost at 10 s) to the 20th generation (less than 8 s).

Figure 5 shows the relationship between the impedance controller damping, ζ, and the rise time of the coupled system dynamics. The colors represent the controller's gain B. Once Fig. 5a, b, and c represent the 1st, 2nd, and 3rd generations, it is possible to notice that the values higher than 400 N s/m disappear after the 4th generation (Fig. 5d). The values of B do not cross the border of 400 N s/m until the end of the experiment, represented by (Fig. 5f). Classical control theory states that, for a second order system, which was modeled by a second order plus a first order, smaller

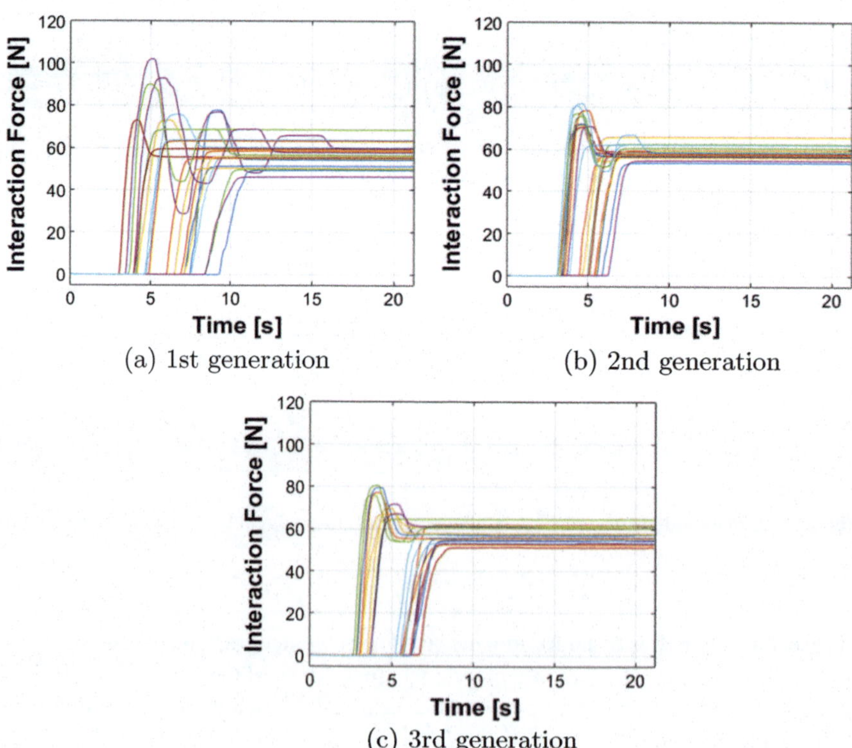

Fig. 4 Interaction forces vector for each individual versus time

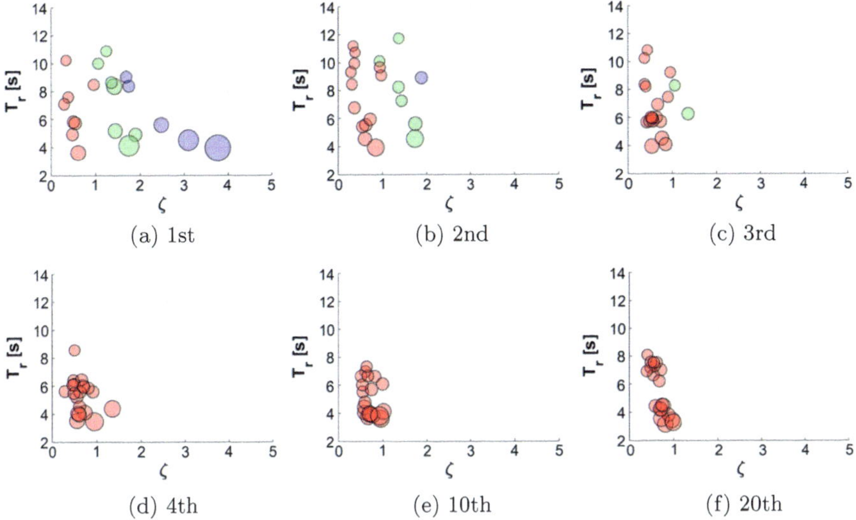

Fig. 5 Plot of $T_r \times \zeta$ for the indicated generations, where: red stands for 100 N s/m $\leq B <$ 400 N s/m; green, 400 N s/m $\leq B <$ 700 N s/m; and blue, 700 N s/m $\leq B <$ 1000 N s/m

damping values (ζ) would lead to smaller rise times. However, looking at Fig. 5, as the damping decreases, rise time increases, what suggests that the real resulting system is not a second order, but higher. Also, the friction is not taken into account at any step in this work's model, although it is a major factor for contact modeling [10].

Another interesting observation that may be discussed is about the analysis of all individuals together generated over the 20 generations of the optimization, where each blue dot is one impedance controller, displayed at Fig. 6. This approach is interesting because, while the data is resulting from an optimization process, each controller results in a respective time metric, for this set of environment+tool+robot.

Two tendencies are possible to note about the controller and the metrics T_r and TV. Plotting the latter values versus the stiffness (Fig. 6b), K, one may notice that lower the oscillations (small TV) are obtained via higher values of stiffness. In the other hand, looking for small values of T_r (Fig. 6a), they require smaller values of the stiffness. The steady state error (Fig. 6c) does not have a clear behavior due to stiff contact with admittance control in industrial robots [21], but it tends to increase with the very high values of stiffness ($K >$ 3000 N/m).

From these two curves, two trends can be observed: T_r leads to a linear fitting curve, which is denoted by the red dotted line (7); and TV is better described by an exponential fitting (8).

$$T_r(K) = 0.00264K + 0.7903 \tag{7}$$

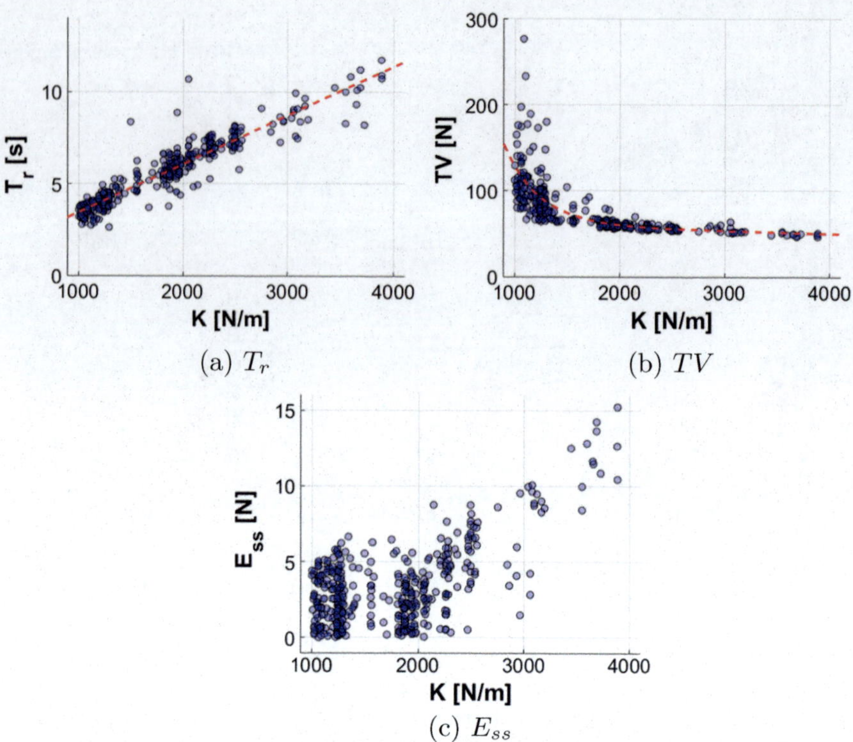

Fig. 6 All individuals recorded over the optimization: metrics versus stiffness

$$TV(K) = 1376 * exp(-0.003006K) + 64.71 * exp(-6.424 * 10^{-5}K) \quad (8)$$

The last analysis consists of the means and standard deviations for each controller parameter. Figure 7 shows the tendency for each gain over the generations, where B and K show a convergence around 204.7 N s/m and 1640.8 N/m respectively, between the 5th and 20th generations, respectively. M, however, does not display a clear convergence since has smaller limits compared to the other metrics, also, small changes in M leads to high terms due to multiplication with acceleration.

4 Conclusion

A hardware-in-the-loop optimization was implemented for better coupled dynamics between robot and environment. Results display that it is possible to implement this technique without worrying about non-modeled dynamics of nonlinearities, since the GA is capable of deal with them. Our method does not consider environment characteristics, which makes an easier implementation. Also, although the fact that

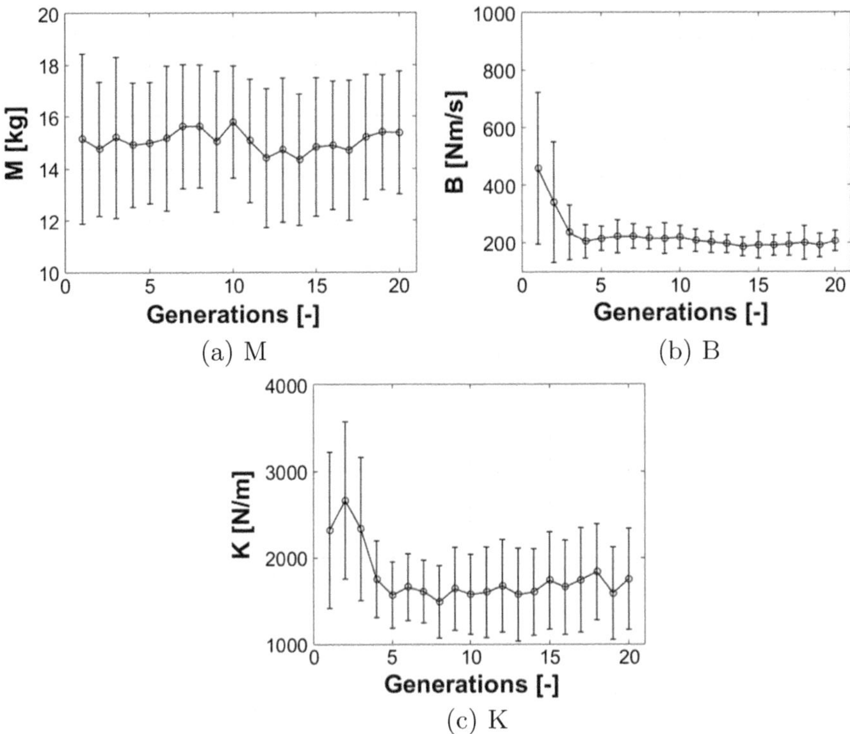

Fig. 7 Controller mean value and standard deviation for parameters versus generations

the whole experiment took around 2.5 h to finish, 5th generation took approximately 40 min to complete, this means that the tendency for B and K were reached quickly.

An important consideration is about the B parameter: all individuals converge for the region below 400 N s/m, what means a softer controller for the limits chosen for these experiments. Only 4 generations were needed to higher values of B disappear. Moreover, higher values of ζ should lead to slower dynamics, instead, it is reaching a faster behavior. This indicates that the coupled system is not a second order, as the supposed earlier.

Next steps of this work are to run the experiment for different environments, checking the robustness of the method. Also, the fitted curves are useful to enhance a theoretical model over a contact situation for this system setup, therefore some studies may be disposed on this intention.

Acknowledgements We would like to acknowledge FAPESP (processes 2013/07276-1, 2015/07484-9 and 2015/24343-0), EMBRAER (cooperation project GSI-1168-15), CNPq and CAPES. Also, we would like to thank professor Maira M. Silva for her support over the development of this technique.

References

1. Boyd, S., Barratt, C.: Linear Controller Design: Limits of Performance. Technical report, Stanford University Stanford United States (1991)
2. Buerger, S.P., Hogan, N.: Complementary stability and loop shaping for improved human robot interaction. IEEE Trans. Robot. **23**(2), 232–244 (2007). https://doi.org/10.1109/TRO.2007.892229
3. Caurin, G.A.P., Siqueira, A.A.G., Andrade, K.O., Joaquim, R.C., Krebs, H.I.: Adaptive strategy for multi-user robotic rehabilitation games. In: Conference Proceedings: Annual International Conference of the IEEE Engineering in Medicine and Biology Society, vol. 2011, pp. 1395–8 (2011). https://doi.org/10.1109/IEMBS.2011.6090328
4. Coutinho, F., Cortesão, R.: Environment stiffness estimation with multiple observers. In: IECON Proceedings (Industrial Electronics Conference), pp. 1537–1542 (2009). https://doi.org/10.1109/IECON.2009.5414740
5. Coutinho, F., Cortesao, R.: Comparison of position and force-based techniques for environment stiffness estimation in robotic tasks. In: IEEE International Conference on Intelligent Robots and Systems, vol. 1, pp. 4933–4938 (2012). https://doi.org/10.1109/IROS.2012.6386161
6. Deb, K., Pratap, A., Agarwal, S., Meyarivan, T.: A fast and elitist multiobjective genetic algorithm: NSGA-II. IEEE Trans. Evol. Comput. **6**(2), 182–197 (2002). https://doi.org/10.1109/4235.996017
7. Erickson, D., Weber, M., Sharf, I.: Contact stiffness and damping estimation for robotic systems. Int. J. Robot. Res. **22**(1), 41–57 (2003). https://doi.org/10.1177/0278364903022001004
8. Ficuciello, F., Romano, A., Villani, L., Siciliano, B.: Cartesian impedance control of redundant manipulators for human-robot co-manipulation. In: IEEE International Conference on Intelligent Robots and Systems, pp. 2120–2125. IEEE (2014). https://doi.org/10.1109/IROS.2014.6942847
9. Fryman, J., Matthias, B.: Safety of industrial robots: from conventional to collaborative applications. In: 7th German Conference on Robotics; Proceedings of ROBOTIK 2012, pp. 1–5 (2012)
10. Gilardi, G., Sharf, I.: Literature survey of contact dynamics modelling. Mech. Mach. Theory **37**(10), 1213–1239 (2002). https://doi.org/10.1016/S0094-114X(02)00045-9
11. Hogan, N.: Impedance control: an approach to manipulation: Parts I–III. Trans. ASME J. Dyn. Syst. Measur. Control **107**(1), 1–7 (1985)
12. Jayender, J., Patel, R.V., Nikumb, S.: Robot-assisted catheter insertion using hybrid impedance control. In: Proceedings of the IEEE International Conference on Robotics and Automation, vol. 2006, pp. 607–612. IEEE (2006). https://doi.org/10.1109/ROBOT.2006.1641777
13. Jhan, Z.Y., Lee, C.H.: Adaptive impedance force controller design for robot manipulator including actuator dynamics. Int. J. Fuzzy Syst. **19**(6), 1739–1749 (2017). https://doi.org/10.1007/s40815-017-0358-2
14. Lahr, G.J.G., Soares, J.V.R., Garcia, H.B., Siqueira, A.A.G., Caurin, G.A.: Understanding the implementation of impedance control in industrial robots. In: LARS 2016 13rd Latin American Robotics Symposium, Recife, PE (2016)
15. Lange, F., Bertleff, W., Suppa, M.: Force and trajectory control of industrial robots in stiff contact. In: 2013 IEEE International Conference on Robotics and Automation, pp. 2927–2934. IEEE (2013). https://doi.org/10.1109/ICRA.2013.6630983
16. Langsfeld, J.D., Kabir, A.M., Kaipa, K.N., Gupta, S.K.: Integration of planning and deformation model estimation for robotic cleaning of elastically deformable objects. IEEE Robot. Autom. Lett. **3**(1), 352–359 (2018). https://doi.org/10.1109/LRA.2017.2749280
17. Li, E.: The robotic impedance controller multi-objective optimization design based on Pareto optimality. In: Lecture Notes in Computer Science (including subseries Lecture Notes in Artificial Intelligence and Lecture Notes in Bioinformatics), vol. 9773, pp. 413–423. Springer International Publishing (2016). https://doi.org/10.1007/978-3-319-42297-8_39

18. Love, L., Book, W.: Environment estimation for enhanced impedance control. In: Proceedings of the IEEE International Conference on Robotics and Automation (ICRA), vol. 2, no. (3), pp. 1854–1859 (1995). https://doi.org/10.1109/ROBOT.1995.525537
19. Mehdi, H., Boubaker, O.: Impedance controller tuned by particle swarm optimization for robotic arms. Int. J. Adv. Robot. Syst. **8**(5), 93–103 (2011). https://doi.org/10.5772/45692
20. Ogata, K.: Modern Control Engineering, 5th edn. Prentice-Hall, New Jersey (2010)
21. Ott, C., Mukherjee, R., Nakamura, Y.: Unified impedance and admittance control. In: 2010 IEEE International Conference on Robotics and Automation, pp. 554–561. IEEE (2010). https://doi.org/10.1109/ROBOT.2010.5509861
22. Rao, S.S.: Engineering Optimization. Wiley (2009). https://doi.org/10.1080/03052150500066646

SDRE Trajectory Tracking Control for a Hovercraft Autonomous Vehicle

Ana Paula Pagotti, Elvira Rafikova and Marat Rafikov

Abstract A hovercraft is an amphibious vehicle lifted by a propeller that causes the effect of an air cushion between the vehicle and the surface. This way, a hovercraft becomes a fast and versatile vehicle to be used in different kinds of tasks such as rescues, environmental monitoring and coast guard patrolling. This paper presents the control problem formulation in order to track a reference trajectory of a hovercraft dynamical model. For this purpose the SDRE (State Dependent Riccati Equation) control method is applied to the model of this dynamical system. The nonlinear control problem is formulated in order to minimize the cost functional. Numerical simulations are performed using Matlab®, so that, the equations of the system and the reference are integrated to generate data about the position, orientation and velocities of the hovercraft. The results confirm that the control method succeeded in controlling the hovercraft in all proposed cases.

Keywords Hovercraft autonomous vehicle · State-dependent Riccati equation Nonlinear control

1 Introduction

A hovercraft is an amphibious vehicle which floats on a cushion of pressurized air supplied by one or more fans and contained inside a cavity on a flexible structure known as skirt [1]. Kuznetsov in [2] says that these fans create a pressure distribution in the opposite direction of the gravitational force, thus minimizing the

A. P. Pagotti (✉) · E. Rafikova · M. Rafikov
Universidade Federal do ABC, Av. dos Estados, 5001, Santo André, SP, Brazil
e-mail: ana.pagotti@ufabc.edu.br

E. Rafikova
e-mail: elvira.rafikova@ufabc.edu.br

M. Rafikov
e-mail: marat.rafikov@ufabc.edu.br

A. de T. Fleury et al. (eds.), *Proceedings of DINAME 2017*, Lecture Notes in Mechanical Engineering, https://doi.org/10.1007/978-3-319-91217-2_23

vehicle contact with the ground or water. This causes a ground effect related to the navigation surface. The ground effect creates a thin air layer that lubricates the base of the vehicle, avoiding the contact with the surface of navigation and consequently reducing the drag forces. It provides the vehicle not only excellent performance on rough surfaces but also the high speed that other conventional marine vehicles cannot achieve [3].

Nonetheless, to plan some missions using Hovercrafts it is necessary to control the trajectory of the vehicle. To do this some controllers of high performance must be applied as pointed in [4]. Finally Hovercrafts present some non-holonomic constraints of movement that don't allow the application of simple feedback control laws [5].

In mobile robotics the vehicle trajectory tracking control can be separated in two different problems: the stabilization of the final position and the trajectory tracking of a reference [6]. The first problem is related to the Brockett's sufficiency condition because of the non-holonomic constraints of movement, that is: any continuous control law couldn't ensure the asymptotic convergence of the vehicle to its resting configuration. And the second problem refers to the determination of a position or a reference trajectory where the robot should be stabilized by the minimization of the error between this reference and the actual trajectory of the system. To solve that problem, a suitable mechanism of control is a state feedback control presented by Rafikova et al. in [7]. A way to control this sort of systems is using a sub-optimal method of control, known as SDRE—(State Dependent Riccati Equation), the main advantage of wich is that it's not necessary to linearize the system.

Then, the objective of the present paper is to formulate the tracking control trajectory problem of a Hovercraft, which has non-holonomic constraints, applying the SDRE control method. In the section Methodology it is shown how the dynamical system is treated to allow the use of the control method, writing it in an error space-state system.

Numerical simulations were performed using Matlab®, so that the equations of the error space-state form, dynamical model of the vehicle and the reference were integrated to generate data about the position, orientation and velocities of the Hovercraft. In this section it will also be presented the graphs about these simulations that provide visualization of the behavior of the dynamical system during a determined interval of time, including the trajectory tracking of the vehicle.

2 Methodology

In this section it is formulated the control problem studying the mathematical dynamic model of the vehicle and the reference chosen. After that, it is applied the SDRE method of control at the error space-state system.

2.1 Modelling of the Dynamical System

The chosen hovercraft model can be found in [8], it considers the dynamics of a sub-actuated hovercraft, which possesses 3 degrees of freedom and only two actuators, two propellers, in this case. The body shape is considered circular and symmetric with respect to x and y direction, as shown in Fig. 1. The damping terms are neglected. This hovercraft dynamic model is obtained through the study of a ship dynamic model developed in [9, 10]. The equations are of the following form:

$$\begin{aligned}
\dot{\mathbf{x}} &= \mathbf{u}\,cos(\varphi) - \mathbf{v}\,sin(\varphi) \\
\dot{\mathbf{y}} &= \mathbf{u}\,sin(\varphi) + \mathbf{v}\,cos(\varphi) \\
\dot{\varphi} &= r \\
\dot{\mathbf{u}} &= \mathbf{vr} + \boldsymbol{\tau}_{\mathbf{u}} \\
\dot{\mathbf{v}} &= -\mathbf{ur} - \beta\mathbf{v} \\
\dot{r} &= \tau_r
\end{aligned} \tag{1}$$

where x, y and φ refer to the position and orientation of the hovercraft regarding an Earth-fixed referential and **u**, **v**, and **r** refer to the linear velocities of surge, sway and angular velocity of the vehicle. The derivatives of **u**, **v**, and **r** correspond to the linear surge acceleration of the vehicle, the linear lateral or sway acceleration of the vehicle and angular acceleration of the vehicle, respectively. The term $\boldsymbol{\tau}_{\mathbf{u}}$ is the linear acceleration in surge direction and τ_r is the rotational acceleration. The quotient between damping coefficient and mass is represented by β.

As shown in Fig. 1 the propellers are positioned symmetrically to the axis that passes through the vehicle center of mass and are actuated by two motors actuated independently.

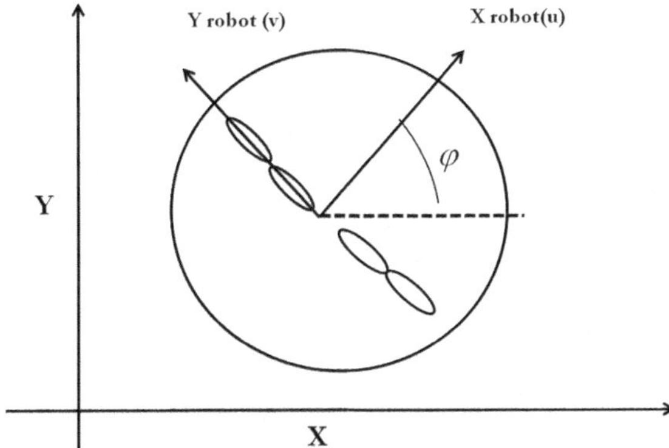

Fig. 1 Schematic representation of the vehicle

2.2 SDRE Control Method

The State Dependent Riccati Equation control method is a suboptimal control of a nonlinear system (2) through the minimization of the functional (3) as shown in [11, 12].

$$\begin{aligned} \dot{\mathbf{x}} &= \mathbf{f}(\mathbf{x}) + \mathbf{B}(\mathbf{x})\mathbf{u} \\ \mathbf{x} &= \mathbf{x_0} \end{aligned} \tag{2}$$

$$\mathrm{J}[\mathbf{u}] = \frac{\mathbf{1}}{\mathbf{2}} \int_{t_0}^{\infty} \left(\mathbf{x}^{\mathbf{T}}\mathbf{Q}(\mathbf{x})\mathbf{x} + \mathbf{u}^{\mathbf{T}}\mathbf{R}(\mathbf{x})\mathbf{u}\right).dt \tag{3}$$

Placing a system (2) in a state dependent form though a non unique parameterization:

$$\dot{\mathbf{x}} = \mathbf{A}(\mathbf{x})\mathbf{x} + \mathbf{B}(\mathbf{x})\mathbf{u} \tag{4}$$

It is possible to obtain a controller in a form:

$$\mathbf{u} = -\mathbf{R}^{-1}(\mathbf{x})\mathbf{B}^{\mathbf{T}}(\mathbf{x})\mathbf{P}(\mathbf{x})\mathbf{x} \tag{5}$$

Solving a state dependent algebraic Riccati equation:

$$\mathbf{P}(\mathbf{x})\mathbf{A}(\mathbf{x}) + \mathbf{A}^{\mathbf{T}}(\mathbf{x})\mathbf{P}(\mathbf{x}) - \mathbf{P}(\mathbf{x})\mathbf{B}(\mathbf{x})\mathbf{R}^{-1}(\mathbf{x})\mathbf{B}^{\mathbf{T}}(\mathbf{x})\mathbf{P}(\mathbf{x}) + \mathbf{Q}(\mathbf{x}) = 0 \tag{6}$$

2.3 Control Problem Formulation

Consider the vehicle system (1) and a reference system of the form:

$$\begin{aligned} \dot{\mathbf{x}}_{\mathbf{r}} &= \mathbf{u_r}\, cos(\varphi_r) - \mathbf{v_r}\, sin(\varphi_r) \\ \dot{\mathbf{y}}_{\mathbf{r}} &= \mathbf{u_r}\, sin(\varphi_r) + \mathbf{v_r}\, cos(\varphi_r) \\ \dot{\varphi}_r &= r_r \\ \dot{\mathbf{u}}_{\mathbf{r}} &= \mathbf{v_r}\mathbf{r}_r + \boldsymbol{\tau}_{\mathbf{ur}} \\ \dot{\mathbf{v}}_{\mathbf{r}} &= -\mathbf{u_r}\mathbf{r_r} - \beta\mathbf{v_r} \\ \dot{r}_r &= \tau_{rr} \end{aligned} \tag{7}$$

An error is defined as the difference between the (1) and (7):

$$\begin{bmatrix} e_1 \\ e_2 \\ e_3 \\ e_4 \\ e_5 \\ e_6 \end{bmatrix} = \begin{bmatrix} x - x_r \\ y - y_r \\ \varphi - \varphi_r \\ u - u_r \\ v - v_r \\ r - r_r \end{bmatrix} \tag{8}$$

Substituting and deriving (8) the system yields:

$$\begin{aligned}
\dot{e}_1 &= \left[\left(\frac{-1+cos(e_3)}{e_3}\right)(\mathbf{u_r}cos(\varphi_r) - \mathbf{v_r}sin(\varphi)_r) - \left(\frac{sin(e_3)}{e_3}\right)(\mathbf{u_r}sin(\varphi_r) + \mathbf{v_r}cos(\varphi_r))\right]e_3 + \\
&\quad + [cos(e_3)cos(\varphi_r) - sin(e_3)sin(\varphi_r)]e_4 - [sin(e_3)cos(\varphi_r) - cos(e_3)sin(\varphi_r)]e_5 \\
\dot{e}_2 &= \left[\left(\frac{-1+cos(e_3)}{e_3}\right)(\mathbf{u_r}sin(\varphi_r) + \mathbf{v_r}cos(\varphi_r)) + \left(\frac{sin(e_3)}{e_3}\right)(\mathbf{u_r}cos(\varphi_r) - \mathbf{v_r}sin(\varphi))\right]e_3 + \\
&\quad + [sin(e_3)cos(\varphi_r) - cos(e_3)sin(\varphi_r)]e_4 + \left[cos(e_3)cos(\varphi_r) - sin(e_3)sin(\varphi)_r\right]e_5 \\
\dot{e}_3 &= e_6 \\
\dot{e}_4 &= (e_6 + r_r)e_5 + \mathbf{v_r}e_6 + \mu_1 \\
\dot{e}_5 &= -(e_6 + r_r)e_4 - \beta e_5 - \mathbf{u_r}e_6 \\
\dot{e}_6 &= \mu_2
\end{aligned} \tag{9}$$

Then, the system in the error model could be written in the matrix form:

$$\begin{bmatrix} \dot{e}_1 \\ \dot{e}_2 \\ \dot{e}_3 \\ \dot{e}_4 \\ \dot{e}_5 \\ \dot{e}_6 \end{bmatrix} = \mathbf{A}(\mathbf{e}) \begin{bmatrix} e_1 \\ e_2 \\ e_3 \\ e_4 \\ e_5 \\ e_6 \end{bmatrix} + \mathbf{B}(\mathbf{e}) \begin{bmatrix} 0 \\ 0 \\ 0 \\ \mu_1 \\ 0 \\ \mu_2 \end{bmatrix} \tag{10}$$

where the matrices **A(e)** and **B(e)** are:

$$\mathbf{A}(\mathbf{e}) = \begin{bmatrix} 0 & 0 & a_1 & a_3 & -a_4 & 0 \\ 0 & 0 & a_2 & a_4 & a_3 & 0 \\ 0 & 0 & 0 & 0 & 0 & 1 \\ 0 & 0 & 0 & 0 & a_5 & \mathbf{v_r} \\ 0 & 0 & 0 & -a_5 & -\beta & -\mathbf{u_r} \\ 0 & 0 & 0 & 0 & 0 & 0 \end{bmatrix} \mathbf{B}(\mathbf{e}) = \begin{bmatrix} 0 & 0 \\ 0 & 0 \\ 0 & 0 \\ 1 & 0 \\ 0 & 0 \\ 0 & 1 \end{bmatrix} \tag{11}$$

and the coefficients of the matrix **A(e)** are:

$$a_1 = \left[\left(\frac{-1+\cos(e_3)}{e_3}\right)(\mathbf{u_r}\cos(\varphi_r) - \mathbf{v_r}\sin(\varphi)) - \left(\frac{\sin(e_3)}{e_3}\right)(\mathbf{u_r}\sin(\varphi_r) + \mathbf{v_r}\cos(\varphi_r))\right]$$
$$a_2 = \left[\left(\frac{-1+\cos(e_3)}{e_3}\right)(\mathbf{u_r}\sin(\varphi_r) + \mathbf{v_r}\cos(\varphi_r)) + \left(\frac{\sin(e_3)}{e_3}\right)(\mathbf{u_r}\cos(\varphi_r) - \mathbf{v_r}\sin(\varphi))\right]$$
$$a_3 = \cos(e_3 + \varphi_r)$$
$$a_4 = \sin(e_3 + \varphi_r)$$
$$a_5 = (e_6 + r_r) \tag{12}$$

Then the state dependent algebraic Riccati equation (13) should be solved each time step which is possible by numeric non-iterative approaches by using the stable eigenvectors of the Hamiltonian matrix.

$$\mathbf{P}(\mathbf{e})\mathbf{A}(\mathbf{e}) + \mathbf{A}^{\mathbf{T}}(\mathbf{e})\mathbf{P}(\mathbf{e}) - \mathbf{P}(\mathbf{e})\mathbf{B}(\mathbf{e})\mathbf{R}^{-1}(\mathbf{e})\mathbf{B}^{\mathbf{T}}(\mathbf{e})\mathbf{P}(\mathbf{e}) + \mathbf{Q}(\mathbf{e}) = 0 \tag{13}$$

to obtain the control vector:

$$\mathbf{u} = -\mathbf{R}^{-1}(\mathbf{e})\mathbf{B}^{\mathbf{T}}(\mathbf{e})\mathbf{P}(\mathbf{e})\mathbf{e} \tag{14}$$

3 Numerical Simulations

After the control problem formulation, the system (11) was solved numerically. For this purpose the numeric integration using the fourth order Runge-Kutta integrator (ode45) in Matlab® is used. Function *lqr* is used to solve the control problem. This command tests the controllability of the system and performs the numeric solution of the algebraic Riccati equation according to [13]. Computation cost of this solution is low and adequate to real-time systems. The initial conditions of the problem were presented in the figures subtitle, the parameter $\beta = 1.2$ was adopted as it is proposed in [4] for all calculations. The system was simulated applying the initial condition of surge direction $\tau_{ur} = 0.1$ to the straight line trajectory. To the circular trajectory it was set to $\tau_{ur} = 0.28$.

The matrices **Q(e)** and **R(e)** to solve the *lqr* were chosen as

$$\mathbf{Q}(\mathbf{e}) = \begin{bmatrix} 5 & 0 & 0 & 0 & 0 & 0 \\ 0 & 5 & 0 & 0 & 0 & 0 \\ 0 & 0 & 5 & 0 & 0 & 0 \\ 0 & 0 & 0 & 5 & 0 & 0 \\ 0 & 0 & 0 & 0 & 5 & 0 \\ 0 & 0 & 0 & 0 & 0 & 5 \end{bmatrix} \mathbf{R}(\mathbf{e}) = \begin{bmatrix} 10 & 0 \\ 0 & 10 \end{bmatrix} \tag{15}$$

Table 1 Physical parameters of the hovercraft

Parameter	Dimension
Mass	5.5 kg
J (rotational inertia)	0.047 kg m^2
l (moment arm)	0.123 m
d_v	5.5 kg/s
d_w	0.41 kg/s
Depth	25.4 cm
Width	35.6 cm
Height	18.1 cm

The choice of (15) is empiric and these values are selected to ensure stability and to minimize the overshoot of the system. The physical parameters of the Hovercraft are presented in the Table 1 and were obtained in [14].

3.1 Linear Trajectory

Here, the purpose is that the hovercraft tracks the trajectory of a virtual reference that was set as a straight line with the orientation of 45°. The initial conditions to the Hovercraft are $[x, y, \varphi, \mathbf{u}, \mathbf{v}, \mathrm{r}] = [5, -5, 1, 0, 0, 0]$ and for the reference are $[x_r, y_r, \varphi_r, \mathbf{u}_r, \mathbf{v}_r, \mathrm{r}_r] = [0, 0, \pi/4, 0.5, 0, 0]$. The simulations of tracking a linear trajectory in Fig. 2a shows the errors related to the position and orientation of the hovercraft converging to zero in less than 10 s, which means that after this time it is expected that the hovercraft will be following the same trajectory that the reference with the same orientation. In Fig. 2b the error related to the velocities of the hovercraft is presented and converges to zero after 12 s.

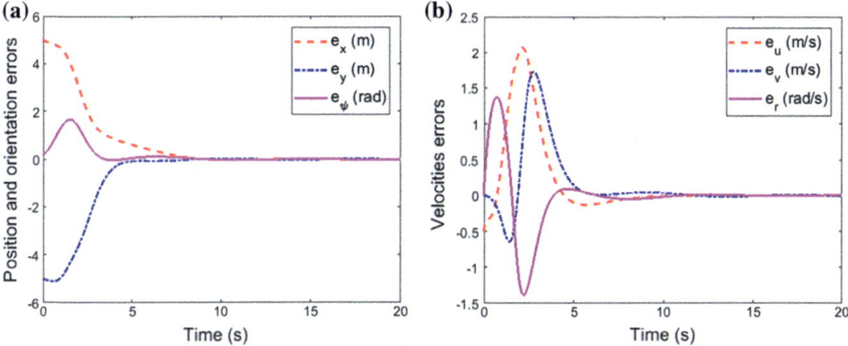

Fig. 2 **a** Simulation of the error for position and orientation. **b** Simulation of the error for velocities

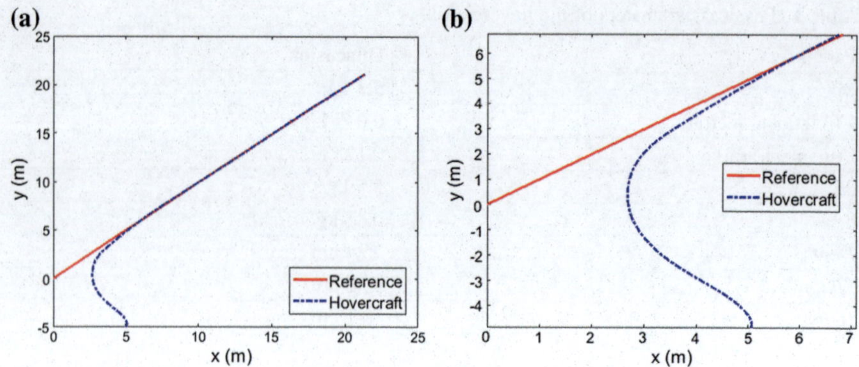

Fig. 3 **a** Trajectories of the vehicle and the reference. **b** The detail of initial convergence

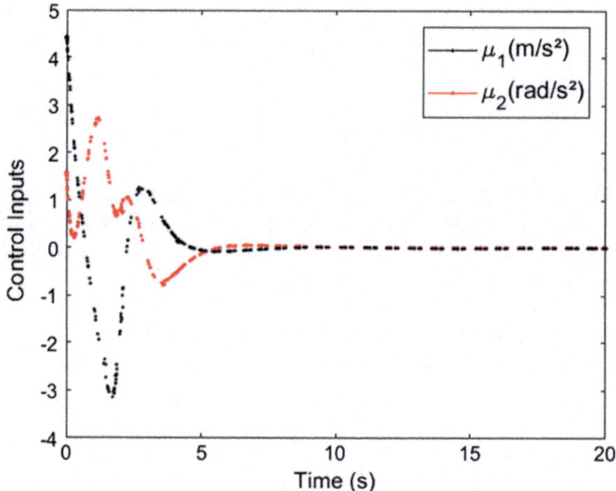

Fig. 4 Control inputs of the hovercraft

In the Fig. 3a both hovercraft and reference trajectories are presented on the plane. It is possible to see that the hovercraft follows the desired trajectory of a straight line with the inclination of 45°. In the Fig. 3b it is presented a zoomed detail of the convergence of the hovercraft to the trajectory. From Fig. 3b it is clear that the hovercraft trajectory starts at different set of initial conditions than a reference and converges successfully towards it.

The control inputs are shown in Fig. 4. The black line represents the control input related to the linear acceleration applied to the system and the red line represents the time evolution of the control input related to the angular acceleration of the system. As it can be seen, after 10 s the control inputs stabilize becoming constant and converging to the reference values as the trajectory converges to the reference set previously.

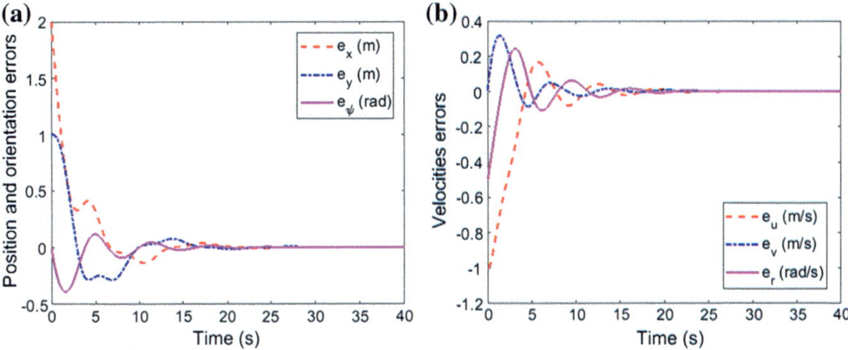

Fig. 5 **a** Error states for position and orientation of the system, performing circular reference. **b** Error states for velocities performing circular reference

3.2 Circular Trajectory

Simulations were performed to achieve tracking of a circular reference trajectory using this algorithm, adjusting only a few parameters. The parameters were adjusted as follows:

$$\begin{aligned} &\dot{\varphi}_r = r_r = const \\ &\varphi_r = r_r \cdot t + c \\ &\mathbf{u_r} = const \\ &\mathbf{v_r} = 0 \\ &\dot{r}_r = \tau_r = 0 \end{aligned} \tag{16}$$

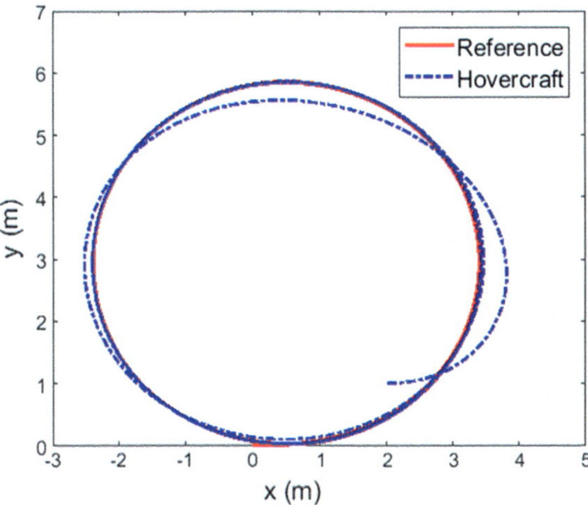

Fig. 6 Trajectory tracking of the hovercraft

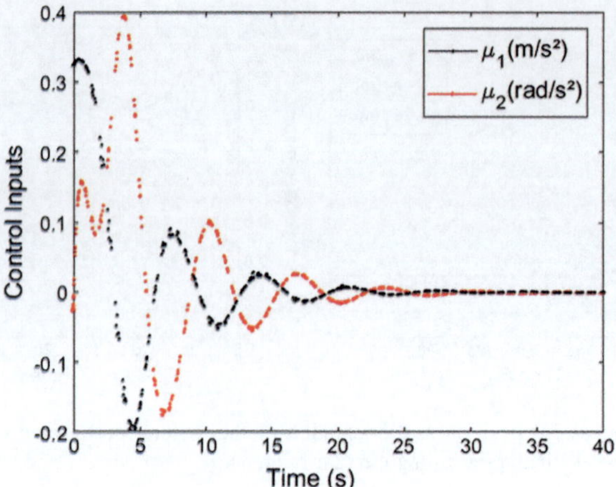

Fig. 7 Control inputs of the hovercraft

It means that to track a circle trajectory the surge velocity ($\mathbf{u}_r$) is constant, the sway velocity ($\mathbf{v}_r$) is null, and the angular velocity ($\dot{\varphi}_\mathbf{r}$) is also constant. Then, it's possible to determine the angle of orientation of the reference at each time of the simulation using the second equation of (16). Initial conditions for the hovercraft trajectory were set as $[x, y, \varphi, \mathbf{u}, \mathbf{v}, \mathrm{r}] = [2, 1, 0, 0, 0, 0]$ and for the reference trajectory as $[x_r, y_r, \varphi_r, \mathbf{u}_r, \mathbf{v}_r, \mathrm{r}_r] = [0, 0, 0.001, 1, 0, 0.5]$.

Figure 5a presents the error states related to position and orientation of the vehicle with respect to time. The system starts to converge to zero in less than 15 s and completely after 23 s. Thus hovercraft tracks the circular trajectory successfully after this time.

Figure 5b present the error systems related to the velocities of the system in forward ($\boldsymbol{u}$), lateral ($\boldsymbol{v}$) and rotational (r) directions of the system. As it is seen from this figure, the velocity error converges to zero after 25 s.

Figure 6 shows the trajectory by the hovercraft in blue and the reference trajectory in red. From this figure it is clear that the hovercraft controlled trajectory converges into a reference with success.

The applied control inputs on the hovercraft for the circular path are shown in Fig. 7. The black line represents the control input related to the linear acceleration applied to the system and the red line represents the time evolution of the control input related to the angular acceleration of the system. As it is seen, after 23 s they stabilize becoming constant and converging to the reference values.

4 Conclusions

In this work the reference trajectory tracking control problem is formulated in which a dynamical system representing a hovercraft vehicle is tracked by the control effort toward a desired trajectory in a general form. The control method Stated Dependent Riccati Equation is applied to the solution of this problem and numerical simulations are presented regarding two basic trajectory regimes: a linear desired trajectory and a circular trajectory. By switching the reference velocities and some initial conditions and parameters it is possible to obtain different set of desired trajectories and track the main vehicle system towards them. The SDRE methods deals with the solution of the algebraic state-dependent Riccati equation by numerical methods based on the Hamiltonian matrix eigenvectors, which permits a step-by-step solution of the control problem and is suitable to real-time applications. The numerical simulations validate the successful trajectory tracking and the effectiveness of the control method.

References

1. Hinchey, M.J., Sullivan, M.A.: On hovercraft overwater heave stability. J. Sound Vib. **163**, 261–273 (1993). https://doi.org/10.1006/jsvi.1993.1163
2. Kuznetsov, N.G.: The problem of forward motion of a hovercraft: asymptotic analysis of a solution. J. Math. Sci. Russ. **159**, 91–104 (2009). https://doi.org/10.1007/s10958-009-9429-8
3. Chung, J., Jun, T.: Optimization of an air cushion vehicle bag and finger skirt using genetic algorithms. Aerosp. Sci. Technol. **8**, 219–229 (2004). https://doi.org/10.1016/j.ast.2003.11.002
4. Morales, R., Sira-Ramirez, H., Somolinos, J.A.: Linear active disturbance rejection control of the hovercraft vessel model. Ocean Eng. **96**, 100–1081 (2015). https://doi.org/10.1016/j.oceaneng.2014.12.031
5. Brockett, R.W.: Asymptotic stability and feedback stabilization In: Differential Geometric Control Theory, pp. 181–191. Birkhauser, MA (1983)
6. Rafikova, E., Kurka, P.R.G.: Controle de Trajetória de um Robô Móvel Autônomo. In: Paulo Sausen, Airam Sausen. (Org.). Pesquisas Aplicadas em Modelagem Matemática, 1 ed., vol. 2, pp. 313–336. Editora Unijui, Ijuí (2012)
7. Rafikova, E., Kurka, P.R.G., Rafikov, M., Gafurov, S.A.: On linear and nonlinear trajectory tracking control for nonholonomic integrator. J. Dyn. Vibroacoust. **1**, 1–6 (2014)
8. Fantoni, I., Lozano, R., Mazenc, F., Pettersen, K.Y.: Stabilization of a nonlinear under actuated hovercraft. In: 38° Conference on Decision and Control, Phoenix, Arizona USA, pp. 2533–2538, December 1999
9. Fossen, T.I.: Guidance and Control of Ocean Vehicles. Wiley, UK (1994)
10. Pettersen, K.Y., Nijmeijer, H.: Global practical stabilization and tracking for an underactuated ship—a combined averaging and backstepping approach. In: Proceedings IFAC Conference on System Structure and Control, Nantes, France, pp. 59–64, July 1998
11. Cloutier, J.R.: State-dependent Riccati equation techniques: an overview. In: Proceedings of the American Control Conference, Albuquerque, New Mexico, pp. 932–936, June 1997

12. Rafikov, M., Rafikova, E., Rinaldo, G.: Synchronization of the mobile robot to a chaotic trajectory. J. Appl. Nonlinear Dyn. **6**, 379–385 (2016). https://doi.org/10.5890/jand.2016.09.005
13. Arnold, W.F., III, Laub, A.J.: Generalized eigenproblem algorithms and software for algebraic Riccati Equations. In: Proceedings of the IEEE, vol. 72, pp. 1746–1754 (1984)
14. Aguiar, P.A., Cremean, L., Hespanha, J.P.: Position tracking for nonlinear underactuated hovercraft: controller design and experimental results. In: Proceedings of the 42nd IEEE Conference on Decision and Control Maui, Hawaii, USA, pp. 3858–3863, December 2003

Method for Controlling Stick-Slip Vibrations in Slender Drilling Systems

Guilherme Sampaio and Hans I. Weber

Abstract Systems actuated trough a flexible shaft poses a big challenge to control strategies as the actuator is not connected directly to the end effector, causing propagation effects as well as an energy accumulation and dissipation in the shaft. This paper focuses on the top driven drilling system used in the oil and gas industry. In these systems, all kind of vibrations are found: longitudinal deformations (bit bouncing), flexional (rubbing), and torsional (stick-slip). This paper is about the torsional deformation of the highly flexible string modeled as a 20 DOF Lumped parameters system. A method for reducing stick-slip vibrations is presented and its results analyzed. The investigation includes the development of a reduced scale test rig adequate for torsional vibrations under damping. Results from the mathematical model and experimental tests are then compared.

Keywords Stick-slip · Torsional vibrations · Friction · Control · Drilling

1 Introduction

Top driven drilling used in the oil and gas industry is one of the most investigated application of systems driven by a highly flexible shaft. These systems pose a big challenge to control strategies, as the actuator is not linked directly to the end effector, causing propagation effects as well as an energy accumulation and dissipation in the shaft. These drilling systems are composed by a top drive linked to the drill bit through hundreds of meters of steel pipes. Drilling is one of the most expensive parts of oil prospecting and involves many risks of accidents, even though the methods in use are still very much based on trial and error experiences. Linear control theories

G. Sampaio (✉) · H. I. Weber
Laboratótio de Dinâmica e Vibrações, PUC-Rio, R. Marqus de S. Vicente 225, Rio de Janeiro, RJ, Brazil
e-mail: guirsp@gmail.com

A. de T. Fleury et al. (eds.), *Proceedings of DINAME 2017*, Lecture Notes in Mechanical Engineering, https://doi.org/10.1007/978-3-319-91217-2_24

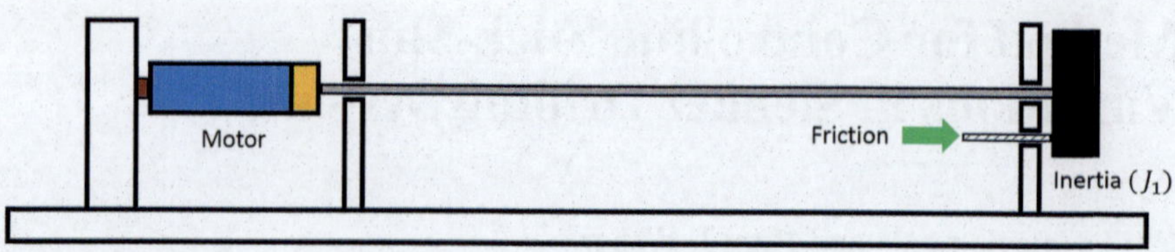

Fig. 1 Test setup

for example, PID have little success on real drilling applications because there are lots of uncertainties present, friction with the well, friction between rock and bit, etc. This paper will present a numerical and experimental study of a control technique that aims to reduce torsional vibrations maintaining a constant speed at the bit in a simple model of an oil drilling rig. A 2 m long test setup Fig. 1 composed by a DC motor, an elongated flexible shaft and a driven inertia that simulates the bit of a drilling structure, was constructed to test in the lab the controller in the presence of uncertainties and sensor noises.

The torsional system was modeled as a 20 degrees of freedom (DOF) flexible shaft. The contact of the pin that simulates the contact between bit and rock in the real system, is modeled as sum of a Coulomb static friction coefficient, a dynamic coefficient and a viscous friction, dependent of the angular speed.

The problem of modelling the torsional dynamics of a flexible shaft can be approached in different ways, the most common in literature are a simple torsional spring or spring-damper [1], lumped parameters or a finite element discretization of the shaft.

Khulief et al. [2] analyze self-excited stick-slip oscillations in drillstrings using a proposed dynamic model where the equation of motion of the rotating drillstring is derived using Lagrangian approach in conjunction with the finite element method and analyses torsional-bending and axial-bending nonlinear couplings.

Navarro-Lopez and Cortes [3] developed a lumped parameter model to investigate the influence of sliding motion on self-excited stick-slip oscillations and bit sticking phenomena. Hopf bifurcations were used to investigate the range of rotary speeds where the undesired torsional vibrations of the drillstring happen.

Rudat and Dashevskiy [4] present in the article, an innovative model based stick-slip control system using a lumped parameters model with parameters identified from real world applications trough Newton Gauss method and extended Kalman filter. The key idea of this paper is to run simulations on an embedded system down hole and transmit the updated model parameters in a lower bitrate trough mud pulses, an established technology. It shows that lumped masses model can reproduce nonlinear dynamics of drilling and shows, with experiments done in field tests, the effectiveness of the proposed approach.

Bayliss et al. [5] analyze a basic pole placement controller design for a Single Input Single Output (SISO) linear model of a drilling system, but recursively evaluated based on an online Recursive Least Squares (RLS) identification of the open-loop plant parameters. It presents a discussion on system architecture implications, and the simulation results with and without adaptive stick/slip mitigation method. The presented method in this paper relies on accurate measurements of the speed on Bit.

Ritto et al. [6] analyze the dynamics of a horizontal drill-string modeling uncertainty on the frictional force and how uncertainties on the frictional forces propagate through the system. A stochastic field with exponential correlation function is used to model the frictional coefficient.

Kreuzer and Steidl [7] present in the paper a method for controlling these vibrations by exactly decomposing the drill string dynamics into two traveling waves traveling in the direction of the top drive and in the direction of the drill bit. Authors state that by using two angular sensors placed 5 m away from each other it is possible to characterize traveling torsional waves in the drillstring and therefore use the top drive to eliminate them.

Kapitaniak et al. [8] experimentally investigates drillstring vibrations using a vertical reduced scale drilling rig. In the test setup presented by authors, one can investigate torsional, compression and helical bucking vibrations. The proposed experiment uses a 10 mm diameter steel drillstring and rotational speeds up to 54 RPM. The article describes methods used to obtain the mechanical properties of the setup as well as the use of finite element models to represent it.

2 Mathematical Model

A model with a 20 DOF Lumped parameters flexible shaft Fig. 2, was chosen to be used for the simulations in this paper. This model was chosen for its simplicity yet being capable of including the inertia of the shaft and dissipation from internal friction. This model also uses a complete DC motor model with electrical and mechanical parts.

The DC motor is modeled by the equations of the mechanical Eq. (1) and electrical parts Eq. (2) as well as the torque constant k_t that is the relation between T_m and i. All the parameters used for the motor were the ones obtained by [9].

$$J_m \frac{d^2\theta}{dt^2} = T_m - b_m \frac{d\theta}{dt} \tag{1}$$

$$L\frac{di}{dt} = -Ri + V - e \tag{2}$$

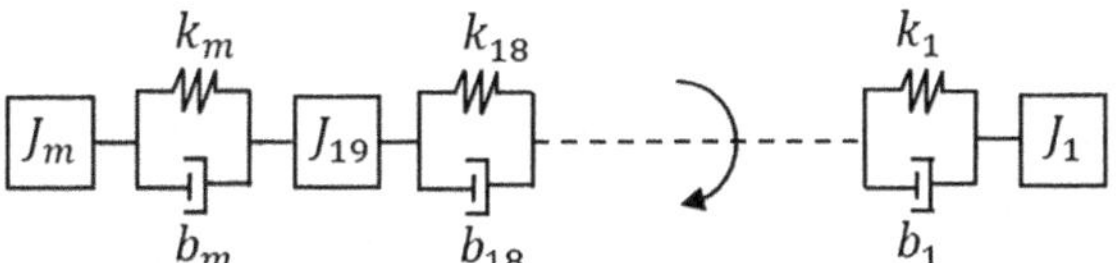

Fig. 2 Lumped parameters flexible shaft

In the lumped parameters model, each element or DOF is an elementary inertia-damper-spring system modeled as:

$$I\frac{d^2\theta}{dt^2} = -k\theta - b\frac{d\theta}{dt} \tag{3}$$

Writing the differential equations in Eq. (3) for each sub system and assembling them in a matrix form it becomes:

$$\bar{M}\ddot{\theta} + \bar{B}\dot{\theta} + \bar{K}\theta = \bar{\tau} \tag{4}$$

where θ is the state vector representing the angular displacements of the lumped masses, $\bar{\tau}$ is the external torques vector, the mass matrix ($\bar{M}$), the damping matrix ($\bar{B}$), and the spring matrix ($\bar{K}$) are:

$$\bar{M} = \begin{bmatrix} J_1 & & & & \\ & J_2 & & & \\ & & \ddots & & \\ & & & J_{19} & \\ & & & & J_m \end{bmatrix} \bar{B} = \begin{bmatrix} b_1 & -b_1 & & & \\ -b_1 & (b_1 + b_2) & -b_2 & & \\ & -b_2 & \ddots & & \\ & & & (b_{18} - b_m) & -b_m \\ & & & -b_m & b_m \end{bmatrix} \tag{5}$$

$$\bar{K} = \begin{bmatrix} k_1 & -k_1 & & & \\ -k_1 & (k_1 + k_2) & -k_2 & & \\ & -k_2 & \ddots & & \\ & & & (k_{18} - k_{19}) & -k_{19} \\ & & & -k_{19} & k_{19} \end{bmatrix} \tag{6}$$

This way, the mechanical lumped parameters of the drill string can be written substituting Eqs. (5) and (6) in Eq. (4). The model parameters were obtained experimentally, Table 1 shows the numerical values.

Table 1 Model parameters used on simulations

Properties	Value	Unit
String length (L)	1.7	m
String diameter (mm)	3	mm
Total inertia of J_1	0.01555819	kg m^2
String stiffness (K)	0.2548	N m/rad
Moment of inertia of motor (J_m)	0.37×10^{-3}	kg m^2
Armature inductance (L_{DC})	1.10×10^{-3}	H
Armature resistance (R_{DC})	0.33	Ω
Torque constant (K_t)	0.12	N m/A
Speed constant (K_e)	6.02×10^{-2}	V/(rad/s)

2.1 Torque on Bit Formulation

The contact between bit and rock is modeled, according to Armstrong-Hlouvry et al. [10] by the sum of a Coulomb static friction coefficient, a dynamic coefficient and a viscous friction, dependent of the angular speed. This contact appears in the model on the inertia J_1.

$$T_r = (T_C + (T_{brk} - T_C) \cdot exp(-c_v|\omega|))sign(\omega) + f\omega \qquad if|\omega| \geq \omega_{th} \tag{7}$$

and:

$$T_r = \omega \frac{(f\omega_{th} + (T_C + (T_{brk} - T_C) \cdot exp(-c_v\omega_{th})))}{\omega_{th}} \qquad if|\omega| < \omega_{th} \tag{8}$$

where: T_r is the friction torque, T_C is the Coulomb friction torque, T_{brk} is the static friction torque, c_v is the dynamic friction coefficients, ω is the angular speed, f is the viscous friction coefficient and ω_{th} is the velocity of threshold that is in the order of 10^{-4} included to avoid numerical problems (Fig. 3).

Table 2 shows the numerical values of friction used in this paper.

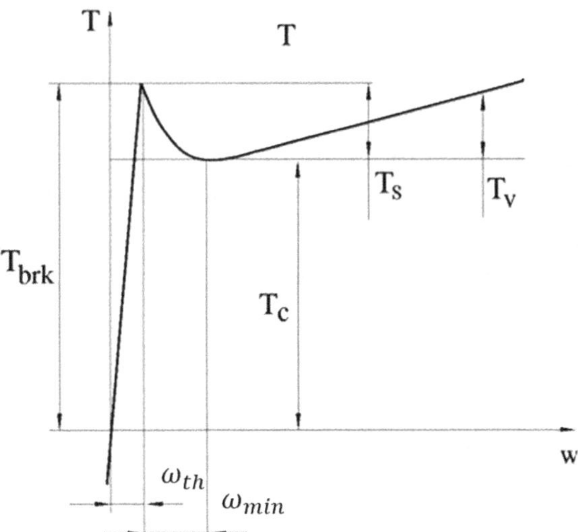

Fig. 3 Friction torque versus rotational speed

Table 2 Friction parameters

Parameter	Value	Unit
Breakaway friction coefficient (μ_{brk})	0.5	–
Coulomb friction coefficient (μ_c)	0.47	–
Normal force on bit (N_f)	26	N
Viscous friction coefficient (b)	0.001	$\frac{\text{N m}}{\text{rad/s}}$

3 Controlling the System with a Torque Source

By analyzing the structure of the friction law Eq. (8) and the results of the simulations, an investigation was started to analyze if adding a torque source to J_1 (a kind of downhole motor), it could be possible to modify the stick slip phenomenon. Using this supposition an analysis was made to verify if it is possible to mitigate the stick slip by controlling the torque on J_1 with the use of a DC motor.

A recent study from Shor et al. [11], showed that the propagation effects on torsional vibrations are important for the implementation of torsional vibrations mitigation techniques, which led to suppose that the phase of the proposed control for an imposed torque on J_1 should be important for the results. Therefore, the lumped masses system described in Sect. 2 was simulated with two torque sources, both DC motors referred as "Motor" and "J_1 Motor" in Fig. 4. The simulations started at $t = 0$ s with angular displacement and speed of the drill string being zero. At $t = 0$ the top drive motor is started at 2 rad/s, and around $t = 9$ s the energy accumulated in the drill string is enough to overcome the static friction force and the stick slip phenomenon begins. At $t = 15$ s a second DC motor attached to J_1 is energized applying a torque of aprox. −0.29 N m to J_1. This method only observes the output (angular speed at J_1) to start the motor in J_1, then this control is an open loop scheme, only applying a constant torque on J_1. The torque applied on J_1 therefore is: 0 N m for $t < 15$ s and $t > 30$ s, and −0.29 N m between $t = 15$ s and $t = 30$ s. Results are shown in Figs. 5 and 6.

As described by Shor et al. [11], the delay effects from the propagation of torsional vibrations along the drill string must be considered for the control structure of the

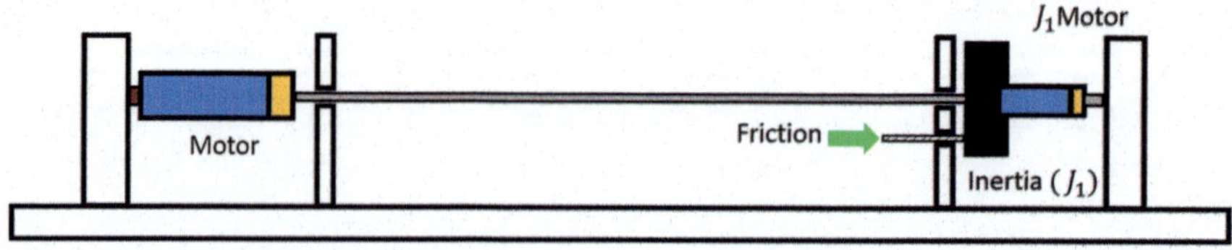

Fig. 4 Mechanical model with motor on J_1

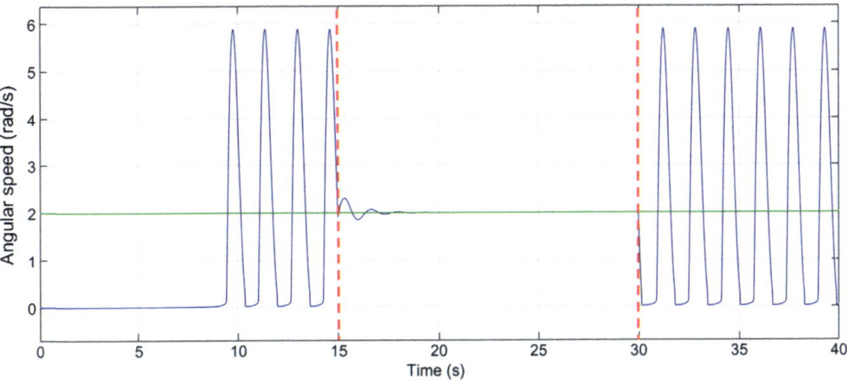

Fig. 5 Angular speed on top drive (green) and J_1 (blue)

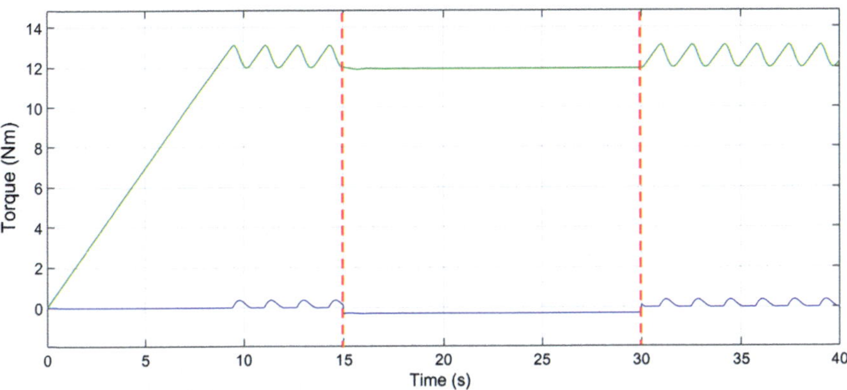

Fig. 6 Torque on top drive (green) and J_1 (blue)

problem. To prove that the developed mathematical model is capable of representing these effects, the system described in Figs. 5 and 6, was also simulated for a torque on J_1 being applied from $t = 14$ s to $t = 30$ s.

Results in Figs. 7 and 8 show that if the torque on J_1 is applied at a wrong moment it will have no effect on the stick-slip, only adding a small disturbance on the angular speed when it is applied. Simulations show that this approach to mitigate the stick slip only works if the torque on J_1 is applied when the angular speed of J_1 is decreasing, i.e. $\ddot{\theta}_1 < 0$.

In order to test this supposition, the same system was simulated again with the beginning of application of torque in J_1 at $T = 13.25$ s.

Simulation results in Figs. 9 and 10 show that the closer to the top drive speed is to the speed on J_1 when the torque on J_1 is applied, the better results are obtained. This is valid considering that the torque on J_1 is applied when the angular speed of J_1 is decreasing, i.e. $\ddot{\theta}_1 < 0$.

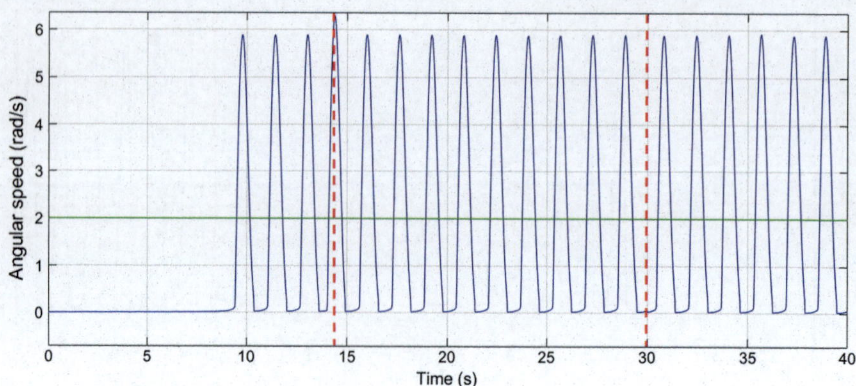

Fig. 7 Angular speed on top drive (green) and J_1 (blue)

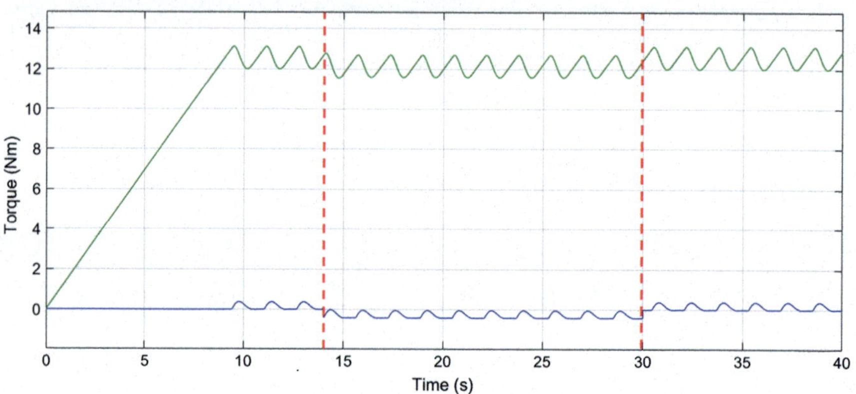

Fig. 8 Torque on top drive (green) and J_1 (blue)

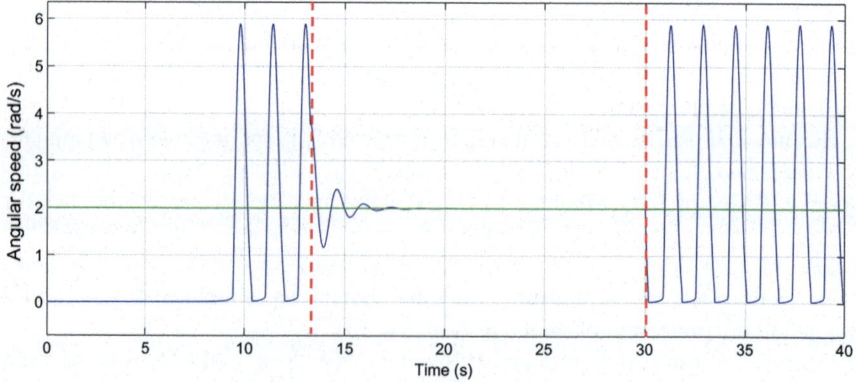

Fig. 9 Angular speed on top drive (green) and J_1 (blue)

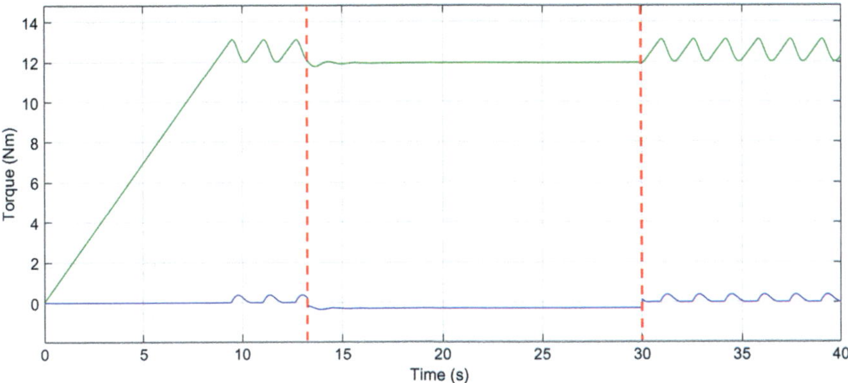

Fig. 10 Torque on top drive (green) and J_1(blue)

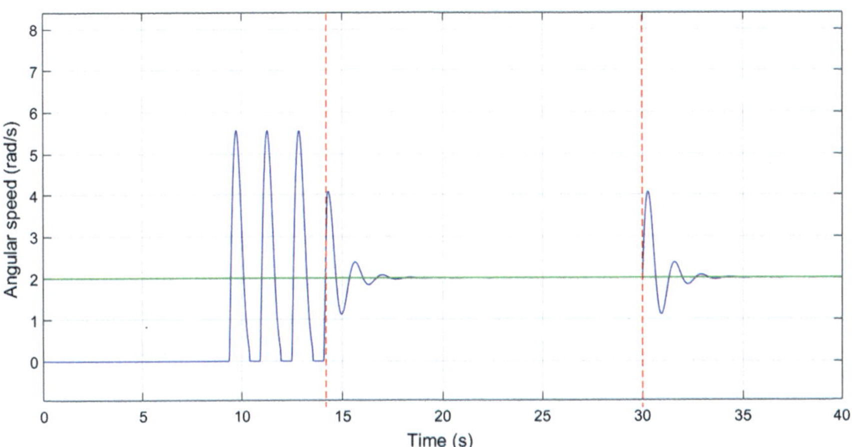

Fig. 11 Angular speed on top drive (green) and J_1 (blue)

One should note that despite this strategy is effective and shows good results, it depends on a precise measure of the speed on the bit (J_1) which limits its application for real life oil drilling problems with the existing technologies of bottom hole measurement available.

If, on the other hand, a positive torque is applied when the system is accelerating i.e. when the angular speed of J_1 is increasing, i.e. $\ddot{\theta}_1 > 0$, the results stay the same, the stick-slip is eliminated, but in this case another phenomenon is observed. When the motor in J_1 is turned off, the system has a torsional perturbation, but it does not come back to a stick-slip behavior. Figure 11 shows the angular speed of the top drive and J_1, and Fig. 12 shows the torque of both motors. As in the previous case, it is worth to note that the torque applied by the motor in J_1 is much smaller than the one from the top drive.

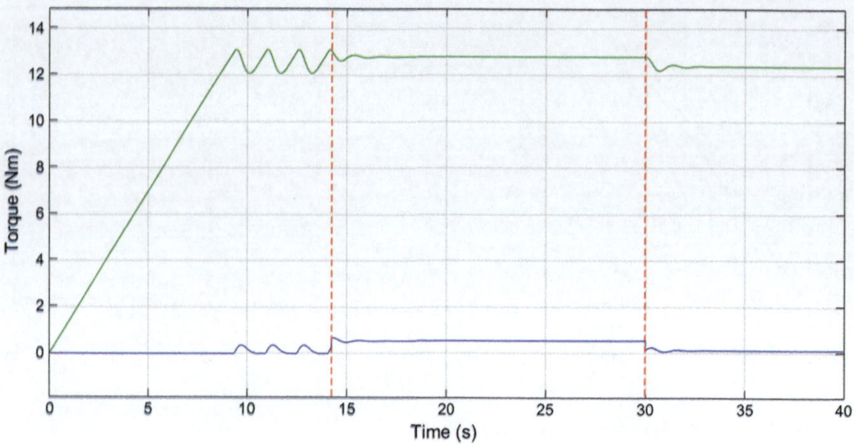

Fig. 12 Torque on top drive (green) and J_1 (blue)

4 Experimental Tests

In order to perform the experimental tests, a test bench was made Fig. 13. This apparatus is proposed to study only the rotational and torsional dynamics of the system, for that reason the motor and the inertia are mounted on bearings so that the setup can be used in an horizontal position, making it easy to operate. The drill string is 1.7 m long and is made of a 3 mm diameter steel rod. The DC motor is mounted on two ball bearings, so the torque applied by the motor is obtained through a force measured by a load cell positioned at a known distance from the motor. This force, the normal force of the brake on the inertia, and the rotational speed of the motor and of the inertia, are measured and these data are acquired by a National Instruments cDAQ system. A complete description of this experimental setup is shown in [9].

Table 3 shows the numerical values measured on the experimental setup used in this work.

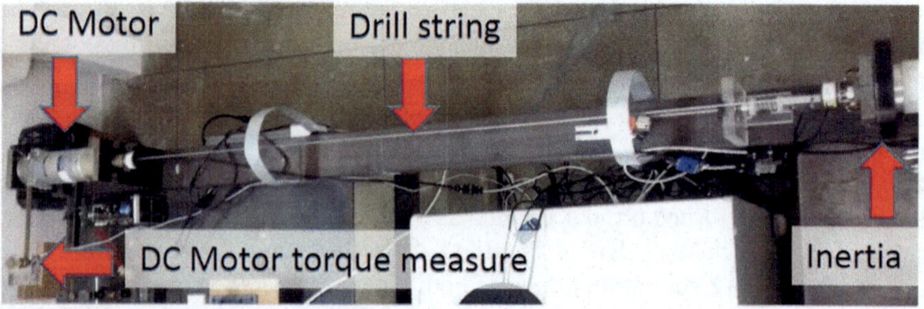

Fig. 13 Experimental setup

Table 3 Mechanical parameters of drillstring

Properties	Value	Unit
String length (L)	1.7	m
String density (ρ)	7850	kg/m^3
String diameter (mm)	3	mm
Young modulus (E)	210	GPa
Poisson ratio (ν)	0.3	–
Inertia of J_1	0.01555	kg m^2

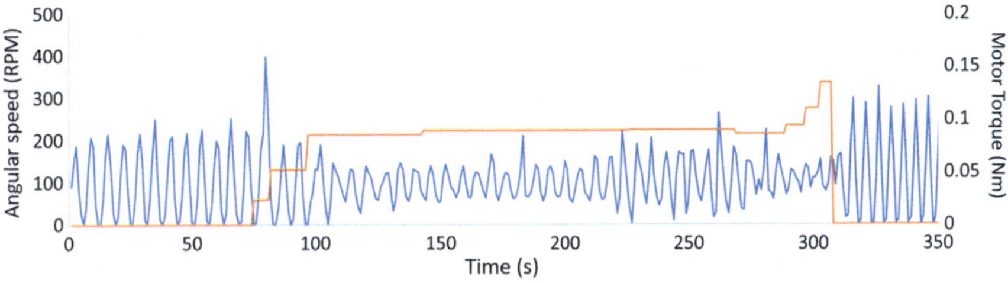

Fig. 14 Experimental results for torque applied on J_1 in green and speed at J_1 in orange

Figure 14 shows the results obtained with the apparatus shown in Fig. 13. In blue, on the left axis is the angular speed in RPM of J_1, in orange on the right axis is the amplitude of the torque in N m applied by the DC motor attached to J_1. Results shown are very similar to the ones obtained with the numerical model Fig. 5. The torsional vibration on the experimental test rig is not completely eliminated due to noise of the sensors and to non-modeled imperfections of the apparatus. But this approach shows that it can eliminate the stick-slip during the period the DC motor is being used.

5 Conclusions

It was shown that it is possible to mitigate the stick-slip phenomenon in the presented test setup and in the model, by applying a small torque on the bit, orders of magnitude lower than the one applied by the main motor, i.e. top drive.

Despite the construction challenges, the results presented in this paper show that the presence of a torque source on the bit of the drilling column (a kind of bottom hole motor, or mud motor) has a very important role on the torsional vibrations of the column. The presence of the bottom motor is almost not studied by the dynamics and vibrations community.

References

1. Patil, P.A., Teodoriu, C.: Model development of torsional drillstring and investigating parametrically the stick-slips influencing factors. J. Energy Resour. Technol. **135**(1), 013103 (2013)
2. Khulief, Y.A., Al-Sulaiman, F.A., Bashmal, S.: Vibration analysis of drillstrings with self-excited stickslip oscillations. J. Sound Vib. **299**(3), 540–558 (2007)
3. Navarro-Lopez, E.M., Corts, D.: Avoiding harmful oscillations in a drillstring through dynamical analysis. J. Sound Vib. **307**(1), 152–171 (2007)
4. Rudat, J., Dashevskiy, D.: Development of an innovative model-based stick/slip control system. In: SPE/IADC Drilling Conference and Exhibition. Society of Petroleum Engineers (2011)
5. Bayliss, M.T., Panchal, N., Whidborne, J.F.: Rotary steerable directional drilling stick/slip mitigation control. IFAC Proc. Vol. **45**(8), 66–71 (2012)
6. Ritto, T.G., Escalante, M.R., Sampaio, R., Rosales, M.B.: Drill-string horizontal dynamics with uncertainty on the frictional force. J. Sound Vib. **332**(1), 145–153 (2013)
7. Kreuzer, E., Steidl, M.: Controlling torsional vibrations of drill strings via decomposition of traveling waves. Arch. Appl. Mech. **82**(4), 515–531 (2012)
8. Kapitaniak, M., et al.: Unveiling complexity of drillstring vibrations: experiments and modelling. Int. J. Mech. Sci. **101**, 324–337 (2015)
9. de Paula, G.R.S.: Dynamics and control of stick-slip and torsional vibrations of flexible shaft driven systems applied to drillstrings. D.Sc. Thesis, Rio de Janeiro, Brazil (2017)
10. Armstrong-Hlouvry, B., Pierre, D., De Wit, C.: A survey of models, analysis tools and compensation methods for the control of machines with friction. Automatica **30**(7), 1083–1138 (1994)
11. Shor, R.J., Pehlivanturk, C., Acikmese, B., van Oort, E.: Propagation of torsional vibrations in drillstrings: how borehole geometry affects transmission and implications on mitigation techniques. In: ICoEV (2015)

Control of Multiple Mobile Robots in Dynamic Formations

Guilherme Rinaldo, Elvira Rafikova and Marat Rafikov

Abstract This work deals with the control of multiple mobile robots in trajectory, while maintaining a formation, through the use of State-Dependent Riccati Equation control method. Three robots with differential drive are used in a scheme in which one is considered the leader and the other two are considered followers. By changing formation parameters, this work seeks to achieve two different formations, V-shaped and Echelon formation, very common in the military field. Simulations are performed using LabVIEW, demonstrating the successful application of the control method in mobile robot tracking problems while maintaining formations.

Keywords Multiple robot control · Leader-follower problem · SDRE control
Nonlinear control

1 Introduction

The control of multiple mobile robots in formation is a theme that became the target of several recent studies [1–3], due to its large number of applications, both civilian and military. The idea is to control a robot so that it converges to a defined path while maintaining a formation with other robots.

Among the various approaches for the control of multiple robots in formations, the most common ones are behavior-based methods and leader-follower. Behavior-based control is the assignment of complex behaviors for each robot, such as speed synchronization, the ability to avoid collisions and centralization of the

G. Rinaldo (✉) · E. Rafikova · M. Rafikov
Center for Engineering, Modeling and Social Science, Federal University of ABC,
Avenida dos Estados 5000, Santo Andre, SP 092210-170, Brazil
e-mail: guilherme.rinaldo@ufabc.edu.br; guilherme.rinaldo@gmail.com

E. Rafikova
e-mail: elvira.rafikova@ufabc.edu.br

M. Rafikov
e-mail: marat.rafikov@ufabc.edu.br

A. de T. Fleury et al. (eds.), *Proceedings of DINAME 2017*, Lecture Notes
in Mechanical Engineering, https://doi.org/10.1007/978-3-319-91217-2_25

formation [3]. The leader-follower method used in this work, one or more robots are designated as leaders, while the others, which are called followers, receive information about the leader's position and converge to its trajectory while maintaining a desired distance and orientation relative to it [4–9]. To adjust its position in respect to the leader, the follower robot calculates track points based on the position and velocity information obtained from the leader. This method is known for simplifying the inclusion of new robots to the group, thus making the system easily scalable [8]. In [9] was proposed a reminiscent of the leader-follower approach that allowed a robot to follow a chaotic trajectory using the State Dependent Riccati Equation (SDRE) control method to achieve synchronization.

One approach to the control of multiple robots in formations is the tracking method, which requires a group of interconnected robots to reach a target or a reference path simultaneously. This approach has been widely studied [10] and can be used to solve a large set of problems, such as navigation while maintaining formation problems, rendezvous problems and dynamic trajectories tracking problems, like the one presented in this work [9–13]. Similar complex problems of cooperative tracking with dynamic references, using different control methods, were addressed in [14–16].

The present work deals with the control of a team of three mobile robots to solve a leader-follower dynamic trajectory tracking problem. The novelty of this work lies on the use of the SDRE control method to design a controller capable of making the robots navigate synchronized with one another in order to achieve and maintain two different formations. Computational simulations are shown to validate the results and the efficiency of the SDRE method, which requires low computational cost, in mobile robots tracking problems.

2 Methodology

2.1 Robotic Model

This study uses mobile robots with differential drive and rear support. The model has two identical wheels, positioned in the same axis, restricted to move only around it and independently controlled by motors. The rolling of the wheels is considered pure, so that there are no drift during movement. Such features have a restrictive nature, preventing the movement of the system in some directions at an initial moment, characterizing it as nonholonomic.

The system analysis is carried in a kinematic level. It is assumed that the center of mass is located in the geometric center of the robot, hence eliminating possible unbalances and consequent Coriolis and centripetal acceleration effects (Fig. 1).

The robot's position in the X and Y axes can be described by the projection of the robot's linear velocities on these axes. The angular velocity of the robot

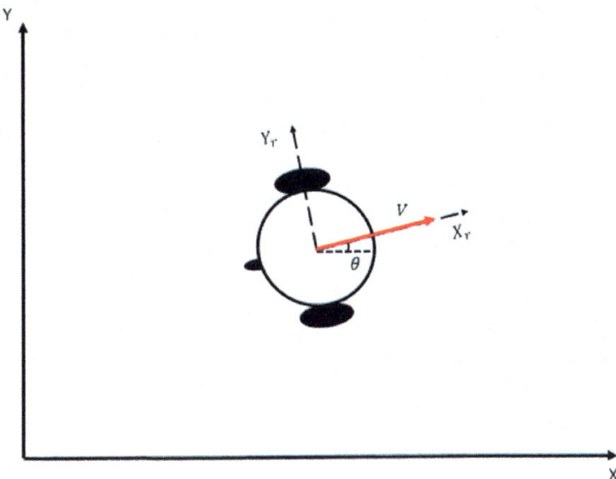

Fig. 1 Model of a differential drive robot

corresponds to the variation of the rotation angle θ over time, where θ is the angle between the global reference axis X and the local reference axis X_r of the robot.

$$\begin{aligned} \frac{dx}{dt} &= \dot{x} = \mathbf{v}\cos(\theta) \\ \frac{dy}{dt} &= \dot{y} = \mathbf{v}\sin(\theta) \\ \frac{d\theta}{dt} &= \dot{\theta} = \boldsymbol{\omega} \end{aligned} \tag{1}$$

Thus, the robot's kinematic model can be described as:

$$\begin{aligned} \dot{x}_i &= \mathbf{v_i}\cos(\theta_i) \\ \dot{y}_i &= \mathbf{v_i}\, sin(\theta_i) \\ \dot{\theta}_i &= \boldsymbol{\omega}_i \end{aligned} \tag{2}$$

where i = 1, 2, 3, …, n and i = 1 represents the leader robot, $\dot{x}$ and $\dot{y}$ describe the robot's velocity on the Cartesian coordinate system and $\dot{\theta}$ indicates its rotation in respect to the X axis, or in other words, the orientation of the robots on the plane.

2.2 *The Problem of Tracking of Multiple Mobile Robots*

The problem of tracking of multiple mobile robots consists on the stabilization of a system with two or more identical robots in a reference trajectory. The reference trajectory, to which all robots converge, is described as being a robot itself, called reference robot, whose kinematic model is identic to Eq. (2).

$$\begin{aligned} \dot{x}_r &= \mathbf{v_r} \cdot \cos(\theta_r) \\ \dot{y}_r &= \mathbf{v_r} \cdot sin(\theta_r) \\ \dot{\theta}_r &= \boldsymbol{\omega}_r \end{aligned} \tag{3}$$

The error vector $\mathbf{e(t)}$ is now introduced as:

$$\mathbf{e_i}(t) = \begin{bmatrix} x_i - x_{ri} - d_{xi} \\ y_i - y_{ri} - d_{yi} \\ \theta_i - \theta_{ri} - d_{\theta i} \end{bmatrix} \tag{4}$$

with i = 1, 2, 3, …, n.

The coordinates x_i, y_i and θ_i represents the position of a given robot, x_{ri}, y_{ri} and θ_{ri} represents the position of the reference robot and d_{xi}, d_{yi} and $d_{\theta i}$ are formation parameters, which determine the layout of the robots' formation. If these constants are null, the controlled robot shall converge to the reference trajectory, if not, the robot will converge to a distance on the X and Y axes and rotate by an angle θ, in respect to the leader robot. These distances and orientations therefore defines the layout of the formation.

Considering Eqs. (2) and (4), for any pair leader-follower, the deviation system can be described as:

$$\begin{aligned} \dot{e}_{1i} &= \mathbf{v_r} \cdot \cos(\theta_r) \cdot \cos(e_{3i}) - \mathbf{v_r} \cdot \sin(\theta_r) \cdot \sin(e_{3i}) + u_{1i} \cdot \cos(\theta_r) \cdot \cos(e_{3i}) \\ &\quad - u_{1i} \cdot \sin(\theta_r) \cdot \sin(e_{3i}) - \mathbf{v_r}\cos(\theta_r) \\ \dot{e}_{2i} &= \mathbf{v_r} \cdot \sin(\theta_r) \cdot \cos(e_{3i}) - \mathbf{v_r} \cdot \sin(e_{3i}) \cdot \cos(\theta_r) + u_{1i} \cdot \sin(\theta_r) \cdot \cos(e_{3i}) \\ &\quad + u_{1i} \cdot \cos(\theta_r) \cdot \sin(e_{3i}) - \mathbf{v_r} \cdot \sin(\theta_r) \\ \dot{e}_{3i} &= u_{2i} \end{aligned} \tag{5}$$

where i = 2, …, n. The control efforts u_{1i} and u_{2i} are components of the suboptimal controller u_i [12].

The system can be rewritten in matrix form as:

$$\dot{\boldsymbol{e}}_i = \boldsymbol{A}_i(e_{3i})e_i + B_i(e_{3i})\boldsymbol{u}_i \tag{6}$$

in which:

$$\boldsymbol{A}_i(e_{3i}) = \begin{bmatrix} 0 & 0 & \frac{V_1 \cos(\theta_{ri})(\cos(e_{3i}) - 1) - V_1 \sin(\theta_{ri}) \sin(e_{3i})}{e_{3i}} \\ 0 & 0 & \frac{V_1 \sin(\theta_{ri})(\cos(e_{3i}) - 1) - V_1 \cos(\theta_{ri}) \sin(e_{3i})}{e_{3i}} \\ 0 & 0 & 0 \end{bmatrix} \tag{7}$$

$$\boldsymbol{B}_i(e_{3i}) = \begin{bmatrix} \cos(e_{3i}) \cos(\theta_{ri}) - \sin(\theta_{ri})\sin(e_{3i}) & 0 \\ \sin(\theta_{ri}) \sin(e_{3i}) + \cos(e_{3i}) \cos(\theta_{ri}) & 0 \\ 0 & 1 \end{bmatrix} \tag{8}$$

$$\boldsymbol{u}_i = \begin{bmatrix} u_{1i} \\ u_{2i} \end{bmatrix} \tag{9}$$

2.3 The Problem of Tracking of Multiple Mobile Robots

Let the dynamic nonlinear system be described as:

$$\dot{\mathbf{x}} = f(x) + \mathbf{B}(x)\mathbf{u} \tag{10}$$

with f(0) = 0.

Using Optimal Control Theory, the goal is to find the optimal control **u** that minimizes our cost functional **J[u]**:

$$J[u] = \int_0^{\infty} \left[\mathbf{Q}(x) + \mathbf{u}^T\mathbf{R}\mathbf{u}\right] dt \tag{11}$$

For **Q(x)** continuous and positive-definite.

According to [17], the solution of this problem can be found through solving the Hamilton-Jacobi-Bellman equation associated to this system. Since the solution of the HJB equation is usually hard to find, one can approximate it using a State Dependent Riccati Equation of the form:

$$\mathbf{P}(\boldsymbol{x})\mathbf{A}(\boldsymbol{x}) + \mathbf{A}(\boldsymbol{x})^T\mathbf{P}(\boldsymbol{x}) - \mathbf{P}(\boldsymbol{x})\mathbf{B}(\boldsymbol{x})\mathbf{R}(\boldsymbol{x})^{-1}\mathbf{B}(\boldsymbol{x})^T\mathbf{P}(\boldsymbol{x}) + \mathbf{Q}(\boldsymbol{x}) = 0 \tag{12}$$

Equation (12) results in a suboptimal controller, however, it makes it much easier problem to be solved, compared to the resolution through the HJB equation. The control law is given by:

$$\mathbf{u} = -\mathbf{R}^{-1}\mathbf{B}^T\mathbf{P}(x) \cdot \mathbf{x} \tag{13}$$

where **P(x)** is the solution to the Riccati equation.

The system can now be written as:

$$\dot{\mathbf{x}} = f(x) + \mathbf{B}(x)\mathbf{u}, \quad f(0) = 0 \tag{14}$$

where $f(x) = \mathbf{A}(x)\mathbf{x}$.

This is known as State-Dependent Coefficients form. The matrices $\mathbf{A}(x)$ and $\mathbf{B}(x)$ are system state functions and thus coefficients in the Riccati equation. For any solution obtained this way, the SDRE method comes down to the resolution of a LQR at each sampling instant [17]. In this work, the computational time needed to compute each instant is 0.1 s. To ensure the existence of the SDRE controller, $\mathbf{A}(x)$ must be a controllable parameterization of the nonlinear system for a given region if $[\mathbf{A}(x), \mathbf{B}(x)]$ is controllable for every $\mathbf{x}$ of that region [17, 18].

3 Results and Numerical Simulations

The concept of mobile robots formation can be considered as physical distances in respect to the leader or adjacent agents that each robot must respect. One may then introduce parameters in the x and y coordinates of each robot in order to ensure that these distances, and therefore, the formation are maintained. In this work, formation parameters called dx and dy have been introduced. These parameters act on each follower robot ensuring a predetermined distance from the leader in each axis (Fig. 2).

Simulations were carried using the software LabVIEW, which provides a good set of tools for working with real-time simulations and an easy integration with hardware for practical tests. The Follower 1, Leader and Follower 2 were positioned, initially, on the points (0,5; 2,5), (2,5; 2,5) and (4,5; 2,5) on the XY plane. The main function of the program is to define the initial conditions of the robots, obtain the appropriate information of the leader and apply the proposed control on the followers, alongside with the desired positioning restrictions. Equation (12) was solved for each instant of time using the LQR function, which is an implementation of a non-iterative method. Therefore the whole system is suited for real-time

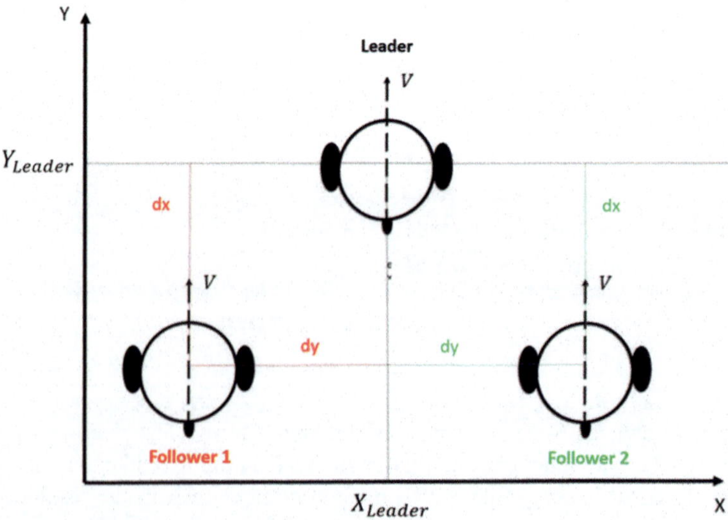

Fig. 2 Formation parameters

applications. The control efforts are limited to ±2 m/s and ±0.6 rad/s to avoid saturation of the robot's actuators.

For the application of the control method the weighted matrices **Q** and **R** of the Riccati equation are design parameters for the controller and thus can be chosen arbitrarily. These matrices have a direct effect on the time and smoothness of system's convergence. In the present work, for simplicity, these matrices were chosen as:

$$\mathbf{Q} = \begin{bmatrix} 10 & 0 & 0 \\ 0 & 10 & 0 \\ 0 & 0 & 10 \end{bmatrix} \tag{15}$$

$$\mathbf{R} = \begin{bmatrix} 1 & 0 \\ 0 & 1 \end{bmatrix} \tag{16}$$

Figures 3, 4, 5 and 6 refer to simulations of the robots navigating in V-shaped formation with desired linear velocity of 0.2 m/s, and desired null angular velocity.

Without angular velocity the V-shaped formation moves forward, as seen in Fig. 3. The followers respect the distance of 2 m on the X axis and 2 m in the Y axis as defined by *dx* and *dy* parameters. Figure 4a shows the control efforts of the leader, where U1 is the control effort on the linear velocity of the robot and U2 is the control on the angular speed. Both reflect the achievement of the desired velocities of 0.2 m/s and 0 rad/s initially set.

Figures 5a and 6a show the fast stabilization on the control efforts of the followers, converging to 0.2 m/s and 0 rad/s in approximately 1 s. Figures 4b, 5b and 6b show the stabilization of the errors of each robot. It is observed that the errors of the leader converge to zero almost immediately and the errors of the followers take about 10 s to converge to zero.

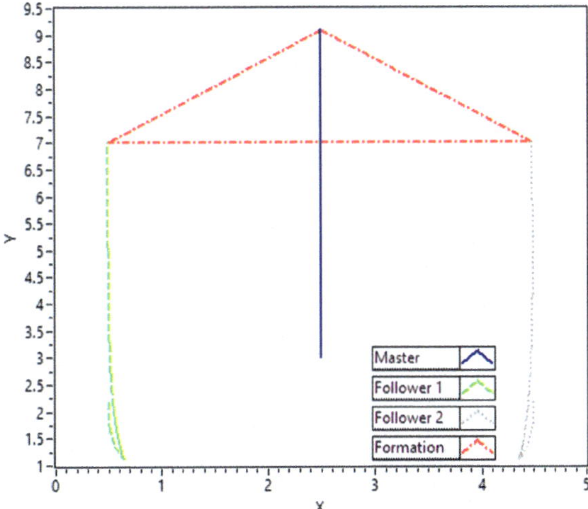

Fig. 3 Robots in V-shaped formation

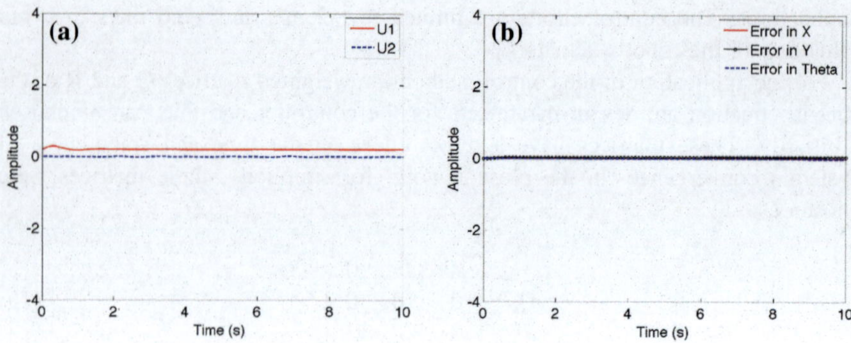

Fig. 4 **a** Control efforts of the leader robot; **b** errors of the leader robot

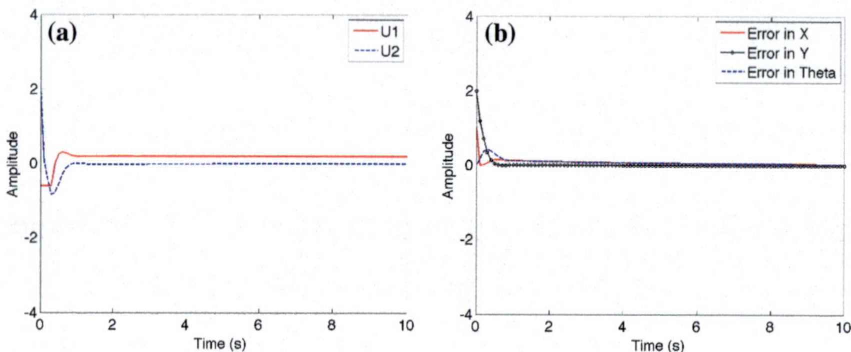

Fig. 5 **a** Control efforts of follower 1; **b** errors of follower 1

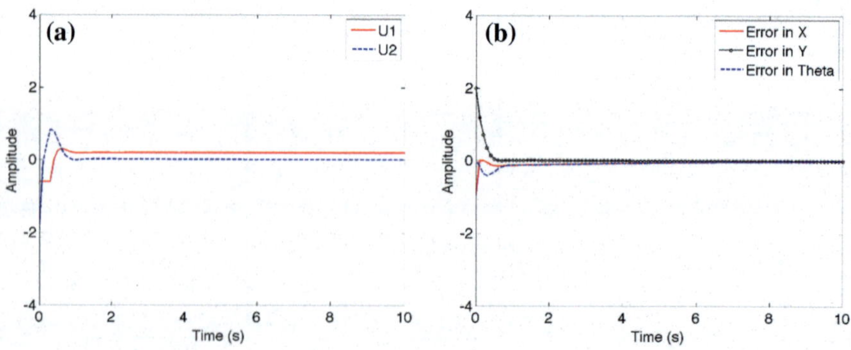

Fig. 6 **a** Control efforts of follower 2; **b** errors of follower 2

Figures 7, 8, 9 and 10 refer to simulations of the robots navigating in Echelon formation. Initially the robots had a linear velocity of 0.2 m/s and null angular velocity. After 15 s, an angular velocity of 0.1 rad/s was introduced in order to make the robot formation turn left.

Figure 7a shows the robots achieving Echelon formation and Fig. 7b depicts the formation turning left due to the positive angular velocity. Figures 8a shows that the control efforts of the leader stabilize immediately to 0.2 m/s and Fig. 8b shows that the errors of leader also converge to zero immediately.

Figures 9a and 10a show that the control efforts of the followers take a bit longer than the leader to stabilize, about 6 s for Follower 1 and 3 s for Follower 2 Both of them converge to 0.2 m/s and 0 rad/s. After 15 s the angular velocity is increased and the control effort U2 of both robots converge to 0.1 rad/s. Figures 9b and 10b show that the errors of the follower robots converge to zero after approximately 16 s for Follower 1 and 8 s for Follower 2.

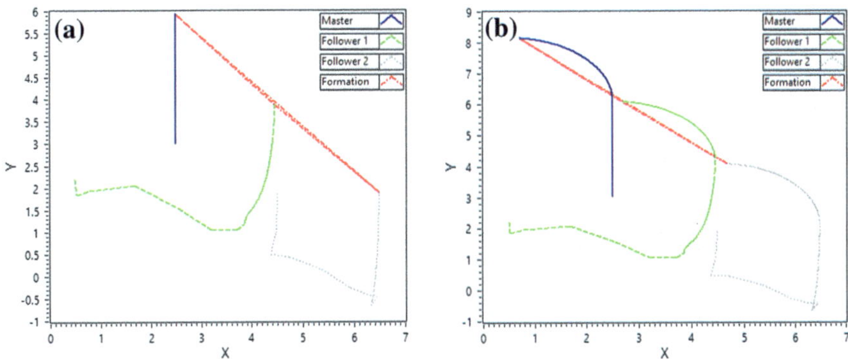

Fig. 7 Robots in Echelon formation

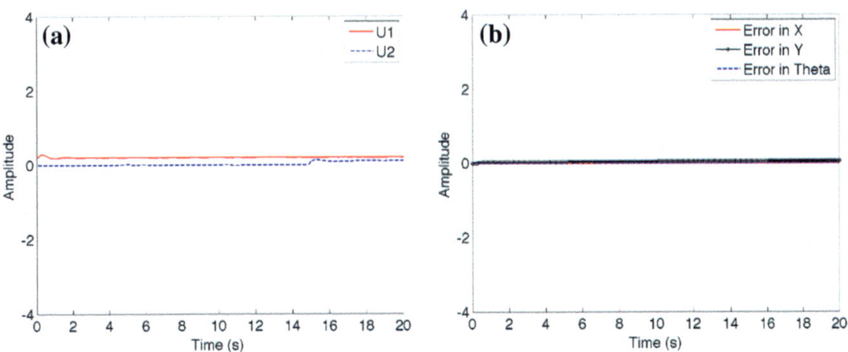

Fig. 8 **a** Control efforts of the leader robot; **b** errors of the leader robot

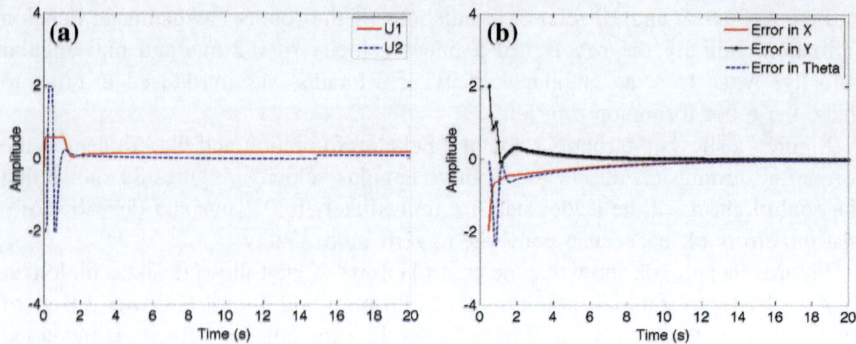

Fig. 9 **a** Control efforts of follower 1; **b** errors of follower 1

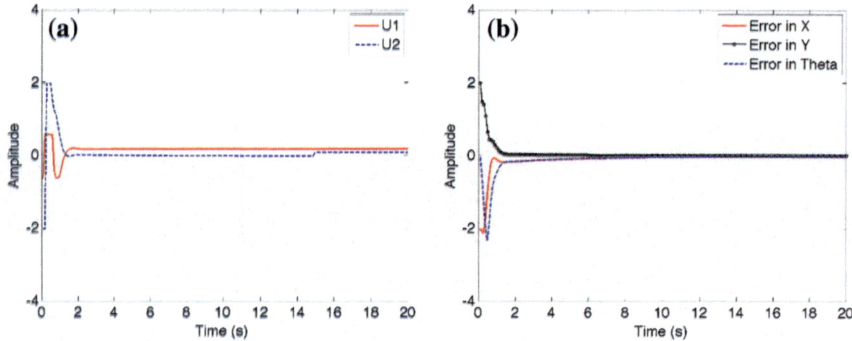

Fig. 10 **a** Control efforts of follower 2; **b** errors of follower 2

4 Conclusions

The SDRE control method was highly effective for the control of multiple mobile robots in formation, as it takes into account the non-linearities of non-holonomic dynamic system that is the robot with differential drive. The computational cost of the control application is low because can be approximated by the Linear Quadratic Regulator instead of the Hamilton-Jacobi-Bellman equation. The proposed formations were achieved and maintained in all the simulated trajectories. The stabilization of errors and control efforts proved to be fast, less than 20 s, for both cases. In more sophisticated robotic platforms, this stabilization time can be decreased. Real time engineering applications are viable because of the small computational time needed to solve the controller at each sampling instant and the feasible control efforts. Future works on this subject could include practical experiments to ratify the efficiency of the SDRE control method.

Acknowledgements The first author would like to thank CAPES for the support on this work.

References

1. Chi, T., Zhang, C., Song, Y., Feng, J.: A strategy of multi-robot formation and obstacle avoidance in unknown environment. In: 2016 IEEE International Conference on Information and Automation (ICIA), pp. 1455–1460. Ningbo (2016)
2. Feng, J., Zhang, C., Song, Y., Chi, T.: A multi-robot dynamic formation scheme based on rigid formation. In: 2016 IEEE International Conference on Information and Automation (ICIA), pp. 1450–1454. Ningbo (2016)
3. Yang, L., Li, J.: A behavioral multi-robot formation control approach in obstacle environments. In: 2016 Chinese Control and Decision Conference (CCDC), pp. 6767–6771. Yinchuan (2016)
4. Meng, Z., Ren, W., Cao, Y., You, Z.: Leaderless and leader-following consensus with communication and input delays under a directed network topology. IEEE Trans. Syst. Man, Cybern. Part B (Cybernetics) **41**(1), 75–88 (2011)
5. Xiao, H., Chen, C.L.P.: Leader-follower multi-robot formation system using model predictive control method based on particle swarm optimization. In: 2017 32nd Youth Academic Annual Conference of Chinese Association of Automation (YAC), pp. 480–484. Hefei (2017)
6. Park, B.S., Yoo, S.J.: Adaptive leader-follower formation control of mobile robots with unknown skidding and slipping effects. Int. J. Control Autom. Syst. **13**, 587–594 (2015)
7. Consolini, L., Morbidi, F., Prattichizzo, D., Tosques, M.: Leader-follower formation control of nonholonomic mobile robots with input constraints. Automatica **44**(5), 1343–1349 (2008)
8. Dai, Y., Lee, S.-G.: The leader-follower formation control of nonholonomic mobile robots. Int. J. Control Autom. Syst. **10**(2), 350–361 (2012)
9. Rafikova, E., Rafikov, M., Rinaldo, G.: Synchronization of the mobile robot to a chaotic trajectory. J. Appl. Nonlinear Dyn. **5**(3), 325–335 (2016)
10. Siciliano, B., Sciavicco, L., Villani, L., Oriolo, G.: Robotics: Modelling, Planning and Control, 1st edn. Springer Publishing Company, Incorporated (2008)
11. Tang, F., Si, B., Ji, D.: A prey-predator model for efficient robot tracking. In: 2017 IEEE International Conference on Robotics and Automation (ICRA), pp. 3568–3574. Singapore (2017)
12. Ma, L., Yao, Y., Wang, M.: The optimizing design of wheeled robot tracking system by PID control algorithm based on BP neural network. In: 2016 International Conference on Industrial Informatics—Computing Technology, Intelligent Technology, Industrial Information Integration (ICIICII), pp. 34–39. Wuhan (2016)
13. Wen, G., Duan, Z., Chen, G., Yu, W.: Consensus tracking of multi-agent systems with Lipschitz-type node dynamics and switching topologies. IEEE Trans. Circuits Syst. I: Regular Papers, **61**(2), 499–511 (2014); Wang, J., Xin, M.: Multi-agent consensus algorithm with obstacle avoidance via optimal control approach. In: American Control Conference (ACC), pp. 2783–2788 (2011)
14. Chung, S.-J., Slotine, J.J.: Cooperative robot control and concurrent synchronization of lagrangian systems. IEEE Trans. Robot. **25**(3), 686–700 (2009)
15. Khoo, S., Xie, L., Man, Z.: Robust finite-time consensus tracking algorithm for multirobot systems. IEEE/ASME Trans. Mechatron. **14**(2), 219–228 (2009)
16. Sun, D., Mills, J.K.: Adaptive synchronized control for coordination of two robot manipulators. IEEE Int. Conf. Robot. Auto. **1**(4), 498
17. Mracek, C., Cl.outier, J.: Control designs for the nonlinear benchmark problem via the state-dependent Riccati equation method. Int. J. Robust Nonlinear Control **8**(4–5), 401–433 (1998)
18. Çimen, T.: Systematic and effective design of nonlinear feedback controllers via the state-dependent Riccati equation (SDRE) method. Annual Rev. Control **34**(1), 32–51 (2010)

An Application of the Lurie Problem in Hopfield Neural Networks

Rafael Fernandes Pinheiro and Diego Colón

Abstract The goal of this work is to present applications of recent results in the Lurie problem, also known as the absolute stability problem, to Hopfield neural networks, aiming its stability analysis. We show how to obtain the mathematical model of a neural network, in terms of differential equations, and present simulations for some examples. We give special attention to networks with multiple inputs and outputs, and point future directions of research to be followed.

Keywords Lurie problem · Hopfield neural network · Absolute stability

1 Introduction

Since the middle of the last century, Lurie problem was an important topic of research in automatic control, specially in aircraft control systems. The Lurie problem appeared in 1947, because of an aircraft control problem, and currently it is considered solved for the case of a single control, e.g. Popov criterion. However, for multiple controls the Lurie problem is not solved completely. Today, it reached complex areas that are targets of several studies, such as the human brain. The Lurie problem [12], named in honor of the mathematician A. I. Lurie, has made its history with significant collaboration of researchers in mathematics and engineering, and in subareas as theory of stability and control, as well as laying the foundation for a new and important area, which is Robust Control.

Many names have worked in this problem, and certainly left significant contributions. Initially, between the 50s and 60s, we can mention Aizerman [1], with the well known Aizerman conjecture, and Krasovskii [8], Popov [17], and Kalman [6].

R. F. Pinheiro (✉) · D. Colón
Automation and Control Laboratory, Telecom and Control Department,
University of São Paulo, Av. Prof. Luciano Gualberto, Travessa 03, No. 158, CEP, São Paulo, SP 05508-900, Brazil
e-mail: rafael.pinheiro@usp.br

D. Colón
e-mail: diego@lac.usp.br

A. de T. Fleury et al. (eds.), *Proceedings of DINAME 2017*, Lecture Notes in Mechanical Engineering, https://doi.org/10.1007/978-3-319-91217-2_26

The Lurie problem took a bigger leap from 1980, when it began to appear works that linked the problem to other areas and with other approaches such as neural networks by Forti et al. [3] and Kaskurewicz and Bhaya [7]; convex approach Gapski and Geromel [4]; and chaos and chaos synchronization by Liao and Yu [11].

The Hopfield network was first proposed in 1984 by Hopfield [5], and it is a nonlinear network with significant interest to engineering and physics, besides neuroscience (more specifically, neurodynamics). There is an important relationship between the Lurie problem and Hopfield neural networks, which will become clear in the following [10]. The Hopfield network has been applied in several areas like biology [13] and to solve optimization problems [3, 7].

This article is divided as follows: In Sect. 2 we will proceed with a re-presentation of the Lurie problem and Hopfield neural network, seeking to bring in a simpler approach, however broad, its main concepts. In Sect. 3, we present the Hopfield Neural Network and its mathematical model. In Sect. 4, it is shown how the Lurie problem can be applied to the Hopfield network, focusing on the stability analysis.

2 The Lurie Problem and Absolute Stability

In 1944, Lurie was imbued to solve a problem of stability of the automatic control system of an aircraft [12]. Based on this system and the analysis of the stability of the null solution of the system (that is the equilibrium point), Lurie did not managed to find necessary and sufficient conditions to global asymptotic stability of the system.

The basic structure of a linear time-invariant control system is generally based on the feedback, as presented in Fig. 1, where $G_a(s)$ is the transfer function that represents the system to be controlled and $G_c(s)$ is the controller. A system's transfer function is a mathematical model which is an operational method to express the ordinary differential equation that relates the output variable to the input variable [14].

Regarding the aircraft control problem, the closed-loop control system in Fig. 1 can be compared to the pilot verifying if the real route of the aircraft σ matches the desired route r and actuating in the control surfaces and the propeller in order to achieve this goal. The role of the controller $G_c(s)$ would be the pilot role.

A much more realistic model of the aircraft control would be the one presented in Fig. 2. In fact, it has been found that the mathematical model of the control surfaces may have nonlinearities, such as saturation. As u represents the deflection of the

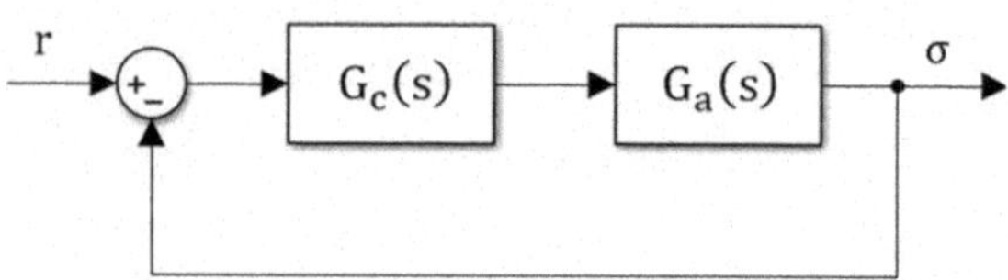

Fig. 1 Aircraft control system

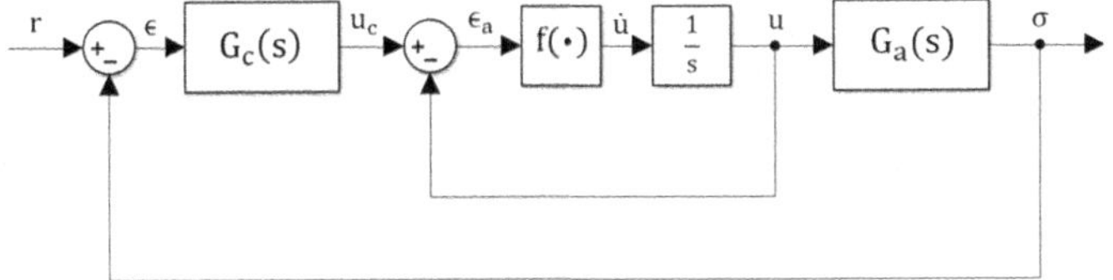

Fig. 2 Aircraft control system with rudder control

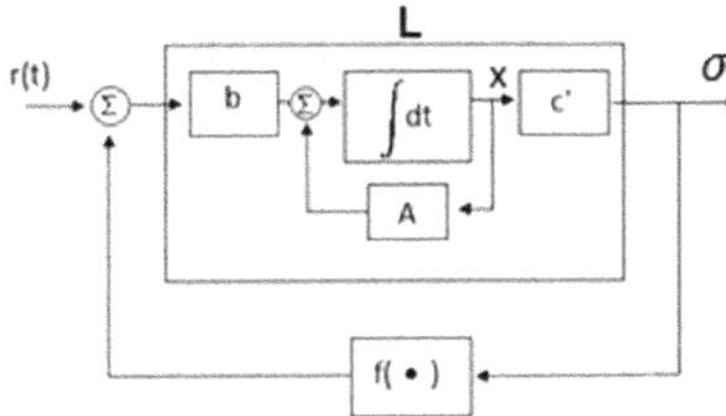

Fig. 3 Block diagram of the type Lurie system

rudder, the rudder control is intended to keep the deflection u_c determined by the controller $G_c(s)$ until it is fixed the deviation ϵ direction of the aircraft.

What Lurie did was to separate the linear part and the nonlinear part of the system (calculations can be checked in [18]). In Fig. 3 it is shown a general form of this separation, where the linear dynamics is represented by the system L, and the function $f(\sigma)$ in the feedback, represents a nonlinearity.

In general, the nonlinearity $f(\sigma)$ is a continuous function restricted to the first and third quadrants of the plane, that belongs to one of the following families:

$$\begin{aligned} F_{(0,k]} &:= \{f | f(0) = F_{[0,k]} := \{f | f(0) = 0, 0 \leq f(\sigma) \leq k\sigma, \sigma \neq 0\}, \\ F_{(0,k)} &:= \{f | f(0) = F_{[0,k)} := \{f | f(0) = 0, 0 \leq f(\sigma) < k\sigma, \sigma \neq 0\}, \\ F_{[k_1,k_2]} &:= \{f | f(0) = 0, k_1\sigma \leq f(\sigma) \leq k_2\sigma, \sigma \neq 0\}, \\ F_{\infty} &:= \{f | f(0) = 0, f(\sigma) > 0, \sigma \neq 0\}. \end{aligned}$$

In Fig. 4, we have some graphical representations of this kind of nonlinearity. Considering $r(t) = 0$, the diagram in the Fig. 3 is expressed by the following system of differential equations, which was known in the literature as a Lurie system [11]:

$$\begin{cases} \dot{x} = Ax + bf(\sigma) \\ \sigma = c^T x \end{cases} \tag{1}$$

where $x \in R^n$ is the state vector, and $b, c \in R^n$, $A \in R^{n \times n}$ are fixed matrices and $\sigma f(\sigma) > 0$.

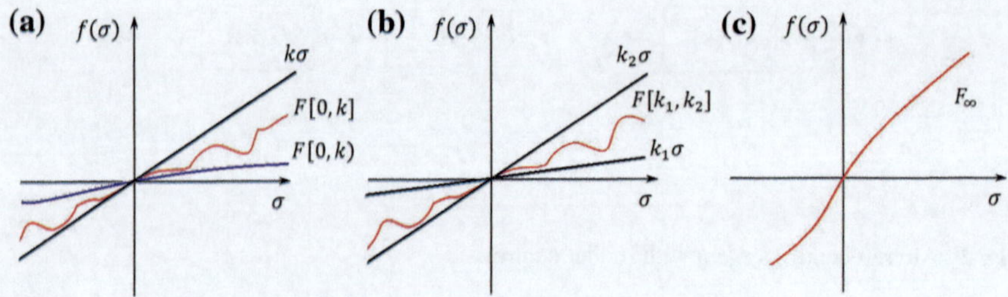

Fig. 4 Functions types f: (a) $F[0,k]$, $F[0,k)$; (b) $F_{[k_1,k_2]}$; (c) F_∞

We can now state the Lurie problem in the following question [11]: **What are the necessary and sufficient conditions for the equilibrium point of the system** (1) **be globally asymptotically stable**? In a more formal way, we can use the following definition:

Definition 1 If the zero solution of the system (1) is globally asymptotically stable for $f(\sigma) \in F_\infty$, then we say that the zero solution of the system (1) is absolutely stable for F_∞.

Thus, considering the system (1) and the Definition 1, we have the formulation for Lurie Problem with a single control and may also be formulated for $F_{(0,k]}$, $F_{(0,k)}$, $F_{[k_1,k_2]}$ and F_∞.

Finally, the Lurie problem can be extended to the case of multiple controls (inputs), i.e. with m nonlinear controls [11]:

$$\begin{cases} \dot{x} = Ax + \sum_{j=1}^{m} b_j f_j(\sigma_j) \\ \sigma_j = c_j^T x = \sum_{i=1}^{n} c_{ij} x_i, \quad j = 1, \dots, m, \end{cases} \tag{2}$$

where $A \in R^{n\times n}, x = (x_1, \dots, x_n)^T, b_j = (b_{1j}, \dots, b_{nj})^T, c_j = (c_{1j}, \dots, c_{nj})^T, f_j \in F_\infty := \{f : f(0) = 0, f(\sigma)\sigma > 0, \sigma \neq 0, f(\sigma) \in C[(-\infty, +\infty), R^1]\}$, $Re\lambda(A) \leq 0$.

Notice that the functions f_j can be not exactly known, but belongs to an uncertain family, which explains why Lurie can be considered one of the founders of robust control. In fact, this is one of the difficulties of the Lurie problem, the other being the fact that the variables are not separated. To work around this problem, we can make a coordinate transformation in (2) that will be presented in the following.

Without loss of generality, we assume that the set of line vectors $c_i = (c_{i1}, \dots, c_{in})$ $(i = 1, 2, \dots, m)$, are linearly independent, and with the purpose of separating the variables, by a transformation [9] the system (2) can be transformed in the following:

$$\dot{y} = \tilde{A}y + \sum_{j=n-m+1}^{n} \tilde{b}_j \tilde{f}_j(y_j), \tag{3}$$

or:

$$\dot{y}_i = \sum_{j=1}^{n} \tilde{a}_{ij} y_j + \sum_{j=n-m+1}^{n} \tilde{b}_{ij} \tilde{f}_j(y_j). \tag{4}$$

3 Hopfield Neural Network and Its Relationship with Lurie Problem

The Hopfield neural network (HNN) is a neural network model which was proposed by Hopfield in 1984 [5], which is part of an area known as neurodynamics. This is the area that study Artificial Neural Networks (ANN) as nonlinear dynamical systems with an emphasis on stability problems. With the publication of Hopfield's article, research on ANN got a jump and the Hopfield network started to be a model of an associative memory, with the main purpose to restore a pattern stored in response to the presentation of an incomplete or distorted version of this pattern. The HNN was applied in various areas, as presented in Braga et al. [2], such as: implementation of an identification systems of military target used in aircrafts B-52, user authentication system, oil exploration, prediction in the financial market, recognition of faces and autonomous navigation control vehicles (ALVINN vehicles). Currently, it is being used in optimization problems in association with absolute stability.

HNN can be represented by the following system of nonlinear differential equations

$$C_i \frac{du_i}{dt} = -\frac{u_i}{R_i} + \sum_{j=1}^{n} T_{ij} V_j + I_i, \qquad i = 1, 2, \ldots, n, \tag{5}$$

where $u \in R^n$, $R \in R^n$, $T \in R^{n \times n}$, $I \in R^n$, and $V_j = g(u_i)$, and $g : R \to [0, 1]$ is continuously differentiable and monotonically increasing, or $g_i'(u_i) > 0$.

The continuous model of HNN can be thought as an electric circuit (for more details [15]). The functions g are called activation functions and usually are sigmoidal functions.

Observe that these functions g are compatible with the nonlinear functions of the Lurie problem, which means that, based in [10], a HNN can be considered a special case of Lurie type system with multiple controls. This fact allows for analysis using the theory of Lurie systems, that is, the theory of absolute stability. Thus, considering the Eq. (5) after some changes of variables, it can take the following form:

$$\frac{dx_i}{dt} = -d_i x_i + \sum_{j=1}^{n} b_{ij} f_j(x_j). \tag{6}$$

In relation to the function f, we have $f \in F_\infty$. Comparing the equations in (6) with the equations in (4)

$$\dot{y}_i = \sum_{j=1}^{n} \tilde{a}_{ij} y_j + \sum_{j=n-m+1}^{n} \tilde{b}_{ij} \tilde{f}_j(y_j).$$

We observe that the Hopfield neural network is a special case of a type Lurie system with multiple control, where $\tilde{a}_{ij} = 0$, $i \neq j$, $\tilde{a}_{ii} = -d_i < 0$, and $m = n$.

Therefore, with this relationship between HNN and type Lurie systems the theory and methodology of Lurie type systems can promote the study of absolute stability of neural networks. So, it makes sense to ask what are the necessary and sufficient conditions for a equilibrium point of (6) to be globally asymptotically stable, that is, to be absolutely stable.

4 An Application of the Lurie Problem to Hopfield Neural Network

We use two theorems due to Liao [9], which are results which provide sufficients conditions for absolute stability of the Lurie problem with multiple controls described by the Eq. (4). Then the same can be applied to the Eq. (6). The proof of these theorem can be found in [9, 16].

For the first theorem, consider the following relationship:

$$\alpha_{ij} = \begin{cases} \tilde{a}_{ij}, & i = 1, \ldots, n, \quad j = 1, \ldots, n-m, \\ \tilde{b}_{ij}, & i = 1, \ldots, n, \quad j = n-m+1, \ldots, n. \end{cases}$$

Consider also the Kronecker delta:

$$\delta_{ij} = \begin{cases} 1, \; i = j, \\ 0, \; i \neq j. \end{cases}$$

Theorem 1 *The zero solution of the (4) or (6) is absolutely stable if the conditions of any of the following set are satisfied:*

(1) $\tilde{a}_{ii} < 0$ *for* $i = 1, \ldots, n$, *and the square matrix of order n, where* $((-1)^{\delta_{ij}} |\tilde{a}_{ij}|)$ *is Hurwitz.*

(2) There is a constant $k_l > 0$ *for* $l = n-m+1, \ldots, n$ *such that*

$$\begin{cases} k_l \tilde{b}_{ll} \leq \tilde{a}_{ll}, \quad k_l |\tilde{b}_{il}| < |\tilde{a}_{il}|, \textit{for} \quad l = n-m+1, \ldots, n, \; i = 1, \ldots, n, \quad i \neq l, \\ \qquad \textit{or} \\ k_l \tilde{b}_{ll} < \tilde{a}_{ll}, \quad k_l |\tilde{b}_{il}| \leq |\tilde{a}_{il}|, \textit{for} \quad l = n-m+1, \ldots, n, \; i = 1, \ldots, n, \quad i \neq l, \end{cases}$$

or

(1') $\tilde{a}_{ii} < 0$ *for* $i = 1, \ldots, n$, *and the square matrix of order n, where* $((-1)^{\delta_{ij}} |\alpha_{ij}|)$ *is Hurwitz.*

(2') *There is a constant* $k_l > 0$ *for* $l = n - m + 1, \ldots, n$ *such that*

$$\begin{cases} k_l \tilde{a}_{ll} \leq \tilde{b}_{ll}, \quad k_l |\tilde{a}_{il}| < |\tilde{b}_{il}|, \textit{for} \quad l = n - m + 1, \ldots, n, \; i = 1, \ldots, n, \quad i \neq l, \\ \qquad\qquad \textit{or} \\ k_l \tilde{a}_{ll} < \tilde{b}_{ll}, \quad k_l |\tilde{a}_{il}| \leq |\tilde{b}_{il}|, \textit{for} \quad l = n - m + 1, \ldots, n, \; i = 1, \ldots, n, \quad i \neq l, \end{cases}$$

Example 1 Let us consider the neural network with two neurons:

$$\begin{cases} \dot{u}_1 = -u_1 - 2g_1(u_1) + 1.5g_2(u_2) + I_1 \\ \dot{u}_2 = -u_2 + 1.5g_1(u_1) - 2g_2(u_2) + I_2 \end{cases}$$

where $g_i(u_i) = tanh(u_i), R_i = C_i = 1, I_i = 0, i = 1, 2, T_{ii} = -2, T_{12} = T_{21} = 1.5$.

We have: $\begin{bmatrix} \tilde{a}_{11} & \tilde{a}_{12} \\ \tilde{a}_{21} & \tilde{a}_{22} \end{bmatrix} = \begin{bmatrix} -1 & 0 \\ 0 & -1 \end{bmatrix}, \alpha_{ij} = \begin{bmatrix} \tilde{b}_{11} & \tilde{b}_{12} \\ \tilde{b}_{21} & \tilde{b}_{22} \end{bmatrix} = \begin{bmatrix} -2 & 1.5 \\ 1.5 & -2 \end{bmatrix}$.

We will apply the theorem to see if this equilibrium point is absolutely stable. Applying the theorem's second set of conditions, we have:

The condition 1' of the theorem is satisfied, because $\tilde{a}_{ii} < 0$ and the matrix $((-1)^{\delta_{ij}} |\alpha_{ij}|)$ has eigenvalues equal to -3.5 and -0.5 so it is Hurwitz.

In relation to condition 2', we have:

$$\text{for} \quad l = 1, \quad i = 2,$$

$$k\tilde{a}_{11} \leq \tilde{b}_{11} \rightarrow k_1(-1) \leq (-2), \quad k_1 |\tilde{a}_{21}| < |\tilde{b}_{21}| \rightarrow k_1 |0| < |1.5|;$$

$$\text{for} \quad l = 2, \quad i = 1,$$

$$k\tilde{a}_{22} \leq \tilde{b}_{22} \rightarrow k_2(-1) \leq (-2), \quad k_2 |\tilde{a}_{12}| < |\tilde{b}_{12}| \rightarrow k_2 |0| < |1.5|.$$

Taking $k_1 = k_2 = 2$ the condition 2' is satisfied, and the equilibrium point is absolutely stable. This also means that the equilibrium point is unique.∘

In the following, we present a simulation for Example 1 using Matlab/Simulink. In Fig. 5, we have the time response of the states u_1 and u_2 with initial conditions $(u_1^0, u_2^0) = (5, -5)$ and $(u_1^1, u_2^1) = (-2, 3)$ respectively. The Figure shows that the solutions converge for zero solution. Figure 6 shows the phase portrait for other initial states.

The next theorem is related to multiple inputs Lurie type problems.

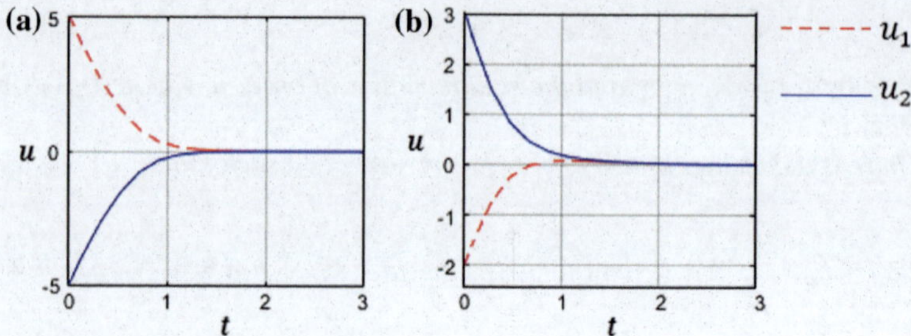

Fig. 5 Temporal response

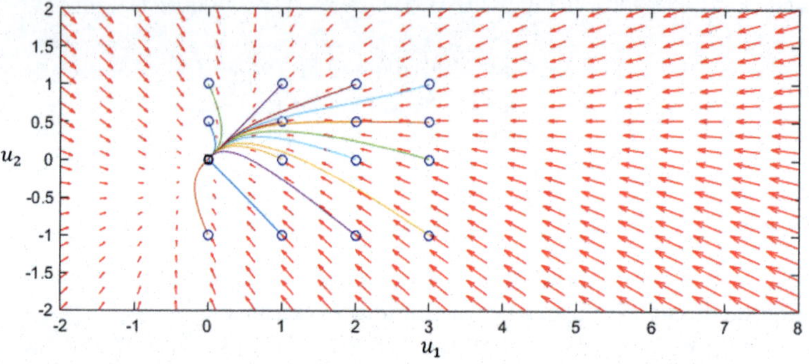

Fig. 6 Phase portrait

Theorem 2 *Suppose that*

1. $\tilde{A} = (\tilde{a}_{ij})_{n\times n}$ *is a Hurwitz matrix,*
2. *There is a constant* $r_j \geq 0\,(j = 1, \ldots, n-m)$, $r_j > 0\,(j = n-m+1, \ldots, n)$ *Such that*

$$\begin{cases} r_j\tilde{a}_{jj} + \sum_{i=1, i\neq j}^{n} r_i|\tilde{a}_{ij}| \leq 0, \quad j = 1, \ldots, n-m, \\ r_j\tilde{a}_{jj} + \sum_{i=1, i\neq j}^{n} r_i|\tilde{a}_{ij}| < 0, j = n-m+1, \ldots, n, \\ r_j\tilde{b}_{jj} + \sum_{i=1, i\neq j}^{n} r_i|\tilde{b}_{ij}| \leq 0, j = n-m+1, \ldots, n, \end{cases}$$

or

$$\begin{cases} r_j\tilde{a}_{jj} + \sum_{i=1, i\neq j}^{n} r_i|\tilde{a}_{ij}| \leq 0, \quad j = 1, \ldots, n-m, \\ r_j\tilde{a}_{jj} + \sum_{i=1, i\neq j}^{n} r_i|\tilde{a}_{ij}| \leq 0, j = n-m+1, \ldots, n, \\ r_j\tilde{b}_{jj} + \sum_{i=1, i\neq j}^{n} r_i|\tilde{b}_{ij}| < 0, j = n-m+1, \ldots, n. \end{cases}$$

Then, the zero solution of the (4) or (6) is absolutely stable.

The following in an example with three neurons in which is used theorem two.

Example 2 Now, let us consider the neural network with three neurons:

$$\begin{cases} \dot{u}_1 = -u_1 \ -3g_1(u_1) - 2g_2(u_2) - 2g_3(u_3) + I_1 \\ \dot{u}_2 = -u_2 \ +3g_1(u_1) - 3g_2(u_2) + g_3(u_3) + I_2 \\ \dot{u}_3 = -u_3 \ +g_1(u_1) + 2g_2(u_2) - 2g_3(u_3) + I_3, \end{cases}$$

where $g_i(u_i) = tanh(u_i), \quad I_i = 0, \quad i = 1, 2, 3.$

We have: $\begin{bmatrix} \tilde{a}_{11} & \tilde{a}_{12} & \tilde{a}_{13} \\ \tilde{a}_{21} & \tilde{a}_{22} & \tilde{a}_{23} \\ \tilde{a}_{31} & \tilde{a}_{32} & \tilde{a}_{33} \end{bmatrix} = \begin{bmatrix} -1 & 0 & 0 \\ 0 & -1 & 0 \\ 0 & 0 & -1 \end{bmatrix}$, and $\begin{bmatrix} \tilde{b}_{11} & \tilde{b}_{12} & \tilde{b}_{13} \\ \tilde{b}_{21} & \tilde{b}_{22} & \tilde{b}_{23} \\ \tilde{b}_{31} & \tilde{b}_{32} & \tilde{b}_{33} \end{bmatrix} = \begin{bmatrix} -3 & -2 & -2 \\ 3 & -3 & 1 \\ 1 & 2 & -2 \end{bmatrix}$,

In this example, we take $m = n = 3$, then we easily find that condition 1 is satisfied. In the condition 2 we use the inequalities:

$$\begin{cases} r_j\tilde{a}_{jj} + \sum\limits_{i=1, i\neq j}^{n} r_i|\tilde{a}_{ij}| < 0, j = n - m + 1, \ldots, n, \\ r_j\tilde{b}_{jj} + \sum\limits_{i=1, i\neq j}^{n} r_i|\tilde{b}_{ij}| \leq 0, j = n - m + 1, \ldots, n. \end{cases}$$

In the first inequality, we have:

$$\begin{cases} r_1\tilde{a}_{11} + r_2|\tilde{a}_{21}| + r_3|\tilde{a}_{31}| < 0 \\ r_2\tilde{a}_{22} + r_1|\tilde{a}_{12}| + r_3|\tilde{a}_{32}| < 0 \\ r_3\tilde{a}_{33} + r_1|\tilde{a}_{13}| + r_2|\tilde{a}_{23}| < 0. \end{cases}$$

Replacing the values, we found that r_1, r_2 and r_3 must be greater than zero.
In the second inequality, we have:

$$\begin{cases} r_1\tilde{b}_{11} + r_2|\tilde{b}_{21}| + r_3|\tilde{b}_{31}| \leq 0 \\ r_2\tilde{b}_{22} + r_1|\tilde{b}_{12}| + r_3|\tilde{b}_{32}| \leq 0 \\ r_3\tilde{b}_{33} + r_1|\tilde{b}_{13}| + r_2|\tilde{b}_{23}| \leq 0. \end{cases} \rightarrow \begin{cases} -3r_1 + 3r_2 + r_3 \leq 0 \\ -3r_2 - 2r_1 + 2r_3 \leq 0 \\ -2r_3 - 2r_1 + r_2 \leq 0. \end{cases}$$

It's easy to check that $r_1 = 5$, $r_2 = 1$ and $r_3 = 1$ satisfy the inequality. Therefore, we conclude that the system of Example 2 is absolutely stable.

We have in Figs. 7 and 8 the simulations that validate Theorem 2. Thus, in the Example 2, similarly to the 1, we find that the systems in question have only one

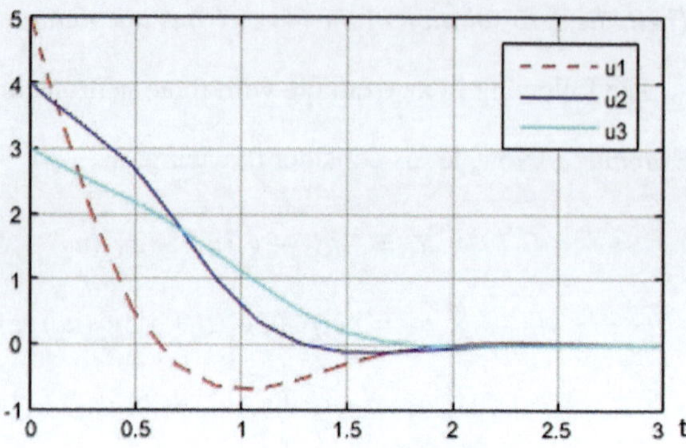

Fig. 7 Temporal response

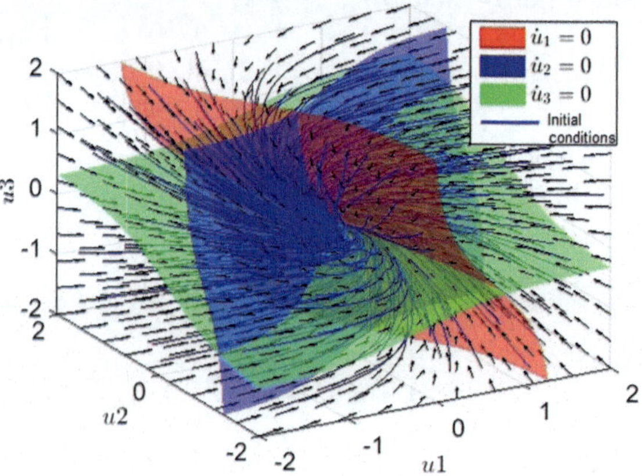

Fig. 8 Phase portrait of the Example 2

equilibrium point, which is the origin, being asymptotically stable. This, therefore, leads to absolute stability.

We could also use Theorem 2 to establish a stability criterion for neural networks. For example, let's determine for which α and β values the system below is absolutely stable:

$$\begin{cases} \dot{u}_1 = -u_1 - 3g_1(u_1) + \alpha g_2(u_2) - 2g_3(u_3) \\ \dot{u}_2 = -u_2 + 3g_1(u_1) - 3g_2(u_2) + \beta g_3(u_3) \\ \dot{u}_3 = -u_3 + g_1(u_1) + 2g_2(u_2) - 2g_3(u_3). \end{cases}$$

We can set the values of r_1, r_2 and r_3 as obtained in Example 2, i.e., $r_1 = 5$, $r_2 = 1$ and $r_3 = 1$, so:

$$-15 + |\alpha| + 1 \leq 0 \rightarrow |\alpha| \leq 14 \rightarrow -14 \leq \alpha \leq 14,$$

$$-3r_2 - 2r_1 + 2r_3 \leq 0 \rightarrow |\beta| \leq 6.5 \rightarrow -6.5 \leq \beta \leq 6.5.$$

5 Conclusion

We conducted a presentation of the Lurie problem in a simple and detailed language in relation to that normally found in the literature. An important fact is that the Lurie problem is considered solved for a single control, but it remains open to multiple controls.

It was verified that the Hopfield network is a particular case of a Lurie system with multiple control, where it became clear that it is feasible to use the Lurie problem theory in Hopfield neural networks.

Possible directions of research include the exploration of the fact that the Lurie Problem is a particular type of robust control problems, so one can also work on controller designs in order to obtain robustness stability and performance for the networks. Also interesting is the situation where there is no absolute stability for Lurie type system. In such situations, we will have questions to answer such as: are there periodic solutions, are there limits cycles, can the system present chaos? what are the conditions for such situations?

References

1. Aizermam, A.: On the effect of nonlinear functions of several variables on the stability of automatic control systems (in Russian). Autom. i Telemekh. **8**(1) (1947)
2. Braga, A.P., Carvalho, A., Ludemir, T.B.: Artificial Neural Networks: theory and applications (in Portuguese). 2nd edition. LTC (2007)
3. Forti, M., Manetti, S., Marini, M., Liberatore, S.: On absolute stability of neural networks. IEEE Trans. Circuits Syst. **6**, 241–244 (1994)
4. Gapski, P.B., Geromel, J.C.: A convex approach to the absolute stability problem. IEEE Trans. Automat. Contr. **25**, 613–617 (1994)
5. Hopfield, J.J.: Neurons with graded response have collective computational properties like those of two-state neurons. Proc. Natl. Acad. Sci. **81**, 3088–3092 (1984)
6. Kalman, R.E.: Liapunov functions for the problem of Lurie in automatic controls. Proc. Natl. Acad. Sci **49**(8), 201–205 (1963)
7. Kaskurewicz, E., Bhaya, A.: Comments on necessary and sufficient condition for absolute stability of neural network. IEEE Trans. Circuits Syst. **42**, 497–499 (1995)
8. Krasovskii, N.N.: On the stability of the solutions of a system of two differential equations (in Russian). Prikl. Mat. i Mekh. **17**(6) (1953)
9. Liao, X.: Absolute Stability of nonlinear Control Systems. Kluwer Academic China Science Press, Beijing (1993)
10. Liao, X., Xu, F., Yu, P.: Absolute stability of Hopfield neural network. In: Wang, J., Yi, Z., Zurada, J.M., Lu, B.L., Yin, H. (eds.) Advances in Neural Networks—Lecture Notes in Computer Science, vol. 3971, pp. 249–254. Springer, Berlin (2006)
11. Liao, X., Yu, P.: Absolute Stability of Nonlinear Control. 2nd edition. Springer (2008)
12. Lurie, A.I., Postnikov, V.N.: On the theory of stability of control systems (in Russian). Prikl. Mat. i Mekh. **8**(3), 246–248 (1944)
13. Monteiro, L.H.A.: Dynamic Systems (in Portuguese). 2nd edition. Physics Bookstore (2006)
14. Ogata, K.: Modern Control Engineering. 4nd edition. Prentice Hall (2002)
15. Pinheiro, R.F.: The Lurie problem and applications to neural networks (in Portuguese). Master's thesis, University of São Paulo, Mathematics and Statistics Institute (2015)

16. Pinheiro, R.F., Colon, D.: An application of Lurie problem in Hopfield neural networks. In: Fleury, A.T., Rade, D.A., Kurka, P.R.G. (eds.) Proceedings of the 17th International Symposium on Dynamic Problems of Mechanics. ABCM, São Paulo (2017)
17. Popov, V.M.: Absolute stability of nonlinear systems of automatic control. Remote Control **22**(8), 857–875 (1961)
18. Rasvan, V.: Systems with monotone and slope restricted nonlinearities. Tatra Mt. Math. Publ. **48**, 165–187 (2011)

Part V
Nonlinear Dynamics

Rotations of the Parametric Pendulum Excited by a Reciprocating Motion with a View on Energy Harvesting

Franco E. Dotti, Florencia Reguera and Sebastián P. Machado

Abstract In this article the nonlinear dynamics of a parametric pendulum considering a reciprocating excitation is addressed. The interest in the study of this kind of forcing lies in its wide use in machines and industrial equipment including a crank-rod mechanism. The work aims at the further development of pendulum devices for energy harvesting. In this context, the study is focused on pendulum rotations, which are highly energetic. Although reciprocating excitation is similar to the classic sinusoidal excitation, a different and more complex rotational behavior is observed and more rotating attractors are found as new rotation zones arise in the space of forcing parameters. It is shown that the existence of these additional rotating attractors, which depend on crank-rod ratio and the amount of damping, increases the possibilities of energy extraction.

Keywords Parametric pendulum · Reciprocating motion · Rotating attractors
Energy harvesting

F. E. Dotti (✉) · S. P. Machado
Facultad Regional Bahía Blanca, Grupo de Investigación en Multifísica Aplicada, Universidad Tecnológica Nacional, Bahía Blanca, Argentina
e-mail: fdotti@frbb.utn.edu.ar

S. P. Machado
e-mail: smachado@frbb.utn.edu.ar

F. E. Dotti · F. Reguera · S. P. Machado
Consejo Nacional de Investigaciones Científicas y Técnicas, Buenos Aires, Argentina
e-mail: florencia.reguera@uns.edu.ar

F. Reguera
Universidad Nacional del Sur, Departamento de Ingeniería, Área Hidráulica, Bahía Blanca, Argentina

A. de T. Fleury et al. (eds.), *Proceedings of DINAME 2017*, Lecture Notes in Mechanical Engineering, https://doi.org/10.1007/978-3-319-91217-2_27

1 Introduction

Energy harvesting from the parametric pendulum is a topic of growing interest for scientists and engineers [1–4], due to the high kinetic energy available in its rotational motion. The basic idea of the devices consists of a pendulum with a vertical motion induced by an ambient energy source. If stable rotations of the pendulum can be reached, a generator attached to the axis of rotation could extract electrical energy. Being the parametric pendulum a problem of escape from a potential well [5], rotations are required because they represent the most energetic motion [6]. Although conceptually simple, this technology is still at a laboratory stage mostly due to the complex nonlinear dynamics of the system. Two sources of ambient vibrations are mainly considered as external excitation: vibrating machines and the motion of the sea waves. In both cases, rotations are possible only in some forcing scenarios. But while sea waves present a stochastic behavior, machine vibrations are generally of harmonic nature, with a consequent high degree of predictability. This is an important feature in the design of suitable pendulum harvesters because the physical dimensions of the system can be defined in terms of the forcing parameters. The goal of the design always is to improve the ability of reaching rotational motion.

In this work, reciprocating motion is regarded as external excitation. This motion is interesting because it can be found in a wide range of industrial machines, including engines and pumps, where a crank-rod system is used. Reciprocating motion is similar to sinusoidal, which is the classical excitation in literature, but slightly more complex [7]. The study of differences and similarities among these two excitations is interesting since many experimental devices aimed to the study of the classic parametric pendulum employ a crank-rod mechanism due to its simplicity [2, 8, 9].

The article is organized as follows. After this introduction, the governing equation of the system under study is presented. The central part of the paper is devoted to the exploration of rotatory dynamics of the pendular system, including an overview of rotating responses, a parametric study and an integrity analysis of the basins of rotations. Finally, the main conclusions of the study are summarized and discussed.

2 The Parametric Pendulum Under Reciprocating Excitation

The governing differential equation of the parametrically excited pendulum of Fig. 1 can be set up by using Lagrange's equation for single-DOF non conservative systems, and its derivation can be easily followed in any classic book of nonlinear dynamics [5, 10]. It is a second-order ordinary differential equation given by

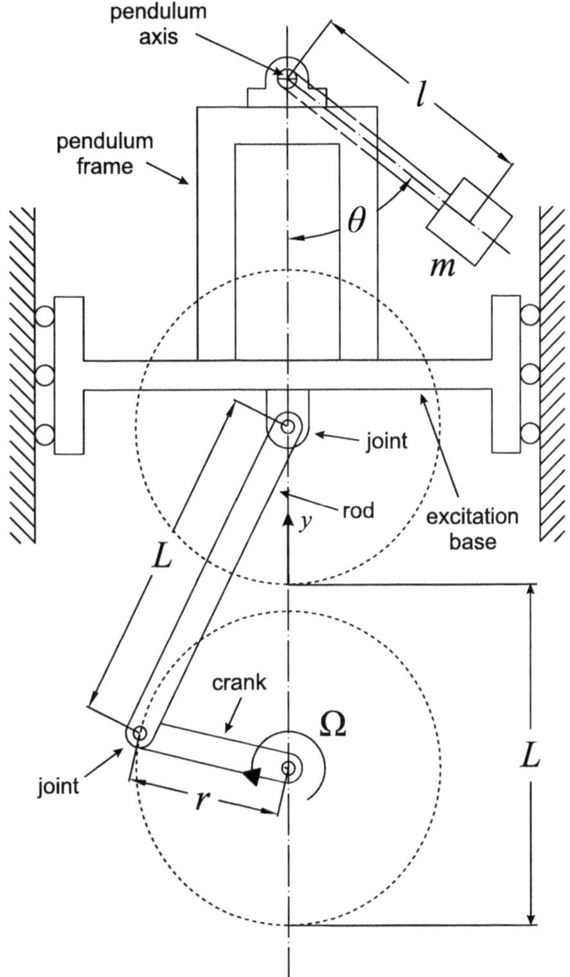

Fig. 1 The parametric pendulum excited by a reciprocating motion in vertical direction

$$ml^2 \frac{d^2\theta}{d\tau^2} + c\frac{d\theta}{d\tau} + ml\left(\frac{d^2y}{d\tau^2} + g\right) \sin\theta = 0 \tag{1}$$

where m is the mass of the pendulum bob, l the distance between the center of gravity and the pendulum axis, c the viscous damping coefficient, τ the time, g the acceleration of gravity, y the vertical displacement of the pendulum system, and θ is the angle positively measured anticlockwise from the hanging position. A reciprocating motion provided by a crank-rod system [7] constitutes the imposed motion y to the pendulum device. This is shown in Fig. 1. The connecting joint between rod and crank rotates at a constant frequency Ω, following a circumferential trajectory. Thus, the displacement of that joint projected horizontally or vertically is sinusoidal in time. However, the angle between the rod and the vertical direction is continuously changing during the cycle of motion. Therefore the linear motion of the upper end of the rod is more complex than a simple sine function. Such excitation gives to the pendulum system the following displacement

$$y = r(1 - \cos \Omega\tau) + L\left(1 - \sqrt{1 - \lambda^2 \sin^2 \Omega\tau}\right) \tag{2}$$

where r is the crank radius, L is the length of the rod and the crank-rod ratio is $\lambda = r/L$.

Now, introducing (2) into (1), a non-dimensional equation is obtained as

$$\ddot{\theta} + \beta\dot{\theta} + \left(R \cos \omega t + \lambda^3 R \frac{\Lambda_3}{\Lambda_1^3} + \lambda R \frac{\Lambda_2}{\Lambda_1} + 1\right) \sin \theta = 0 \tag{3}$$

where the following definitions have been made

$$\begin{gathered} \omega_0 = \sqrt{\tfrac{g}{l}}, \quad t = \omega_0 \tau, \quad \beta = \tfrac{c}{ml^2 \omega_0}, \quad \omega = \tfrac{\Omega}{\omega_0}, \quad R = \tfrac{r\omega^2}{l}, \\ \Lambda_1 = \sqrt{1 - \lambda^2 \sin^2 \omega t}, \quad \Lambda_2 = \cos^2 \omega t - \sin^2 \omega t, \quad \Lambda_3 = \cos^2 \omega t \cdot \sin^2 \omega t \end{gathered} \tag{4}$$

The superimposed dot in (3) means derivation with respect to dimensionless time t. The magnitudes R, ω and β are non-dimensional parameters associated to the forcing amplitude, forcing frequency and damping, respectively. Depending on λ, R, ω and β, and initial conditions θ_0 and $\dot{\theta}_0$, different steady states of the system can be obtained [11]. These responses include: rest position, oscillations, rotations and chaos.

3 Exploring Rotating Attractors

3.1 Overview

The dynamics of rotating attractors is explored, for different configurations of the parametric pendulum under reciprocating excitation. Equation (3) is solved numerically by employing a dimensionless simulation time of $t_s = 2500$, with the purpose of ensure steady state responses. Control spaces, bifurcation diagrams and basins of attraction are constructed, based on extensive numerical simulations, considering different settings of the control parameters λ, R, ω and β. To avoid transients, the first $t_d = 2300$ are discarded in the construction of all the diagrams.

Steady state rotations are classified in four categories [12]: pure rotations, oscillating rotations, straddling rotations and large amplitude rotations. Pure rotations have a very significant attribute: the angular velocity always keeps the same sign ($\dot{\theta} > 0$ or $\dot{\theta} < 0$). This ensures no change in the direction of rotation, implying no oscillatory motion of any kind. Pure rotations exist in conjugate pairs: clockwise

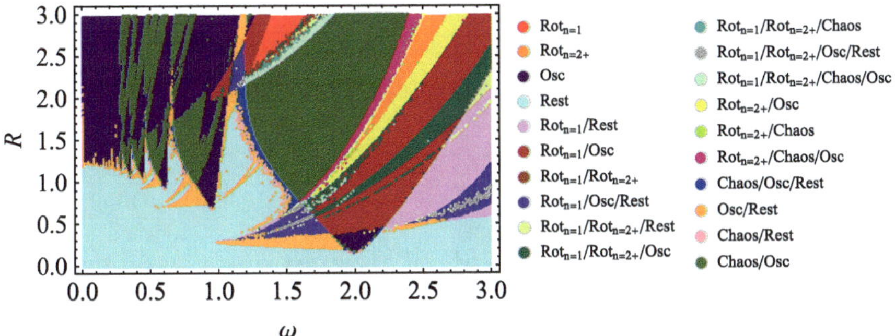

Fig. 2 Control space R-ω showing the physical responses of a system with $\lambda = 0.126$ and $\beta = 0.1$. $\text{Rot}_{n=1}$ means "pure rotations of period 1" while $\text{Rot}_{n=2+}$ means "pure rotations of period 2 or higher"

and anticlockwise [13]. A pure rotation is highly energetic, being the desired motion for energy harvesting purposes. In this article, pure rotations are regarded as synonymous of rotations, while the other categories are considered merely as oscillations.

For a given set (R, ω), the coexistence of periodic and chaotic solutions is possible, evidencing the nonlinear nature of the system. This coexistence depends on initial conditions. As an example, the control space R-ω of Fig. 2 shows all the possible steady states. This map is constructed as follows: for each fixed pair (R, ω), several simulations are performed employing different initial conditions, which produces different dynamical patterns; the topology of all these patterns is computed to give the color classification of the corresponding point (R, ω) in the control space. It can be seen that, for low excitation amplitudes, the rest position is the commonest solution. As R increases, oscillations, rotations and tumbling chaos appear. Rotations are the dominant type of stable solutions in the main resonance zone ($\omega = 2$), but for most of the scenarios they coexist with other responses. Besides, there is a wide range of the control space where rotations are not possible, irrespective of the initial conditions.

3.2 *Influence of the Crank/Rod Ratio* λ

In Fig. 3, control spaces R-ω are presented for different values of λ, with fixed damping of $\beta = 0.1$.

Figure 3a corresponds to the classic parametric pendulum (sinusoidal forcing), which can be recovered from (3) by setting $\lambda = 0$. For low λ (say $\lambda \lesssim 0.3$, Fig. 3b, c), the bifurcational behavior is topologically similar to the classic system. For higher λ (Fig. 3d), an additional rotation zone appears due to the significance of λ-terms in (3). These additional rotating attractors are studied by means of the bifurcation

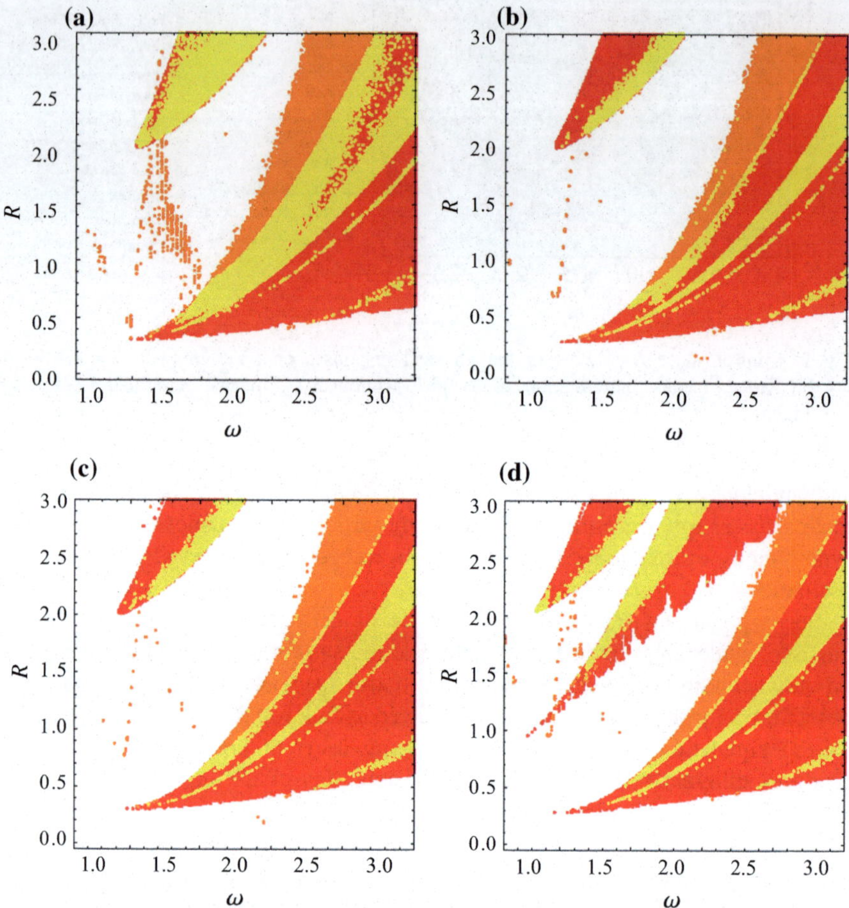

Fig. 3 Control space *R*-ω for the purely rotating attractors (clockwise and anticlockwise) with β = 0.1 and: **a** $\lambda = 0$ (classic parametric pendulum); **b** $\lambda = 0.126$; **c** $\lambda = 0.185$; **d** $\lambda = 0.356$. (•): period-1 rotations; (•): period-2 or higher rotations; (•): coexisting period-1 and period-2 or higher rotations

diagrams of Fig. 4. Figure 4a, b are constructed by fixing ω in the control space of Fig. 3d and plotting Poincaré points of the steady state response (a sampling time 2 π/ω is employed). In Fig. 4a, the main resonance (ω = 2) is studied. Up to $R \approx 1.31$, the system is topologically similar to the classic parametric pendulum (see [11] for an equivalent bifurcation diagram with $\lambda = 0$): two period-1 symmetric rotations appear at a saddle-node bifurcation ($R \approx 0.42$), then undergo a period-doubling cascade ($R \approx 1.07$) and vanish at a crisis scenario ($R \approx 1.31$). Then, after a narrow strip of tumbling chaos, a period-6 oscillation (actually a large amplitude rotation) appears as the only stable solution for a relatively broad range of *R*, until it also vanishes in a crisis. At $R \approx 1.58$ tumbling chaos take place. The additional rotating attractors appear at $R \approx 2.31$, maintaining as solutions of the physical system up to $R = 3.5$ and above. Two minor period-3 rotating attractors are born at $R \approx 2.89$, but they soon vanish in a crisis at $R \approx 2.9$, after a rapid period-doubling cascade. Rotations and tumbling chaos coexists with the inverted pendulum solution from $R \approx 3.35$ on. Bifurcation diagram of Fig. 4b shows the three pairs of rotating attractors which can be obtained for sufficiently high values of λ and $\beta = 0.1$. Attractors appearing at $R \approx 0.36$ and $R \approx 2.44$ exist in the classic parametric pendulum, while attractors at $R \approx 1.88$ are exclusive of the reciprocating excitation. Figure 4c is associated to the control space of Fig. 3b, i.e. a low-λ scenario. As expected, additional rotation zones cannot be found in such situation, with a bifurcational behavior similar to the classic parametric pendulum.

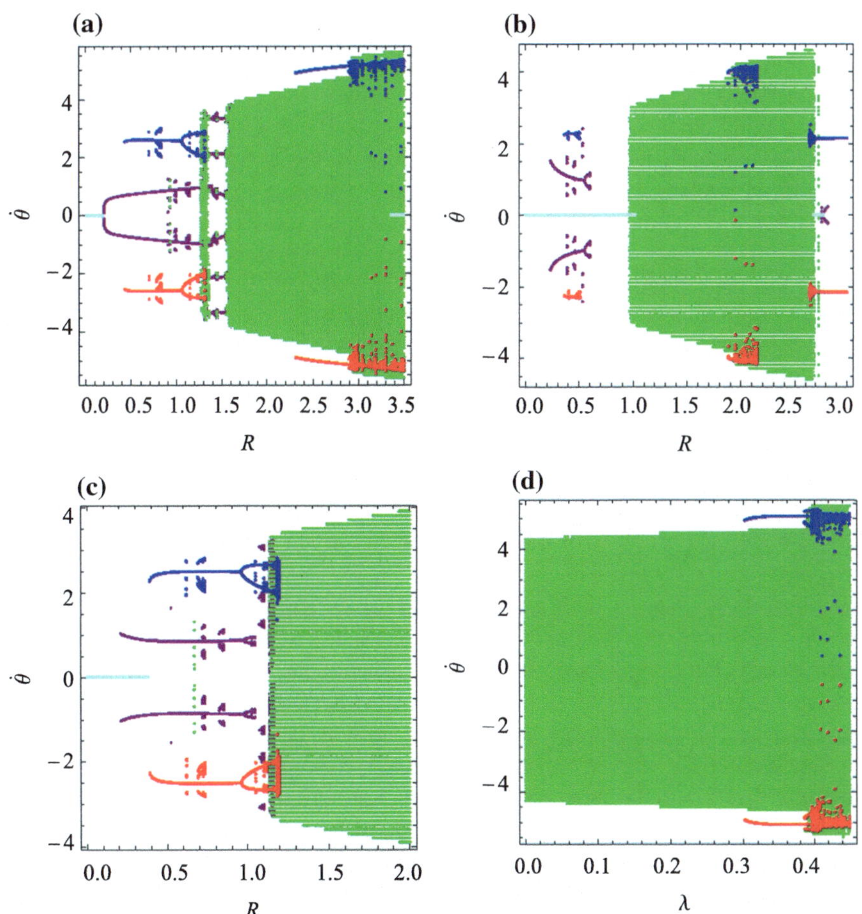

Fig. 4 Bifurcation diagram of the non-dimensional angular velocity for $\beta = 0.1$ and: **a** $\omega = 2$, $\lambda = 0.356$; **b** $\omega = 1.45$, $\lambda = 0.356$; **c** $\omega = 1.82$, $\lambda = 0.126$; **d** $R = 2.7$, $\omega = 2$. (•): clockwise rotations, (•): anticlockwise rotations, (•): rest, (•): oscillations, (•): tumbling chaos

Figure 3c shows a bifurcation diagram of λ, obtained by fixing R, ω and β in such a way to ensure the existence of the additional rotational attractors for a high value of λ (say $\lambda = 0.356$ as in Fig. 3d). As expected from Fig. 3a–c, there are not stable rotating solutions for low λ, but as λ is increased rotations appear at a saddle-node bifurcation ($\lambda \approx 0.303$). Now, considering $\lambda = 0.4$ as an upper practical limit of mechanical systems, results of Fig. 4d seem to indicate that an almost extreme value of λ is needed to ensure the existence of those additional rotating attractors. This is correct for $\beta = 0.1$, but not for lower amounts of damping, as shown in the next subsection.

3.3 *Influence of the Damping Parameter β*

The previous study was conducted assuming a fixed damping of $\beta = 0.1$. Besides speeding numerical integration, this choice allows us to compare our results with those in many other works of literature [6, 11–14]. But it has been demonstrated [6] that damping must be of $\beta < 0.1$ to ensure a viable energy extraction. Thus, a scenario with lower damping must be studied.

For $\lambda = 0$, it is known that a change in damping moves the control space R-ω downwards or upwards, by decreasing or increasing β respectively [8, 12–15] In fact, it has been pointed [8] that an increase of the excitation amplitude (R in our system) is equivalent to a decrease of β and vice versa. A more complex damping behavior was found for reciprocating excitation. This is evidenced in Fig. 5. Control space of Fig. 5a can be compared with that of Fig. 3b, since for both cases $\lambda = 0.126$ but with different β. From this comparison it is clear that with a decrement of β, an additional rotation zone appears, just as happened when the parameter λ was increased (Fig. 3a–d). The bifurcation diagram of Fig. 5b shows that as β decreases with fixed R, ω and λ, a pair of rotational attractors arise at a saddle node bifurcation ($\beta \approx 0.118$). This saddle node is the same of Fig. 4b but projected on the β-$\dot{\theta}$ plane instead of the λ-$\dot{\theta}$ plane. An imaginary motion picture of the bifurcation diagram in Fig. 5b as λ decreases should show the saddle node moving left, until the rotating attractors completely vanish when $\lambda = 0$, leaving behind only tumbling chaos. In conclusion, with an adequate (not necessarily extremely high) value of the crank-rod ratio λ, the additional rotating attractors exist for a range of β where energy extraction is feasible [6].

Finally, Fig. 5b suggests that the rotational response at low β deserves some attention. For $\beta = 0.01$, most of the rotations has period-1 (some period-4 motions are observed). However, steady states can be preceded by long periods of transient tumbling chaos [16]. In such cases, a very large simulation is required to avoid the transient. Figure 6 shows an example where a non-dimensional time of $t_d = 10{,}000$ must be discarded to obtain a steady period-1 rotation. This phenomenon explains the "blurred" rotating attractor of Fig. 5b and also the intermittencies of Figs. 3d and 5a.

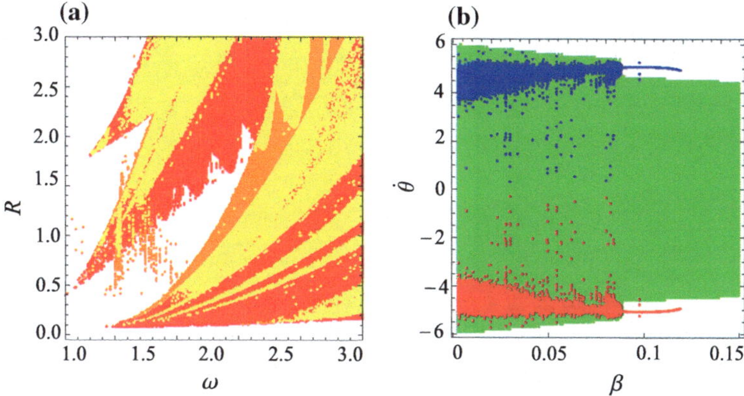

Fig. 5 **a** Control space R-ω for the purely rotating attractors (clockwise and anticlockwise) with $\lambda = 0.126$, β = 0.027. (•): period-1 rotations; (•): period-2 or higher rotations; (•): coexisting period-1 and period-2 or higher rotations. **b** Bifurcation diagram of the non-dimensional angular velocity for $R = 2.7$, ω = 2, $\lambda = 0.356$. (•): clockwise rotations, (•): anticlockwise rotations, (•): rest, (•): oscillations, (•): tumbling chaos

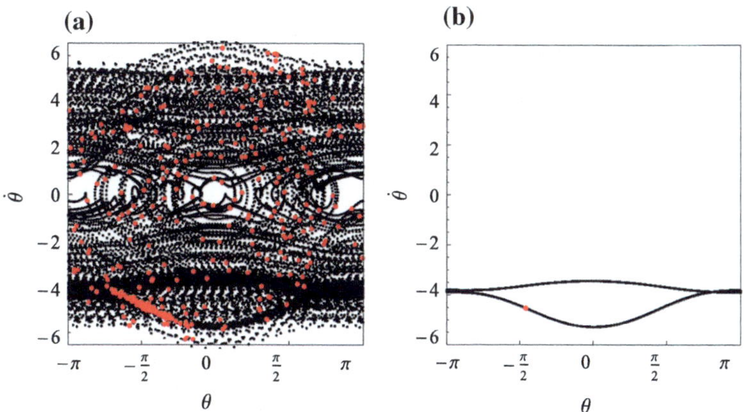

Fig. 6 Phase portraits and Poincaré sampling for $R = 2.7$, ω = 2, $\lambda = 0.356$ and β = 0.01. Initial conditions θ = 2 and $\dot{\theta} = -1.88$. Simulation time $t_s = 45{,}000$. **a** Discarded time $t_d = 7500$, transient tumbling chaos is present. **b** Discarded time $t_d = 10{,}000$, transient tumbling chaos is avoided and period-1 rotation is obtained. (•): pendulum response, (•): Poincaré points

3.4 Robustness and Probability of Rotations

After establishing the parameter settings where rotations are possible, the dynamics of the basins of attraction must be studied. This is necessary to know how difficult it is to achieve a steady state rotation, and how predictable could be this motion.

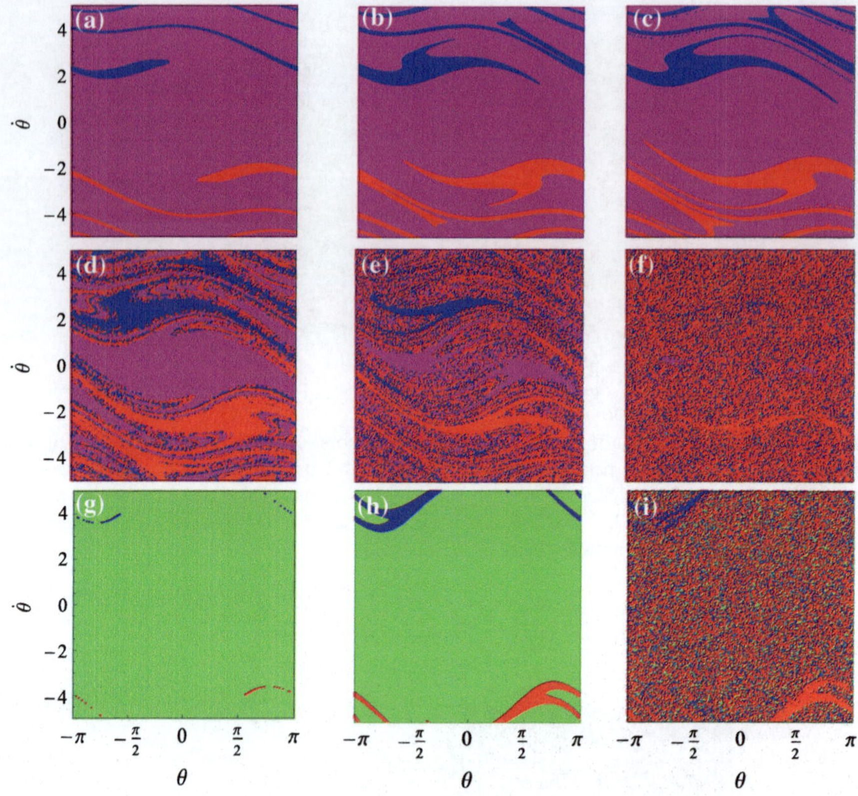

Fig. 7 Basin sequence for ω = 2, λ = 0.356, β = 0.1 and: **a** $R = 0.45$, born of period-1 rotations; **b** $R = 0.6$, basin of rotations grows; **c** $R = 0.65$, fractal erosion starts; **d** $R = 0.75$ and **e** $R = 1$, progress of erosion; **f** $R = 1.3$, basins of oscillations and rotations almost vanish; **g** $R = 2.22$, born of period-3 rotations; **h** $R = 2.7$, basin of rotations grows; **i** $R = 3$, new erosion. (•): clockwise rotations, (•): anticlockwise rotations, (•): oscillations, (•): tumbling chaos

The dynamics of the basins of rotation is followed, as R increases. Figure 7 shows basins of attraction associated to the bifurcation diagram of Fig. 4a. The birth of rotations at a saddle node bifurcation ($R \approx 0.425$) is observed and, as R increases, the basin of rotations grows (Fig. 7a, b). After the homoclinic tangency [5], the basin boundary initiates its fractalization: *fractal fingers* sweep across the basin of oscillations, leading to the erosion of the entire basin (Fig. 7c–e). At $R = 1.3$ (Fig. 7f) the basin of rotations is almost fully eroded; there is a high *final state sensitivity*: small variations of the initial conditions modifies the attractor ultimately chosen [5]; rotating chaos is present [8, 13], but it is about to be replaced by tumbling chaos. At $R \approx 1.32$ there is a crisis and then, for a broad range of R, tumbling chaos is the only stable attractor. At $R \approx 2.22$ a new basin of rotations appears inside the chaotic basin (Fig. 7g). This basin evolves until it fractalizes from $R \approx 2.75$ on. At $R = 3$, both the basin of rotations and tumbling chaos are

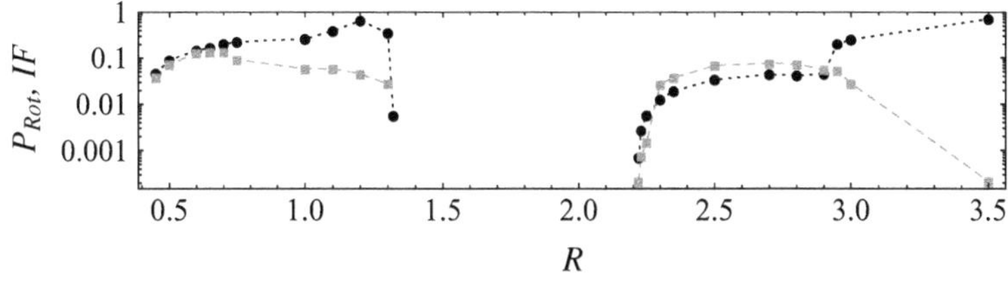

Fig. 8 Integrity and probabilistic analysis of rotations as R is increased, for $\omega = 2$, $\lambda = 0.356$ and $\beta = 0.1$. (■): probability of rotations, P_{Rot}; (•): integrity factor, *IF*

eroded. Rotations are of period-1 and they evidence long chaotic transients, as discussed in the previous subsection.

A visual inspection of Fig. 7 allows qualitative observations on the interaction among attractors. But with a view on energy harvesting, a quantitative evaluation of the robustness is required. For this purpose, the *integrity factor* (*IF*) is considered, which is defined in 2D as the normalized radius of the largest circle entirely belonging to a basin [17]. As rotations are studied regardless of its period, the *IF* can be thought as a measure of sensitivity of rotations to initial conditions: with a high *IF* (*IF* $\rightarrow$ 1), a small variation of the initial conditions also produces a steady rotation; meanwhile, with a low *IF* (*IF* $\rightarrow$ 0), the opposite happens. Besides, since initial conditions usually cannot be accurately determined in practice, it is important to know what happens given an unknown initial state. Thus P_{Rot} is defined as the probability of occurrence of rotations, for random initial conditions into a given range. Being *IF* and P_{Rot} two normalized magnitudes, they can be compared directly.

Results of the robustness/probabilistic analysis are presented in Fig. 8. Up to fractalization ($R \approx 0.65$, see Fig. 7c), P_{Rot} and *IF* give increasing values since rotations are confined into their basin. The peak of robustness is at $R \approx 0.6$ ($IF = 0.129$). As erosion evolves, the *IF* decays as the final state sensitivity increases, but P_{Rot} keeps growing since more initial conditions give rotations (Fig. 7d, e). At $R = 1.2$ the basin of oscillations is almost fully eroded and there is not possible in practice to predict the direction of rotations. P_{Rot} reaches a maximum: $P_{Rot} = 0.66$. Right before the crisis ($R \approx 1.32$), rotations vanish and P_{Rot} fall dramatically. At the crisis, a few initial conditions produce rotations: $IF = 0$ and $P_{Rot} \approx 0$. Similar behavior is observed for the rotational attractors at $R \approx 2.22$, which are exclusive of reciprocating excitation. The peak probability is $P_{Rot} = 0.71$ ($R = 3.5$), as 71% of the initial conditions produce rotations.

It is interesting to note that for some settings of the parameters it could be $P_{Rot} = 1$. This means that all initial states produce rotations. Actually, Fig. 9 has a "red zone" where all the responses are rotational with period-1. Due to the erosion of the basin, final state sensitivity is high; but the choice is reduced to clockwise or anticlockwise rotations, thus only direction of motion is unpredictable.

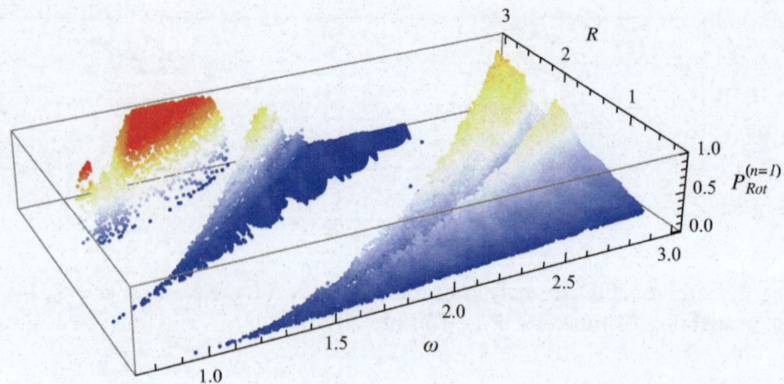

Fig. 9 Probability of period-1 rotations given random initial conditions ($\lambda = 0.356$)

4 Conclusions

The dynamics of the parametric pendulum with a reciprocating excitation was addressed with a view on energy harvesting. A rich dynamic behavior is elucidated, with substantial differences with respect to the classic sinusoidal forcing. Crank/rod ratio and viscous damping are crucial for rotational dynamics of the system: with a sufficiently high crank/rod ratio and/or a sufficiently low damping, new rotating attractors appear which are impossible with a sinusoidal excitation. These attractors exist for ranges of damping where energy extraction is feasible, more that more excitation scenarios allow rotational motion, increasing the possibilities for energy extraction.

The structural stability of the attractors and probabilities of obtaining rotations with unknown initial conditions were studied. It is shown that both robustness and probability of rotations grow until fractal erosion of the phase portrait starts. After this, robustness decays due to fractal erosion, but probability keeps growing since more initial conditions produce rotations. This means that rotations are easy to obtain but direction of rotation is difficult to predict due to a high final state sensitivity. The first is good for energy harvesting purposes, while the second should not lead to great difficulties in practical applications: rotations are desired, regardless of their direction.

A main conclusion of this work is that rotations are reachable and predictable with an adequate configuration of forcing and damping parameters. These parameters are closely related to the design of a suitable pendulum harvester. As excitation source is commonly known, damping depends only on the pendulum system and must be measured. Of course, a low friction is desired in energy harvesting applications.

Acknowledgements The authors would like to thank the support of Secretary of Science and Technology and Secretary of International Relations of UTN, CONICET, ANPCyT and the Engineering Department of UNS.

References

1. Wiercigroch, M.: A New Concept of Energy Extraction from Waves via Parametric Pendulor. UK Patent Application (2010)
2. Yurchenko, D., Alevras, P.: Dynamics of the *N*-pendulum and its application to a wave energy converter concept. Int. J. Dynam. Control **1**(4), 290–299 (2013)
3. Najdecka, A., Narayanan, S., Wiercigroch, M.: Rotary motion of the parametric and planar pendulum under stochastic wave excitation. Int. J. Nonlin. Mech. **71**, 30–38 (2015)
4. Reguera, F., Dotti, F., Machado, S.: Rotation control of a parametrically excited pendulum by adjusting its length. Mech. Res. Commun. **72**, 74–80 (2016)
5. Thompson, J., Stewart, H.: Nonlinear dynamics and chaos, 2nd edn. Wiley, West Sussex, England (2002)
6. Nandakumar, K., Wiercigroch, M., Chatterjee, A.: Optimum energy extraction from rotational motion in a parametrically excited pendulum. Mech. Res. Commun. **43**, 7–14 (2012)
7. Rattan, S.: Theory of machines. Tata McGraw Hill, New Delhi, India (2009)
8. Leven, R., Pompe, B., Wilke, C., Koch, B.: Experiments on periodic and chaotic motions of a parametrically forced pendulum. Physica D **16**, 371–384 (1985)
9. Dotti, F., Reguera, F., Machado, S.: Damping in a parametric pendulum with a view on energy harvesting. Mech. Res. Commun. **81**, 11–16 (2017)
10. Thomsen, J.: Vibrations and stability. Springer, Berlin, Germany (2003)
11. Dotti, F., Reguera, F., Machado, S.: A review on the nonlinear dynamics of pendulum systems for energy harvesting from ocean waves. In: Proceedings of the 1st Pan-American Congress on Computational Mechanics—PANACM 2015, pp. 1516–1529, Buenos Aires, Argentina (2015)
12. Garira, W., Bishop, S.: Rotating solutions of the parametrically excited pendulum. J. Sound Vib. **263**, 233–239 (2003)
13. Clifford, M., Bishop, S.: Rotating periodic orbits of the parametrically excited pendulum. Phys. Lett. A **201**, 191–196 (1995)
14. Horton, B., Sieber, J., Thompson, J., Wiercigroch, M.: Dynamics of the nearly parametric pendulum. Int. J. Nonlin. Mech. **46**, 436–442 (2011)
15. Xu, X., Wiercigroch, M.: Approximate analytical solutions for oscillatory and rotational motion of a parametric pendulum. Nonlin. Dyn. **47**, 311–320 (2007)
16. Szemplinska-Stupnicka, W., Tyrkiel, E., Zubrzycki, A.: The global bifurcations that lead to transient tumbling chaos in a parametrically driven pendulum. Int. J. Bifurcat. Chaos **10**(9), 2161–2175 (2000)
17. Lenci, S., Rega, G.: Experimental versus theoretical robustness of rotating solutions in a parametrically excited pendulum: a dynamical integrity perspective. Physica D **240**, 814–824 (2011)

Part VI
Vehicle Dynamics and Multibody Systems

Analysis of Control Strategies for Autonomous Motorcycles Stabilization and Trajectories Tracking

Marília Maurell Assad, Marco Antônio Meggiolaro and Mauro Speranza Neto

Abstract Autonomous vehicles—defined as vehicles with carrying capacity of persons or property without the use of a human driver—are an interesting and recent problem, with increasing studies in the last 20 years. Regarding this type of vehicles, a less explored option is the motorcycle: apart from the difficulties inherent in making a vehicle move independently, autonomous motorcycles have to be able to remain stable at any speed and trajectory. This work's main object of study is a small-scale electric motorcycle; represented by a linear model through a multibody approach: its four rigid bodies—wheels, chassis, handlebar and fork—have separately a characteristic behavior and together they influence the dynamics of each other. This approach results in lower order models, easier to simulate and to apply classical or modern control strategies. The two-wheeled vehicle is considered an inverted pendulum with a mobile base and other simplifications are proposed, as constant displacement speed or small steering and yaw angles. Since this vehicle is naturally unstable, to ensure a follow-up course without overturning it is necessary to apply an adjusted control signal; once the autonomous system studied will not have the presence of a mechanical counterbalance, there remains only the steering as a control strategy. Thus, this work analyzes the dynamic characteristics of the zero track vehicles and verifies the validity of different stability and path tracking control strategies of a motorcycle using as input only the steering of the handlebar.

Keywords Autonomous vehicles · Motorcycle dynamics · Stability control Trajectory control

M. M. Assad (✉) · M. A. Meggiolaro · M. S. Neto
Pontifical Catholic University, Rio de Janeiro, RJ 22451-900, Brazil
e-mail: marilia.assad@gmail.com

M. A. Meggiolaro
e-mail: meggi@puc-rio.br

M. S. Neto
e-mail: msn@puc-rio.br

A. de T. Fleury et al. (eds.), *Proceedings of DINAME 2017*, Lecture Notes in Mechanical Engineering, https://doi.org/10.1007/978-3-319-91217-2_28

1 Introduction

Motorcycles are an interesting study subject due to its unique dynamics: inherently unstable, they can achieve stability without a rider at certain speeds or settings, thanks to its geometry and gyroscopic effect. With respect to the mathematical modeling of this system, the most complete approach to describe its dynamic behavior is the multibody, where the motorcycle is interpreted as a combination of rigid bodies—including the wheels, bumpers, chassis, handlebar and fork driver. Each body has a separate characteristic behavior, but when combined they influence the dynamics of one another.

An alternative step to obtain a simplified multibody model is to consider only the main components of the vehicle, such as the driver, motorcycle's main body, wheels and front fork. This approach results in lower order models which are, therefore, easier to simulate and implement control strategies [1, 3, 7]. Other simplifications include considering the two-wheeled vehicle as an inverted pendulum mobile base, constant displacement speed or small steering and yaw angle [5]. This simpler multibody approach to the motorcycle model was also chosen in this paper, since it easily allows to apply classical and modern control strategies [6, 8].

Regarding autonomous vehicles, a topic increasingly in vogue in recent years, it is important to comprehend the human driving behavior in order to successfully mimic it. The control exercised by motorcyclists is complex and divided into two main categories: the steering torque, applied at the motorcycle handlebar, and the driver's body roll movement, which cause a double reverse pendulum effect on the vehicle dynamics.

The effects of each controller is clear in a curve trajectory: in motorcycle competitions, the driver tilts its body in the same direction of the curve, which results in a torque that changes the rear wheel rotational axis, causing an arched trajectory. Examples of this application are an actuated inverted pendulum [3] and a rotating mass [8] reproducing the torque given by the driver.

The driving system also plays an important role in the motorcycle dynamics: to successfully perform a curve with only the steering torque, it is necessary to initially move the motorcycle's handlebar in the opposite direction of the desired movement and then steer it in the right path, a phenomenon known as *counter-steering* [2]. Guiding the vehicle in the opposite direction causes a centrifugal force that tilts the motorcycle in the desired direction, enabling a curved path.

Interestingly, the steering control has the same operating principle of the body roll, without requiring a mass to cause the motorcycle inclination. Given that this work's object of study is a small-scale electric motorcycle, only the first category of control was applied, i.e., the stabilization and path tracking control are obtained solely by the steering torque, disregarding the driver's body roll torque.

2 Motorcycle Dynamics

The motorcycle model, based on [5], is derived from the Whipple bicycle model: the vehicle is composed of two structures, or frames, joined by a revolution joint on the handlebars, with each frame supporting a free rotating axisymmetric wheel. The driver's body is not considered in this paper; the front frame has uniformly distributed mass and wheels provide purely rotational movement, i.e., the sideslip and deformation of tires is not accounted in the model.

Roll, steer and yaw freedoms are allowed; longitudinal velocity is considered constant. To reduce the complexity of the model, the four rigid bodies have been simplified to two sets: the front frame includes the handlebar, fork and front wheel assembly; the rear frame is the main and rear wheel frame. The main system parameters are shown in Fig. 1; it is important to highlight the origin of the coordinate system—at the contact point between the rear wheel and the floor—and its direction, with the z axis positive in favor of gravity.

In order to validate the proposed model and test the suggested control strategies, an electric motorcycle was modeled using *SolidWorks*. The Duratrax450 is a 1/5 small-scale motorcycle, actuated by a *brushless* motor, which controls the vehicle speed, and steered by a servomotor, as seen in Fig. 2. The motorcycle had its four main systems measured and weighted and the computational model, portrayed in Fig. 2, gave its moments of inertia; the result is in Table 1.

With this data, is possible to combine all four bodies into two main frames: the front frame, consisting of the front wheel and the steering body, and the rear frame, including the rear wheel and main body. Meijaard et al. [5] details this variables' manipulation, which makes explicit some important relations, such as the gyroscopic effect, i.e., torque about one axis due to angular acceleration about the other.

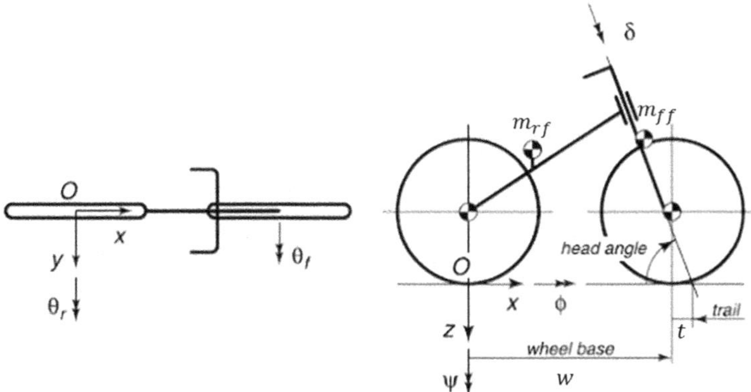

Fig. 1 Schematic of motorcycle's main parameters, based on [5]

Fig. 2 Test apparatus and computational model by *SolidWorks*

The motorcycle model is then completely defined by 19 design parameters, including the vehicle's velocity; Table 2 provides additional variables, dependent on main variables shown in Table 1, whose development is explicit in [5].

Given that the motorcycle is considered to be freely rolling forward on the *xy* plane, with constant speed, and approximately parallel to the global *x* axis—i.e., small roll, yaw and steer angles, as well as small displacement on the *y* axis—the governing linear equations of motion are the ones referring to two lateral degrees of freedom: the rightward lean of the rear frame (ϕ) and rightward steer of handlebars (δ). The forces acting on the system are: gravity in each center of mass, the ground reactions to the front wheel and the torque applied to the steering system, T_δ, considered positive when moving the handlebars to the right, and applied negatively on the rear frame. The lean and steer equations are described in detail in [5]; the dynamical system may also be written in state-space representation; the numerical result to Table 1 data is in Eq. 1.

$$\begin{bmatrix} \ddot{\phi} \\ \ddot{\delta} \\ \dot{\phi} \\ \dot{\delta} \end{bmatrix} = \begin{bmatrix} -0.93v & -3.5v & 91.0 & -30.0v^2 - 2.7 \\ 8.1v & -6.4v & -26.0 & 12.0v^2 + 100.0 \\ 1 & 0 & 0 & 0 \\ 0 & 1 & 0 & 0 \end{bmatrix} \begin{bmatrix} \dot{\phi} \\ \dot{\delta} \\ \phi \\ \delta \end{bmatrix} + \begin{bmatrix} -177.0 \\ 1.5 \times 10^3 \\ 0 \\ 0 \end{bmatrix} T_\delta \tag{1}$$

It is known that, for certain speed ranges, the motorcycle is able to remain stable without the aid of a driver; thus it is interesting to analyze the vehicle stability with zero input torque. The four poles of this system can be divided in three categories: the smaller real value is associated with capsize mode, in which the motorcycle falls sideways, with the roll and steering angles slowly increasing; imaginary eigenvalues are associated with the weave mode, in which the rear frame oscillates around the steering body; the last eigenvalue, real and with larger module, is known as the wobble mode, in which the steering shaft oscillates and aligns quickly with the motorcycle body in the direction of its movement, in a tractrix-like movement. The weave and wobble modes are significant at high speeds (up to 10 m/s), since these oscillatory effects can lead to serious instability.

Table 1 Duratrax450's individual bodies' experimental data

Parameter	Symbol	Value
Rear wheel		
Center of mass (m)	$(x_{rw} \quad y_{rw} \quad z_{rw})$	$(0 \quad 0 \quad -0.061)$
Mass (kg)	m_{rw}	0.69
Moments of inertia (kg m^2)	$[J_{rw_{xx}} \quad J_{rw_{yy}} \quad J_{rw_{zz}}]$	$[5.04 \quad 9.16 \quad 5.04]\ 10^{-4}$
Main body		
Center of mass (m)	$(x_m \quad y_m \quad z_m)$	$(0.14 \quad 0 \quad -0.11)$
Mass (kg)	m_m	1.19
Moments of inertia (kg m^2)	$\begin{bmatrix} J_{m_{xx}} & 0 & J_{m_{xz}} \\ 0 & J_{m_{yy}} & 0 \\ J_{m_{xz}} & 0 & J_{m_{zz}} \end{bmatrix}$	$\begin{bmatrix} 20.02 & 0 & -4.01 \\ 0 & 41.92 & 0 \\ -4.01 & 0 & 29.61 \end{bmatrix} 10^{-4}$
Steering body		
Center of mass (m)	$(x_s \quad y_s \quad z_s)$	$(0.27 \quad 0 \quad -0.13)$
Mass (kg)	m_s	0.13
Moments of inertia (kg m^2)	$\begin{bmatrix} J_{s_{xx}} & 0 & J_{s_{xz}} \\ 0 & J_{s_{yy}} & 0 \\ J_{s_{xz}} & 0 & J_{s_{zz}} \end{bmatrix}$	$\begin{bmatrix} 4.17 & 0 & 1.94 \\ 0 & 4.58 & 0 \\ 1.94 & 0 & 1.89 \end{bmatrix} 10^{-4}$
Front wheel		
Center of mass (m)	$(x_{fw} \quad y_{fw} \quad z_{fw})$	$(0.31 \quad 0 \quad -0.065)$
Mass (kg)	m_f	0.12
Moments of inertia (kg m^2)	$[J_{fw_{xx}} \quad J_{fw_{yy}} \quad J_{fw_{zz}}]$	$[1.63 \quad 3.19 \quad 1.63]\ 10^{-4}$
Rear frame		
Center of mass (m)	$(x_{rf} \quad y_{rf} \quad z_{rf})$	$(0.09 \quad 0 \quad -0.087)$
Mass (kg)	m_{rf}	1.88
Moments of inertia (kg m^2)	$\begin{bmatrix} J_{rf_{xx}} & 0 & J_{rf_{xz}} \\ 0 & - & 0 \\ J_{rf_{xz}} & 0 & J_{rf_{zz}} \end{bmatrix}$	$\begin{bmatrix} 3.41 & 0 & 2.19 \\ 0 & - & 0 \\ 2.19 & 0 & 12.23 \end{bmatrix} 10^{-3}$
Front frame		
Center of mass (m)	$(x_{ff} \quad y_{ff} \quad z_{ff})$	$(0.29 \quad 0 \quad -0.10)$
Mass (kg)	m_{ff}	0.25
Moments of inertia (kg m^2)	$\begin{bmatrix} J_{ff_{xx}} & 0 & J_{ff_{xz}} \\ 0 & - & 0 \\ J_{ff_{xz}} & 0 & J_{ff_{zz}} \end{bmatrix}$	$\begin{bmatrix} 8.27 & 0 & 0.28 \\ 0 & - & 0 \\ 0.28 & 0 & 4.09 \end{bmatrix} 10^{-4}$

Given the maximum speed of the electric motorcycle is of 15 m/s, the numerical result to the eigenvalues variation with speed increase is in Fig. 3. The analysis of eigenvalues position in an uncontrolled configuration shows that the Duratrax450 is not self-stabilizing, since its weave mode never reach the stable region; this phenomenon is due to, mainly, its geometric design, with a steering body much lighter than the rear frame.

Table 2 Duratrax450's multibody experimental data

Total mass (kg)	m_t	2.13	Wheel base (m)	w	0.31
Caster angle (rad)	ε	0.49	Trail (m)	t	0.028
Gravity (m/s^2)	g	9.81			
Global moment of inertia (kg m^2)					
J_{xx}	2.11×10^{-2}	J_{xz}	2.41×10^{-2}	J_{zz}	4.83×10^{-2}
Front frame moment of inertia (kg m^2)					
$J_{\varepsilon x}$	5.11×10^{-4}	$J_{\varepsilon z}$	6.64×10^{-4}	$J_{\varepsilon\varepsilon}$	5.27×10^{-4}
Global center of mass (m)					
x_t	0.11	z_t	−0.089	f	0.079
Gyroscopic coefficients					
S_f	3.80×10^{-3}	S_t	2.46×10^{-2}	S_u	1.99×10^{-2}

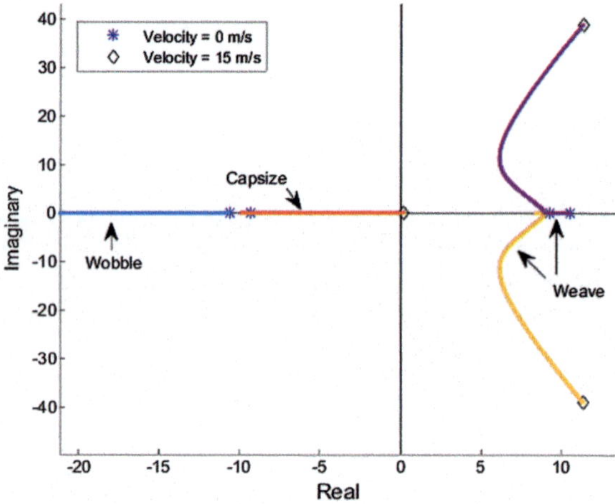

Fig. 3 Poles variation with speed increase

3 Stabilization Control Strategies

The first step to control an autonomous motorcycle is to balance it so it will not fall down. In this work, two strategies to keep the motorcycle straight are analyzed: an ideal state feedback control and a state feedback control with an observer, since not all state variables are measurable.

3.1 State Feedback

Considering all state variables ideally measurable, the state space model output of the motorcycle systems becomes the identity matrix, and the input control is proportional to all state variables through a gain vector. To verify if the state feedback can be made in this new system, the controllability matrix must have full rank; this system, for any speed range, is completely controllable.

Since the capsize and wobble eigenvalues are already stable—negative real poles—the main objective of the state feedback is to change the complex part of the weave eigenvalues. Table 3 brings the gain vector for three speeds and Figs. 4 and 5 illustrate the simulation results, considering an initial roll speed of 0.5 rad/s; is worth mentioning that the maximum torque provided by the servomotor is 0.32 N m.

3.2 State Feedback with Observer

Experimentally, not all state variables are available for feedback: only the steer angle (δ), given by the servomotor, and the roll angular speed ($\dot{\phi}$), measured by a gyrometer, are directly measurable. Therefore, Eq. 2 brings the observed state, where l is the observer gain vector, and the input control is a state feedback of the observed state. To verify if the full order state observer can be made in this new system, the observability matrix must have full rank; again, for any speed range, the system is completely observable.

$$\dot{\tilde{\mathbf{X}}} = \mathbf{A}\tilde{\mathbf{X}} + \mathbf{lC}\left(\mathbf{X} - \tilde{\mathbf{X}}\right) - \mathbf{B}\left(\mathbf{K}\tilde{\mathbf{X}}\right) \tag{2}$$

Since the observer and state feedback gain are independent, it is possible to choose two different sets of desirable eigenvalues; the poles of the observer are chosen so that its response is faster than the system's. For a fair comparison of the performance of the control strategies, the feedback poles will follow the previous method's design, meaning the feedback gain will be the same in Table 3. Table 4 brings the observer gain vector for three speeds and Figs. 6 and 7 illustrate the simulation results, considering an initial roll speed of 0.5 rad/s only to the measured state space.

4 Trajectory Control Strategy

The second step to control an autonomous motorcycle is to make it follow a desired trajectory. Since the previous model only considered the roll and steer angles and speeds, it needs to be adjusted to include the yaw angle (ψ) and lateral speed ($\dot{y}$). The yaw angular speed and rear wheel lateral velocity are, respectively, in Eq. 3; considering small displacements and angles. The new state space model adds these two equations to Eq. 1; the desired motorcycle's trajectory is to make a lane change of one meter, with zero initial conditions.

Table 3 State feedback gain to three different speeds

Speed (m/s)	Desired eigenvalues	State feedback gain
5	$[-0.68 \quad -3.1\pm24.0i \quad -42.0]$	$[-8.2\times10^{-5} \quad 8.3\times10^{-3} \quad -4.3\times10^{-2} \quad 0.35]$
10	$[-0.18 \quad -4.6\pm52.0i \quad -82.0]$	$[1.2\times10^{-5} \quad 1.2\times10^{-2} \quad -2.9\times10^{-2} \quad 0.98]$
15	$[-0.1 \quad -6.4\pm80.0i \quad -122.0]$	$[1.2\times10^{-5} \quad 1.7\times10^{-2} \quad -2.8\times10^{-2} \quad 2.07]$

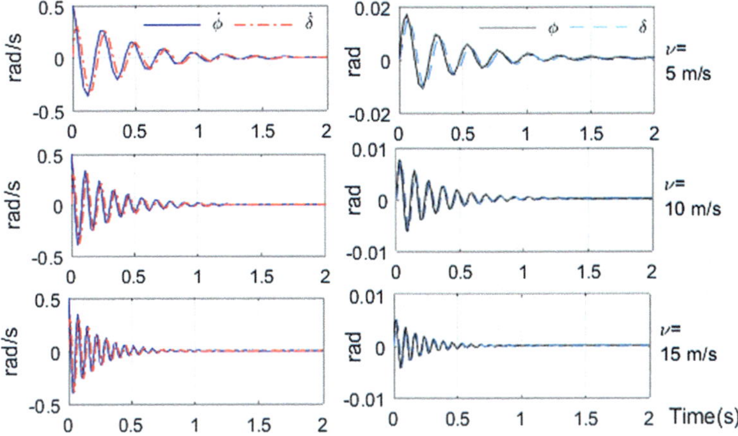

Fig. 4 Simulation at 5, 10 and 15 m/s with state feedback

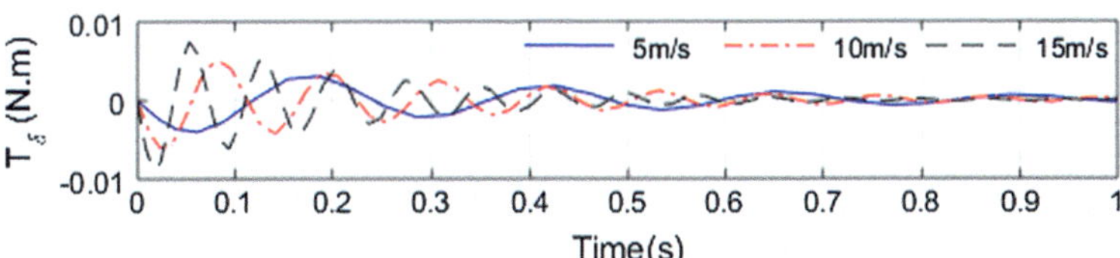

Fig. 5 Torque input at different speeds with state feedback

$$\begin{aligned} \dot{\psi} &= (\cos(\varepsilon)/w)(t\dot{\delta} + v\delta) \\ \dot{y} &= v\psi \end{aligned} \tag{3}$$

4.1 State Feedback

Considering all new state variables ideally measurable and the path desired, the input control becomes the desired trajectory minus a proportional gain to the state variables. Unlike the previous strategies, the desired eigenvalues are the same to every speed simulated, [−1 −5 −10 −15 −20 −25], in order to guarantee an equal time response at different speeds, making the controller performance independent of the longitudinal velocity.

Table 5 brings the gain vector for three speeds and Fig. 8, a comparison between the path traveled in each velocity; Figs. 9 and 10 illustrate the simulation results. The countersteering phenomenon appears on simulation, in which the motorcycle is

Table 4 Observer gain to three different speeds

Speed (m/s)	Desired eigenvalues	Observer gain vector
5	$[-3.39 \quad -15.7\pm121.9i \quad -210.8]$	$\begin{bmatrix} 5.22 & -650.36 & 27.84 & -42.80 \\ -773.25 & 22.03 & -32.06 & 203.89 \end{bmatrix}^T$
10	$[-0.89 \quad -22.8\pm262.4.0i \quad -410.1]$	$\begin{bmatrix} -26.25 & -1890 & 12.26 & -16.95 \\ -3040 & 1190 & -18.40 & 409.79 \end{bmatrix}^T$
15	$[-0.51 \quad -32.1\pm398.7i \quad -612.0]$	$\begin{bmatrix} -44.75 & -2960 & 9.85 & -9.87 \\ -6840 & 2750 & -12.06 & 611.86 \end{bmatrix}^T$

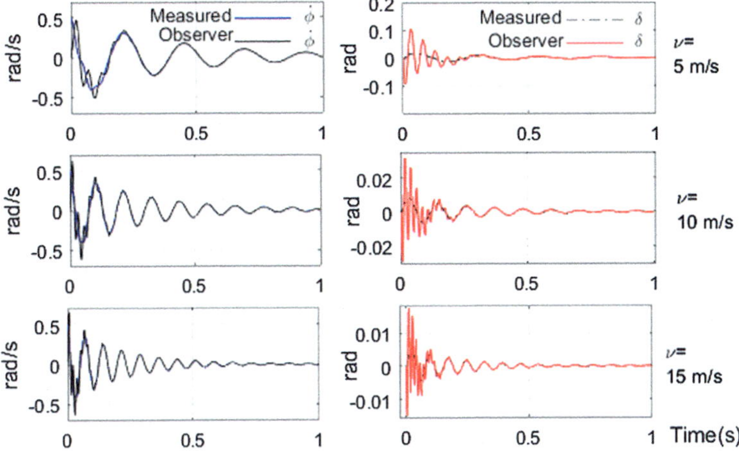

Fig. 6 Simulation at 5, 10 and 15 m/s with state feedback and observer

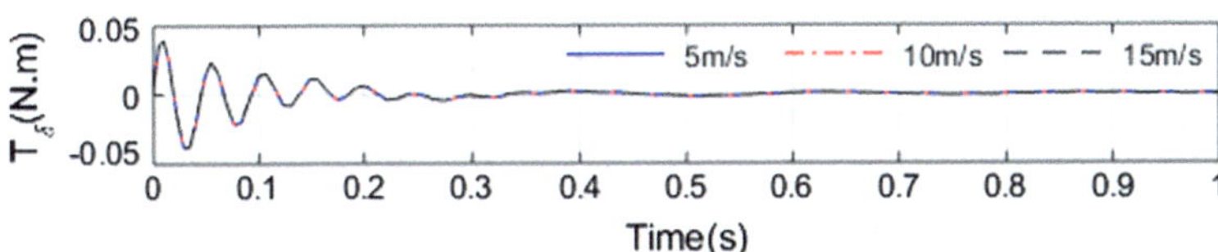

Fig. 7 Torque input at different speeds with state feedback and observer

initially moves to the opposite side to which you want to move. The adjustment of controller's gains has respected the system torque limit, which translates to a higher stabilization time than expected.

4.2 Linear Quadratic Regulator

At last, the linear quadratic regulator control strategy is applied to the path tracking problem. The steering torque is still proportional to state variables, but the control gains aim to minimize a cost function. The parameters of matrices Q and R are designed to penalize deviations on the state variables and control signal; since all variables are considered independent, they became diagonal weights with higher gains to the roll angle, steer angle and lateral displacement. The values were adjusted after simulations in order to guarantee a maximum path error of 10%. Equation 4 brings the matrices used in all speeds simulated, where it resulted in different closed loop poles.

Table 5 State feedback gain to three different speeds

Speed (m/s)	State feedback gain
5	$\left[-3.98\times10^{-3}\quad 2.55\times10^{-2}\quad -0.31\quad 1.11\quad -0.28\quad -3.91\times10^{-2}\right]$
10	$\left[4.47\times10^{-2}\quad 7.01\times10^{-3}\quad -9.75\times10^{-2}\quad 2.03\quad -0.14\quad -9.88\times10^{-3}\right]$
15	$\left[7.6\times10^{-2}\quad -1.34\times10^{-2}\quad -5.42\times10^{-2}\quad 3.10\quad -9.50\times10^{-2}\quad -4.30\times10^{-3}\right]$

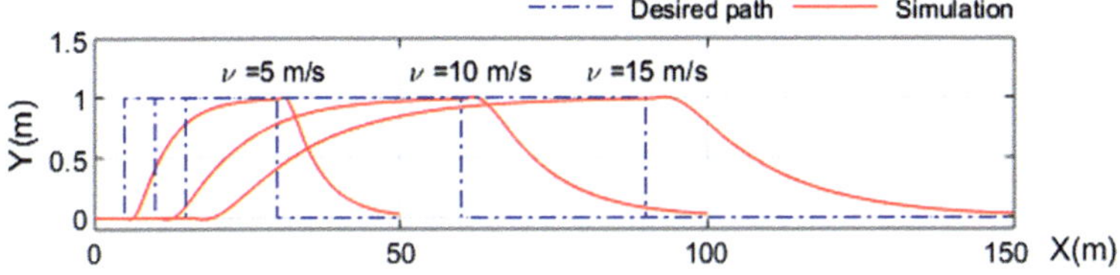

Fig. 8 Comparison between trajectory control with different speeds and state feedback control

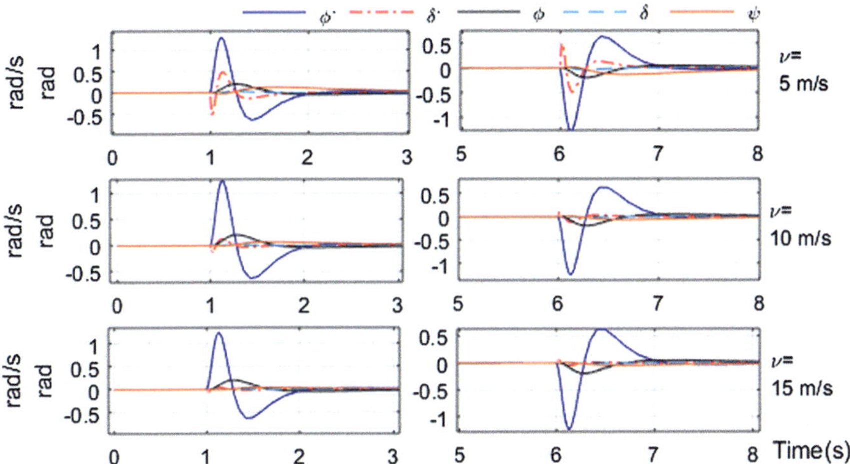

Fig. 9 Simulation at 5, 10 and 15 m/s with state feedback control

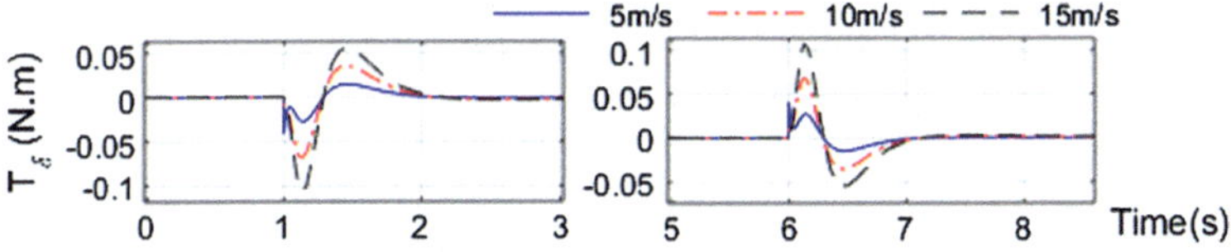

Fig. 10 Torque input at different speeds with state feedback

$$Q_{\dot{\phi}} = 1 \quad Q_{\dot{\delta}} = 1 \quad Q_{\phi} = 5 \quad Q_{\delta} = 5 \quad Q_{\psi} = 1 \quad Q_{y} = 10 \quad \mathbf{R} = [0.5] \tag{4}$$

Finally, Table 6 brings the closed loop poles and control gain vector; the resulting paths are explicit on Fig. 11. Figures 12 and 13 bring with more details the behavior of other state variables: the roll and steer speeds are higher than the previous simulation, with maximum amplitude of 2 rad/s; roll, steer and yaw angles are all under 0.3 rad, maintaining the small angle hypothesis; the torque control,

Table 6 Linear quadratic regulator to three different speeds

Speed (m/s)	Closed loop eigenvalues	State feedback gain
5	$\begin{bmatrix} -3.2 & -1.7\pm2.4i \\ -2000 & -22.0\pm15.0i \end{bmatrix}$	$[-2.07 \quad 1.19 \quad -15.03 \quad 26.71 \quad -17.11 \quad -4.47]$
10	$\begin{bmatrix} -3.2 & -1.7\pm2.4i \\ -2000 & -42.0\pm33.0i \end{bmatrix}$	$[-1.84 \quad 1.23 \quad -12.31 \quad 51.48 \quad -32.85 \quad -4.47]$
15	$\begin{bmatrix} -3.2 & -1.7\pm2.4i \\ -2000 & -62.0\pm50.0i \end{bmatrix}$	$[-1.75 \quad 1.24 \quad -11.51 \quad 77.44 \quad -48.67 \quad -4.47]$

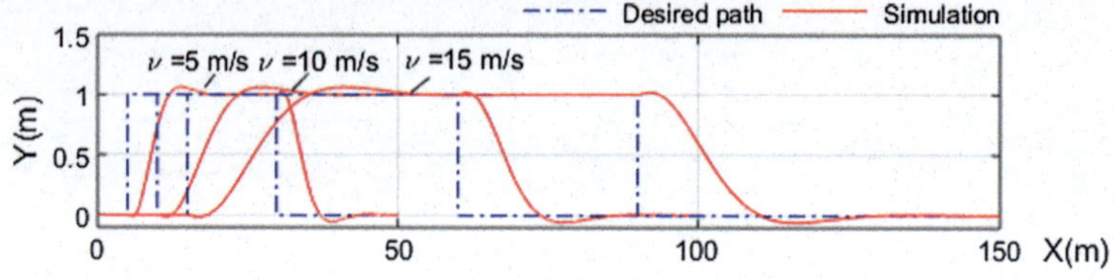

Fig. 11 Comparison between trajectory control at 5, 10 and 15 m/s with LQR control

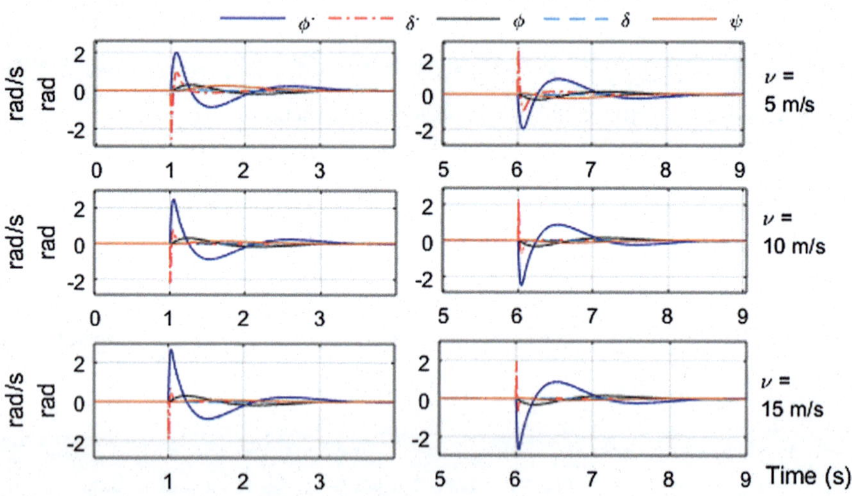

Fig. 12 Simulation at 5, 10 and 15 m/s with LQR control

however, reaches its limit in an infinitesimal time, which could be problematic on an experimental setting. Nevertheless, the system has a faster stabilization, with minimum overshoot.

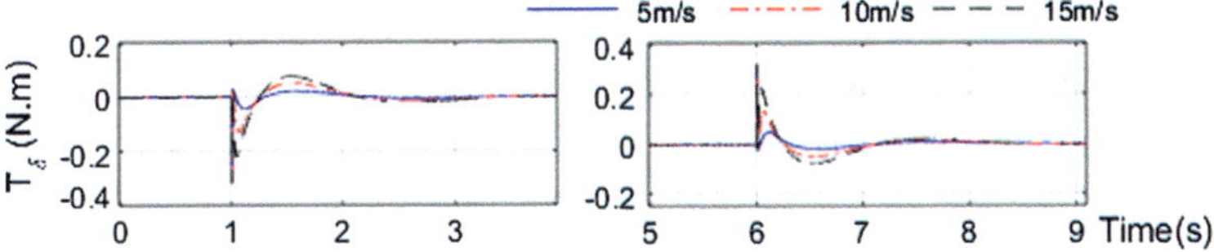

Fig. 13 Torque input at different speeds with LQR control

5 Conclusions

In this work the linearized model of a small-scale electric motorcycle was developed through a multibody approach, considering its main four rigid bodies—wheels, chassis, handlebar and fork. Simplifications were brought into the system to facilitate its analysis: the two-wheeled vehicle is considered an inverted pendulum with a mobile base, with constant longitudinal velocity and small steering, roll and yaw angles.

Four control strategies are proposed to act on the system natural instability and replace a human driver. Since the small scale motorcycle considered in this work does not have the presence of a mechanical counterbalance, there remains only the steering system as control input.

Numerical simulations with the complete system were carried out to prove the modern control strategies' validity: a state space feedback controller was capable to successfully keep the motorcycle upright; since only two of the state variables are measurable by the instrumentation system, another space feedback was proposed, also successful, with an state observer to estimate the missing variables values and optimize the control algorithm.

At last, a state feedback and linear quadratic regulator were proposed to make the motorcycle follow a desired trajectory, i.e., a lane change path. The state space model was adjusted to include two new variables regarding the lateral dynamics of the system and simulations confirmed the validity of these strategies.

As future works, the authors intend to investigate other control strategies as well as to improve the dynamic motorcycle model, so as to make lateral displacement, yaw and roll motions of the bicycle closer to reality.

References

1. Åström, K.J., Klein, R.E., Lennartson, A.: Bicycle dynamics and control: adapted bicycles for education and research. Control Syst. IEEE **25**(4), 26–47 (2005)
2. Fajans, J.: Steering in bicycles and motorcycles. Am. J. Phys. **68**(7), 654–659 (2000)
3. Keo, L., et al.: Experimental results for stabilizing of a bicycle with a flywheel balancer. In: 2011 IEEE International Conference on Robotics and Automation (ICRA), IEEE, pp. 6150–6155 (2001)

4. Martins, G.N., Speranza Neto, M., Meggiolaro, M.A.: 2016, Dynamic models of bicycles and motorcycles using power flow approach, CONEM 2016. IX Congresso Nacional de Engenharia Mecânica, Fortaleza, Brazil (2016)
5. Meijaard, J.P., et al.: Linearized dynamics equations for the balance and steer of a bicycle: a benchmark and review. In: Proceedings of the Royal Society of London A: Mathematical, Physical and Engineering Sciences. vol. 463. No. 2084. The Royal Society (2007)
6. Sharp, R.S.: Motorcycle steering control by road preview. J. Dyn. Syst. Meas. Contr. **129**(4), 373–381 (2007)
7. Tanaka, Y., Murakami, T.: A study on straight-line tracking and posture control in electric bicycle. IEEE Transactions on Industrial Electronics, vol. 56, no. 1, pp. 159–168 (2009)
8. Yi, J., et al.: Trajectory tracking and balance stabilization control of autonomous motorcycles. In: Proceedings 2006 IEEE International Conference on Robotics and Automation, 2006. ICRA 2006. IEEE, pp. 2583–2589 (2006)

Modelling and Simulation of the Rolling Dynamics of a Tractor-Trailer Truck Vehicle

Ricardo Sampaio and Flavio Celso Trigo

Abstract In this work, a 6 degree-of-freedom analytical model that describes the rolling dynamics of an articulated heavy-duty vehicle composed by a tractor and a trailer was developed and numerically simulated. The behaviour of a typical medium-duty truck-trailer set was evaluated in two common traffic manoeuvres, namely, performing a constant steering wheel angle curve and a change of direction, both at constant velocity (36 km/h). In the latter case, in order to simulate a sudden change, steering amplitude ranging from zero to 30° was imposed to the front wheels through a step function during 2 s. Results of the first trial revealed that the tractor roll angle presents a peak of 2.6° and achieves a steady value of 0.8° respectively at 4 and 20 s after the beginning of the manoeuvre. Roll angle amplitudes of the trailer, spanning from −0.015 to 0.023° (a negative value means the vehicle is leaning inwards at the curve), revealed an oscillatory characteristic, before reaching a stable value of 0.0025° in about 15 s. In the second trial, roll angles of both tractor and trailer reach maximum values higher than those of the previous test (respectively 11.5 and −0.5°); in addition, the oscillatory movement of the trailer was enhanced, since positive and negative roll angles alternated five times until both units return to a steady (zero roll angle) condition after about 20 s from the beginning of the manoeuvre. Those results are compatible with the expected behaviour of actual vehicles, thus suggesting that the proposed model can be tailored to include other truck-trailer configurations and manoeuvres.

Keywords Articulated vehicle dynamics · Roll dynamics modelling
Multibody system dynamics · Tractor-trailer numerical simulation

R. Sampaio · F. C. Trigo (✉)
University of São Paulo, São Paulo, SP 05508-000, Brazil
e-mail: trigo.flavio@usp.br

R. Sampaio
e-mail: ricsampa@gmail.com

A. de T. Fleury et al. (eds.), *Proceedings of DINAME 2017*, Lecture Notes in Mechanical Engineering, https://doi.org/10.1007/978-3-319-91217-2_29

1 Introduction

Truck vehicles play an important role in current freight transportation networks. Particularly in Brazil, according to data from 2008 [1], road vehicles are used to carry 58% of the total amount of freights. An usual manner to enhance the load capacity of those vehicles is to employ a tractor unit to haul one or several trailer and/or semi-trailer units. The difference between a trailer and a semi-trailer is related to the kind of coupling between the hauled units to the towing vehicle and among themselves. When the coupling allows transferring part of the load from the hauled unit to the tractor or to another hauled unit (normally, through the coupling of a fifth wheel on the tractor and a device called kingpin on the hauled unit, see Fig. 1a, b), the towed unit is called a semi-trailer; otherwise, it is a trailer. A consequence of the previous definition is that trailers have at least two axles. A typical 3-axle trailer is shown in Fig. 1c. However, regardless of the kind of coupling, those vehicles are known as articulated trucks.

Due to their weight and dimensions, articulated trucks are prone to present dynamical instability even at normal operating conditions, i.e., forward travel at constant speed. Those instabilities may be caused by small perturbations on track (holes, bumps, sudden friction changes, uneven surfaces) or natural causes, such as wind gusts, that may lead to the potentially dangerous situations of jack-knifing, trailer swing, truck/trailer rollover, and flutter [6]. One of the possible outcomes of the above-cited situations is the rolling of the trailer/semi-trailer, tractor or of both. In the literature, there are two main sets of models that attempt to describe and

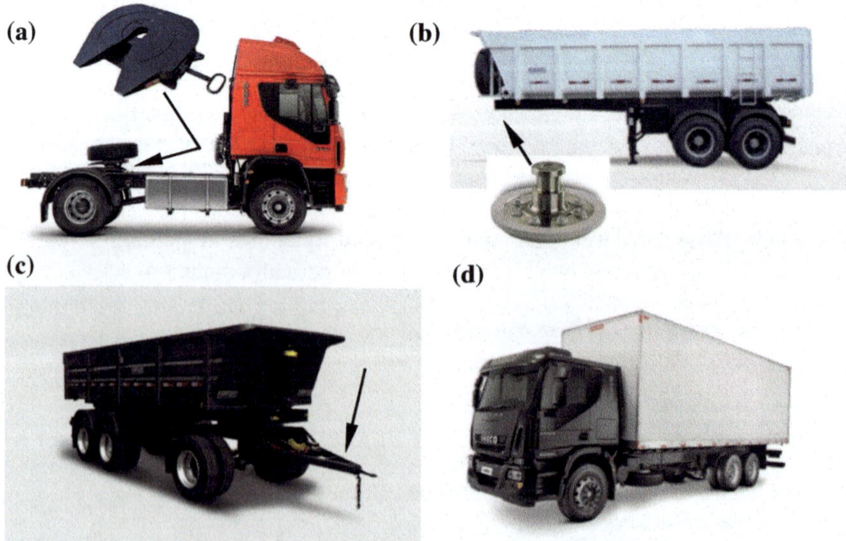

Fig. 1 **a** Fifth wheel and tractor truck [2, 3]; **b** kingpin and semi-trailer [2, 4]; **c** trailer unit with coupling device [5]; **d** box truck [3]

predict the dynamical behaviour of articulated trucks, mostly for the purpose of supporting the design of passive or active control systems capable of avoiding unstable conditions: The ones whose equations of motion are obtained directly from the application of the theorems of mechanics [7–9] and those that rely on multibody dynamics simulation software, in which the whole model is numerically built and evaluated as, for instance, the works by [10, 11].

To this end, a common feature of purely theoretical models is the use of simplifying hypotheses such as jack-knifing does not imply in rolling [9], or identical roll angles of tractor and trailer [7, 8]. On the other hand, multibody numerical models such as that from [10, 11], despite considering several degrees of freedom and handling both holonomic and non-holonomic constraints, may represent a challenge for the control designer, given the universe of possible parameter combinations.

Considering the above rationale, the present work aims at the development of a 6-degree-of-freedom analytical model to describe the rolling dynamics of an articulated tractor-trailer vehicle, and evaluating its performance through numerical simulations in two standard traffic manoeuvres, namely, constant radius curve and sudden change in direction manoeuvre, both at constant velocity. The proposed approach poses a contribution since, unlike [7–9], the rolling of each unit is considered as a separate degree-of-freedom, thus allowing one to evaluate the dynamical interaction between trailer and tractor under normal or perturbed operational conditions.

2 Materials and Methods

This section aims at presenting the theoretical background and the hypotheses used in the development of the analytical model of the vehicles. Firstly, only the tractor was considered, in order to qualitatively check the adequacy of the adopted hypotheses, especially those concerning tyre models. As the results were satisfactory, with physical responses consistent with the proposed manoeuvres, the model was extended to the entire vehicle, taking into account the trailer unit. All simulations were performed with the assistance of Matlab (R2012a version) and Mathematica (version 9.0).

2.1 Tractor Vehicle Model

The model to be developed is based on Chen and Tomizuka's [7] box truck (Fig. 1d), which is adapted to haul a trailer unit. Equations of motion are obtained through the Lagrangean formalism [12], and they include tractor roll and yaw. A schematic representation of the vehicle, as well as the reference frames and their coordinate systems used in the development of the model, are shown in Fig. 2.

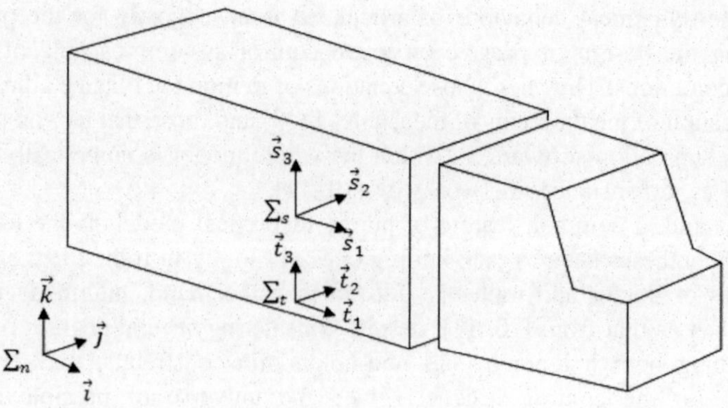

Fig. 2 Box truck (tractor vehicle) and reference frames

The system of coordinates Σ_t is rigidly attached to the unsprung element of the vehicle in such a way that the unit vector $\vec{t}_1$ is on the rolling axis, and that the direction of $\vec{t}_3$ intercepts the vehicle center of mass. On the other hand, system of coordinates Σ_s is fixed to centre of mass (sprung mass), with axes initially parallel with those of system Σ_t. Finally, Σ_n is the inertial reference frame.

The three systems of coordinates are related by transformation Eqs. (1) and (2), in which the functions sine and cosine are respectively represented by $s(x)$ and $c(x)$.

$$\{\vec{n}\} = \begin{pmatrix} c(\varphi) & -s(\varphi) & 0 \\ s(\varphi) & c(\varphi) & 0 \\ 0 & 0 & 1 \end{pmatrix} \{\vec{t}\} \tag{1}$$

$$\{\vec{t}\} = \begin{pmatrix} 1 & 0 & 0 \\ 0 & 1 & -\theta \\ 0 & \theta & 1 \end{pmatrix} \{\vec{s}\} \tag{2}$$

In Eq. (1), φ stands for the *yaw angle of the unsprung element in relation to the inertial reference frame*, whereas θ is *the rolling angle of the sprung mass as observed from the reference frame attached to the unsprung element*. The hypothesis of small angular displacements $(\sin(\theta) \approx \theta,\ \cos(\theta) \approx 1)$ was employed in Eq. (2). This way, it is possible to write expressions for the kinetic energy and potential function for the vehicle centre of mass (CM) in relation to the inertial reference frame. To begin with, one expresses the position of the CM in relation to the inertial reference frame:

$$\begin{aligned} \vec{p}_{CM/n} &= \vec{p}_{CM/t} + \vec{p}_{t/n} \\ \vec{p}_{CM/n} &= z\vec{t}_3 + h\vec{s}_3 + x_n\vec{i} + y_n\vec{j} \end{aligned} \tag{3}$$

In Eq. (3), x_n and y_n are the displacements of the origin of the unsprung element coordinate system in relation to the inertial reference frame, z is its height, and h is the height of the CM to the rolling axis. Differentiating Eq. (3) with respect to time, one obtains the velocity of the CM according to Eq. (4),

$$\{\vec{v}_{CG/n}\} = \left\{ \begin{array}{c} (h\dot{\varphi}\theta + \dot{x}_n c(\varphi) + \dot{y}_n s(\varphi))\vec{s}_1 \\ (-h\theta - \dot{x}_n s(\varphi) + \dot{y}_n c(\varphi))\vec{s}_2 \\ (\dot{x}_n \theta s(\varphi) - \dot{y}_n \theta c(\varphi))\vec{s}_3 \end{array} \right\}, \tag{4}$$

in which the derivative of the unit vectors of the Σ_s base was calculated using the rotation vector of the sprung mass,

$$\Omega_{s/n} = \dot{\theta}\vec{s}_1 + \dot{\varphi}\theta\vec{s}_2 + \dot{\varphi}\vec{s}_3 \tag{5}$$

The kinetic energy of the tractor unit can be obtained using

$$\begin{gathered} T_c = \tfrac{1}{2} m_c (\vec{v}_{CM/n} \cdot \vec{v}_{CM/n}) + \tfrac{1}{2} \overrightarrow{\Omega}_{s/n} I \overrightarrow{\Omega^t}_{s/n} \Rightarrow \\ T_c = \tfrac{1}{2} m_c \Big[(h\dot{\varphi}\theta + \dot{x}_n c(\varphi) + \dot{y}_n s(\varphi))^2 + (-h\theta - \dot{x}_n s(\varphi) + \dot{y}_n c(\varphi))^2 \\ + (\dot{x}_n \theta s(\varphi) - \dot{y}_n \theta c(\varphi))^2 \Big] + \tfrac{1}{2} (I_x \dot{\theta}^2 + I_y (\dot{\varphi}\theta)^2 + I_z \dot{\varphi}^2), \end{gathered} \tag{6}$$

where m_c(kg) is the mass and $I(\mathrm{kg\,m^2})$ is the central inertia tensor. The potential energy function, on the other hand, is given by

$$V_c = m_c g h (c(\theta) - 1), \tag{7}$$

in which the hypothesis that the sprung mass swings about the vehicle rolling axis was considered. From Eqs. (6) and (7), one writes the Lagrangean

$$L_c = T_c - V_c \tag{8}$$

and, through the Euler-Lagrange equations,

$$\frac{d}{dt}\left(\frac{\partial L}{\partial \dot{q}_i}\right) - \frac{\partial L}{\partial q_i} = Q_{q_i}, \quad i = 1, \ldots, 4, \tag{9}$$

the equations of motion can be written for the generalised coordinates $\boldsymbol{q} = [q_1, q_2, q_3, q_4]^T = [x_n, y_n, \theta, \varphi]^T$ and for the generalised non-conservative forces $\boldsymbol{Q} = \left[Q_{q_1}, Q_{q_2}, Q_{q_3}, Q_{q_4}\right]^T = \left[Q_x, Q_y, Q_\theta, Q_\varphi\right]^T$ associated to those coordinates. In this model, the generalised non-conservative forces are those acting on the tyres, as depicted in Fig. 3, in which F_{lj} and F_{aj}, $j = 1, \ldots, 4$, respectively stand for the longitudinal and lateral forces. The forces acting at the suspension elements (orthogonal to the tyre forces, not shown in the figure) are represented by F_{sj}.

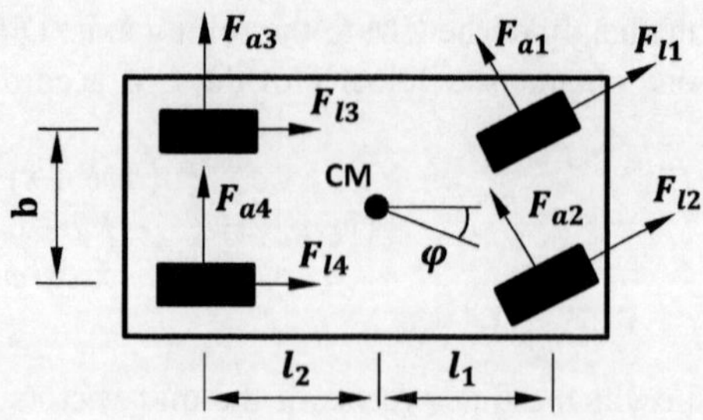

Fig. 3 Lateral and longitudinal forces at the tyres of the tractor truck

For the computation of the non-conservative generalised forces, one first describes F_{lj}, F_{aj}, F_{sj} in coordinates of the Σ_n, as follows:

$$F_{xj} = F_{lj}c(\varphi) - F_{aj}s(\varphi) \tag{10}$$

$$F_{yj} = F_{lj}s(\varphi) + F_{aj}c(\varphi) \tag{11}$$

Since the scope of this work is the analysis of the lateral dynamics of the vehicle, all the man oeuvres are performed at constant velocity, i.e., $F_{lj} = 0, \ j = 1, 2, 3, 4$ [13]. Lateral forces, on the other hand, follow the model by [14], whose main assumption is that, at low speeds, their magnitude is proportional to the slip angle; moreover, the proportionality constant is the so called cornering stiffness (CS). Using δ to represent the *steering angle at the font tyres*, the slip angles for tyres $j = 1, \ldots, 4$ of Fig. 3 are given by:

$$\alpha_1 = \delta - arctg\left(\frac{\dot{y}_t + l_1\dot{\varphi}}{\dot{x}_t - \frac{b}{2}\dot{\varphi}}\right) \tag{12}$$

$$\alpha_2 = \delta - arctg\left(\frac{\dot{y}_t + l_1\dot{\varphi}}{\dot{x}_t + \frac{b}{2}\dot{\varphi}}\right) \tag{13}$$

$$\alpha_3 = -arctg\left(\frac{\dot{y}_t - l_2\dot{\varphi}}{\dot{x}_t - \frac{b}{2}\dot{\varphi}}\right) \tag{14}$$

$$\alpha_4 = -arctg\left(\frac{\dot{y}_t - l_2\dot{\varphi}}{\dot{x}_t + \frac{b}{2}\dot{\varphi}}\right) \tag{15}$$

Another hypothesis that holds for low speeds is $arctg(x) \simeq x$; therefore, lateral forces are obtained by directly multiplying the slip angles and the CS, the latter admitted the same for all tyres. Suspension is modelled as a spring-viscous linear damping system. Thus,

$$F_s = k\epsilon + d_c\dot{\epsilon}, \tag{16}$$

with k representing spring stiffness and d_c denoting viscous damping coefficient, whereas ϵ stands for the displacement of the suspension, measured from the equilibrium position. Their values, for each tyre, are given by

$$\epsilon_1 = \epsilon_3 = -\frac{b}{2}\theta \tag{17}$$

$$\epsilon_2 = \epsilon_4 = -\frac{b}{2}\theta \tag{18}$$

Given those hypotheses and constraints, the generalised forces, considered applied at points whose position is defined by vector $\vec{p}_j$ at each tyre j, $j = 1, \ldots, 4$, are obtained according to

$$Q_{q_i} = \sum_{j=1}^{4} F_{a_j} \frac{\partial(\vec{p}_j \cdot \vec{\imath})}{\partial q_i} + \sum_{j=1}^{4} F_{l_j} \frac{\partial(\vec{p}_j \cdot \vec{\jmath})}{\partial q_i} + \sum_{j=1}^{4} F_{s_j} \frac{\partial(\vec{p}_j \cdot \vec{k})}{\partial q_i}, \quad i = 1, \ldots, 4. \tag{19}$$

2.2 Tractor and Trailer Vehicle Model

In this section, an analytical model for the complete vehicle is developed. Its schematic representation is in Fig. 4, which includes reference frames Σ_U and Σ_r respectively fixed at the trailer rolling axis (unsprung element) and at the trailer sprung mass. Trailer attitude is described by the *yaw angle* η, measured relatively to Σ_t, and by the *roll angle*, λ, *of the trailer sprung mass*, in relation to Σ_U. Rotation matrices among coordinate systems attached to reference frames Σ_U, Σ_r, and Σ_t are given by

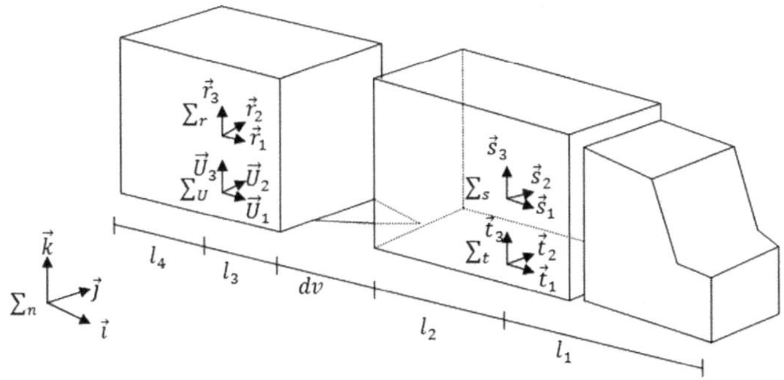

Fig. 4 Tractor and trailer vehicles and reference frames

$$\{\vec{U}\} = \begin{pmatrix} c(\eta) & s(\eta) & 0 \\ -s(\eta) & c(\eta) & 0 \\ 0 & 0 & 1 \end{pmatrix} \{\vec{t}\} \quad \{\vec{r}\} = \begin{pmatrix} 1 & 0 & 0 \\ 0 & 1 & \lambda \\ 0 & -\lambda & 1 \end{pmatrix} \{\vec{U}\} \tag{20}$$

In order to compute the kinetic energy of the trailer, one differentiates the position vector of its centre of mass with respect to time to obtain the velocity vector:

$$\frac{d}{dt}(\vec{p}_{CG_2}) = \frac{d}{dt}(\vec{p}_{CG_1} - l_2\vec{s}_1 - h\vec{s}_3 - (dv + l_3)\vec{r}_1 + h_r\vec{r}_3) \tag{21}$$

$$\begin{aligned}\{\vec{v}_{CG_2}\} = {} & [(h\dot{\varphi}\theta + \dot{x}c(\varphi) + \dot{y}s(\varphi))c(\eta) + (-h\dot{\theta} - \dot{x}s(\varphi) + \dot{y}c(\varphi))s(\eta) + (\dot{x}s(\varphi) - \dot{y}c(\varphi)) \\ & (-s(\eta)\theta^2) - l_2\dot{\varphi}s(\eta) - l_2\dot{\varphi}\theta^2 s(\eta) - h\dot{\varphi}\theta c(\eta) + h\dot{\theta}s(\eta) + h_r\dot{\eta}\lambda]\vec{r}_1 \\ & + [(h\dot{\varphi}\theta + \dot{x}c(\varphi) + \dot{y}s(\varphi))(-s(\varphi)) + (-h\dot{\theta} - \dot{x}s(\varphi) + \dot{y}c(\varphi))(c(\eta) + \theta\lambda) \\ & + (\dot{x}s(\varphi) - \dot{y}c(\varphi))(-\theta^2 c(\eta) + +\theta\lambda) - l_2\dot{\varphi}(c(\eta) + \theta\lambda) + l_2\dot{\varphi}\theta(-\theta c(\eta) + \lambda) \\ & + hs(\eta) + h\dot{\theta}(c(\eta) + \theta\lambda) - (dv + l_3)\dot{\eta} - h_r\dot{\lambda}]\vec{r}_2 \\ & + [(h\dot{\varphi}\theta + \dot{x}c(\varphi) + \dot{y}s(\varphi))(\lambda s(\eta)) + (-h\dot{\theta} - \dot{x}s(\varphi) + \dot{y}c(\varphi))(-\lambda c(\eta) + \theta) \\ & + (\dot{x}s(\varphi) - \dot{y}c(\varphi))(\theta^2\lambda c(\eta) + \theta) - l_2\dot{\varphi}(-\lambda c(\eta) + \theta) + l_2\dot{\varphi}\theta(\theta\lambda c(\eta) + 1) \\ & - h\lambda s(\eta) + h\dot{\theta}(-\lambda c(\eta) + \theta) + (dv + l_3)\dot{\eta}\lambda]\vec{r}_3\end{aligned} \tag{22}$$

Furthermore, considering that m_r(kg) and I_r(kg m^2) are the mass and the central inertia tensor of the trailer unit, its kinetic energy is written as

$$T_r = 0.5 m_r \vec{v}_{CG_2} \cdot \vec{v}_{CG_2} + 0.5 \overrightarrow{\Omega}_{s/n} I_r \overrightarrow{\Omega^t}_{s/n} \tag{23}$$

Under the same assumption of Sect. 2.1, i.e., the sprung mass swings about the rolling axis, the potential energy of the trailer is

$$V_r = m_r g h_r (c(\lambda) - 1) \tag{24}$$

The last step to be taken before building the equations of motion for the whole vehicle concerns the determination of the non-conservative forces at the trailer tyres, as shown in Fig. 5, and the forces at the trailer suspension.

Following the model proposed by [14] for the computation of forces at the tyres according to Eqs. (10) and (11), briefly described in the previous section, the slip angles at low speeds can be approximated by

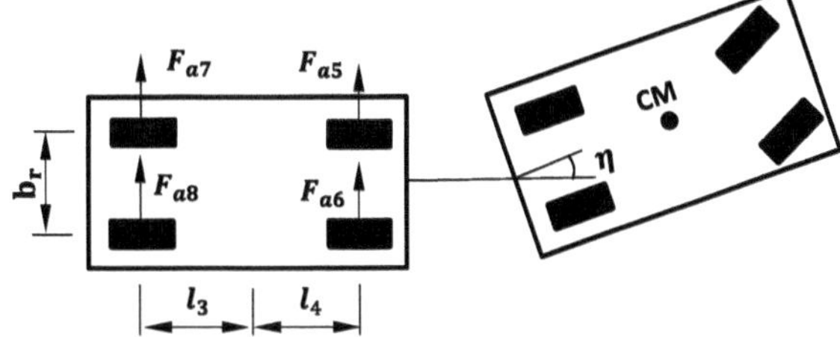

Fig. 5 Lateral forces at the trailer tyres

$$\alpha_j = arctg\left(\frac{\vec{v}_j \cdot \vec{U}_2}{\vec{v}_j \cdot \vec{U}_1}\right), j = \{5, 6, 7, 8\}, \tag{25}$$

in which $\vec{v}_j$ is the velocity of the trailer's j-th tyre, in coordinates of the Σ_U frame, obtained as follows:

$$\begin{cases} \vec{v}_j = \vec{v}_U + \dot{\eta}\vec{U}_3 \wedge (\vec{p}_j - \vec{p}_U) \\ \vec{v}_U = \vec{v}_v + \dot{\eta} \wedge \vec{U}_3\Big(-(dv + l_3)\vec{U}_1\Big) \\ \vec{v}_v = (\dot{x}_n c(\varphi) + \dot{y}_n s(\varphi))\vec{t}_1 + (-\dot{x}_n s(\varphi) + \dot{y}_n c(\varphi))\vec{t}_2 + \dot{\varphi}\vec{t}_3 \wedge (-l_2\vec{t}_1). \end{cases} \tag{26}$$

Trailer suspension forces are calculated according to the linear spring-viscous damping model of Eq. (16), described in Sect. 2.1, as functions of the displacements

$$\epsilon_5 = \epsilon_7 = -\frac{b_r}{2}; \quad \epsilon_6 = \epsilon_8 = \frac{b_r}{2}\lambda. \tag{27}$$

Finally, the Lagrangean for the combined tractor-trailer vehicle can be written as

$$L = T_c + T_r - (V_c + V_r), \tag{28}$$

and the equations of motion follow from Eq. (9) for the vector of generalised coordinates $\boldsymbol{q} = [q_1, q_2, q_3, q_4, q_5, q_6]^T = [x_n, y_n, \theta, \varphi, \eta, \lambda]^T$, that includes the degrees of freedom from both units, and from Eq. (18), for the non-conservative generalised forces, with the summations $j = 1, \ldots, 8,$ and $i = 1, \ldots, 6$. Both the algebraic derivation of the equations of motion and their simulation were performed with the aid of the software Mathematica.

Table 1 Tractor vehicle parameters

Param.	m_c (kg)	h (m)	z (m)	l_1 (m)	l_2 (m)	d_c (N s/m)
Value	8400	0.5	1.2	2.5	1.5	9080
Param.	b (m)	k (N/m)	Ix (kg m^2)	Iy (kg m^2)	I_z (kg m^2)	CS (N/rad)
Value	2.0	5×10^5	12,447	65,735	65,735	14,330

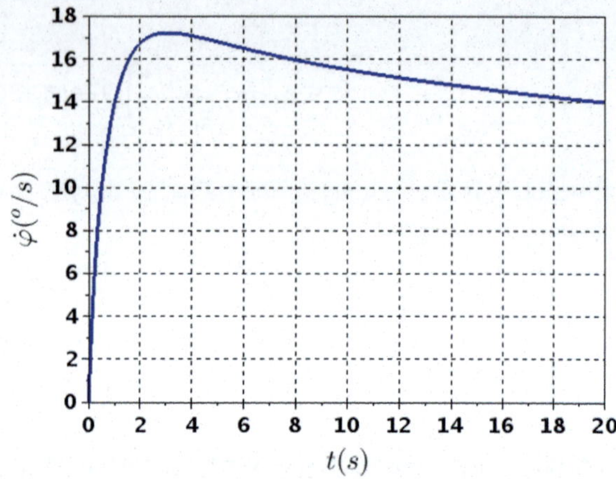

Fig. 6 Tractor yaw rate versus time for the constant steering wheel angle manoeuvre

3 Results and Discussion

Firstly, the dynamic behaviour of the tractor unit without the trailer was evaluated in a simulated curve at a constant velocity of 36 km/h. In order to perform the manoeuvre, a constant steering angle of 4° is imposed at the front wheels through a step function. Simulation parameters of a medium-weight commercial vehicle, given in Table 1, were obtained from [7].

Results for this simulation are depicted in Figs. 6 and 7, from which it is possible to assert that the response of the dynamical model is coherent with the expected behaviour of a real vehicle in a similar manoeuvre. The initial gradient of the yaw rate (Fig. 6) is justified, since the vehicle is suddenly forced to detour from a straight line path until achieving a steady value ($d\varphi/dt \simeq 15$ °/s), which corresponds to the constant steering input. The roll angle and roll rate (Fig. 7) are also compatible with the input condition: from the beginning of the manoeuvre, the vehicle leans outwards at the curve, as evidenced by the values of the roll angle θ, whereas, at the same time, roll rate $d\theta/dt$ presents oscillations that vanish once the steady condition is reached, at $\theta \simeq 0.9°$, after about 20 s.

Then, the dynamical response of the complete vehicle was obtained for two manoeuvres that frequently occur in normal traffic conditions, namely, performing a curve with the same characteristics as the one previously described, and a sudden change of direction from a straight line path, both at constant velocity (36 km/h). In

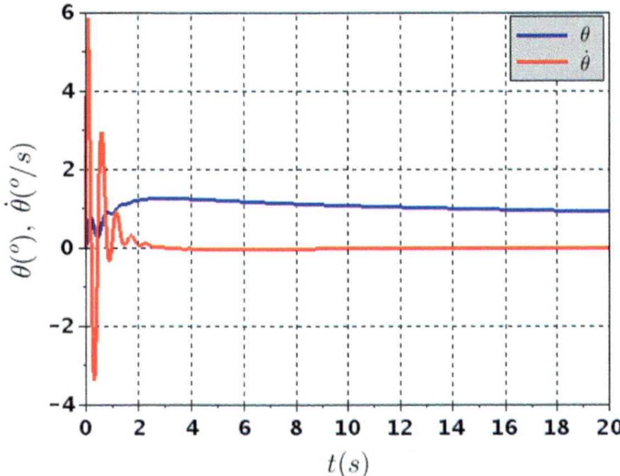

Fig. 7 Tractor roll angle and roll rate versus time for the constant steering wheel angle manoeuvre

Table 2 Trailer vehicle parameters

Param.	m_r^* (kg)	h_r (m)	z_r (m)	$l_3 = l_4$ (m)	CS_r (N/rad)
Value	4200	0.5	1.2	2.5	1.5
Param.	b_r (m)	k_r (N/m)	Ix (kg m^2)	Iy (kg m^2)	I_z (kg m^2)
Value	2.0	5×10^5	6624	32,868	32,868

Values with an asterisk were adapted from [7] for the specific purpose of this work

the latter case, a 30° amplitude steering angle was imposed at the front wheels of the tractor through a step function. Parameters for this simulation can be found in Table 2, in which trailer values are identified by the "r" subscript.

Temporal evolutions of the tractor and trailer roll amplitudes for the constant steering wheel angle manoeuvre are depicted in Figs. 8 and 9. It is possible to

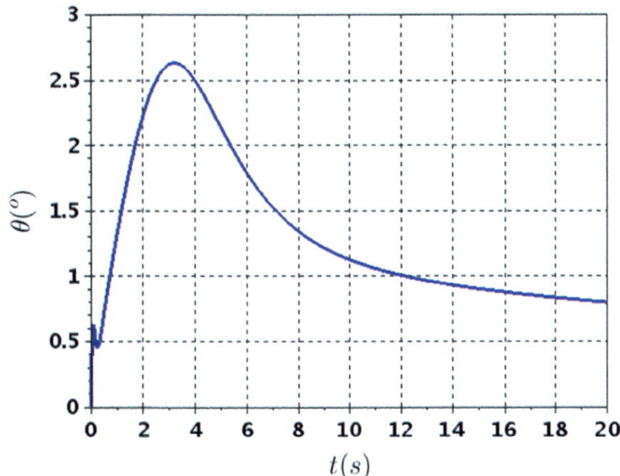

Fig. 8 Tractor roll angle versus time for the constant steering wheel angle manoeuvre

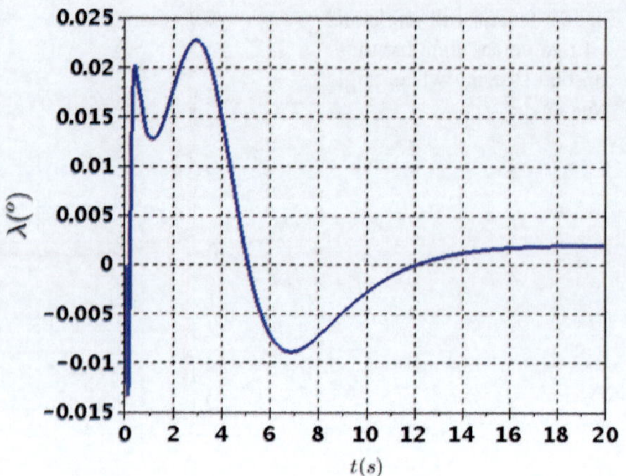

Fig. 9 Trailer roll angle versus time for the constant steering wheel angle manoeuvre

realize that the trailer, constrained to the tractor unit, imposes an inertial resistance to the rolling tendency of the latter when the convoy achieves a steady condition. This assertion is corroborated by comparing results from Figs. 7 and 8. In Fig. 7, the steady roll angle for the tractor is $\theta \simeq 0.9°$ whereas, for the convoy, $\theta \simeq 0.8°$. However, during the transient that follows the imposition of the step function, the tractor roll amplitude reaches a peak value of about 2.6° (Fig. 8), against about 1.2° (Fig. 7) when the same vehicle was simulated without the trailer. The roll angle of the trailer initially accompanies the movement of the tractor and, due to the action of the suspension, oscillates during the transient phase until stabilizing after about 20 s, with the unit leaning outwards, as expected.

Results of the simulated sudden change of direction from a straight line path are depicted in Figs. 10 and 11. Although the step steering amplitude of 30° was imposed during 2 s, starting at the abscissa $t = 3$ s in Fig. 10, the whole convoy

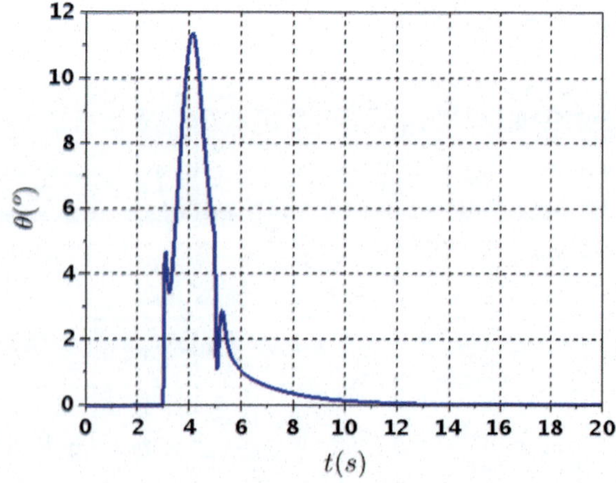

Fig. 10 Tractor roll angle versus time for the change of direction manoeuvre

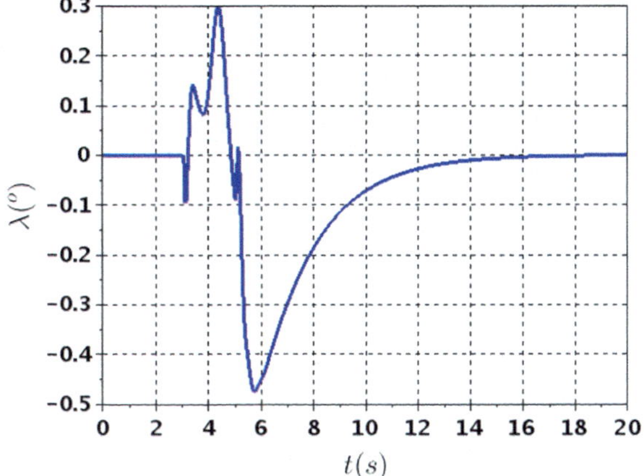

Fig. 11 Trailer roll angle versus time for the change of direction manoeuvre

took about 20 s to fully stabilize at the new heading. The tractor leaning outwards and the trailer oscillating behaviour (Fig. 11) were compatible with the characteristics of the manoeuvre. It must be noted, though, that the peak of roll amplitudes of both vehicles, when compared to those obtained in the constant steering wheel angle condition, were considerably higher, namely, about 4 and 10 times, respectively for the tractor and trailer.

Likewise, it is possible to infer that the combined inertial effects of both units imposes a limiting condition on the safe operation of the convoy, since roll angle amplitude for the tractor unit alone (Fig. 7) was about 1/2 times that of the same vehicle constrained to the trailer, as shown in Fig. 8. In practice, the maximum constant velocity at which the manoeuvres are performed must be smaller for the convoy, which is, again, a result that supports the behaviour of similar vehicles under actual driving conditions.

4 Conclusions

In this work, a six-degree of freedom model of the rolling dynamics of a convoy vehicle was analytically obtained through the Lagrangean formalism. In order to cope with the complexity of the six nonlinear second-order coupled differential equation, the model was linearized for small rolling angles, thus helping its numerical integration. Still, simulated constant radius curve and sudden change of direction manoeuvres, both at constant velocity, exhibited results compatible to those observed in actual vehicles under similar operating conditions, thus stating the efficacy of the proposed model. Altogether, the transient response of the convoy evidenced the influence of the trailer on the behaviour of the tractor unit, since the peaks of roll amplitudes of the latter, especially in the change direction manoeuvre, were considerably higher, thus suggesting its susceptibility to experience unstable behaviour during that phase.

It should be pointed out that, due to the coherent results presented despite its simplicity, the model here developed might be used in parallel with some commercial multibody dynamics software in order to perform a cross-check on the outcomes of simulated conditions on convoy vehicles, since even recent publications rely solely on results provided by the above mentioned proprietary packages. Moreover, it is the intention of the authors to include combined effects of rolling and longitudinal dynamics (jackknife) in the analytical model and to analyse the overall stability, a key issue, for instance, in the development of active control systems.

References

1. Banco Nacional de Desenvolvimento Econômico e Social, O transporte rodoviário de Carga e o Papel do BNDES, BNDES, Brasil (2008)
2. Copeza: http://www.copeza.com.br. Accessed 3 Nov 2016
3. Iveco: http://www.iveco.com. Accessed 12 Sept 2015
4. Rodoclara: http://www.rodoclara.com.br. Accessed 3 Nov 2016
5. Rodocentro: http://www.rodocentroms.com.br. Accessed 3 Nov 2016
6. Vieira, J.L.M.: Estudo de dirigibilidade de veículos longos combinados, Dissertação (Mestrado). Departamento de Engenharia Mecânica, EESC-USP, São Carlos, SP, Brasil (2010)
7. Chen, C., Tomizuka, M.: Lateral control of commercial heavy vehicles. Veh. Syst. Dyn. **33** (6), 391–420 (2000)
8. Gäfvert, M., Lindgärde, O.: A 9-DOF tractor-semitrailer dynamic handling model for advanced chassis control studies. Veh. Syst. Dyn. **41**(1), 51–82 (2004)
9. Luijten, M.F.J.: Lateral dynamic behaviour of articulated commercial vehicles. Master Thesis, Eidhoven University of Technology, Eidhoven, The Netherlands (2010)
10. Zhou, S., Zhang, S.: Assessing the effect of chassis torsional stiffness on tractor semi-trailer rollover. Appl. Math. Inf. Sci. **7**(2), 633–637 (2013)
11. Canale, A.C., Alvarenga, G.S., Viveros, H.P.: Modelagem e análise de um veículo articulado utilizando a técnica dos multicorpos de Simmechanics em Matlab/Simulink®, Minerva **6**(3), 279–286 (2009)
12. Greenwood, D.T.: Classical Dynamics. Dover Books on Physics, USA (1997)
13. Gillespie, T.D.: Fundamentals of Vehicle Dynamics. Society of Automotive Engineers, USA (1992)
14. Baraket, Z., Facnher, P.: Representation of truck tire properties in braking and handling studies: the influence of pavement and tire conditions on frictional characteristics. UMITRI-89-33, University of Michigan, USA (1989)

Use of Integrated Control to Enhance the Safety of Vehicles in Run-Off-Road Scenarios

Abel Castro, Rafael B. Chaves, Georg Rill and Hans I. Weber

Abstract In this work, an integrated vehicle control system (IC) is tested in run-off-road scenarios. The integrated approach was employed in order to coordinate vehicle control systems, i.e. the Anti-Lock Brake System (ABS), Four-wheel Steering (4WS) and the Electronic Stability Program (ESP). To perform a run-off-road maneuver, a fuzzy virtual test driver was designed. By receiving the lateral position of an obstacle and the vehicle's relative yaw angle, the virtual test driver is capable of following a reference trajectory. Furthermore, to test the performance of the standalone controllers, i.e. ABS, ESP and 4WS, individual maneuvers are performed using a multibody vehicle model. The vehicle without any coordination between the control systems is used as reference. For the simulation results, it is concluded that the IC improves the vehicle stability and maneuverability in comparison with the non-integrated approach.

Keywords Integrated control · Multibody vehicle model · Fuzzy control Run-off-road scenarios

1 Introduction

In spite of the decreasing number of single vehicle accidents (SVA) during the last decade, it still represents a high percentage of road fatalities as shown on the left side of Fig. 1. Due to this fact, there is a constant concern of the society about the mobility and traffic safety. Therefore, the development of control systems that can reduce the number of road fatalities is of high priority for the automotive industry and engineers. Moreover, since the introduction of the Anti-lock Braking System (ABS) in 1978 by Bosch, the number of electronic systems to assist drivers and improve the vehicle safety is increasing. For instance, the Electronic Stability Program (ESP) is

A. Castro (✉) · R. B. Chaves · H. I. Weber
Pontifical Catholic University of Rio de Janeiro, Rio de Janeiro, Brazil
e-mail: abel.arrieta@aluno.puc-rio.br

R. B. Chaves · G. Rill
Regensburg University of Applied Sciences, Regensburg, Germany

A. de T. Fleury et al. (eds.), *Proceedings of DINAME 2017*, Lecture Notes in Mechanical Engineering, https://doi.org/10.1007/978-3-319-91217-2_30

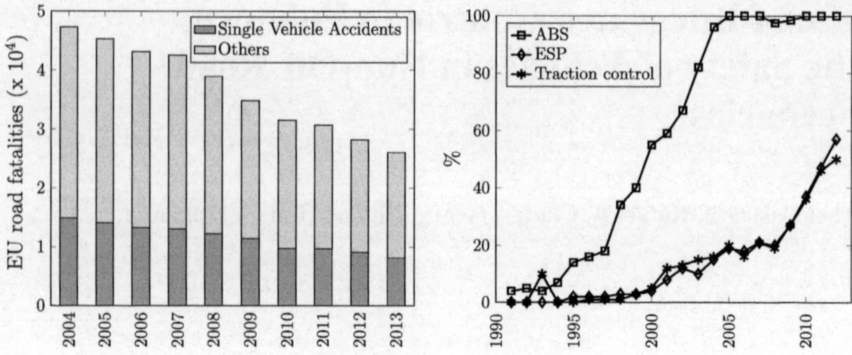

Fig. 1 **Left:** Road fatalities in Europe. **Right:** Safety systems using by top 50 selling cars

capable of reducing in 49% SVA [4]. Consequently, many companies are fitting its best selling cars with safety technologies [10] as depicted on the right side of Fig. 1.

In instability scenarios, the vehicle does not follow the driver's commands. In these cases, the active safety systems should actuate in order to return the vehicle to a maneuverability condition [1, 2]. This task is done by exploiting the tire limits. Nevertheless, if these controllers try to use the tire potential at the same time, this may lead to a conflict that could become a worse scenario for the passengers. Therefore, it is important to design a system to integrate these active safety devices in order to improve the vehicle's stability. In a previous study [3], it was shown that the integration of the Active Front Steering (AFS) and the ESP can improve the vehicle lateral stability in critical scenarios. In this work, an integration control strategy (IC) was designed in order to coordinate the Four-Wheel Steering System (4WS), the ABS and the ESP. Finally, to prove the benefits of the IC system against the non-integrated approach, simulations using a fully nonlinear and three-dimensional vehicle model in run-off-road scenarios were performed.

2 Road Vehicle and Tire Modeling

2.1 *Fully Non-linear Road Vehicle Model*

In order to describe the complete vehicle dynamics, the multibody approach is employed. Using this method, it is possible to model the vehicle by subsystems as presented in [7, 8]. In this work, nine rigid bodies are used, i.e. 4 knuckles, 4 wheel-tires and 1 chassis, see Fig. 2. The equations of motion are obtained using the Jourdain's Principle, also called the method of virtual power. Finally, the vehicle dynamics is characterized by a set of nonlinear first order differential equations.

$$K(y)\,\dot{y} = z\,, \quad M(y)\,\dot{z} = q(y,z,s,u)\,, \quad \dot{s} = f(y,z,s,u), \tag{1}$$

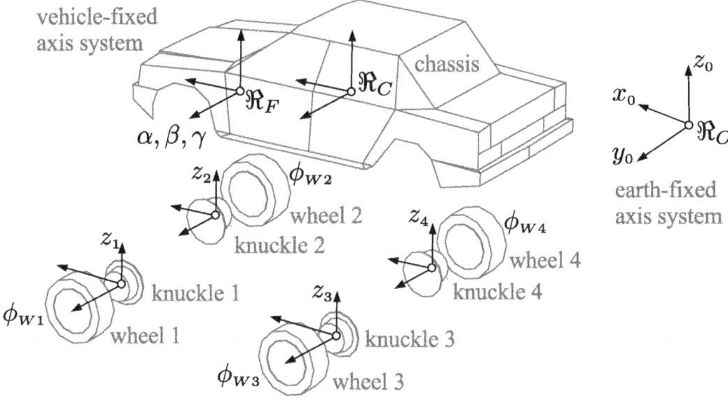

Fig. 2 Rigid bodies and coordinate axis systems of a multibody vehicle model

where y is a vector that collects the generalized coordinates of the vehicle and K is the kinematic matrix used to define an appropriate vector of generalized velocities z. The mass matrix of the multibody vehicle model is denoted by M. The vector of generalized forces q is a function that depends of the input u and additional states s. Furthermore, the vector s collects the internal states of the dynamic force elements and tire deflections respectively.

2.2 *Simple Handling Model*

For control design, the desired response of the vehicle to the driver inputs is required. Generally, the simple handling vehicle model is used as reference, see Fig. 3. This model captures some important features of the vehicle, e.g. yaw rate $\dot{\psi}$ and the sideslip angle β, that are relevant to analyze its stability.

Figure 3 shows the degrees of freedom (DOF), i.e. the lateral and yaw motions, and assumptions of the simple handling model. These DOF are described by:

$$m\left(v\omega + |v|\dot{\beta}\right) = F_{yf} + F_{yr} \qquad \text{and} \qquad \Theta\dot{\omega} = a_1 F_{yf} - a_2 F_{yr} \tag{2}$$

Furthermore, the tire lateral forces are described as function of the cornering stiffness c_{si} and the lateral slips s_{yi} $(i = f, r)$ as follows:

$$F_{yi} = c_{si} s_{yi} \qquad s_{yf} = -\beta - \frac{a_1}{|v|}\omega + \frac{v}{|v|}\delta_f \qquad s_{yr} = -\beta + \frac{a_2}{|v|}\omega + \frac{v}{|v|}\delta_r \tag{3}$$

Finally, rearranging and simplifying (2) and (3), the equations of motion are:

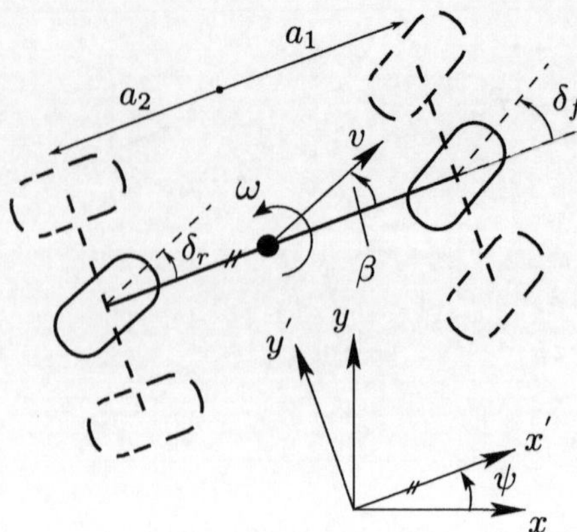

- tires operates in the linear region.
- only planar motion.
- no driving or braking is applied i.e, $\dot{v} = 0$.
- δ_f represents the mean of the front steering angles.
- δ_r represents the mean of the rear steering angles.
- δ_f, δ_r and β are assumed to be small.

Fig. 3 Simple handling vehicle model and list of assumptions

$$\begin{bmatrix} \dot{\beta} \\ \dot{\omega} \end{bmatrix} = \begin{bmatrix} -\frac{c_{sf}+c_{sr}}{m|v|} & \frac{a_2 c_{sr} - a_1 c_{sf}}{m|v||v|} - \frac{v}{|v|} \\ \frac{a_2 c_{sr} - a_1 c_{sf}}{\Theta} & -\frac{a_1^2 c_{sf} + a_2^2 c_{sr}}{\Theta|v|} \end{bmatrix} \begin{bmatrix} \beta \\ \omega \end{bmatrix} + \begin{bmatrix} \frac{v}{|v|}\frac{c_{sf}}{m|v|} & \frac{v}{|v|}\frac{c_{sr}}{m|v|} \\ \frac{v}{|v|}\frac{a_1 c_{sf}}{\Theta} & -\frac{v}{|v|}\frac{a_2 c_{sr}}{\Theta} \end{bmatrix} \begin{bmatrix} \delta_f \\ \delta_r \end{bmatrix} \tag{4}$$

where the time derivative of the yaw angular velocity ω and the sideslip angle β are the output states of the model. The cornering stiffness at the front and rear axle are represented by c_{sf} and c_{sr} respectively. The inertia properties of this model are represented by its mass m, the moment of inertia around the z-axis Θ and the distances a_1 and a_2 from its center of gravity (COG) to the front and rear axle respectively. Finally, the vehicle velocity v, the steering angles δ_f and δ_r are inputs of the simple handling vehicle model.

2.3 TMeasy Tire Model

The tire model easy-to-use or TMeasy [9], is a semi-empirical tire model that uses a small number of parameters to characterize the tire-road contact forces and torques. In addition, TMeasy has a good balance between accuracy and computational processing time. In normal driving maneuvers, e.g. acceleration and deceleration in a curve, the longitudinal slip s_x and lateral slip s_y occur at the same time. Therefore, the combination of tire slips and thus, the longitudinal and lateral forces should be handled by the tire model. In order to consider the contribution of the longitudinal and lateral slip on the combined slip, TMeasy performs a slip normalization process:

$$s = \sqrt{\left(\frac{s_x}{\hat{s}_x}\right)^2 + \left(\frac{s_y}{\hat{s}_y}\right)^2} = \sqrt{\left(s_x^N\right)^2 + \left(s_y^N\right)^2}, \tag{5}$$

where s_x^N and s_y^N are the normalized slips. Furthermore, the normalizing factors $\hat{s}_x$ and $\hat{s}_y$ take into account the longitudinal and lateral force characteristics and are defined as follows:

$$\hat{s}_i = \frac{s_i^M}{s_x^M + s_y^M} + \frac{F_i^M/dF_i^0}{F_x^M/dF_x^0 + F_y^M/dF_y^0} \quad \text{where} \quad i = x, y \tag{6}$$

Similar to the curve of longitudinal and lateral forces, the combined force $F = F(s)$ can be defined by their characteristic parameters dF^0, s^M, F^M, s^S and F^S, and are defined via:

$$\begin{aligned}
dF^0 &= \sqrt{(dF_x^0 \hat{s}_x \cos\phi)^2 + (dF_y^0 \hat{s}_y \sin\phi)^2} \\
s^M &= \sqrt{\left(\frac{s_x^M}{\hat{s}_x}\cos\phi\right)^2 + \left(\frac{s_y^M}{\hat{s}_y}\sin\phi\right)^2} \qquad F^M = \sqrt{\left(F_x^M\cos\phi\right)^2 + \left(F_y^M\sin\phi\right)^2} \\
s^S &= \sqrt{\left(\frac{s_x^S}{\hat{s}_x}\cos\phi\right)^2 + \left(\frac{s_y^S}{\hat{s}_y}\sin\phi\right)^2} \qquad F^S = \sqrt{\left(F_x^S\cos\phi\right)^2 + \left(F_y^S\sin\phi\right)^2}.
\end{aligned} \tag{7}$$

The angular function ϕ is used to guarantee a smooth transition from the longitudinal and lateral force to the combined one. Finally, the longitudinal and lateral result forces are derived from the combined force as follows:

$$F_x = F\cos(\phi) \text{ and } F_y = F\sin(\phi), \text{ where: } \cos(\phi) = \frac{s_x^N}{s} \text{ and } \sin(\phi) = \frac{s_y^N}{s}. \tag{8}$$

Figure 4 shows, on the left side the tire friction limits and on the right side the mutual influence of longitudinal and lateral forces (computed by TMeasy) for a standard commercial tire.

3 Control Design

The ABS, ESP and 4WS control design is explained in this section. For all the simulations performed, a fullsize car was used. Characteristic parameters of this vehicle are defined in Table 1. The road and off-road scenarios are characterized by coefficients of friction of $\mu = 1.0$ and $\mu = 0.4$ respectively. In addition, the road surface is located 5 cm above the off-road surface and an obstacle is placed at a distance of 150 m in front of the vehicle. Avoiding this obstacle will force the vehicle to go off-road. Finally, the driver model was designed using fuzzy logic and simple rules

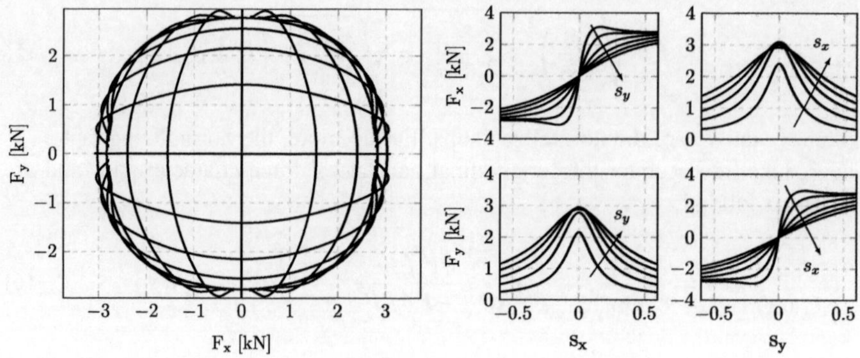

Fig. 4 Left: Tire friction limits. **Right:** Longitudinal and lateral forces

Table 1 Overall characteristic parameters of a fullsize car

Parameter	Value	Units
Mass	2127.8	kg
Inertia at COG	$\begin{bmatrix} 585.9 & 0.0 & 2.4 \\ 0.0 & 3086.4 & 0.0 \\ 2.4 & 0.0 & 3358.3 \end{bmatrix}$	kg m^2
Height of COG	0.55	m
Distance from COG to front/rear axle	1.50/1.40	m
Track width front/rear	1.53/1.52	m
Suspension front/rear	Double wishbone	–
Tires front/rear	P265/40 R18	–

were employed in order to follow a path defined by the road center line. The lateral error and the current orientation of the vehicle are the inputs of the virtual driver.

3.1 Anti-lock Braking System—ABS

The main feature of an ABS model is the capacity to control the slip, avoiding the wheels from locking up, in a region in which the longitudinal tire force is close to the maximum. Therefore, the vehicle's maneuverability and handling stability are retained even during full braking maneuvers [6].

The longitudinal force at each wheel depends of the coefficient of friction μ and wheel load F_z. In this work, μ is assumed to be unknown, for this reason the longitudinal tire slip s_x was taken as control variable and it is defined as follows:

$$s_x = -\frac{(v_x - \Omega r_D)}{|\Omega| r_D + v_N}, \tag{9}$$

where v_x is the vehicle velocity, Ω is the wheel angular velocity, r_D is the tire dynamic radius and v_N is a fictitious velocity, introduced to avoid numerical problems.

The proposed ABS model is based on fuzzy logic due to the highly nonlinear wheel dynamics. This model use three rules, i.e. (1) if s_x is low then increase the braking pressure, (2) if s_x is medium then hold the braking pressure and (3) if s_x is high then decrease the braking pressure. In addition, the fuzzy membership functions are triangular and symmetric where the medium member function is centered at $s_x = 0.15$. This value is because, in standard passenger tires, the maximum longitudinal tire force is obtained at longitudinal slip s_x between 0.1 and 0.2.

In order to prove the effectiveness of the ABS model, two scenarios were simulated. In the first one, a straight line braking maneuver was performed. Using the vehicle with the ABS-off configuration, all wheels have been locked instantly as we can see in the left upper plot (dashed lines) of Fig. 5. On the other hand, with the ABS-on configuration, the wheels are still rolling and in consequence, they have small slip values, i.e. 0.1–0.2 (same plot, solid lines). In addition, the vehicle is reaching decelerations up to $\dot{v}/g = 1.0$, see left bottom plot (solid line) of Fig. 5, which is the limit imposed by the coefficient of friction ($\mu = 1.0$). The last scenario was performed in order to observe the gain in stability using the ABS model. In this simulation, the vehicle is driving in a straight trajectory and at $t = 3.6$ s a steering wheel angle of 30° is applied and then, at $t = 4.0$ s, the brakes are triggered. With the ABS-off, the wheels have been locked again. Therefore, the vehicle follows a straight trajectory because the front wheels are not capable to generate neither lateral nor longitudinal forces. In the case of the vehicle with the ABS-on configuration, it follows the driver's intentions because the front wheels are not locked completely and therefore they can generate lateral forces, see the right multiframe shot of Fig. 5.

3.2 Electronic Stability Program—ESP

The main objective of this system is to assist the driver in critical driving situations, e.g. to avoid an unexpected obstacle on the road [6]. Before the ESP can respond to a critical driving situation, it is necessary to analyze the current state of the vehicle. This analysis takes into account in what direction is the driver steering and the direction in which the vehicle is moving. In order to determine these information, ESP uses the difference between the actual yaw rate ω and the one obtained from the linear vehicle model ω_d (Eq. 4), see Fig. 6. This difference give to the ESP the necessary data to trigger the ABS in order to apply the required brake torque to the selected wheel and then, a compensatory yaw moment T is generated.

In Fig. 7, two simulations are compared. It is possible to distinguish that the torque frequency applied by ESP to the outer front wheel is high when the vehicle is on off-road $t = 5.75 \rightarrow 8$ s. This happens because there is a μ-split condition when the

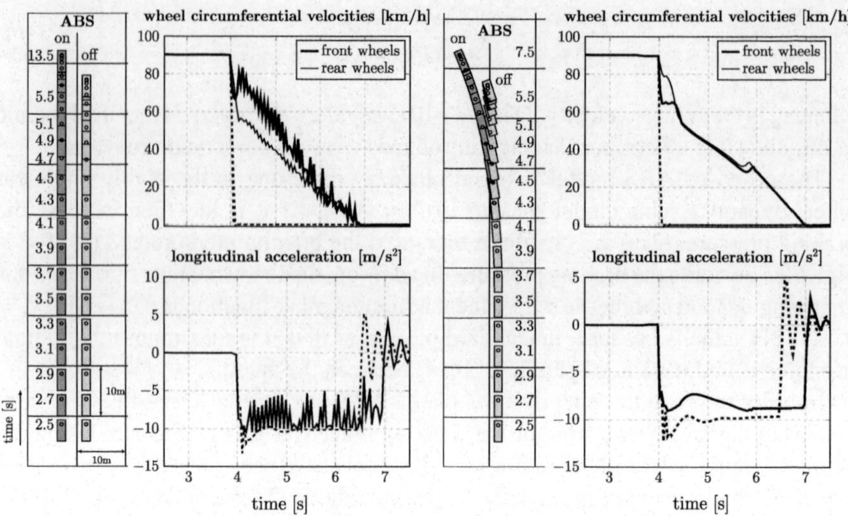

Fig. 5 Braking in a straight line and in a turn (— ABS-on, - - - ABS-off)

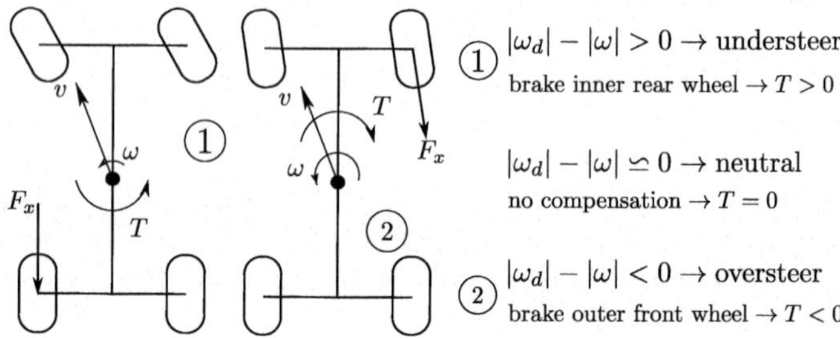

Fig. 6 ESP control process

vehicle goes off-road and comes back to the road again and this requires a compensation torque (longitudinal force due to the braking torque × track width of the axle relative to the selected wheel) to maintain the vehicle's stability. In this period, the yaw angular velocity (left upper plot), lateral acceleration (right upper plot) and the overall yaw moment (left bottom plot) are maintained in a safe range for the ESP. Finally, ESP assisted the driver to maintain the desired path as indicated by the multiframe shot at the very left of Fig. 7.

3.3 Four-Wheel Steering System—4WS

The 4WS was introduced by Nissan in its Skyline model in the late of 1985. The main advantages of this system are: improving the maneuverability at low speeds and the lateral stability at high speeds. In [5], it is concluded that a simple

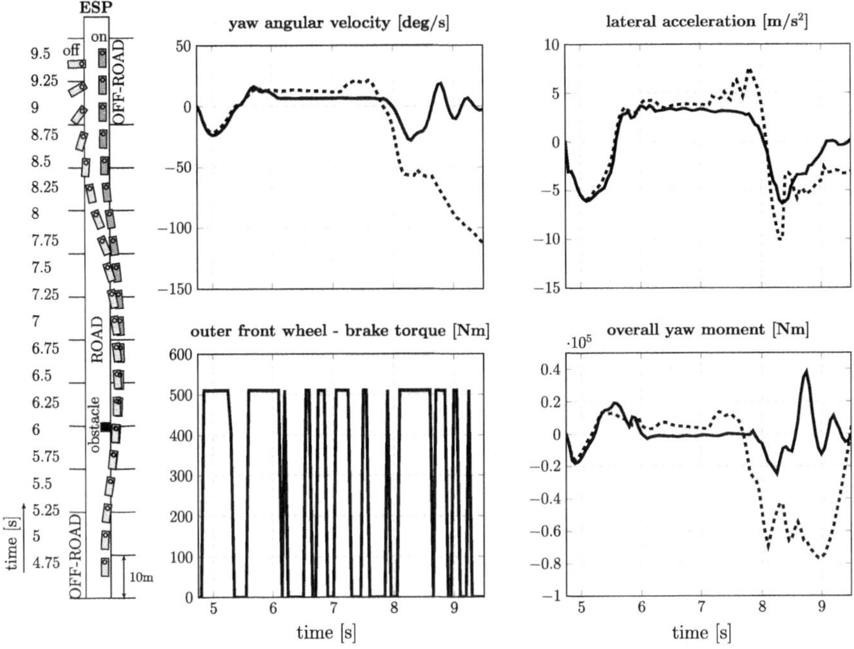

Fig. 7 Trajectory and main states of the vehicle: — ESP-on, - - - ESP-off

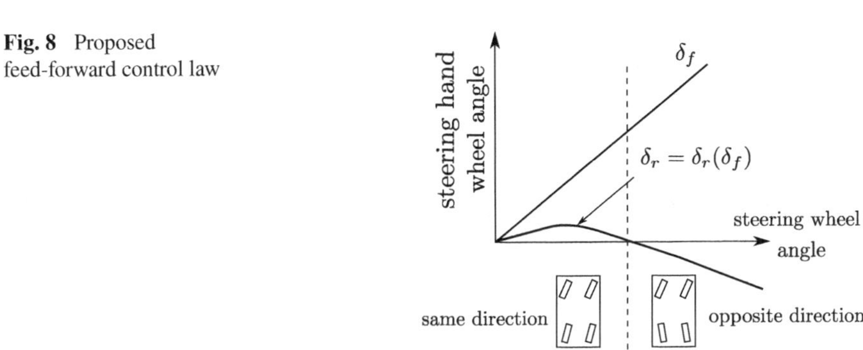

Fig. 8 Proposed feed-forward control law

feed-forward control can improve the vehicle lateral stability. This controller monitors the front steering angle and depending of this value a rear wheel angle is imposed. In this work, a value of 200° ($\delta_f \approx 11.5°$) for the steering hand wheel angle was chosen, see Fig. 8.

Figure 9 shows a simulation of the vehicle with 4WS-on and 4WS-off configuration. From $t \approx 5$–7 s, there is a slight improvement on the vehicle stability with the 4WS-on. A large enhancement can be noticed between $t \approx 7$–9.5 s. In this interval, the yaw angular velocity, lateral acceleration and the overall yaw moment are main-

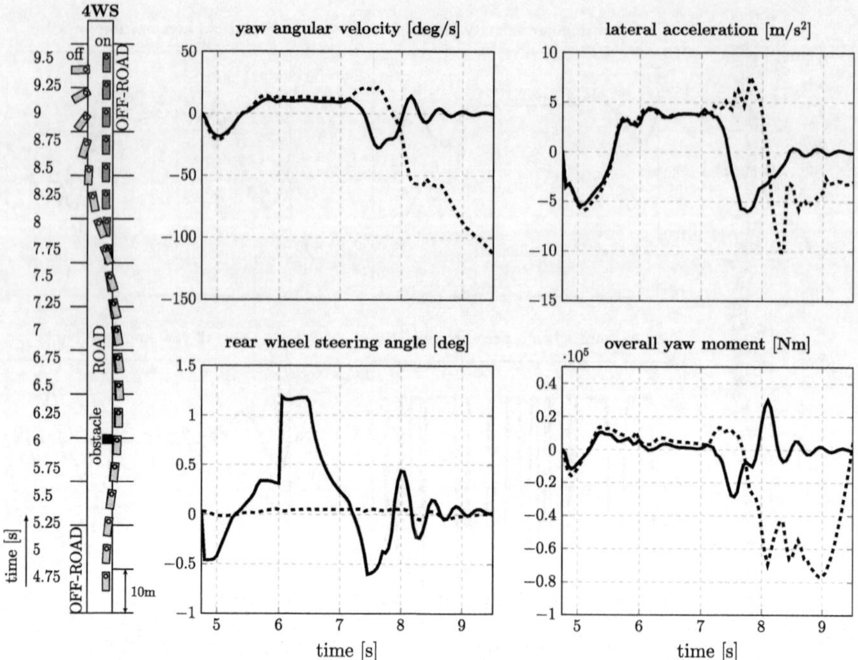

Fig. 9 Trajectory and main states of the vehicle: — 4WS-on, - - - 4WS-off

tained in a safe range due to the 4WS. However, the 4WS system also produces undesirable oscillations on the rear wheels. This behavior is because the virtual driver model is trying to maintain the desired trajectory.

4 Integrated Control

In order to prove the benefits of the integrated vehicle control, two simulation were performed. In the first one, a vehicle with the ESP-on and 4WS-on configuration, i.e. without any integration between them is used, this system will be defined as ESP+4WS. For the second simulation, a rule to avoid the conflict between ESP and 4WS was defined. This rule consists in limiting the use of the braking torque when the rear wheels are steered by a certain angle. In other words, the ESP is applied only when the rear steer angles are bellow a boundary limit. The short name of this integrated control system is IC and it is defined as follows:

$$\text{IC} = \text{4WS-on} + \text{ESP-}\begin{cases} \textbf{on}, & \text{if } |\delta_r| \leq \delta_U \\ \textbf{off}, & \text{otherwise} \end{cases} \tag{10}$$

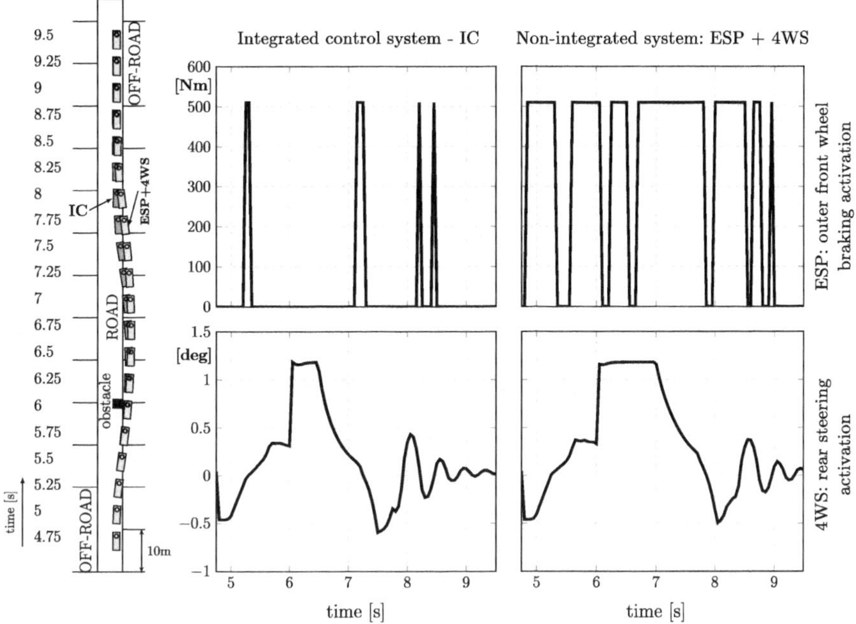

Fig. 10 Multiframe shot and the ESP and 4WS actions of the vehicle with the IC configuration and with the ESP+4WS configuration

where δ_r is the rear steering angle and $\delta_U = 1°$ is the upper bound value.

For a better analysis, the action of each standalone controller, i.e. the ESP and the 4WS, of the IC and ESP+4WS system and the multiframe shot of the vehicle's trajectory are shown in Fig. 10. For the ESP+4WS system (right plots), we can distinguish a conflict between the ESP and 4WS. The reason is because both systems contribute at the same time to the vehicle yaw reaction. In addition, this conflict can be noticed in Fig. 7, where the yaw angular velocity and overall yaw moment are controlled by the ESP without having to steer the rear wheels. This kind of interactions between control systems inside the vehicle can lead to an instability condition. Therefore, in order to avoid this conflict, the rule in Eq. 10 was introduced to limit the functionality of these subsystems. This new strategy is called IC as mentioned above. In the left upper and left bottom plots of Fig. 10, it is noticed that the IC system avoids the conflict between 4WS and ESP in the periods of $t \approx 5.2 \rightarrow 7.2$ s and $t \approx 8.5 \rightarrow 9.5$ s. Furthermore, the lateral deviation of the vehicle using the IC is less than the vehicle equipped with the ESP+4WS system as depicted in the multiframe shot at the very left of Fig. 10.

In Fig. 11, a full comparison between a vehicle equipped with IC and only with ESP is shown. As we can see, in the multiframe shot at the very left of Fig. 11, the lateral deviation of the vehicle with the integrated systems IC is smaller than the vehicle with the ESP-on configuration. This difference is because the 4WS system steers the

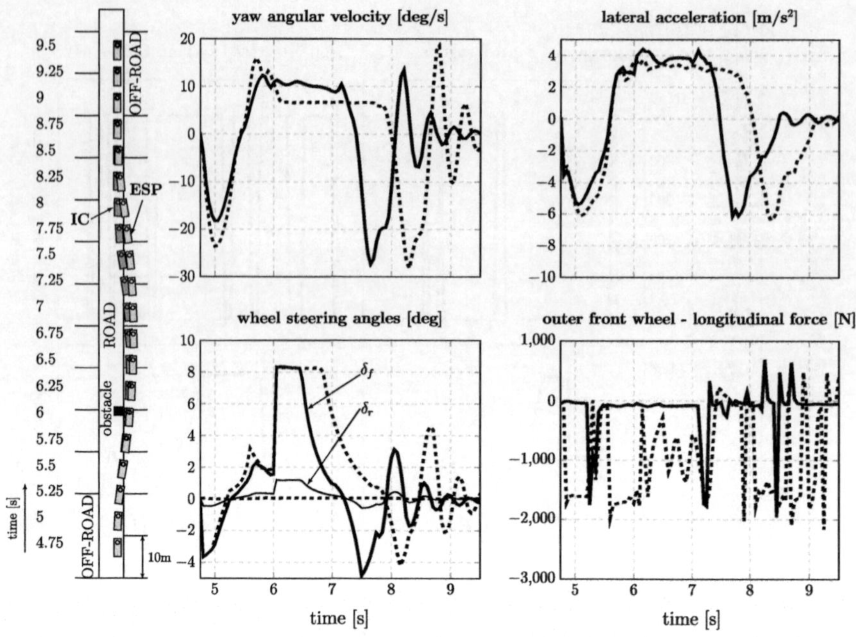

Fig. 11 Trajectory and main states: — IC and - - - ESP-on

rear wheels when the vehicle is in off-road condition, i.e. during $t \approx 5.2 \rightarrow 7.2$ s. The use of rear wheels improves the vehicle lateral stability because it decreases the yaw reaction produced by the front wheels. Therefore, in this period, it is not necessary the use of the ESP in order to produce a compensatory torque and thus, maintain the yaw stability. In the period of $t \approx 8.5 \rightarrow 9.5$ s, an oscillation on the vehicle is produced by the driver model, because it tries to maintain the vehicle close to the center path. In this two periods, the use of a brake torque triggered by the ESP is limited to cases in which the 4WS is not capable to maintain the vehicle's stability. This is easily distinguished because the longitudinal forces at the outer front wheel of the vehicle with the IC system are generated during this periods, i.e. when braking torques are applied. In addition, the vehicle with the IC system use less braking torque actions than with the ESP only, right bottom plot of Fig. 11.

5 Conclusions

The design of individual controllers was done using different methods. The ABS model, based on fuzzy logic, was tested in two scenarios: braking in a straight trajectory and in a curve. The simulations show a good performance of this system, e.g. avoiding wheels from locking up and therefore retaining the vehicle steerability during heavy braking scenarios. The ESP and 4WS models also improve the vehicle stability during the critical driving scenario performed in this work.

Finally, it can be deduced that the direct use of the ESP and 4WS, i.e. the ESP+4WS system, creates conflicts between these systems in the simulated scenario. A simple rule that limits the application of the ESP was defined. From the simulations performed, we can conclude that the proposed integrated control system IC can avoid the conflicts between the ESP and 4WS. Using this strategy, an improvement of the vehicle lateral response and stability was achieved.

References

1. Beal, C.E., Gerdes, J.C.: Model predictive control for vehicle stabilization at the limits of handling. IEEE Trans. Control Syst. Technol. **21**(4), 1258–1269 (2013)
2. Bobier, C.G., Gerdes, J.C.: Staying within the nullcline boundary for vehicle envelope control using a sliding surface. Veh. Syst. Dyn. **51**(2), 199–217 (2013)
3. Castro, A., Rill, G., Weber, H.: Designing an integrated control based on a hierarchical architecture to improve the performance of ground vehicles. In: 23rd ABCM International Congress of Mechanical Engineering (2015)
4. Erke, A.: Effects of electronic stability control (ESC) on accidents: a review of empirical evidence. Accid. Anal. Prev. **40**(1), 167–173 (2008)
5. Furukawa, Y., Yuhara, N., Sano, S., Takeda, H., Matsushita, Y.: A review of four-wheel steering studies from the viewpoint of vehicle dynamics and control. Veh. Syst. Dyn. **18**(1–3), 151–186 (1989)
6. Reif, K.: Fundamentals of Automotive and Engine Technology. Springer, Bosch professional automotive information (2014)
7. Rill, G.: Vehicle modeling by subsystems. J. Braz. Soc. Mech. Sci. Eng. **28**(4), 430–442 (2006)
8. Rill, G.: Road Vehicle Dynamics: Fundamentals and Modeling. CRC Press (2011)
9. Rill, G.: Tmeasy–a handling tire model based on a three-dimensional slip approach. In: Proceedings of the XXIII International Symposium on Dynamic of Vehicles on Roads and on Tracks (IAVSD 2013). Quingdao, China (2013)
10. Vereniging, R.: Mobiliteit in cijfers auto's 2013/2014. Technical Report BOVAG-RAI Mobiliteit (2013)

Yaw Stability Analysis of Articulated Vehicles Using Phase Trajectory Method

André de Souza Mendes, Marko Ackermann, Fabrizio Leonardi and Agenor de Toledo Fleury

Abstract This paper addresses the yaw stability analysis of articulated vehicles using the phase trajectory method. The goal of this work is to ascertain the dynamic conditions that the articulated vehicle can assume without the occurrence of instability events such as jackknife and rollover. The study focuses on the vehicle configuration composed by one tractor unit and a driven unit such as, for instance, a tractor semi-trailer combination. The system consists of a nonlinear tire model and a nonlinear articulated bicycle model with four degrees of freedom. The analysis presented in this paper illustrates the convergence regions of equilibrium points obtained through numerical integration of the equations of motion of the model for different initial conditions in the phase plane. In addition, the changes in the obtained regions are presented as a function of the tractor speed and the position of the articulation point between the two units.

Keywords Vehicle dynamics · Articulated vehicles · Stability analysis · Phase trajectory method · Jackknife

1 Introduction

In heavy weight truck operations, driver and fuel are significant sources of spending. So, in economic terms, the owners seek to transport the maximum amount of cargo with the lowest possible vehicle weight [1]. However, the increase of load capacity

A. de Souza Mendes (✉) · A. de Toledo Fleury
Escola Politécnica da USP, Cidade Universitária, São Paulo, SP, Brazil
e-mail: andremendes@usp.br

A. de Toledo Fleury
e-mail: agenorfleury@usp.br

M. Ackermann · F. Leonardi
Centro Universitário FEI, Av. Humberto de Alencar Castelo Branco, 3972, Assunção, São Bernardo do Campo, SP, Brazil
e-mail: mackermann@fei.edu.br

F. Leonardi
e-mail: fabrizio@fei.edu.br

A. de T. Fleury et al. (eds.), *Proceedings of DINAME 2017*, Lecture Notes in Mechanical Engineering, https://doi.org/10.1007/978-3-319-91217-2_31

implies, in most cases, in the increase of vehicle dimensions which are limited by legislation and structural characteristics of the roads. Very long vehicles are not able to perform tight turns because the rearmost axle tends to move towards the lateral limits of the road not following the path imposed by the front axle. This phenomenon is called offtracking. By splitting long vehicles into several units with shorter wheelbase distances and connecting them through articulations, a higher level of maneuverability can be achieved and tight curves can be made. However, with this vehicle configuration, instability events such as jackknife may occur.

To better understand the yaw instability events associated with articulated vehicles, this paper aims to verify in which dynamic conditions the system, without any input (steering angle, breaking and acceleration), returns to travel in a straight line without the occurrence of any instability phenomenon. Based on this information it is possible to obtain a region in the phase plane that gathers the set of dynamic conditions that satisfy this requirement. Furthermore, the influence of some parameters on the shape of this region should be checked, such as the speed of the tractor and the position of the articulation point.

2 Stability Analysis of Articulated Vehicles

Relatively recent studies present yaw stability analysis of articulated vehicles using linear models [2–4]. The applicability of such models is studied by [5]. The authors verify different linear models of articulated vehicles to be used in stability and dynamic simulation analysis. Dynamic results from single lane changes are compared to those from a nonlinear experimentally validated model developed in the software *TruckSim*. Good agreement was obtained between the linear and nonlinear models for maneuvers with low overall lateral acceleration (0.3 g). When the lateral acceleration exceeds 0.3 g the linear models begin to present significant errors, showing their limitations in reproducing reality. Furthermore, the nonlinear relationship between the lateral force and the slip angle, also known as tire characteristic curve, has a great influence on the dynamic behavior of vehicles [6]. This relationship interferes directly in the stability boundaries of vehicles and its shape depends primarily on the friction coefficient, the vertical load and the longitudinal force of the tire [7]. In an effort to take into consideration the nonlinear behavior of the characteristic curve, many authors use a third order polynomial model [8–10]. An alternative for purely numerical simulation is the semi-empirical model known as *Magic Formula* [7] which is also widely used [11–13].

The stability analysis of nonlinear systems, if not linearized, can be performed using the Lyapunov direct method and the phase trajectories method. The first consists in determining stability based on a fictitious energy function called Lyapunov function. Applications of this method in vehicle systems can be found in [14, 15]. The second method is based on the analysis of the behavior of the orbits of the model states from different initial conditions. Ding et al. [12] use the phase trajectories method to determine the stability regions of a nonlinear model of a tractor-semitrailer combination. Moreover, the authors analyze the variation of the stability region by changing vehicle parameters such as speed and friction coefficient value. However, the model used, although not linear, assumes that the longitudinal speed of the tractor and that of the trailer are equal and constant, which can lead to anomalous results for conditions distant from the equilibrium points. Sun and He [16] use the phase plane to check the stability of nonlinear articulated vehicle model in single lane change maneuvers with open loop sinusoidal steering input. The trajectories of the state variables are compared to the dynamic behavior of a 21 DOF *CarSim* model. Thus, the authors determine the maximum steering value allowed in this maneuver without the occurrence of instability events.

It is possible to notice that there is a trend in the use of nonlinear models of articulated vehicles in stability analysis because, in certain scenarios, linear models are not able to reproduce the vehicle behavior with the necessary accuracy. Moreover, the importance of using an appropriate method for identifying the equilibrium points and regions of stability is evident. A more comprehensive discussion on the present topic can be found in [17].

3 Articulated Vehicle Model

The mathematical model of the tractor-semitrailer combination for yaw stability analysis must be able to represent the dynamic behavior, at least qualitatively, of an actual vehicle for the entire phase plane domain. However, for high values of friction coefficient and high lateral acceleration yaw instability may not occur, resulting instead in rollover event [12]. This type of instability occurs when the lateral acceleration exceeds a certain value, known as rollover threshold. The typical value of this lateral acceleration limit is between 0.2 and 0.5 g for articulated trucks [18]. In contrast, if the friction coefficient is low it is possible to reach the nonlinear region of the tire characteristic curve even under low lateral acceleration. Therefore, for maneuvers at high speed and low friction coefficient the risk of jackknife exists and involves the nonlinear characteristics of the tires. Thus, the tire model used in this paper is the *magic formula* tire model [19] for pure lateral slip, because it is able to adequately represent the tire lateral force for the entire range of slip angles.

The physical model of the vehicle combination is shown in Fig. 1. The points T and S locate the center of gravity (CG) of the tractor and semitrailer, respectively. F, R and M locate the axles of the vehicle and A is the articulation point. The distances a, b and c separate the points F, T, R and A of the tractor and the distances d and e separates the points A, S and M of the semitrailer. The velocity vectors $\mathbf{v}$ and slip angles α receive the subscripts regarding the points to which they are associated.

The modeling of the tractor-semitrailer combination is done considering two rigid bodies moving on a horizontal plane and joined by a single articulation point. Thus, the model has four degrees of freedom and do not take into consideration the roll dynamics. In this case, the rollover threshold is used to determine the occurance of rollover event. The generalized coordinates can be given by x, y, ψ and ϕ. x and y

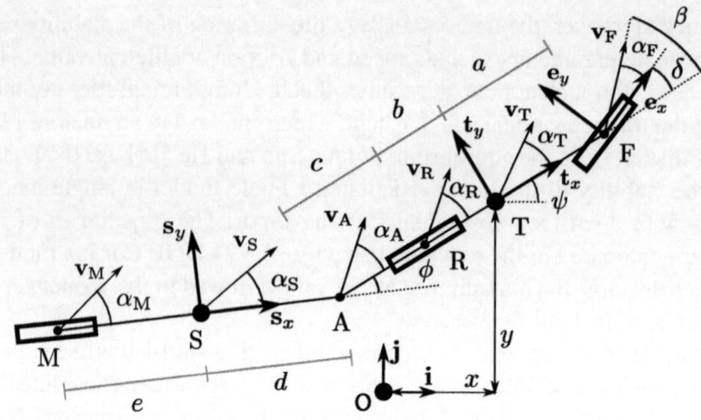

Fig. 1 Single track bicycle model

are the coordinates of the center of gravity of the tractor. ψ is the yaw angle of the tractor and ϕ is the relative yaw angle of the semitrailer. The equations of motion were developed using the Lagrangian approach.

The nonlinear system state equation can be written as

$$\mathbf{M}(\mathbf{x})\,\dot{\mathbf{x}} = \mathbf{f}(\mathbf{x}, \mathbf{u}), \tag{1}$$

where the state vector is

$$\mathbf{x} = \left[\mathrm{x}_1\ \mathrm{x}_2\ \mathrm{x}_3\ \mathrm{x}_4\ \mathrm{x}_5\ \mathrm{x}_6\ \mathrm{x}_7\ \mathrm{x}_8\right]^T = \left[x\ y\ \psi\ \phi\ v_\mathrm{T}\ \alpha_\mathrm{T}\ \dot{\psi}\ \dot{\phi}\right]^T \tag{2}$$

and the input vector is

$$\mathbf{u} = \left[\delta\ F_{x,\mathrm{F}}\ F_{x,\mathrm{R}}\ F_{x,\mathrm{M}}\right]^T. \tag{3}$$

where δ is the steering angle and $F_{x,\mathrm{F}}$, $F_{x,\mathrm{R}}$ and $F_{x,\mathrm{M}}$ are the longitudinal forces at each axle.

The matrix $\mathbf{M}$ is

$$\mathbf{M} = \begin{bmatrix} 1 & 0 & 0 & 0 & 0 & 0 & 0 & 0 \\ 0 & 1 & 0 & 0 & 0 & 0 & 0 & 0 \\ 0 & 0 & 1 & 0 & 0 & 0 & 0 & 0 \\ 0 & 0 & 0 & 1 & 0 & 0 & 0 & 0 \\ 0 & 0 & 0 & 0 & M_{55} & M_{56} & M_{57} & M_{58} \\ 0 & 0 & 0 & 0 & M_{65} & M_{66} & M_{67} & M_{68} \\ 0 & 0 & 0 & 0 & M_{75} & M_{76} & M_{77} & M_{78} \\ 0 & 0 & 0 & 0 & M_{85} & M_{86} & M_{87} & M_{88} \end{bmatrix}, \tag{4}$$

where the elements are given as

$$M_{55} = \left(m_T + m_S\right) \cos\left(\psi + \alpha_{\mathrm{T}}\right) \tag{5}$$
$$M_{56} = -\left(m_T + m_S\right) v_{\mathrm{T}} \sin\left(\psi + \alpha_{\mathrm{T}}\right) \tag{6}$$
$$M_{57} = m_S\left[(b+c)\sin\psi + d\sin(\psi - \phi)\right] \tag{7}$$
$$M_{58} = -m_S d\sin(\psi - \phi) \tag{8}$$
$$M_{65} = \left(m_T + m_S\right) \sin\left(\psi + \alpha_{\mathrm{T}}\right) \tag{9}$$
$$M_{66} = \left(m_T + m_S\right) v_{\mathrm{T}} \cos\left(\psi + \alpha_{\mathrm{T}}\right) \tag{10}$$
$$M_{67} = -m_S\left[(b+c)\cos\psi + d\cos(\psi - \phi)\right] \tag{11}$$
$$M_{68} = m_S d\cos(\psi - \phi) \tag{12}$$
$$M_{75} = -m_S\left[(b+c)\sin\alpha_{\mathrm{T}} + d\sin\left(\alpha_{\mathrm{T}} + \phi\right)\right] \tag{13}$$
$$M_{76} = -m_S\left[(b+c)v_{\mathrm{T}}\cos\alpha_{\mathrm{T}} + d v_{\mathrm{T}}\cos\left(\alpha_{\mathrm{T}} + \phi\right)\right] \tag{14}$$
$$M_{77} = m_S[(b+c)^2 + 2(b+c)d\cos\phi + d^2] + I_T + I_S \tag{15}$$
$$M_{78} = -m_S[(b+c)d\cos\phi + d^2] + I_S \tag{16}$$
$$M_{85} = m_S d\sin\left(\alpha_{\mathrm{T}} + \phi\right) \tag{17}$$
$$M_{86} = m_S d v_{\mathrm{T}}\cos\left(\alpha_{\mathrm{T}} + \phi\right) \tag{18}$$
$$M_{87} = -m_S[d^2 + (b+c)d\cos\phi] + I_S \tag{19}$$
$$M_{88} = \left(m_S d^2 + I_S\right). \tag{20}$$

The vector function $\mathbf{f}$ is given as

$$\mathbf{f} = \left[v_{\mathrm{T}}\cos\left(\psi + \alpha_{\mathrm{T}}\right)\ v_{\mathrm{T}}\sin\left(\psi + \alpha_{\mathrm{T}}\right)\ \dot{\psi}\ \dot{\phi}\ f_5\ f_6\ f_7\ f_8\right]^T, \tag{21}$$

where

$$\begin{aligned} f_5 = {} & F_{x,\mathrm{F}}\cos(\psi + \delta) + F_{x,\mathrm{R}}\cos\psi + F_{x,\mathrm{M}}\cos(\psi - \phi) - F_{y,\mathrm{F}}\sin(\psi + \delta) - \ldots \\ & \ldots - F_{y,\mathrm{R}}\sin\psi - F_{y,\mathrm{M}}\sin(\psi - \phi) - m_S(b+c)\dot{\psi}^2\cos\psi - \ldots \\ & \ldots - m_S d\left(\dot{\psi} - \dot{\phi}\right)^2\cos(\psi - \phi) + \left(m_T + m_S\right)v_{\mathrm{T}}\sin\left(\psi + \alpha_{\mathrm{T}}\right)\dot{\psi} \end{aligned} \tag{22}$$

$$\begin{aligned} f_6 = {} & F_{x,\mathrm{F}}\sin(\psi + \delta) + F_{x,\mathrm{R}}\sin\psi + F_{x,\mathrm{M}}\sin(\psi - \phi) + F_{y,\mathrm{F}}\cos(\psi + \delta) + \ldots \\ & \ldots + F_{y,\mathrm{R}}\cos\psi + F_{y,\mathrm{M}}\cos(\psi - \phi) - m_S(b+c)\dot{\psi}^2\sin\psi - \ldots \\ & \ldots - m_S d\left(\dot{\psi} - \dot{\phi}\right)^2\sin(\psi - \phi) - \left(m_T + m_S\right)v_{\mathrm{T}}\cos\left(\psi + \alpha_{\mathrm{T}}\right)\dot{\psi} \end{aligned} \tag{23}$$

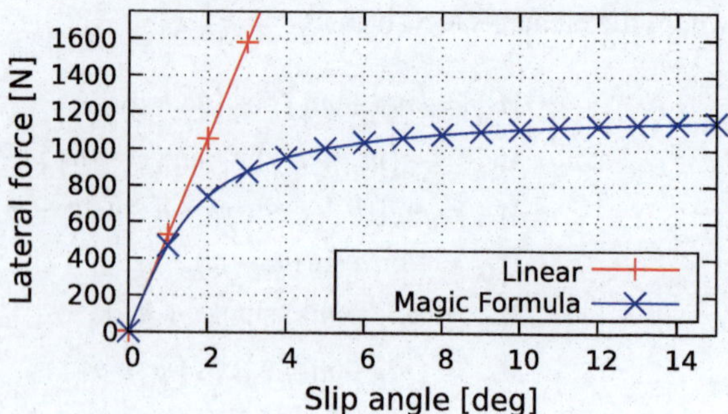

Fig. 2 Characteristic curve of the nonlinear *magic formula* tire model with the parameter values from Table 2, vertical load of $F_z = 4\,\text{kN}$ and road-tire friction coefficient of $\mu = 0.3$. For comparison, the associated linear tire model curve is also plotted

$$
\begin{aligned}
f_7 = {} & F_{x,\text{F}} a \sin\delta + F_{x,\text{M}} (b+c) \sin\phi + F_{y,\text{F}} a \cos\delta - F_{y,\text{R}} b - \dots \\
\dots & - F_{y,\text{M}} \left[(b+c)\cos\phi + (d+e)\right] - m_S (b+c) d \left(\dot{\psi} - \dot{\phi}\right)^2 \sin\phi + \dots \\
\dots & + m_S (b+c) d \dot{\psi}^2 \sin\phi + m_S \left[(b+c) v_\text{T} \cos\alpha_\text{T} + d v_\text{T} \cos\left(\alpha_\text{T} + \phi\right)\right] \dot{\psi}
\end{aligned}
\tag{24}
$$

$$
f_8 = F_{y,\text{M}} (d+e) - m_S (b+c) d \dot{\psi}^2 \sin\phi - m_S d v_\text{T} \cos\left(\alpha_\text{T} + \phi\right) \dot{\psi}. \tag{25}
$$

where $F_{y,\text{F}}$, $F_{y,\text{R}}$ and $F_{y,\text{M}}$ are the lateral forces of each axle given by the *magic formula* tire model [19]. The resulting lateral force (See Fig. 2) is the sum of the contributions of all tires on that axle and the load at each axle is evenly distributed among all tires. For more details about the tire and vehicle models, see [17].

4 Simulation Model

The mathematical models described above are implemented in a open source simulation package called *OpenVD: Open Vehicle Dynamics* [20]. The vehicle parameters of the simulation model are listed in Tables 1 and 2. Thereby, the mass over each axle is given by $m_\text{F} = 6{,}000.0\,\text{kg}$, $m_\text{R} = 10{,}000.0\,\text{kg}$ and $m_\text{M} = 17{,}000.0\,\text{kg}$, which correspond to the weight limits of Brazilian legislation for the number of tires used in each axle. The negative value of c indicates that the articulation point lies within the wheelbase of the tractor.

To illustrate the performance of the simulation model two cases are presented below where two different and arbitrary initial conditions are given to the system with no input. Hence, $\delta = 0\,\text{rad}$, $F_{x,\text{F}} = 0\,\text{N}$, $F_{x,\text{R}} = 0\,\text{N}$ and $F_{x,\text{M}} = 0\,\text{N}$. The integration parameters can be found in Table 3 for both cases.

In Figs. 3 and 4 the vehicle is plotted at different stages of the maneuver. The tractor is represented by the smallest rectangle and the semitrailer by the larger one. The trajectories of each axles are also shown.

In Fig. 3 the vehicle moves sideways and rotates. The relative yaw angle oscillates with decreasing amplitude and converge to zero. Finally, the vehicle travels in a straight line but in a different direction from the initial one. Figure 5a shows the CG

Table 1 Articulated vehicle model—vehicle parameters

Item	Description	Value	Unit
m_T	Mass of the tractor	7,677.0	kg
m_S	Mass of the semitrailer	25,323.0	kg
I_T	Moment of inertia of the tractor	4.61×10^4	kg m^2
I_S	Moment of inertia of the semitrailer	4.52×10^5	kg m^2
a	Distance between points F and T	1.128	m
b	Distance between points T and R	2.422	m
c	Distance between points R and A	−0.31	m
d	Distance between points A and S	4.901	m
e	Distance between points S and M	2.399	m
n_F	Tires at the front axle	2	–
n_R	Tires at the rear axle	4	–
n_M	Tires at the semitrailer axle	8	–
μ	Friction coefficient	0.3	–

Table 2 Articulated vehicle model—*magic formula* tire parameters

Item	Description	Value	Unit
a_0	Shape factor	1.003	–
a_1	Load dependency of friction coefficient	2.014	1/kN
a_2	Friction coefficient level	710.501	–
a_3	Maximum cornering stiffness	5.226×10^3	N/deg
a_4	Load at maximum cornering stiffness	78.877	kN
a_5	Camber sensitivity of cornering stiffness	0.011	1/deg
a_6	Load dependency of E	−0.005	1/kN
a_7	E level	0.670	–

Table 3 Articulated vehicle model—integration parameters

Item	Description	Value (Case 1)	Value (Case 2)	Unit
t	Simulation time	12	15	s
x_0	Initial longitudinal position	0.0	0.0	m
y_0	Initial transversal yaw angle	0.0	0.0	m
ψ_0	Initial yaw angle	0.0	0.0	rad
ϕ_0	Initial relative yaw angle	0.0	0.0	rad
$v_{T,0}$	Initial speed	20.0	20.0	m/s
$\alpha_{T,0}$	Initial vehicle side slip angle	0.3	0.0	rad
$\dot{\psi}_0$	Initial yaw rate	0.25	0.4	rad/s
$\dot{\phi}_0$	Initial relative yaw rate	0.25	0.4	rad/s

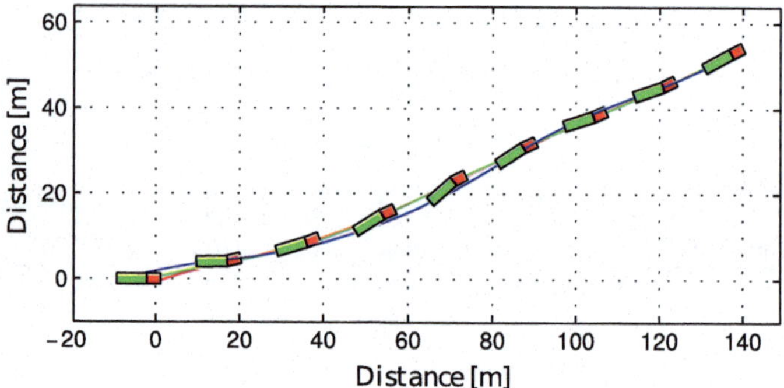

Fig. 3 Successive frames of the articulated vehicle maneuver. Case 1. Without jackknifing (The first 8 s of the maneuver)

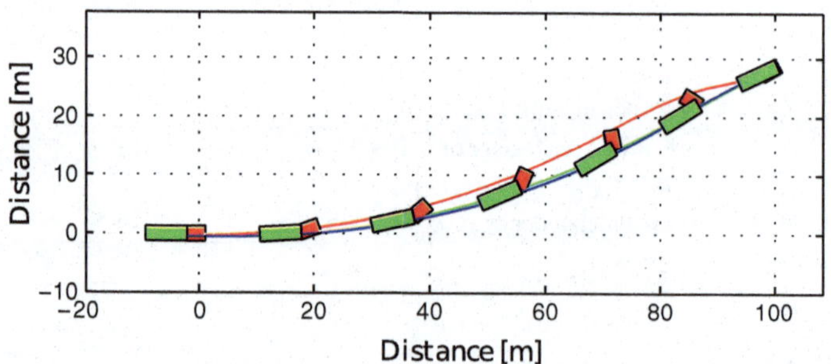

Fig. 4 Successive frames of the articulated vehicle maneuver. Case 2. With jackknifing (The first 6 s of the maneuver)

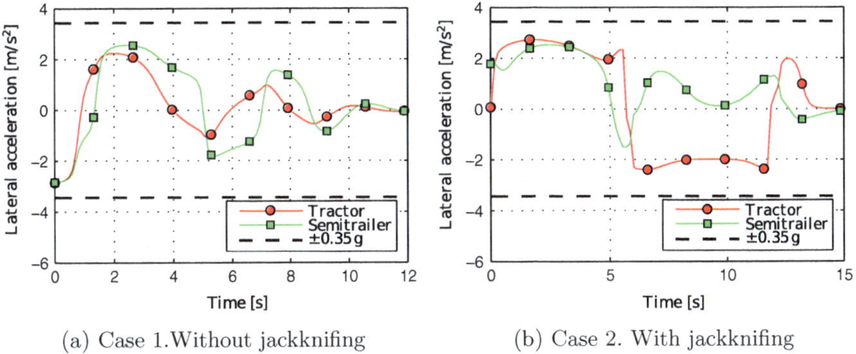

(a) Case 1.Without jackknifing

(b) Case 2. With jackknifing

Fig. 5 Acceleration for both cases

acceleration signal for the first case maneuver. This behavior is considered favorable because there is no jackknife event and the maximum lateral acceleration does not exceed the stipulated rollover threshold (0.35 g).

Figure 4 shows the maneuver in which the lateral force of the tires are not able to prevent the excessive yaw rate of the tractor causing the jackknife. Again, Fig. 5b shows the CG acceleration signal for this maneuver. It is important to note that, in spite of the occurrence of jackknifing, the maximum lateral acceleration is maintained below the rollover threshold during the entire maneuver.

5 Results and Discussion

As discussed so far, the steady-state condition of the system with no input consists in the movement of the vehicle in a straight line with a specific final speed. However, not every final condition of the model is achieved without the occurance of instability events. If α_T converges, for instance, to $+\pi$ rad or $-\pi$ rad the vehicle travels in a straight line but it is evident that the tractor is moving backwards. Therefore, it is possible to trace the convergence region in a phase plane where all the initial conditions converge to the movement of the vehicle in a straight line without the occurrence of jackknife or rollover. Additionally, in the maneuver of interest the value of ϕ should not assume values greater than 90° due to the constructive characteristics of the vehicle. Besides that, the initial relative yaw rate $\dot{\phi}_0$ gets the same value assigned to the initial condition $\dot{\psi}_0$ throughout the phase plane sweep. This feature ensures an initial relative rotation between the two units.

The two most representative yaw-related states, α_T and $\dot{\psi}$, form the phase plane used in this section. All other states and relevant quantities are monitored for assessing the initial conditions. The phase plan was discretized vertically from point $\dot{\psi} = -1.395$ rad/s to the point $\dot{\psi} = 1.395$ rad/s in steps of $\Delta\dot{\psi} = 0.015$ rad/s and in the horizontal direction from the point $\alpha_T = -1.560$ rad to the point $\alpha_T = 1.560$ rad in steps of $\Delta\alpha_T = 0.015$ rad. So, the phase plane grid has 39,083 points. The system is integrated for each initial condition with the vehicle data presented in Tables 1 and 2. The integration time is 20 s with resolution of 0.1 s. The configuration of the states to which the system converges is checked and the initial condition is marked appropriately. The maximum value of the relative yaw angle and the maximum values of the lateral acceleration of the two units are also checked. The lateral acceleration threshold is 0.35 g. If the system converges to the pair $(\alpha_{T,f} = 0, \dot{\psi}_f = 0)$ and the other criteria are also satisfied the initial condition is part of the region of interest. Otherwise it is assigned to one of the adjacent region.

5.1 Varying the CG Velocity

In this section we investigate the variation of the convergence region according to the initial speed of the center of gravity of the tractor $v_{T,0}$. The phase plane sweep is performed for each value of the initial speed. The analyzed speed range goes from $v_{T,0} = 10$ m/s to $v_{T,0} = 30$ m/s in steps of $\Delta v_{T,0} = 2$ m/s, totaling eleven initial speeds. Figure 6 shows the change in the convergence region as a function of the variation of the initial speed of the tractor in the phase plane.

The total area of the convergence region decreases with the increase of the initial speed of the tractor. For positive values of α_T (transversal component of the velocity vector $\mathbf{v}_T$ pointing to the left of the vehicle) the convergence region is, for all speeds, larger when the yaw rate is positive (vehicle turning counter-clockwise) than when the yaw rate is negative (vehicle turning clockwise). This implies that for $\alpha_T > 0$ a small negative initial yaw rate puts the system out of the convergence region. However, the first quadrant region which is larger for smaller initial speeds, has its area significantly reduced as the initial speed increases. Meanwhile, in the fourth

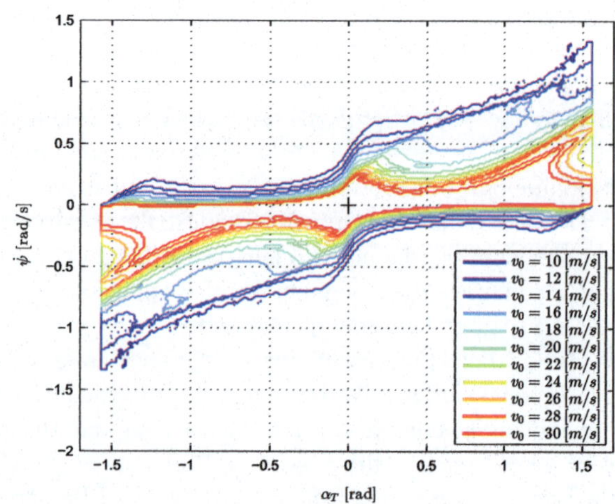

Fig. 6 Convergence region with varying CG velocity. α_T and $\dot{\psi}$ are the side slip angle and yaw rate of the tractor, respectively. Initial conditions inside the convergence region generate maneuvers without the occurance of rollover and jackknife

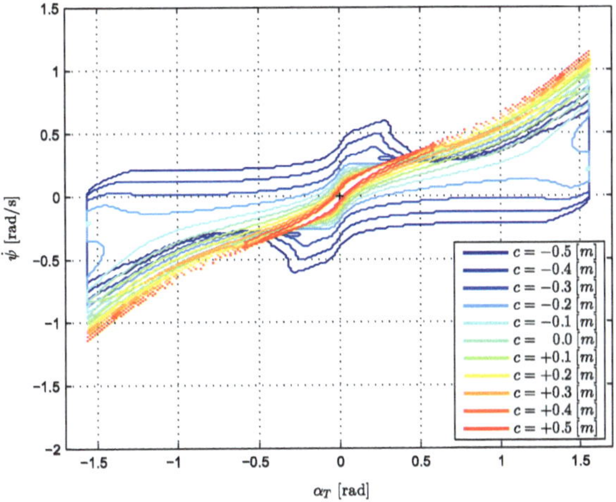

Fig. 7 Convergence region with varying articulation point. α_{T} and $\dot{\psi}$ are the side slip angle and yaw rate of the tractor, respectively. Initial conditions inside the convergence region generate maneuvers without the occurance of rollover and jackknife

quadrant the displacement of the convergence region border is smaller. A similar analysis can be made for negative values α_{T} due to the anti-symmetry of the convergence region.

5.2 *Varying the Articulation Point*

This section shows the variation of the convergence region caused by different logitudinal positions of the articulation point. The values of the quantity that indicates the position of the articulation point in relation to the rear axle of the tractor range from $c = -0.5$ m to $c = +0.5$ m in steps of $\Delta c = 0.1$ m, totaling eleven values. It is important to note that negative values of c indicate that the articulation point lies between the wheelbase of the tractor and positive values indicate that the articulation lies behind the rear axle of the tractor (Fig. 7).

The region with the largest area in the phase plane has the articulation point at $c = -0.5$ m and as the value of c increases the total area decreases until the smallest area at $c = +0.5$ m. Two different phenomenon can be observed with respect to the changes experienced by the region of convergence. The first is the narrowing of the region toward the state $\dot{\psi}$, i.e. for a given state value of α_{T} the set of values of $\dot{\psi}$ contained in the convergence region decreases as the value of c increases. The second phenomenon is the rotation of the convergence region. The region, which for $c = -0.5$ m is approximately horizontal, rotates counter-clockwise and become diagonal when $c = +0.5$ m. A consequence of these rotation is the increase of the range of $\dot{\psi}$ which can contain convergent initial conditions, despite the significant reduction of the total area of the region.

Moreover, as the quantity c increases you can check the fragmentation of the region into several smaller regions, although a larger central region of convergence is maintained. This effect may be due to the phase trajectory method which makes use of the discretization of the phase plane. However, it is also reasonable to assume that the system shows this kind of characteristic.

6 Conclusion

The phase trajectory method allowed the estimation of a convergence region in the phase plane for an articulated vehicle composed by a tractor and a semitrailer. The integration of the system with a initial condition belonging to the interest convergent region produces a maneuver in which the vehicle ends up traveling in a straight line without the occurance of instability events such as jackknife and rollover. Furthermore, the variation of this convergence region is analysed as a function of the tractor speed and the position of the articulation point. In all cases it is possible to observe convergent initial conditions predominantly in the first and third quadrant of the phase plane. As the speed increases and the articulation point moves backwards the convergence region becomes smaller. Particularly, the variation of c produces the narrowing and rotation of the these region.

Acknowledgements These research was supported by the Brazilian research funding agency *Coordenao de Aperfeioamento de Pessoal de Nvel Superior* (CAPES).

References

1. Fancher, P., Winkler, C.: Directional performance issues in evaluation and design of articulated heavy vehicles. Veh. Syst. Dyn. **45**(7-8), 607–647 (2007). https://doi.org/10.1080/00423110701422434
2. Maas, J.W.L.H.: Jackknife stability of a tractor semi-trailer combination. Master's thesis, Technische Universiteit Eindhoven (2007)
3. Hac, A., Fulk, D., Chen, H.: Stability and control considerations of vehicle-trailer combination. SAE Int. J. Passeng. Cars Mech. Syst. **1**, 925–937 (2008). https://doi.org/10.4271/2008-01-1228
4. Luijten, M.F.J.: Lateral dynamic behaviour of articulated commercial vehicles. Master's thesis, Technische Universiteit Eindhoven (2010)
5. Islam, M.M., He, Y., Zhu, S., Wang, Q.: A comparative study of multi-trailer articulated heavy-vehicle models. Proc. Inst. Mech. Eng. Part D: J. Automob. Eng. **229**(9), 1200–1228 (2015). https://doi.org/10.1177/0954407014557053
6. Ervin, R.D., Nisonger, R.L., Mallikarjunarao, C., Gillespie, T.D.: The yaw stability of tractor-semitrailers during cornering. Technical Report, UM-HSRI-79-21-2, The University of Michigan. Highway Safety Research Institute, The University of Michigan. Highway Safety Research Institute (1979)
7. Pacejka, H.B.: Tyre and Vehicle Dynamics, 2nd edn. Butterworth-Heinemann, Oxford (2006)
8. Johnson, D.B., Huston, J.C.: Nonlinear lateral stability analysis of road vehicles using liapunov's second method. In: SAE Technical Paper. SAE International (1984). https://doi.org/10.4271/841057

9. Samsundar, J., Huston, J.C.: Estimating lateral stability region of a nonlinear 2 degree-of-freedom vehicle. In: SAE Technical Paper. SAE International (1998). https://doi.org/10.4271/981172
10. Sadri, S., Wu, C.: Stability analysis of a nonlinear vehicle model in plane motion using the concept of Lyapunov exponents. Veh. Syst. Dyn. **51**(6), 906–924 (2013). https://doi.org/10.1080/00423114.2013.771785
11. Stotsky, A., Hu, X.: Stability analysis of robustly decoupled car steering system with nonlinear tire model. In: Proceedings of the 36th IEEE Conference on Decision and Control, vol. 5, pp. 4750–4755 (1997). https://doi.org/10.1109/CDC.1997.649761
12. Ding, N., Shi, X., Zhang, Y., Chen, W.: Analysis of bifurcation and stability for a tractor semi-trailer in planar motion. Veh. Syst. Dyn. **52**(12), 1729–1751 (2014). https://doi.org/10.1080/00423114.2014.960431
13. Wideberg, J., Dahlberg, E., Svensson, M.: A comparative study of legislation and stability measures of heavy articulated vehicles in different regions. Int. J. Heavy Veh. Syst. **16**(3), 354–361 (2009). https://doi.org/10.1504/IJHVS.2009.027138
14. Yin, G., Jin, X., Qing, Z., Bian, C.: Lateral stability region conservativeness estimation and torque distribution for FWIA electric vehicle steering. Sci. China Technol. Sci. **58**(4), 669–676 (2015). https://doi.org/10.1007/s11431-014-5687-x
15. Yan, Y., Xu, H., Liu, H.: Estimating Vehicle Stability Region Based on Energy Function. Discret. Dyn. Nat. Soc. **2015**, 1–7 (2015). https://doi.org/10.1155/2015/805063
16. Sun, T., He, Y.: Phase-plane analysis for evaluating the lateral stability of articulated vehicles. In: SAE Technical Paper. SAE International (2015). https://doi.org/10.4271/2015-01-1574
17. Mendes, A.S.: Análise de estabilidade em guinada de veículos articulados (in english: Yaw stability analysis of articulated vehicles). Master's thesis, Centro Universitário FEI (2016)
18. Winkler, C.B., Ervin, R.: Rollover of heavy commercial vehicles. Technical Report, UMTRI-99-19, The University of Michigan. Transportation Research Institute, The University of Michigan. Transportation Research Institute (1999)
19. Bakker, E., Pacejka, H.B., Lidner, L.: A new tire model with an application in vehicle dynamics studies. In: SAE Technical Paper. SAE International (1989). https://doi.org/10.4271/890087
20. Mendes, A.S., Meneghetti, D.R.: OpenVD (2017). https://github.com/andresmendes/openvd

The Influence of Inertial Forces on Manual Wheelchair Propulsion

Alberto Amancio Jr., Fabrizio Leonardi, Agenor de Toleto Fleury and Marko Ackermann

Abstract Both experimental and computational studies have contributed to the understanding of the loads during wheelchair propulsion and the factors leading to the incidence of musculoskeletal disorders. However, few studies have addressed the influence of inertial forces on wheelchair propulsion, which are potentially large as upper limb segments undergo large accelerations along the different phases of the propulsion cycle. This study determines and investigates the influence of inertial forces during manual wheelchair propulsion for a subject at two different locomotion velocities. The isolated influence of inertial as well as gravitational forces is determined using a planar model of the upper extremity and an inverse-dynamics approach. The results show that the inertial forces are preponderant even at lower speeds. These findings evidence that quasi-static models are inappropriate to investigate wheelchair propulsion and show the importance of accurate estimation of anthropometric parameters such as segment masses and moments of inertia, which directly affect inertial force estimations in inverse dynamics-based studies of wheelchair propulsion. The results can also help guide investigations on efficient propulsion techniques, as they show that the radial component of the pushrim forces are, to a large extent, determined by inertial effects rather than by an inefficient propulsion technique.

Keywords Wheelchair · Propulsion · Biomechanics

A. Amancio Jr. (✉) · F. Leonardi · M. Ackermann
FEI University, São Bernardo do Campo, Brazil
e-mail: betoamjr@gmail.com

F. Leonardi
e-mail: fabrizio@fei.edu.br

M. Ackermann
e-mail: mackermann@fei.edu.br

A. de Toleto Fleury
Polytechnic School, University of São Paulo, São Paulo, Brazil
e-mail: agfleury@usp.br

A. de T. Fleury et al. (eds.), *Proceedings of DINAME 2017*, Lecture Notes in Mechanical Engineering, https://doi.org/10.1007/978-3-319-91217-2_32

1 Introduction

According to the World Health Organization (WHO), approximately 1% of the world's population needs wheelchairs. Unfortunately, wheelchair locomotion is an inefficient means of locomotion because of the biomechanics of manual wheelchair propulsion [1]. Furthermore, due to the large and repetitive loads on the upper limbs, the incidence of upper extremity pain and injury in long-term users is high [2]. According to [3], the most affected area is the shoulder.

Both experimental and computational studies have contributed to the understanding of the loads during propulsion and the factors leading to the incidence of musculoskeletal disorders [1–5]. Many studies have focused on the reduction of the required demand on the upper limbs through modifications in the wheelchair configuration and in the propulsion technique [3]. Few studies, however, have addressed the influence of inertial forces on wheelchair propulsion, which are potentially large as upper limb segments undergo large accelerations along the different phases of the propulsion cycle [5]. With the increasing accuracy of musculoskeletal models and the growing computational power, computational simulations permit the virtual testing in various scenarios [4, 5].

This study investigates the influence of inertial forces during manual wheelchair propulsion for a subject at two different locomotion velocities (approximately 1.3 and 2.0 m/s). The influence of inertial, gravitational and muscle forces is determined using a planar model of the upper extremity, an inverse-dynamics approach and static optimization. The propulsion patterns are measured in a motion analysis laboratory. The upper limb and wheelchair kinematics are computed from videos acquired by means of a camera and the hand forces by an instrumented pushrim [6].

2 Methods

2.1 Mechanical Model

We employed a planar multibody model of the upper limb composed of two rigid bodies representing forearm and arm (Fig. 1). The shoulder and the elbow are modeled as ideal hinge joints driven by total active joint moments, τ_s and τ_e, respectively, due to the muscles crossing these articulations. The masses, moments of inertia, center of mass locations and segment lengths are estimated using anthropometric data from a scaled OpenSim model [7, 8] for a 1.69 m, 69.5 kg person, which corresponds to the stature and weight of the subject. The adopted generalized coordinates q are the angle between the upper arm and the vertical β, the angle between the forearm and the upper arm α, and the horizontal displacement of the wheelchair and shoulder joint, x, as $q = [x\ \beta\ \alpha]^{\mathrm{T}}$. In the propulsion phase, as in Fig. 1, the hands P are in contact with the pushrims and the contact forces F_x and F_y arise, which are measured by the instrumented wheel.

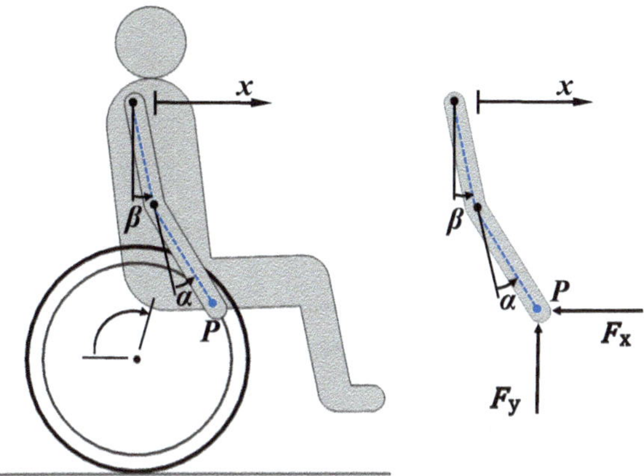

Fig. 1 Upper limb model

The equations of motion of the wheelchair-user system depicted in Fig. 1 were derived using the Newton-Euler Formalism [9] incorporating the hand-rim contact forces into the equations of motion as

$$M\ddot{q} + k(\dot{q}, q) = k_g(q) + G(q)\begin{bmatrix} F_x \\ F_y \end{bmatrix} + H(q)\begin{bmatrix} \tau_s \\ \tau_e \end{bmatrix}, \tag{1}$$

where M is the mass matrix, k is the vector of generalized Coriolis and centrifugal forces, k_g is the vector of generalized forces due to gravity, G transforms the horizontal F_x and vertical F_y components of the handrim force (Fig. 1) in generalized forces and H transforms the shoulder moment τ_s and the elbow moment τ_e in generalized forces.

2.2 *Inverse Dynamics*

The joint moments, τ_s and τ_e, are computed using inverse dynamics from the measured hand contact forces (F_x and F_y) and upper limb kinematics ($q, \dot{q}$ and $\ddot{q}$) as

$$\begin{bmatrix} \tau_s \\ \tau_e \end{bmatrix} = H(q)^{-1}\left(M\,\ddot{q} + k(\dot{q}, q) - k_g(q) - G(q)\begin{bmatrix} F_x \\ F_y \end{bmatrix}\right). \tag{2}$$

2.3 *Musculoskeletal System Model*

In this study, we used the open biomechanics simulation package OpenSim, which provides a platform for the development and simulation of musculoskeletal models [8]. We adopted the upper extremity model by Holzbaur et al. [7]. This model has

21 Hill-type muscle models, four of them biarticular, acting across the shoulder and elbow joints. In this analysis, the tendon was considered stiff and the force-length and force-velocity relationships were adopted from [10]. The force in muscle i is

$$Ft_i = a_i\, fl_i\, fv_i\, \cos(\theta_i) Fiso_i = a_i\, k_i\, Fiso_i, \tag{3}$$

where a_i is the muscle activation, fl_i is the force-length relationship, fv_i is the force-velocity relationship, θ_i is the muscle fibers pennation angle, $Fiso_i$ is the maximal isometric force of muscle i, and k_i is a modulation factor incorporating the force-length and force-velocity relationships.

2.4 Model Integration

It was necessary to integrate the OpenSim musculoskeletal model to the mechanical model of the upper extremity implemented in Matlab. Figure 2 shows all the steps necessary to estimate muscle forces along the propulsion cycle using the OpenSim and Matlab models. From the upper limb kinematics data obtained experimentally, the generalized coordinates and their time derivatives $(t, q(\mathrm{t}), \dot{q}(\mathrm{t})\,\mathrm{and}\,\ddot{q}(\mathrm{t}))$ are computed. The joint angles $\alpha(t)$ and $\beta(t)$ are used as input to the OpenSim program for the computation of the modulation factor of each muscle $k_i(t)$, as well as the muscle moment arm of each muscle with respect to the shoulder $d_{s,i}(t)$ and elbow $d_{e,i}\,(t)$ along the whole propulsion cycle. OpenSim also provides the maximal isometric muscle force for each muscle $Fiso_i$. The collected contact forces on the handrim, F_x and F_y, allow for the computation of the joint moments by inverse dynamics, Eq. (2). Finally, all the information is integrated for the estimation of muscle activations using the static optimization technique [11].

From the moment arms, modulation factors and maximal isometric forces, it is possible to write the shoulder and elbow moments as function of muscle activations as

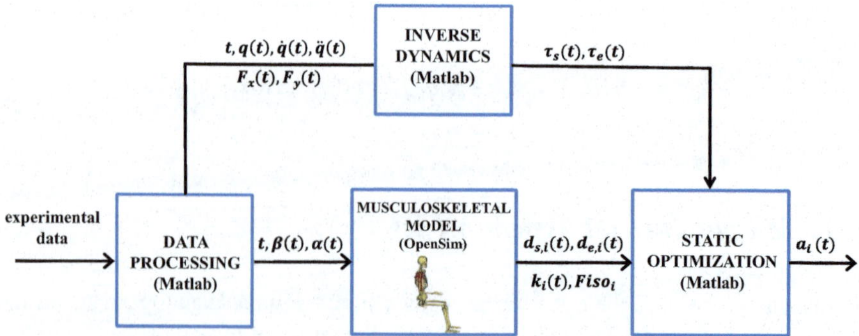

Fig. 2 Block diagram showing steps to estimate muscle forces

$$\begin{bmatrix} \tau_s \\ \tau_e \end{bmatrix} = \begin{bmatrix} d_{O,1}k_1 Fiso_1 & \cdots & d_{O,21}k_{21} Fiso_{21} \\ d_{C,1}k_1 Fiso_1 & \cdots & d_{C,21}k_{21} Fiso_{21} \end{bmatrix} \begin{bmatrix} a_1 \\ \vdots \\ a_{21} \end{bmatrix} = D \begin{bmatrix} a_1 \\ \vdots \\ a_{21} \end{bmatrix}. \tag{4}$$

3 Estimation of Muscle Forces

Equation (4) represents a system of 2 equations for 21 unknown muscle activations, which results in the so-called muscle redundancy problem. Assuming the user propels the wheelchair so as to minimize a given performance criterion, the problem of determining the muscle forces or muscle activations can be formulated as an optimization problem, an approach widely used in the literature and known as static optimization [11]. According to this approach, the muscle forces/activations are determined by solving the optimization problem in each considered instant of time.

In this study, we adopted the sum of the squared muscle activations as the cost function J, which is commonly used in the literature (e.g. [12]), as

$$J = \sum_{i=1}^{21} a_i^2 \tag{5}$$

The optimization problem is subject to physiological lower and upper constraints on the muscle activations, which ensure muscles are not pushing and do not exceed their maximal force application capacity.

4 Determining the Contributions of Inertial, Gravitational and Muscle Forces

In order to estimate the contribution of inertial, gravitational and muscle forces to the handrim forces, we partition the equations of motion, Eq. (1), into three parts [5] as

$$M\ddot{q} + k(q, \dot{q}) = G(q) \begin{bmatrix} F_{x,i} \\ F_{y,i} \end{bmatrix}, \tag{6}$$

$$0 = k_g(q) + G(q) \begin{bmatrix} F_{x,g} \\ F_{y,g} \end{bmatrix}, \tag{7}$$

$$0 = G(q) \begin{bmatrix} F_{x,a} \\ F_{y,a} \end{bmatrix} + H(q) D \begin{bmatrix} a_1 \\ \vdots \\ a_{21} \end{bmatrix}. \tag{8}$$

Note that the sum of these three equations, Eqs. (6)–(8), results in the equations of motion, Eq. (1), if

$$\begin{bmatrix} F_X \\ F_Y \end{bmatrix} = \begin{bmatrix} F_{x,i} \\ F_{y,i} \end{bmatrix} + \begin{bmatrix} F_{x,g} \\ F_{y,g} \end{bmatrix} + \begin{bmatrix} F_{x,a} \\ F_{y,a} \end{bmatrix}, \tag{9}$$

where $F_{x,i}/F_{y,i}$, $F_{x,g}/F_{y,g}$ and $F_{x,a}/F_{y,a}$ are the handrim forces due to the inertial, gravitational and muscle forces, respectively.

From the perspective of the upper limb joint moments, a similar approach allows for the determination of the individual contributions of the inertial, gravitational and handrim contact forces to the joint moments. Here again the equations of motion

$$\begin{bmatrix} \tau_s \\ \tau_e \end{bmatrix} = H(q)^{-1}\left(M\ddot{q} + k(\dot{q}, q) - k_g(q) - G(q)\begin{bmatrix} F_x \\ F_y \end{bmatrix}\right) \tag{10}$$

are partitioned into three parts. The contribution of the inertial forces to the shoulder and elbow moments is

$$\begin{bmatrix} \tau_{s_i} \\ \tau_{e_i} \end{bmatrix} = H(q)^{-1}(M\ddot{q} + k(\dot{q}, q)). \tag{11}$$

The contribution of the gravitational forces to the upper limb joint moments is

$$\begin{bmatrix} \tau_{s_g} \\ \tau_{e_g} \end{bmatrix} = H(q)^{-1}(-k_g(q)). \tag{12}$$

Finally, the contribution of the handrim contact forces to the upper limb joint moments is

$$\begin{bmatrix} \tau_{s_c} \\ \tau_{e_c} \end{bmatrix} = H(q)^{-1}\left(-G(q)\begin{bmatrix} F_x \\ F_y \end{bmatrix}\right) \tag{13}$$

The individual contributions sum up to the total joint moments as

$$\begin{bmatrix} \tau_s \\ \tau_e \end{bmatrix} = \begin{bmatrix} \tau_{s_i} \\ \tau_{e_i} \end{bmatrix} + \begin{bmatrix} \tau_{s_g} \\ \tau_{e_g} \end{bmatrix} + \begin{bmatrix} \tau_{s_c} \\ \tau_{e_c} \end{bmatrix}. \tag{14}$$

5 Experiments

The experimental protocol was approved by a Brazilian Research Ethics Committee (CAAE: 48153015.0.0000.5508). One 26-year old, male, healthy subject was selected and interviewed about the existence of any previous history of musculoskeletal disorder, injury or pain. In the absence of any of the mentioned conditions

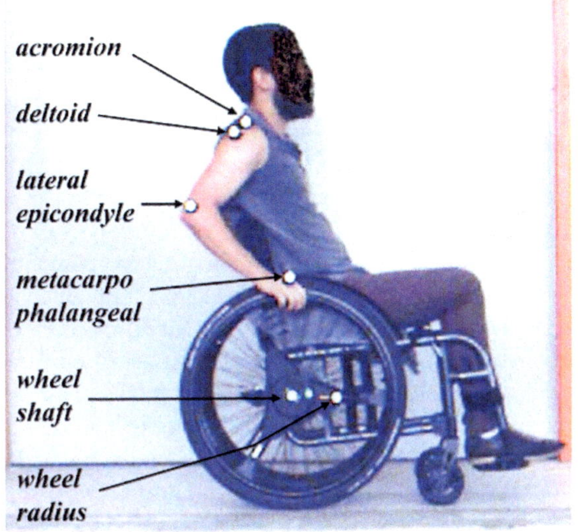

Fig. 3 Reflective markers on the subject and wheelchair

and after an explanation of the experimental procedure, the subject gave his informed consent. The subject's weight, stature and other anthropometric measurements were collected. The experiments consisted of two conditions: (1) comfortable, self-selected locomotion velocity on a level surface along a straight distance of about 7.5 m; and (2) fast, self-selected locomotion on a level surface along a straight distance of approximately 9.5 m. Each test was repeated three times. The subject started from rest and achieved the steady-state condition at about the midpoint of the trajectory. The subject was instructed to keep his trunk motion to a minimum during the trials.

The experiments were performed in the Biomechanics and Motor Control Lab (BMCLab) of the Federal University of ABC, led by Prof. Dr. Marcos Duarte. The subject propelled a manual wheelchair (Kueschall, Compact 2009 SB 400 mm). The left wheel was replaced by the force measurement system SmartWheel [6]. The handrim force data was measured at 240 Hz. Reflective, 25 mm markers were placed on anatomic locations [13] and on the wheelchair's wheel (see Fig. 3). Sagittal plane kinematics was collected with a digital camera (BASLER, scA630) at 120 frames/s.

6 Data Processing

The recorded videos were processed to obtain the 2D, sagittal plane marker trajectories using the software SkillSpector (www.video4coach.com). In order to reduce skin motion artifacts affecting the deltoid marker, the horizontal shoulder joint position x was assumed equal to the horizontal position of the acromion marker. The vertical position of the shoulder joint, in turn, was considered fixed and located 5 cm below the average vertical position of the acromion along the locomotion cycle.

All the data was filtered using a low-pass, fourth-order, zero-lag Butterworth filter with a cut-off frequency of 6 Hz [14, 15]. The sinchronization of the kinematics and handrim force data was performed by comparing the wheel angular position profile provided by the SmartWheel system and the one computed through the wheel markers.

The joint angles β and α (Fig. 1) were computed from the trajectories of the marker positions. The shoulder angle β was obtained from the trajectories of the shoulder joint and the lateral epicondyle marker. The elbow flexion angle α was obtained from the trajectories of the lateral epicondyle and metacarpophalangeal markers. Finally, the first and second time derivatives of the coordinates x, β and α were obtained through finite differences.

7 Results

Figure 4 shows the upper limb kinematics and the handrim contact force along a complete propulsion cycle in a trial performed at an average speed of 1.3 m/s. The blue and red segments represent the arm and the forearm, respectively. The contact force is represented by the black arrows.

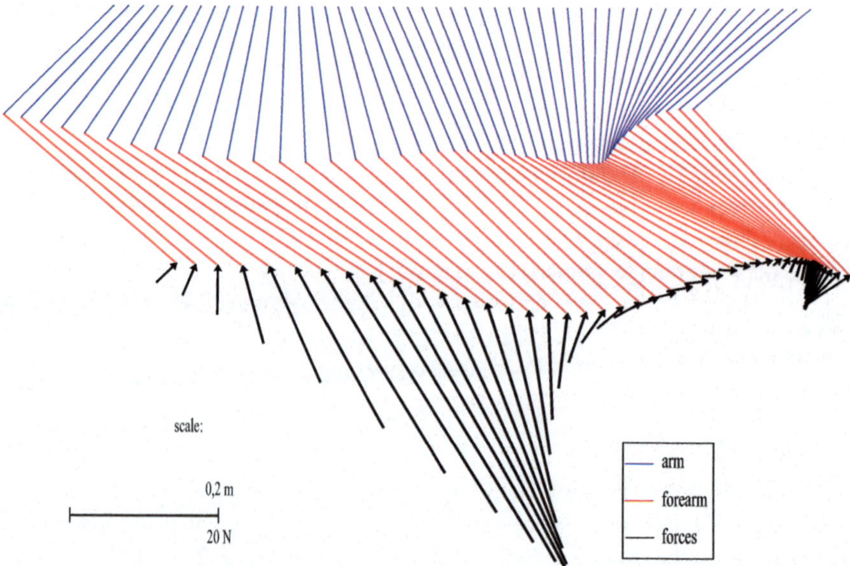

Fig. 4 Graphical representation of upper limb kinematics and handrim contact force along a complete propulsion cycle at an average speed of 1.3 m/s

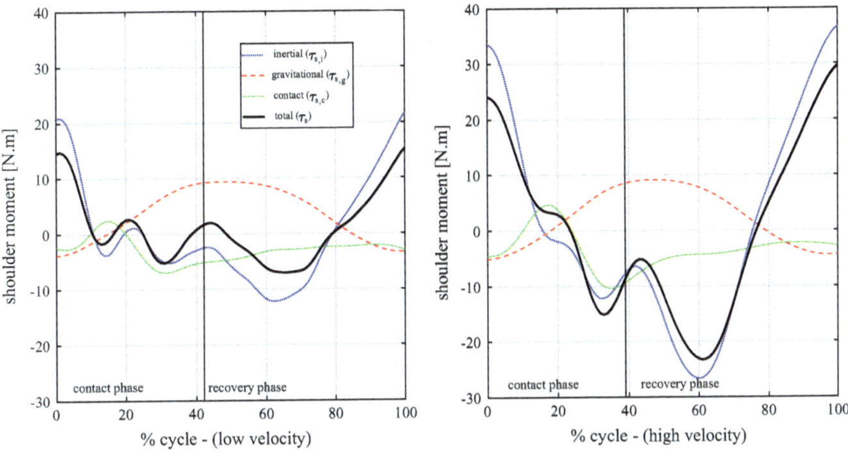

Fig. 5 Individual contributions of the inertial (blue), gravitational (red) and contact (green) forces to the total shoulder moment (black), for the average speeds of 1.3 m/s (left) and 2.0 m/s (right)

Figure 5 shows the individual contributions of the inertial (blue), gravitational (red), and contact (green) forces to the total shoulder moment (black) for both locomotion speeds, 1.3 m/s (on the left) and 2.0 m/s (on the right). The reported results represent the average over the three trials for each locomotion speed.

Figure 6 shows the individual contributions of the inertial (blue), gravitational (red) and contact (green) forces to the total elbow moment (black) for both

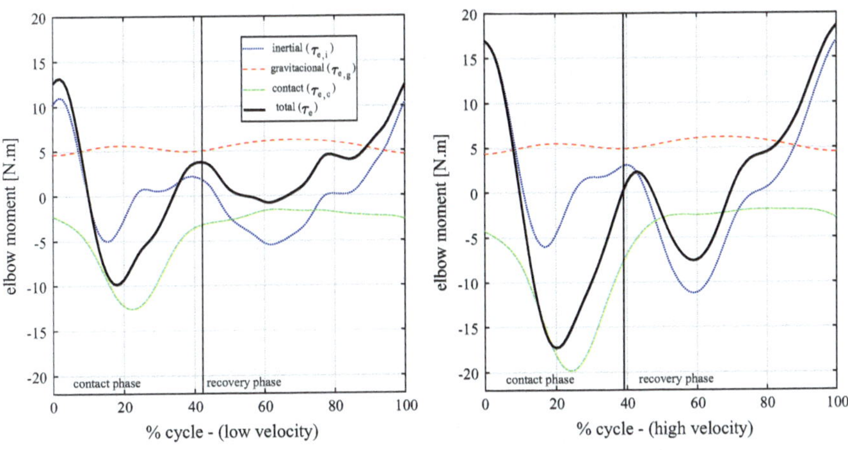

Fig. 6 Individual contributions of the inertial (blue), gravitational (red) and contact (green) forces to the total elbow moment (black), for the average speeds of 1.3 m/s (left) and 2.0 m/s (right)

locomotion speeds, 1.3 m/s (on the left) and 2.0 m/s (on the right). The reported results represent the average over the three trials for each locomotion speed. Note that the contact force contribution is not zero in the recovery phase in which the hand is not in contact with the handrim. This occurs because of measuring errors in the SmartWheel system, which provide non-zero contact values at the recovery phase as shown in Fig. 4.

8 Discussion and Conclusion

The results show that the inertial forces determine, to a great extent, the profile of the total joint moments, specially of the shoulder, even at the lower locomotion speed. While the gravitational contribution remains almost the same in the two locomotion speeds because it depends exclusively on the upper limb configuration, the inertial forces contribution increases considerably at the larger locomotion speed as accelerations of the upper limb segments increase. It is interesting to observe that the contribution of the handrim contact force to the total shoulder moment is small compared to the contribution of the inertial forces, indicating that a large part of the muscular activity at the shoulder is employed to accelerate the joint segments rather than effectively applying propulsion forces on the handrim.

These findings evidence that quasi-static models (e.g. [16]) are inappropriate to investigate wheelchair propulsion, even at relatively low locomotion speeds. Moreover, these results show the importance of accurate estimations of anthropometric parameters such as segment masses and moments of inertia, as well as measurements of segment accelerations which directly affect inertial force estimations in inverse-dynamics-based studies of wheelchair propulsion.

The results can also help guide investigations on efficient propulsion techniques, as they show that the radial component of the handrim forces, Fig. 4, is, to a great extent, determined by inertial and gravitational effects rather than by an inefficient propulsion technique [5]. This corroborates experimental results by Bregman et al. [17], according to which instructing wheelchair users to direct propulsion forces tangentially to the handrim led to increased metabolic cost during wheelchair propulsion, i.e. resulted in a less efficient locomotion.

Responsibility Notice

The authors are the only responsible for the material included in this paper.

Acknowledgements This study was supported by the National Council for Scientific and Technological Development—CNPq through the grant MCTI-SECIS/CNPq 458717/2013-4. We would like to thank Prof. Dr. Marcos Duarte for making his lab and acquisition equipment available for data collection and Kristy Alejandra Godoy Jaimes for helping with data collection.

References

1. Van der Woude, L.H.V., Veeger, H.E.J., Dallmeijer, A.J., Janssen, T.W.J., Rozendaal, L.A.: Biomechanics and physiology in active manual wheelchair propulsion. Med. Eng. Phys. **23**(10), 713–733 (2001)
2. Veeger, H.E.J., Van Der Woude, L.H.V., Rozendal, R.H.: Load on the upper extremity in manual wheelchair propulsion. J. Electromyogr. Kinesiol. **1**(4), 270–280 (1991)
3. Boninger, M.L., Souza, A.L., Cooper, R.A., Fitzgerald, S.G., Koontz, A.M., Fay, B.T.: Propulsion patterns and pushrim biomechanics in manual wheelchair propulsion. Arch. Phys. Med. Rehabil. **83**, 718–723 (2002)
4. Rankin, J.W., Kwarciak, A.M., Richter, W.M., Neptune, R.R.: The influence of wheelchair propulsion technique on upper extremity muscle demand: a simulation study. Clin. Biomech. **27**(9), 879–886 (2012)
5. Ackermann, M., Costa, H.R., Leonardi, F.: A Modeling Framework to Investigate the Radial Component of the Pushrim Force in Manual Wheelchair Propulsion. In: MATEC Web of Conferences (Vol. 35). EDP Sciences (2015)
6. Out-Front, 2015. Out-Front: SmartWheel Overview (Online). http://www.out-front.com/smartwheel_overview.php. Accessed on 22 Sept, 2015
7. Holzbaur, K.R., Murray, W.M., Delp, S.L.: A model of the upper extremity for simulating musculoskeletal surgery and analyzing neuromuscular control. Ann. Biomed. Eng. **33**(6), 829–840 (2005)
8. Delp, S.L., Anderson, F.C., Arnold, A.S., Loan, P., Habib, A., John, C.T., Thelen, D.G.: OpenSim: open-source software to create and analyze dynamic simulations of movement. IEEE Trans. Biomed. Eng. **54**(11), 1940–1950 (2007)
9. Schiehlen, W.: Multibody system dynamics: roots and perspectives. Multibody Sys.Dyn. **1**(2), 149–188 (1997)
10. Schutte, L.M., Rodgers, M.M., Zajac, F.E., Glaser, R.M.: Improving the efficacy of electrical stimulation-induced leg cycle ergometry: an analysis based on a dynamic musculoskeletal model. IEEE Trans. Rehabil. Eng. **1**(2), 109–125 (1993)
11. Erdemir, A., McLean, S., Herzog, W., van den Bogert, A.J.: Model-based estimation of muscle forces exerted during movements. Clin. Biomech. **22**(2), 131–154 (2007)
12. Van der Helm, F.C.: A finite element musculoskeletal model of the shoulder mechanism. J. Biomech. **27**, 551–569 (1994)
13. Boninger, M.L., Cooper, R.A., Robertson, R.N., Rudy, T.E.: Wrist biomechanics during two speeds of wheelchair propulsion: an analysis using a local coordinate system. Arch. Phys. Med. Rehabil. **78**(4), 364–372 (1997)
14. Boninger, M.L., Cooper, R.A., Shimada, S.D., Rudy, T.E.: Shoulder and elbow motion during two speeds of wheelchair propulsion: a description using a local coordinate system. Spinal Cord **36**(6), 418–426 (1998)
15. Finley, M.A., Rasch, E.K., Keyser, R.E., Rodgers, M.M.: The biomechanics of wheelchair propulsion in individuals with and without upper-limb impairment. J. Rehabil. Res. Dev. **41**(3B), 385–395 (2004)
16. Leary, M., Gruijters, J., Mazur, M., Subic, A., Burton, M., Fuss, F.K.: A fundamental model of quasi-static wheelchair biomechanics. Med. Eng. Phys. **34**(9), 1278–1286 (2012)
17. Bregman, D.J.J., van Drongelen, S., Veeger, H.E.J.: Is effective force application in handrim wheelchair propulsion also efficient? Clin. Biomech. **24**, 13–19 (2009)

Numerical Simulation of Track Settlement Using a Multibody Dynamic Software—A Holistic Approach

Alejandro de Miguel, Albert Lau and Ilmar Santos

Abstract A novel and numerical methodology to analyse the train/track dynamic interaction and its influence on the overall track settlement mechanism is presented. This will be achieved by creating an iterative loop that makes possible to assess the condition of the track based on the vehicle forces. The main contribution of this work rests on performing a track degradation analysis considering a regular stretch of railway track. In the first phase, a train/track interaction analysis is developed and assessed by evaluating the contact forces between the wheel and the rail. In a second phase, the forces at each particular support, beneath the rail, are extracted and transformed, by applying a degradation law at the ballast layer, into vertical displacements that in turn are applied as longitudinal level irregularities in the rail. The process is completed by including the updated geometry that enables the further calculations, in a loop mode, considering as many cycles as required.

Keywords Train/track interaction · Multibody simulation · Railway turnouts

1 Introduction

The high performance requirements, together with the increasing demand of the railway transport system is quite evident in many developed countries around the world. As a result, a sustainable development for both passenger and freight traffic, is currently taking place.

It is a priority for the different actors involved in the railway field to accomplish a significant reduction in maintenance costs caused by track quality deterioration and

A. de Miguel (✉) · I. Santos
Department of Mechanical Engineering, Solid Mechanics, Technical University of Denmark, Copenhagen, Denmark
e-mail: almite@mek.dtu.dk

A. Lau
Faculty of Engineering, Department of Civil and Environmental Engineering, Norwegian University of Science and Technology, Trondheim, Norway
e-mail: albert.lau@ntnu.no

A. de T. Fleury et al. (eds.), *Proceedings of DINAME 2017*, Lecture Notes in Mechanical Engineering, https://doi.org/10.1007/978-3-319-91217-2_33

permanent settlements in ballasted track. To achieve such a goal it is fundamental to develop accurate numerical track degradation tools that combined with empirical data, provide an efficient mechanism that allows one to simulate track degradation and plan the proper maintenance works, minimizing the overall maintenance cost of railway tracks.

According to Suiker [1], deterioration of the track is a phenomenon that takes place in two different phases: the short-term and the long-term. Other authors gave important insights by analyzing the permanent settlement of the track when simulating track deterioration, as it is described in Refs. [2–5].

Nguyen et al. [6], in their numerical work, pointed out that the permanent deformation of the track is strongly influenced by the train load and the number of train passages.

Currently, train-track numerical simulation is divided into two main methodologies that provide a good approach that helps to understand the complex behaviour between two systems that interact together. These models, properly calibrated, have the advantage of reducing reliance on experimental campaigns that, in some cases, may be expensive and time-consuming. The first one is the multibody simulation software (MBS) and the second one is the finite element method (FEM). Both of them present pros and cons. For instance, MBS software provide efficient solutions, from the point of view of computational time, by oversimplifying the track modelling. On the other side, FEM codes are able to simulate the interaction between the train and the track by modelling both systems accurately but spending too much computational effort.

This study presents a novel technique that combines, in an efficient way, the capabilities of MBS and FEM methods so the computational advantages and the accurate representation of the track system are kept. This combination makes possible to consider the forces from the train acting on the track and its impact on the track settlement phenomenon. Likewise, the settlements impact on the forces by the train, which in turn will increase the settlement even further.

The mentioned capabilities are taken into account in order to implement a loop process that enables one to consider the track irregularities at different analysis steps.

2 Description of the MBS Model Based on the Euler-Bernoulli Beam Theory

2.1 Dynamic Interaction Through Numerical Approaches

Computer software has developed continuously during last decades, so the capabilities of the programs that enable us to evaluate the dynamic interaction between the train and the track, have growth exponentially. It is expected that numerical tools will gain a lot of importance in the upcoming decades because they reduce significantly reliance on arduous experimental campaigns.

In some works as [7–9], the authors did integrate MBS and FEM methodologies with the goal of overcoming the disadvantages coming from both of them. However, the computational effort remaining when integrating MBS and FEM platforms is still high and this is one of the main reasons that motivates this work.

In the present article a new approach to simulate track degradation, is used. In it, an Euler-Bernoulli beam track (EBT) model is used as the starting point of this research.

2.2 EBT Model

The EBT model described in [10], is a methodology that is implemented in the commercial software GENSYS [11].

A fixed track model is used to simulate the dynamic interaction between the moving masses, representing the train and the track. A continuous multi-span Euler-Bernoulli beam, is elastically constrained at the supports, which aims at simulating the effect of the sleepers. The sleepers are modelled as lump masses with a regular spacing of 60 m. Between the sleepers and the rail a set of spring-damper systems is placed with the purpose of including the effect of the railpads. In the same way, underneath the sleepers a set of spring-damper system is implemented in order to account for the behaviour of the ballast layer.

The dynamic behaviour of the Euler-Bernoulli beam can be characterized by using the Lagrange equation that relates the kinetic energy and the potential energy of the system. By using modal superposition, the vertical deflection of the beam $Y(t)$ is expressed below, given by a linear combination of its first h undamped mode shapes.

$$Y(t) = \sum_{i=1}^{h} \phi_i W_i(t) \tag{1}$$

where ϕ_i denotes the i undamped mode-shape vector and $W_i(t)$ is the modal coefficient of vertical deflection. By substituting Eq. (1) into the energetic expression given by Lagrange equation, one gets a set of i ordinary differential equations, for i varying from 1 to h (the number of undamped modes considered in the analysis). Each equation describes a single degree of freedom model, in the considered modal subspace, since the damping coefficient in this case is assumed to be proportional (Rayleigh viscous damping). The set of ordinary differential equations Eq. (2) can be solved either analytically or by using some of the available numerical integration techniques.

$$\ddot{W}_h + 2\xi_h \omega_h \dot{W}_h + \omega_h^2 W_h = \frac{q}{m_h} \tag{2}$$

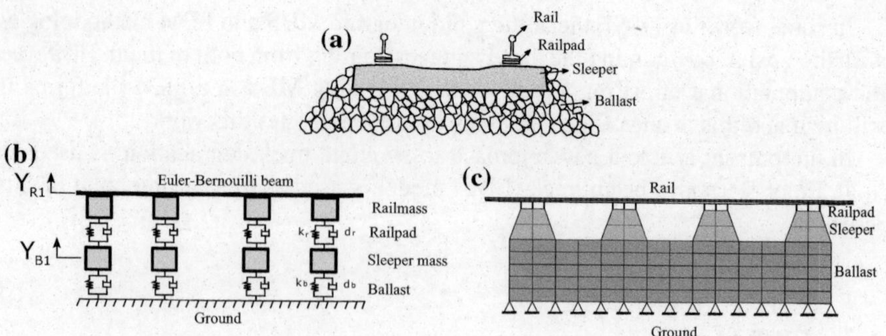

Fig. 1 Track layouts. **a** Physical track system; **b** EBT model; **c** FEM model

where m_h, ξ_h and ω_h refer to the modal mass, damping ratio and modal frequency of the hth mode respectively. Parameter q refers to the external load vector acting on the system.

2.3 *Static and Dynamic Validations of the Track Model*

At this point, it is necessary to validate the EBT model. It is decided to validate the model statically and dynamically, against a more sophisticated FEM model described in [12], that in turn, was also validated against experimental results. A similar validation technique can be found in [6] in which 2-D and 3-D finite element track models were benchmarked.

To achieve the aforementioned goal, a regular stretch of track is created in both the MBS software, where the EBT beam has been implemented, and in the two-dimensional FEM (used in this work as a benchmark). The track components considered in this comparison were the rail, railpads, sleeper and ballast layer, see Fig. 1a. The EBT model, Fig. 1b, consists of an Euler-Bernoulli beam that can be discretized into smaller beam elements. This beam aims at simulating the rail and it is elastically supported on the railpads, which are modelled by a set of spring-damper systems that rest on the sleepers. A set of masses placed every 60 m is used to simulate the sleepers effect. Finally, the ballast layer is modelled in a similar way as the railpads by connecting the sleepers to the ground that is considered to have an infinite stiffness.

An equivalent FEM model is created in the commercial software ANSYS [13]. Similar models were previously used and validated in [12, 14], so they were proved to characterize the track statically and also dynamically. The equivalent FEM model, Fig. 1c, is based on a two-dimensional approach in which rails are modelled using discrete beam elements (BEAM3). The railpads beneath, are modelled using a spring-damper system (COMBIN14). Sleepers and ballast layer are modelled by means of plane stress elements (PLANE182). The type of element and the mechan-

Table 1 Particular elements used to model the track components

Track components	Elements EBT model	Elements FEM model	Nomenclature EBT	Nomenclature FEM
Rail	Beam	Beam	Beam_3	BEAM3
Railpad	Spring-damper	Spring-damper	Plin_36	COMBIN14
Sleeper	Mass	2-D plane stress	M_rigid_6f	PLANE182
Ballast	Spring-damper	2-D plane stress	Plin_36	PLANE182

Table 2 Main parameters used in the simulations

Parameter	Value
Model length (m)	30
Beam discretization (m)	0.01
Sleeper spacing (m)	0.60
Rail bending stiffness (N m^2)	6.11×10^6
Rail mass (kg/m)	60
Railpad stiffness (N/m)	200×10^6
Railpad damping (N s/m)	30×10^3
Sleeper mass (kg)	157.4
Ballast stiffness (N/m)	160×10^6
Ballast damping (N s/m)	500×10^3
Ballast thickness (m)	0.40

ical characteristics used for both models are depicted in Table 1 and Table 2, respectively. Assumed values taken from [6, 15].

Validation of the EBT model was a fundamental issue to check the suitability of the program. It is worth noting that the MBS code has been previously used to analyze interaction forces between the train and the track, particularly at track locations were complex geometries make the analysis of train/track interaction an even harder task. This is the case of railway turnouts. Prediction of contact forces between train wheels and rail is not a trivial task and GENSYS has been successfully benchmarked against other software packages in [16].

Simulations of train/track dynamic interaction have been traditionally carried out in GENSYS considering a moving track system that consists of a set of springs and dampers. The moving track system moves together with the wheelset of the train model. This simplification is valid and provides a rough approach of the locations where the maximum peaks of the wheel/rail contact forces take place in a railway turnout, as addressed in Ref. [17]. This feature together with the demonstrated computational effort efficiency, makes the software appropriate for track optimization analyses. However, the current modelling approach used in GENSYS presents two small difficulties. On one hand, the magnitude of contact forces is not accurately obtained because of the simplifications assumed in the track modelling. On the other

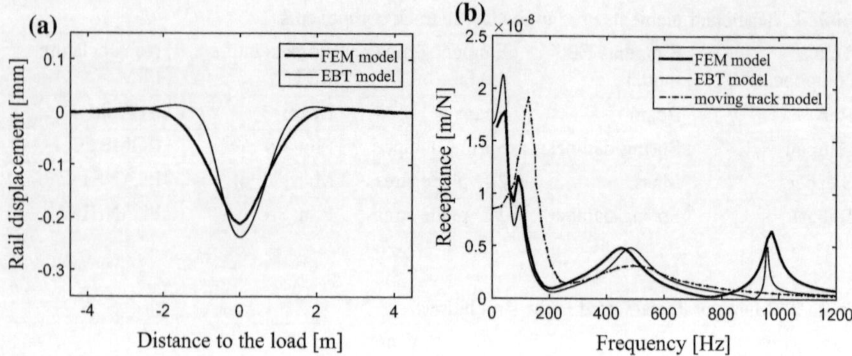

Fig. 2 Validation of the EBT model taking as a reference the FEM model. **a** Static test; **b** dynamic test

hand, the simplifications of the track modelling does not allow one to analyze the effect of the train/track interaction on the layers beneath the rail, thus it is not possible for the program to provide information regarding displacements, velocities, accelerations and the force transmitted to the rest of the track elements.

The latter information is fundamental when a degradation analysis has to be performed; it is together with the implementation of a degradation law the strongest point of this work. In light of the above, it is very important to perform two types of validation, static test and dynamic test. Through the first one, the static displacement is obtained when a vertical point load is applied on the top of the rail, so the proper configuration of the spring set underneath the rails can be verified, see Fig. 2a. The purpose of the second test is to characterize the track system from a dynamic point of view, so the proper configuration of the mass-damper-spring and beam elements of the EBT model is able to provide the main resonant frequencies and mode shapes of the track, as referred in [18]. This test was performed by applying a harmonic load in the middle point between two sleepers. The relationship between the amplitude of displacements and the applied force, at the same point, provides the receptance curve of the track, see Fig. 2b.

2.4 *Validation of the Train/Track Interaction*

Once the static and dynamic validations of the model have been completed, it is necessary to validate the track model when it is coupled with the vehicle model. This validation will verify that a proper train/track interaction is obtained when the vehicle is moving over the track. For this purpose, a simplified model of a train wheelset is created for both FEM and EBT models. The wheelset moves over a regular stretch of track with a 30 m length, as described before. To simplify the analysis, only the rail and the flexibility provided by the railpads will be considered in this assessment.

Table 3 Main parameters used in the wheelset/track simulation

Parameter	Value
Model length (m)	30
Beam discretization (m)	0.01
Railpad spacing (m)	0.60
EB bending stiffness (N m^2)	6.11×10^6
Rail mass (kg/m)	60
Railpad stiffness (N/m)	200×10^6
Railpad damping (N s/m)	30×10^3
Wheelset speed (km/h)	80
Wheelset mass (kg)	1800
Wheel/rail contact stiffness (N/m)	600×10^6
Wheel/rail damping coefficient (N s/m)	600×10^3
Contact stiffness (N/m)	1.43×10^9

The dynamic interaction response obtained in this analysis may be used to prove that the EBT model is able to capture some of the main resonant frequencies of the train/interaction phenomenon.

Mechanical characteristics used to perform the dynamic interaction between the wheelset and the track are listed in Table 3. Assumed values taken from [6, 15].

Through the analysis of the dynamic interaction forces in frequency domain, the main resonant frequencies can be detected in the EBT model. Analyzing Fig. 3a, three main peaks are obtained at 36, 67 and 74 Hz. The first and the last one correspond to the sleeper passing frequency and its second harmonic. The sleeper passing frequency is obtained by dividing the speed of the wheelset by the distance between sleepers. So in this case, the sleeper passing frequency is 37.03 Hz (the second harmonic lies around 74 Hz). The second peak corresponds to a transient behaviour of the wheelset/track interaction, which has a large influence in the initial phase of the contact force. This frequency corresponds to a transient interval in which the vehicle comes into the track model and both systems bounce up and down together. A modal analysis which is carried out in ANSYS provides a frequency, for the aforementioned bouncing vibration mode of 66.60 Hz.

Note that for the case in which the interaction forces, coming from the moving track, are analyzed none of these peaks are captured (dashed line) in Fig. 3a. This implies a substantial improvement of the response given by the former models used in MBS software to analyze dynamic interaction between the vehicle and the track. It should be noted that the computational efficiency of the MBS approaches is also kept in the EBT model, which makes this methodology suitable to perform tedious degradation simulations.

3 Description of the Implemented Degradation Process

3.1 Empirical Laws of Track Degradation

According to Ref. [19], the contribution of the ballast layer settlement may represent up to 70% of overall settlement of the track, see Fig. 3b. For this reason, the ballast layer is one of the track components with the highest influence on the overall track degradation process, which is why in many cases, research works focus on predicting track settlement evolution by only considering deformations at the ballast layer.

In this respect, Selig and Waters [19] came up with different degradation laws based on experimental campaigns. They suggested the following law, given in Eq. (3), to describe permanent ballast settlement:

$$\varepsilon_N = \varepsilon_1 N^{\beta} \tag{3}$$

where ε_1 refers to the permanent deformation obtained during the first loading cycle, β is a constant and N is the number of cycles. According the experimental data obtained by Selig and Waters [19], ε_1 and β take the values of 0.35% and 0.21, respectively.

It has been concluded that the growth rate of permanent track deformation is significantly reduced as the number of loading cycles is increased [20]. The author verified that the first loading cycle causes a very high deformation and afterwards, the deformation of the ballast layer follows a logarithmic function like the one given in Eq. (4).

$$\varepsilon_N = \varepsilon_1 N(1 + C\log(N)) \tag{4}$$

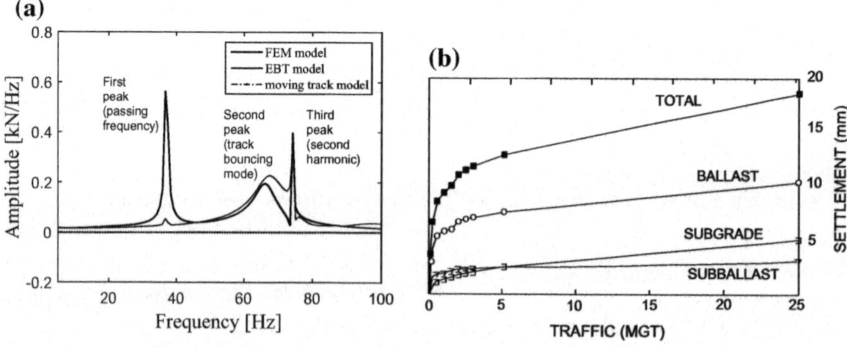

Fig. 3 **a** Interaction forces amplitude in frequency domain for the FEM, EBT and moving track models; **b** contribution of different layers to the total track settlement. Adapted from [19]

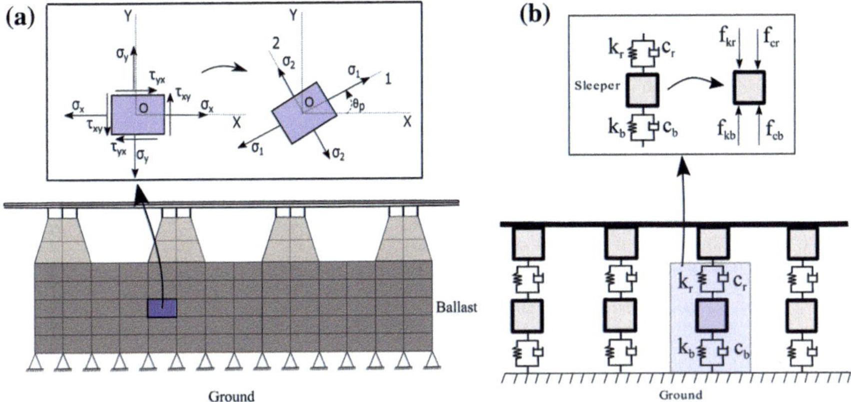

Fig. 4 Stress/force status at the ballast layer given by **a** deviatoric stress; **b** reaction magnitude in the sleeper

where ε_N refers to the permanent deformation of the ballast layer after N loading cycles, C is a constant that, according to the authors, is equal to 0.20 and ε_1 refers to the permanent deformation given by the first loading cycle in Eq. (5):

$$\varepsilon_1 = 0.082(100n_p - 38.2)(\sigma_1 - \sigma_2)^2 \tag{5}$$

where $(\sigma_1 - \sigma_2)$ is the deviatoric stress magnitude (difference between the major and minor principal stresses) acting in the ballast layer, see Fig. 4a. Parameter n_p refers to the porosity of the ballast layer that, in turn, depends on the initial tamping level and on the characteristics of the ballast layer. The latter parameter usually varies between 0.40 and 0.50 according to Ref. [20].

In order to implement this law into a numerical code, it requires the previous assessment of stress levels at the ballast layer, [21]. This unique characteristic makes the degradation law perfect for numerical codes that enable the evaluation of stresses in different materials, such the FEM packages. For this particular case in the EBT model, the law does not match directly into the main capabilities of the program. For that purpose, it would be necessary for the deformation of the ballast layer to be directly connected to the magnitude of a force rather than stresses.

Another degradation law, provided by [2, 22], suggested a similar expression as the one defined in Eq. (4) in order to define, for this particular case, the total ballast settlement. This expression fits perfectly into the capabilities of the EBT model, because unlike Eq. (4), this one takes into account the reaction force on the sleepers, $F = f_{kr} + f_{cr} + f_{kb} + f_{cb}$, to define the degradation of the ballast layer, see Fig. 4b. This is advantageous, as the reaction force on the sleepers is obtained directly from the proposed EBT model. Such degradation law is expressed in Eq. (6).

$$u_N = u_1(1 + Clog(N)) \tag{6}$$

where u_N is the permanent settlement of the ballast layer after N loading cycles. C is a constant that, according to the authors, is equal to 0.43. The settlement during the first loading cycle, u_1, depends on the reaction magnitude in the sleeper F and also on parameters s and a, Eq. (7).

$$u_1 = sF^a \tag{7}$$

where the parameter s takes a constant value of 0.00095 mm/kN [18], and variable vary 0.001–0.0004 mm/kN, depending on the conditions of the track foundation [2]. For variable a, both authors assume a value of 1.60.

3.2 Loop Process to Simulate Track Degradation in the EBT Model

Prediction of railway track settlement is done in this work by implementing an iterative process. To achieve such a goal, two different packages are integrated into a single tool that is able to predict track degradation by implementing the ballast settlements during each iteration. The first package is the EBT vehicle track model, created in the commercially available multi-body simulation tool, GENSYS. Capabilities of this model were already demonstrated in the previous sections. The second package is written in a program developed in the OCTAVE code [23], which contains the algorithm that calculates the track settlement, according to Eqs. (6) and (7).

In the first phase, the vehicle/track interaction using the EBT model is simulated in GENSYS. The passage of the train generates dynamic forces in all the spring-damper sets located underneath the rail. Afterwards, the dynamic forces of the elements (spring and dampers) that converge into the sleeper are stored.

At each time step the sum of all the previous forces is done and the maximum negative force (pointing downwards) is taken and stored again for each sleeper. As a result of this, an array with the forces that have a significant influence in the ballast settlement is generated. The size of the array is equal to the number of sleepers in the EBT model.

In a second phase, the forces will be taken to calculate the ballast settlement during the first loading cycle, Eq. (7). and the permanent settlement (in the vertical direction) of the ballast layer after N cycles, Eq. (6).

Permanent deformation caused by a single train passage is too low, for this reason the degradation analysis will be carried out considering a set of ΔN rather than just one at each iteration. This is an efficient way to avoid calculating the settlement for each load cycle individually.

The values of ballast settlement obtained in OCTAVE are introduced back in the EBT model, as longitudinal level track irregularities in the rail. Location of track irregularities will be right above the sleepers and in the mid-span between two sleepers, so discretization of the irregularities along the track is done every 30 m. A spline-cubic interpolation is used to smooth the track irregularity profiles and also to obtain the magnitude of the irregularities at the mid-span locations.

Once the geometry of the track has been updated, a new dynamic interaction between the train and the track is performed in the EBT model, so the loop process can be extended as much as it is desired.

4 Results

With the purpose of testing the described methodology in which a degradation law is implemented into the EBT model, a simplified train/track interaction example is assessed. Both track and wheelset models used to perform such analysis are similar than those described in the previous sections. The speed of the wheelset is for this particular case 80 km/h.

It is possible to evaluate the increase in contact forces when an initial irregularity is considered. This initial irregularity will create an uneven distribution of contact forces locally surrounding the track defect, compared to other track sections. The different force magnitudes at the sleepers around the irregularity will generate different settlement rates at each sleeper (generally with a higher magnitude of the settlement) and degradation rates will be more accelerated in the vicinities of the irregularity, which in turn causes increased differential settlements around the initial irregularity. Two different scenarios are considered in this assessment. The first one, in which a perfect track geometry without irregularities is considered, and the second one where an initial defect is introduced in the EBT model. In the second scenario, the degradation analysis is carried out by simulating one iteration corresponding to 50,000 train passages. Contact forces between the wheel and the rail are depicted in Fig. 5a.

Figure 5b shows the evolution of the track irregularities that have been obtained when performing the track degradation analysis. The figure also shows track irregularities when two iterations are considered and an initial track defect has been implemented before starting the first simulation. It can be seen how track irregularities grow especially in the vicinity of the initial defect considered in the first simulation, when the vertical force at each individual sleeper is extracted and used to predict the track settlement under each individual sleeper after 50,000 passages. Afterwards, the updated geometry of track irregularity after 50,000 passages is inserted back into the model to predict the subsequent vertical wheel force after another 50,000 passages. This time, the vertical wheel force has higher amplitudes than it did when only the initial irregularity was considered. The vertical force at each individual sleeper is then used to predict the final geometry of the track across the model after a total of 100,000 passages, and the corresponding settlement is shown in Fig. 5b (dashed line).

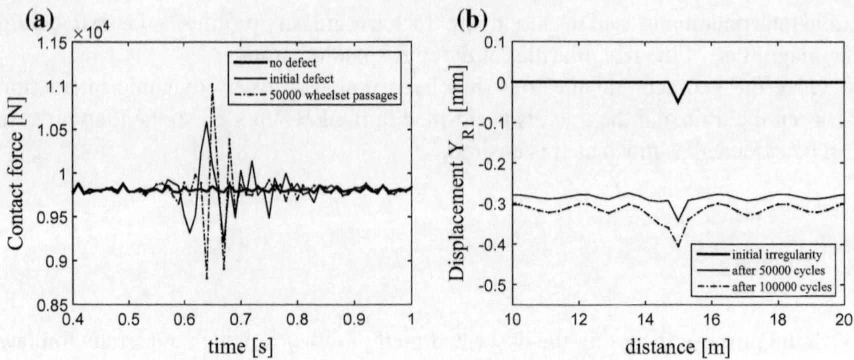

Fig. 5 Degradation analysis results: **a** vertical wheel/rail contact forces; **b** track irregularities

5 Conclusions and Future Aspects

A computational procedure to predict track degradation is described in this work. The goal was to implement an evaluative ballast settlement algorithm in a commercial MBS code, so that the program is able to provide a long-term simulation of the dynamic interaction between the train and the track. The evolution of the track irregularities is thereby obtained by considering a certain number of loading cycles. The advantage of this procedure is the low computational time compared to other similar methodologies that are either implemented in FEM codes or by linking MBS and FEM software as cross-disciplinary tools that are able to predict track degradation iteratively.

This procedure applied for a regular track section can be extended to other critical regions, such as railway turnouts, and transition zones where degradation process are very important and accelerated. At these locations, numerical calculations of interaction forces are a non-trivial task, mainly due to the complexity of the track geometry. MBS software and particularly the EBT model have proven to provide accurate results of train/track interactions with low computational effort. The advantages of implementing a settlement algorithm model have also been shown within this work, as it enables one to evaluate track degradation phenomenon in a simple and efficient manner. Furthermore, the settlement algorithm allows one to include a broad variety of settlement laws and not only the one described in [2, 22]. In this respect, further studies are needed to adapt and make the different degradation laws compatible with the capabilities of the EBT model.

A systematic measurement program would be also needed to provide both a complete calibration of the mechanical properties considered in the EBT degradation methodology and a validation of the numerical results against real onsite data.

Acknowledgements The authors gratefully acknowledge the Innovation Fund Denmark for supporting the INTELLISWITCH project (Grant no. 4109-00003A) within which this work was performed. Special thanks to Jernbaneverket, the Norwegian National Rail Administration. Mr. Ingemar Persson is also gratefully acknowledged for assisting and advising several modelling issues in GENSYS.

References

1. Suiker, A.: The Mechanical Behaviour of Ballasted Railway Tracks, PhD Thesis, Delft Technical University, Delft, The Netherlands (2002)
2. Hunt, H.E.M.: Track Settlement Adjacent to Bridge Abutments. Vehicle Infrastructure Interaction IV, San Diego, USA (1996)
3. Fröhling, R.D.: Deterioration of Railway Track Due to Dynamic Vehicle Loading and Spatially Varying Track Stiffness, PhD Thesis, University of Pretoria, Pretoria (1997)
4. Abdelkrin, M., de Bonnet, G., Buhan, P.: A computational procedure for predicting the long term residual settlement of a platform induced by repeated traffic loading. Comput. Geotech. **30**(6), 463–476 (2003). https://doi.org/10.1016/S0266-352X(03)00010-7
5. Ferreira, P.A., López-Pita, A.: Improving Ballasted High-Speed Railway Track Design for the Reduction of Vibration Levels and Maintenance Needs, TRB 2014, Annual Meeting (2013)
6. Nguyen, K., Goicolea, J.M., Gabaldón, F.: Comparison of dynamic effects of high-speed traffic load on ballasted track using a simplified two-dimensional and full three-dimensional model. Proc. Institut. Mech. Eng. Part F J. Rail Rapid Transit. **228**, 128–142 (2012). https://doi.org/10.1177/0954409712465710
7. Song, M., Noh, H., Choi, C.A.: A new three-dimensional finite element analysis model of high-speed train-bridge interactions. Eng. Struct. **25**, 1611–1626 (2003)
8. Cai, X., Zhao, L., Lau, A., Tan, S., Cui, L.: Analysis of vehicle dynamic behavior under ballasted track irregularities in high-speed railway. Noise Vibrat. Worldwide **46**, 10–17 (2015). https://doi.org/10.1260/0957-4565.46.10.10
9. Kouroussis, G., Florentin, J., Verlinden, O.: Ground vibrations induced by intercity/interregions trains: a numerical prediction based on the multibody/finite element modeling approach. J. Vibrat. Control **22**(20), 4192–4210 (2015). https://doi.org/10.1177/1077546315573914
10. Sun, Y.Q., Cole, C., Spiryagin, M., Dhanasekar, M.: Vertical dynamic interaction of trains and rail steel bridges. Electron. J. Struct. Eng. **13**(1), 88–97 (2013)
11. GENSYS, Users Manual, Release 1509 (2016)
12. Ribeiro, C., Paixão, A., Fortunato, E., Calçada, R.: Under sleeper pads in transition zones at railway underpasses: numerical modelling and experimental validation. Struct. Infrastruct. Eng. **11**(11), 1432–1449 (2014). https://doi.org/10.1080/15732479.2014.970203
13. ANSYS, Academic Research, Release 15.0, Help System, Coupled Field Analysis Guide, ANSYS, Inc. (2015)
14. Paixão, A., Fortunato, E., Calçada, R.: A numerical study on the influence of backfill settlements in the train/track interaction at transition zones to railway bridges. J. Rail Rapid Transit. 1–13 (2015). https://doi.org/10.1177/0954409715573289
15. Zhai, W.M., Wang, K.Y. and Lin, J.H.: Modelling and experiment of railway ballast vibrations. J. Sound Vibrat. 673–683 (2004)
16. Iwnicki, S.: The Manchester benchmarks for rail vehicle simulation. Vehicle Syst. Dynam. **31**(Suppl.) (1999)
17. Kassa, E., Nielsen, J.C.O.: Dynamic interaction between train and railway turnout: full-scale field test and validation of simulation models. Vehicle Syst. Dynam. **46**, 521–34 (2008). https://doi.org/10.1080/00423110801993144

18. De Man, P.: Dynatrack a Survey of Dynamic Railway Tracks Properties and Their Quality, PhD thesis, Delft University of Technology, Delft, The Netherlands (2002)
19. Selig, E., Waters, J.M.: Track Geotechnology and Substructure Management. Thomas Telford Services Ltd., London (1994)
20. Ford, R.: Differential ballast settlement, and consequent undulations in track, caused by vehicle-track interaction. Vehicle Syst. Dynam. Suppl. **24**, 222–223 (1995)
21. Ribeiro, C.: Transiçoes aterro-estrutura em linhas ferrovirias de alta velocidade: Análise experimental e numérica. PhD Thesis. Technical University of Porto, Portugal (2012)
22. Mauer, L.: An iterative track-train dynamic model for calculation of track error growth. Vehicle Syst. Dynam. Suppl. **24**, 209–211 (1995)
23. GNU, Octave, Scientific Programming Language

Part VII
Wave Propagation, Acoustics and Vibroacoustics

Flexural Wave Band Gaps in a 1D Phononic Crystal Beam

Edson Jansen Pedrosa de Miranda Jr. and José Maria Campos dos Santos

Abstract The forced response of flexural waves propagating in a 1D phononic crystal (PC) beam and its band structure are investigated theoretically and experimentally. PC beam unit cell is composed by steel and polyethylene. The study is performed by using six methods, finite element (FE), spectral element (SE), wave finite element (WFE), wave spectral element (WSE), conventional plane wave expansion (CPWE) and improved plane wave expansion (IPWE). Simulated examples of a 1D PC beam considering unit cells of different sizes are analyzed. Forced response results are presented in the form of displacement, transmittance and receptance, and the elastic band structure is investigated using its real and imaginary (attenuation) parts. Numerical and analytical results of all approaches are in a good agreement, except by WFE and FE numerical results in high frequencies. The effect of the amounts of polyethylene on the attenuation constant is studied. Depending on the application, choosing polyethylene quantity correctly is not simple, because it is related to the unit cell size and in which frequency the band gap is opened up. An experiment with a 1D PC beam is proposed and numerical and analytical results can localize the band gap position and width close to the experimental results. A small Bragg-type band gap with low attenuation is observed between 405 and 720 Hz. The 1D PC beam with unit cells of steel and polyethylene presents potential application for vibration control.

Keywords 1D phononic crystal beam · Flexural vibration · Band gaps Vibration control

E. J. P. de Miranda Jr. (✉) · J. M. C. dos Santos
University of Campinas, UNICAMP-FEM-DMC, Rua Mendeleyev, 200, CEP, Campinas, SP 13083-970, Brazil
e-mail: edson.jansen@ifma.edu.br

J. M. C. dos Santos
e-mail: zema@fem.unicamp.br

E. J. P. de Miranda Jr.
Federal Institute of Maranhão, IFMA-NIB-DEP, Rua Afonso Pena, 174, CEP, São Luís, MA 65010-030, Brazil

A. de T. Fleury et al. (eds.), *Proceedings of DINAME 2017*, Lecture Notes in Mechanical Engineering, https://doi.org/10.1007/978-3-319-91217-2_34

1 Introduction

Artificial periodic composites known as phononic crystals (PCs), consisting of a periodic array of scatterers embedded in a host medium, have been quite studied [12]. PCs have received renewed attention since they exhibit band gaps in frequency ranges that include only mechanical (elastic or acoustic) evanescent waves. The physical origin of phononic and photonic band gaps can be understood at micro-scale using the classical wave theory to describe the Bragg and Mie resonances based on the scattering of mechanical and electromagnetic waves propagating within the crystal [11]. PCs have many applications, such as vibration isolation technology [3, 8–10], acoustic barriers/filters [16], noise suppression devices [15] and surface acoustic devices [1].

Most of the studies concerning PCs focused on investigation of bulk mechanical waves [12] and its results have shown that band gaps may appear because of the contrast between physical properties, for instance elastic modulus and density of inserts and matrix. Other important properties that influence band gaps are insert geometry, filling fraction and PC lattice. Band gaps may also be affected by PC physical nature, which can be: solid/solid, fluid/fluid and mixed solid/fluid PCs. Only few studies have focused on 1D PCs [5, 14, 17] and all of them considered solid/solid PCs.

The main purpose of this study is to investigate the Bragg-type band gap formation, band structure, also known as dispersion relation, forced response, and attenuation constant of a 1D PC beam using the finite element (FE), spectral element (SE), wave finite element (WFE), wave spectral element (WSE), conventional plane wave expansion method (CPWE) and improved plane wave expansion (IPWE) methods.

2 Model and Method

Figure 1 sketches a PC beam with a periodic array of unit cells containing two different materials, i.e. steel (blue) and polyethylene (white), where a is the lattice parameter. Each unit cell is composed by 2/3 of steel and 1/3 of polyethylene. It is important to mention that Euler-Bernoulli (EB) beam theory is used. Mathematical formulation of the different methods employed in this study are not provided since the page number limitation, however, references are suggested.

EB beam mathematical formulation using SE and FE methods can be found in Lee [6]. Each unit cell is discretized in three spectral elements using SE method, i.e. each part of unit cell, Fig. 1b, corresponds to one spectral element. However, for FE method, each part of unit cell is discretized, Fig. 1b, in two finite elements. Thus, the global dynamic stiffness matrix can be obtained by the assembly of dynamic stiffness matrices of EB beam elements modeled by SE and FE methods.

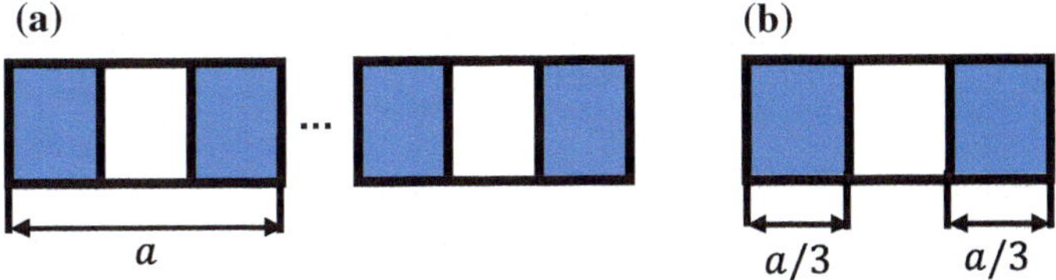

Fig. 1 **a** Schematic representation of a PC beam with N unit cells of steel (blue) and polyethylene (white). **b** PC beam unit cell

Mathematical formulation of WFE method can be found in Mencik [7]. The main difference between WSE and WFE method is the dynamic stiffness matrix used, i.e. the dynamic stiffness matrix can be derived from SE and FE methods, respectively. Modeling just one unit cell of the PC beam is one great advantage of WSE and WFE methods in relation to SE and FE methods.

CPWE, also known as $\omega(k)$ method, is a semi-analytical method used to predict the band structure. CPWE method presents slow convergence, mainly for systems with large property mismatching. To solve this convergence problem, it is used the IPWE method. Cao et al. [4] proposed the IPWE method for PCs and showed that this method provides much more accurate results than CPWE.

3 Results and Discussion

3.1 Numerical Verification

PC EB beam parameters and material properties are summarized in Table 1, where subscripts A and B refer to steel and polyethylene, respectively.

Note that the hysteretic damping, η_A, η_B, also known as loss factors, are included as a complex Young's modulus, $E_A = E_A(1 + i\eta_A)$, $E_B = E_B(1 + i\eta_B)$. PC beam

Table 1 Beam geometric parameters and material properties

Geometry/Property	Value
Unit cell length $(a = 2a_A + a_B)$, a_A, $a_B = \frac{1}{3}a$	0.0424 m
Beam length (WSE, WFE, FE and SE methods) (L)	0.424 m
Beam length (CPWE and IPWE methods) (L)	∞
Number of unit cells (WSE, WFE, FE and SE methods) (N)	10
Circular cross section area $(S = \pi r^2,\ r = 9.45\,\text{mm})$	2.8055×10^{-4} m^2
Young's modulus (E_A, E_B)	21×10^{10} Pa, 0.72×10^9 Pa
Mass density (ρ_A, ρ_B)	7800, 935 kg/m^3
Loss factor (η_A, η_B)	0.0013, 0.01
Second moment of area $(I = \pi r^4/4)$	6.2635×10^{-9} m^4

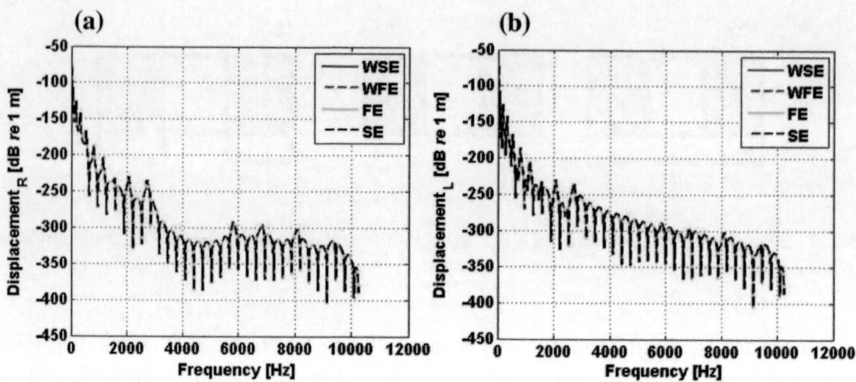

Fig. 2 PC beam displacement of the left (**b**) side, first node, and right, (**a**) side, last node, calculated by WSE, WFE, FE and SE methods

forced response is analyzed regarding a free-free boundary condition and an excitation force as a cosine-shaped pulse on beam left side. Figure 2 shows the PC beam displacement of left (first node) and right (last node) sides.

As can be seen in Fig. 2, there are some regions where resonances do not appear. However, it is difficult to localize exactly the band gaps. To overcome this issue, it is plotted in Fig. 3a, b the frequency response function (FRF) and the transmittance in Fig. 4, defined as the division between last and first node displacements. For FRF, it is chosen the receptance, i.e. the division between displacement of the first or the last nodes, and the force, which gives H_{11} or H_{21}, respectively, also known as point receptance and transfer receptance.

From Figs. 3, 4, it can be seen a band gap created between 2780 and 5798 Hz. This band gap is known as a Bragg-type band gap, because the mechanism involved is the Bragg scattering. Thus, frequency location is governed by Bragg's law,

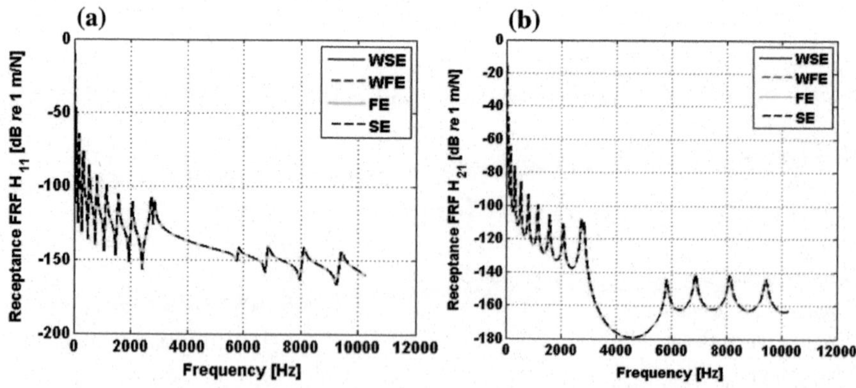

Fig. 3 Point receptance (**a**) and transfer receptance, (**b**) of the PC beam calculated by WSE, WFE, FE and SE methods showing the band gap

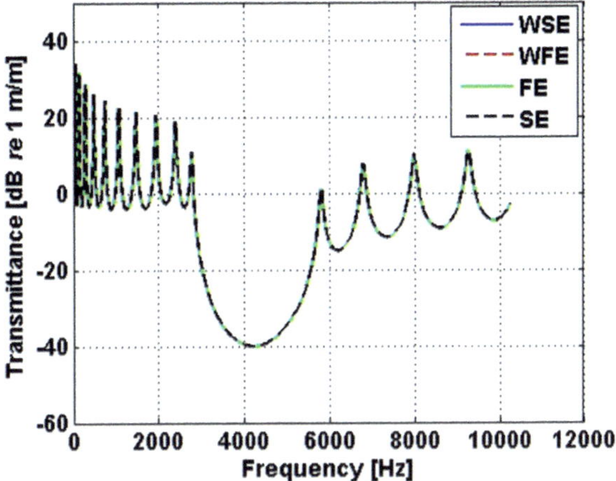

Fig. 4 Transmittance of the PC beam calculated by WSE, WFE, FE and SE methods showing the band gap

$a = n(\lambda/2)$, $(n = 1, 2, 3, \ldots)$, where λ is the wave wavelength in host material. Bragg's law implies that it is difficult to achieve a low frequency Bragg-type band gap in PCs with small size.

In Fig. 5a, it is compared the transmittance for different lattice parameters calculated by WSE method until 10240 Hz. It is considered $a = 0.212$, 0.106, 0.0707, 0.053, which results in $N = 2$, 4, 6, 8, for the fixed beam length in Table 1. WSE is chosen since it is an analytical method. For the first case, $a = 0.212$, $N = 2$, it can be observed four wide Bragg-type band gaps, that is to say 122.5–456.9 Hz, 626.3–2884 Hz, 3174–4439 Hz and 4439–7961 Hz. For the other cases, $a = 0.106$, 0.0707, 0.053, $N = 4$, 6, 8, it can just be seen completely the first Bragg-type band gap between 456.9 Hz–1204 Hz, 1008 Hz–2325 Hz and

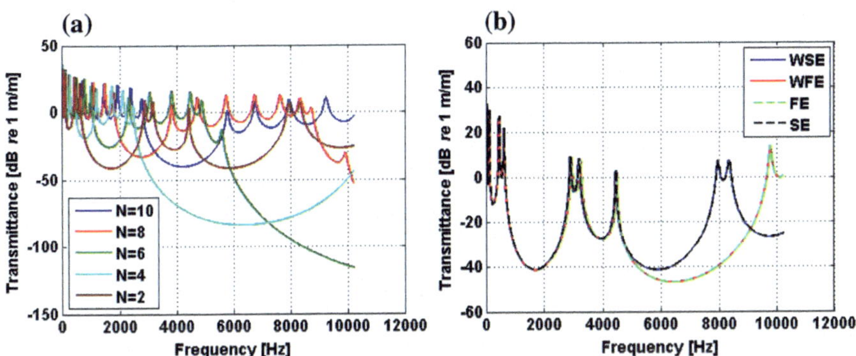

Fig. 5 Transmittance of the PC beam calculated by WSE method for $a = 0.212$, 0.106, 0.0707, 0.053, $N = 2$, 4, 6, 8 (**a**) and transmittance calculated by WSE, WFE, FE and SE methods for $a = 0.212$, $N = 2$ (**b**)

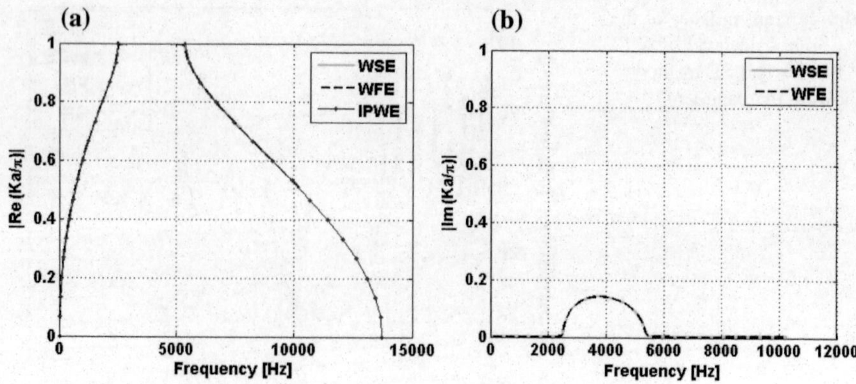

Fig. 6 Elastic band structure of the PC beam considering the data in Table 1. Reduced Bloch wave vector real part (**a**) calculated by WSE, WFE and IPWE methods and its imaginary part calculated by WSE and WFE methods (**b**)

1782 Hz–3849, respectively. Thus, increasing unit cell length for a fixed beam length the Bragg-type band gaps will appear in low frequencies, as expected.

In Fig. 5b it is shown the transmittance for $a = 0.212$, $N = 2$, calculated by WSE, WFE, FE and SE methods. It can be seen that WFE and FE methods do not match with WSE and SE methods in high frequencies, because the unit cell discretization (two finite elements for each part of the unit cell) is not enough in higher frequencies.

Figure 6 illustrates the elastic band structure considering the data in Table 1. Figure 6a shows real part of the reduced Bloch wave vector, ka/π, also known as reduced wavenumber, using WSE, WFE and IPWE methods. Figure 6b shows imaginary part of the reduced Bloch wave vector using WSE and WFE methods. For IPWE and CPWE calculations, it is considered 101 plane waves in the Fourier series expansion. In Fig. 6, it is only shown the first irreducible Brillouin zone (FIBZ) [2], i.e. $[0, \pi/a]$.

Figure 7 shows the comparison between IPWE and CPWE methods inside the first Brillouin zone, i.e. $[-\pi/a, \pi/a]$. In Fig. 7, only the first 10 branches are illustrated. The matching between IPWE and CPWE does not occur only for the high bands even considering a high number of planes waves.

In order to demonstrate the agreement between WSE and IPWE methods for other values of a, we plot in Fig. 8a the elastic band structure real part for $a = 0.212$, 0.106, 0.0707, 0.053, 0.0424, $N = 2$, 4, 6, 8, 10. Note that the curves related to each lattice parameter are not identified by a specific color, but it can be seen the matching. In Fig. 8b, this comparison is done and its behavior is the same discussed in Fig. 5a. WSE method is used to calculate imaginary part of the Bloch wave vector. From Fig. 8b, it can be observed that the attenuation performance of

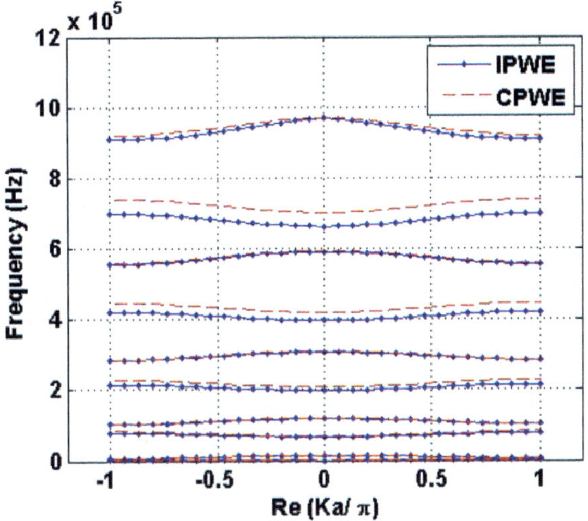

Fig. 7 Real part of the reduced Bloch wave vector calculated by IPWE and CPWE methods

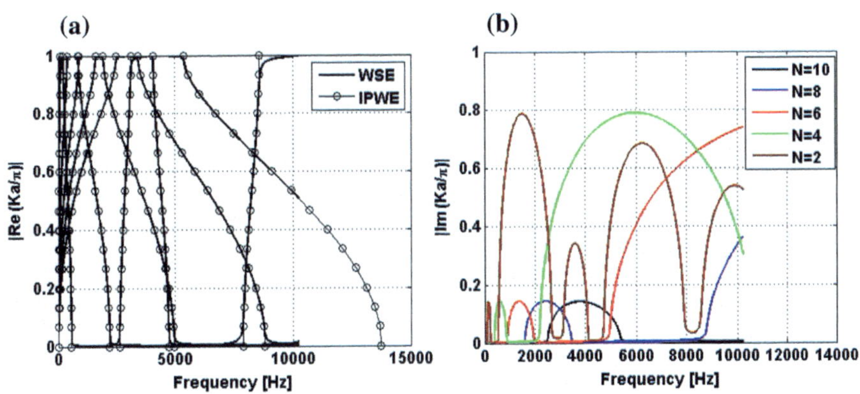

Fig. 8 Elastic band structure of the PC beam considering a = 0.212, 0.106, 0.0707, 0.053, 0.0424, i.e. $N = 2, 4, 6, 8, 10$. Real part of the reduced Bloch wave vector (**a**) calculated by WSE and IPWE methods and the imaginary part of reduced Bloch wave vector calculated by WSE method (**b**)

the Bragg-type band gaps is better for $a = 0.212$, 0.106, because there are more Bragg-type band gaps in low frequencies. The attenuation constant, defined by $\mu = \Im\{k\}a$ is an important information which can be analyzed from the imaginary part of Bloch wave vector. The attenuation constant gives some insight about attenuation of the unit cell, but it can not be confused with beam attenuation.

Figure 9a–e shows the influence of polyethylene quantity (5–95%) on the unit cell attenuation for $a = 0.212$, 0.106, 0.0707, 0.053, 0.0424, $N = 2$, 4, 6, 8, 10, respectively. It is important to mention that until Fig. 8, it is only considered 1/3 (≈33,33%) of polyethylene, as described in Table 1. The polyethylene quantity influence on the unit cell attenuation performance is complicated and varies with

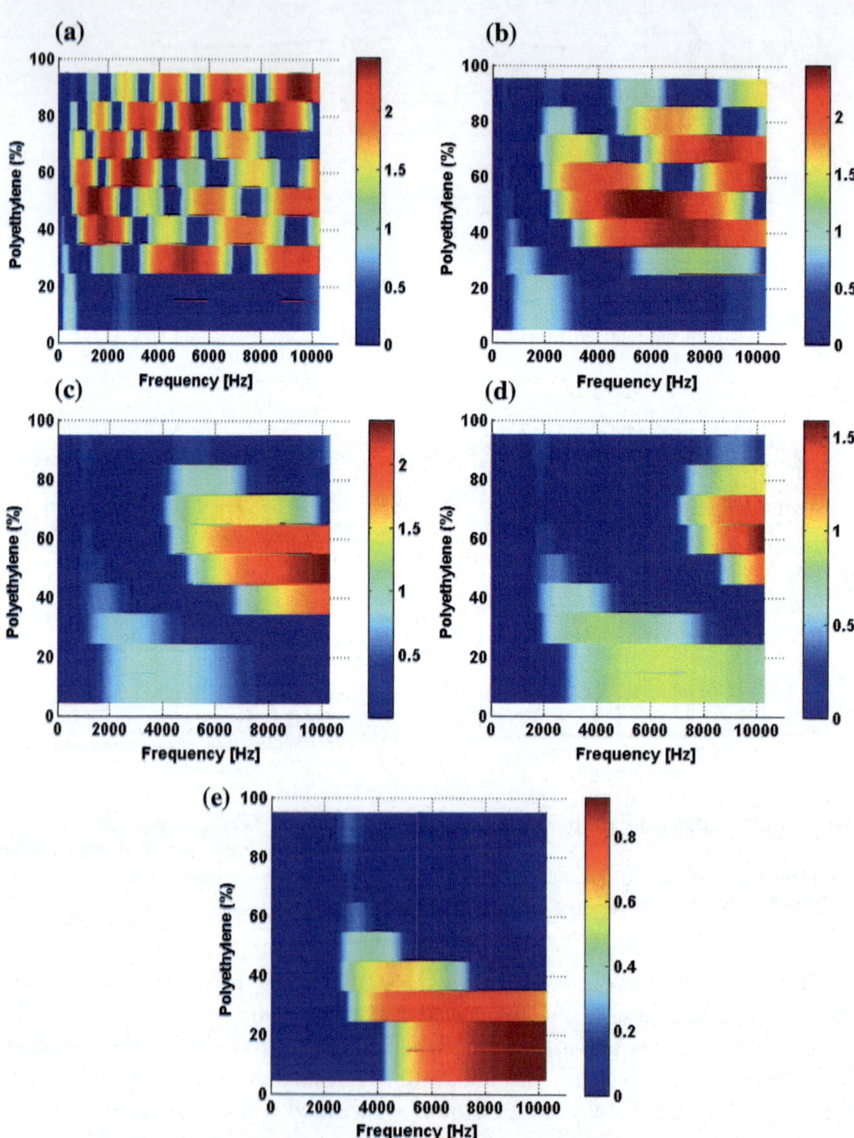

Fig. 9 Attenuation constant surface x-y view of the PC beam unit cell considering $a = 0.212$, 0.106, 0.0707, 0.053, 0.0424, (**a**–**e**), respectively, calculated by WSE method

unit cell length. No attenuation is observed for $a = 0.0424$ until 2500 Hz—Fig. 9e, $a = 0.053$, until 1612 Hz—Fig. 9d, $a = 0.0707$, until 874 Hz—Fig. 9c, $a = 0.106$, until 406 Hz—Fig. 9b, and $a = 0.212$, until 98 Hz—Fig. 9a, independently of the polyethylene quantity. Thus, depending on the application, choosing polyethylene quantity correctly it is not easy, since it is related to the unit cell length and in which frequency the band gap is opened up.

In Fig. 9b–e, there are some regions which present higher attenuation, i.e. between 40%–80%, 50%–60%, 60%–80% and 5%–30% of polyethylene, for $a = 0.106$, 0.0707, 0.053, 0.0424, respectively. In Fig. 9a, it is difficult to identify where is the best attenuation region, because there are many band gaps. However, below 20% of polyethylene is the worst attenuation region.

3.2 Experimental Verification

A PC beam is used to perform an experimental test. The PC beam is similar to the model proposed in Fig. 1 with free-free boundary condition, however, the $a_B = 0.041$ m and $a_A = 0.0325$ m, with $a = 2a_A + a_B = 0.106$ m. The properties are the same described in Table 1, with $N = 4$. The measurement instruments used in the experimental setup are summarized in Table 2. Figure 10 shows the experimental setup with details of the impact hammer and accelerometers positions. By using an impact force excitation applied to the right and left ends of the PC beam, acceleration measurements are taken on PC beam right end.

However, it is chosen to plot the displacement, that is $u = -a_c/\omega^2$, where u is the displacement, a_c is the acceleration measured and ω is the angular frequency. Inertance point and transfer FRFs are measured with 5 averages, with the frequency discretization of 0.625 Hz.

Figure 11a–d illustrates the displacement of the last beam node (right side), the transmittance, and the FRFs H_{11} and H_{21}, respectively. Numerical results present good agreement. However, FE and WFE methods do not match in higher frequencies the analytical methods, as discussed before. Furthermore, there is some mismatching related to the experimental FRFs.

Numerical and analytical results of band gap widths do not match the experimental results, since the numerical model considered may not capture all aspects of

Table 2 Measurement instruments

Instrument	Manufacture and model	Sensitivity	Measure range
Impulse hammer	PCB 86E80	22.5 mV/N	222.0 N (peak)
Accelerometer	KISTLER 8614A500M1	3.46 mV/g	(±5%) 0–12.5 kHz
Data acquisition	LMS SCR05	–	–

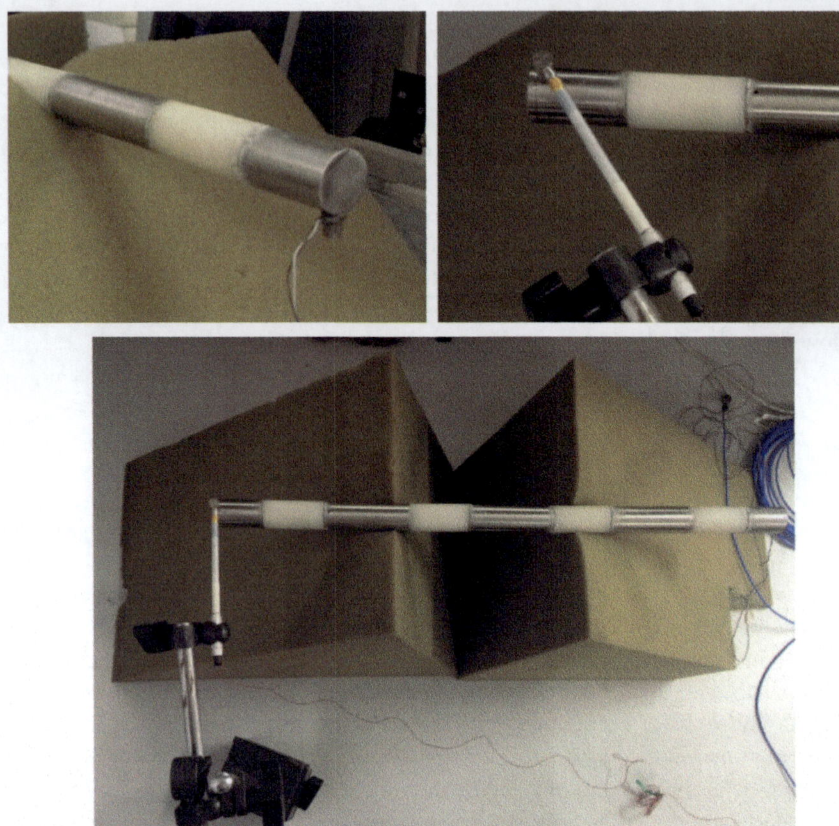

Fig. 10 Experimental setup of the PC beam

PC beam, such as the material used to glue the polymer and metal, material properties may not be exactly the same of Table 1, for instance. In addition, it is used the EB beam theory, possibly considering higher theories, such as Timoshenko beam theory [13], the results may be improved in higher frequencies.

Figure 12 shows the elastic band structure of the PC beam. Band gap widths observed in Fig. 11 can be confirmed in Fig. 12. Furthermore, it may be observed in Fig. 12 a small band gap between 405 and 720 Hz with low attenuation.

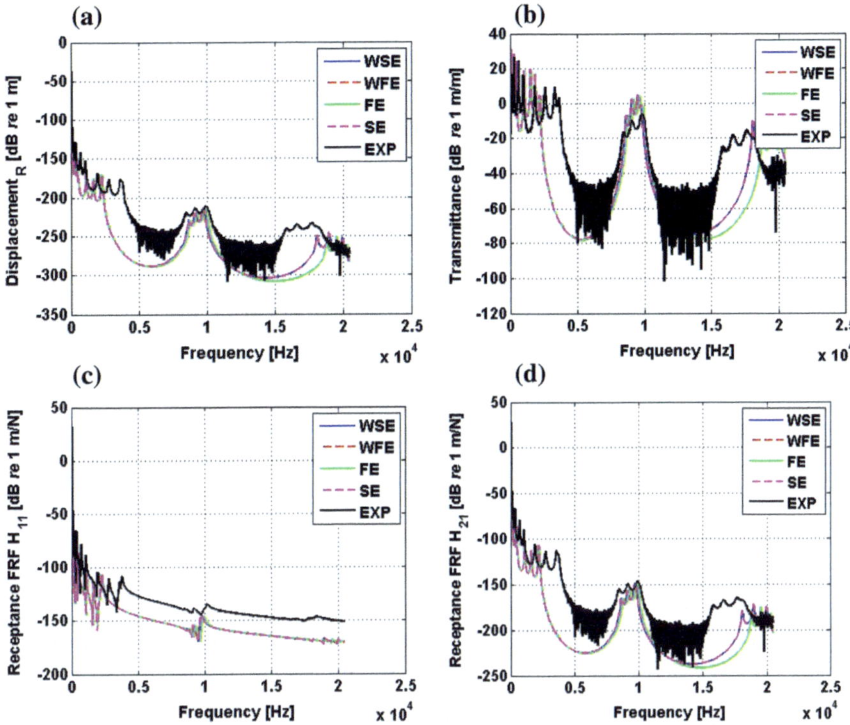

Fig. 11 PC beam displacement of the right (**a**) side, transmittance (**b**), point receptance (**c**) and transfer receptance (**d**) calculated by WSE, WFE, FE, SE methods and measured experimentally

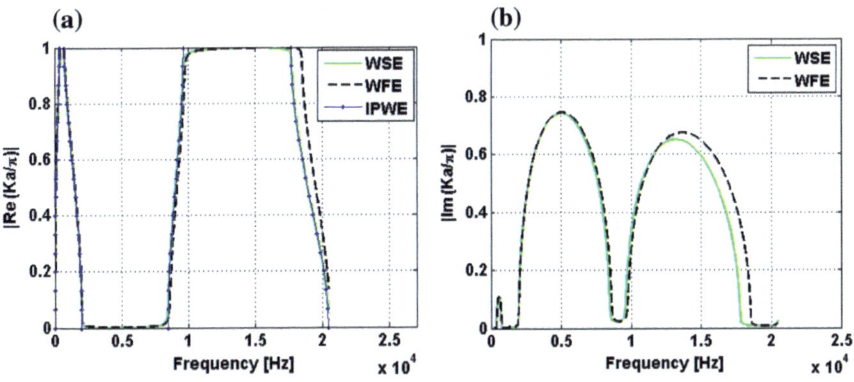

Fig. 12 Elastic band structure of the PC beam. Real part of the reduced Bloch wave vector (**a**) calculated by WSE, WFE and IPWE methods and imaginary part of reduced Bloch wave vector calculated by WSE and WFE methods (**b**)

4 Conclusions

Forced response and elastic band structure of a 1D PC beam are obtained. Forced response is obtained by WFE, WSE, FE and SE methods and a good matching is observed, instead of in high frequencies, where WFE and FE do not match the spectral analytical methods. Real part of elastic band structure is calculated by WFE, WSE and IPWE methods and it is shown a good agreement between them. CPWE presents some disadvantages compared to the IPWE for the higher bands. The influence of unit cell length is also studied and for larger unit cells, Bragg-type band gaps are opened up in low frequencies. The polyethylene quantity into the unit cell is an important variable and its influence on the attenuation constant depends on the unit cell length. In some ranges of frequency no attenuation is observed independently of polyethylene quantity and there are regions that present higher attenuation. Numerical and analytical results present good agreement with experimental results and they can localize the band gap position and width close to the experimental. A small Bragg-type band gap with low attenuation is observed between 405 and 720 Hz. The 1D PC beam with unit cells of steel and polyethylene presents potential application for vibration management.

References

1. Benchabane, S., Khelif, A., Robert, L., Rauch, J.Y., Pastureaud, T., Laude, V.: Elastic band gaps for surface modes in an ultrasonic lithium niobate phononic crystal. Proc. SPIE **6182** (618216), 1–13 (2006)
2. Brillouin, L.: Wave Propagation in Periodic Structures. Dover Publications, New York, USA (1946)
3. Casadei, F., Beck, B.S., Cunefare, K.A., Ruzzene, M.: Vibration control of plates through hybrid configurations of periodic piezoeletric shunts. J. Intell. Mater. Syst. Struct. **23**(10), 1169–1177 (2012)
4. Cao, Y.J., Hou, Z.L., Liu, Y.Y.: Convergence problem of plane-wave expansion method for phononic crystal. Phys. Lett. A **327**, 247–253 (2004)
5. Jian-Yu, F., Dian-Long, Y., Xiao-Yun, H., Li, C.: Coupled flexural-torsional vibration band gap in periodic beam including warping effect. Chin. Phys. B **18**(4), 1–6 (2009)
6. Lee, U.: Spectral Element Method in Structural Dynamics, 1st edn. Wiley, Singapore (2009)
7. Mencik, J.-M.: On the low- and mid-frequency forced response of elastic structures using wave finite elements with one-dimensional propagation. Comput. Struct. **88**, 674–689 (2010)
8. Miranda Jr., E.J.P., dos Santos, J.M.C.: Band structure in carbon nanostructure phononic crystals. Mater. Res. (2017). https://doi.org/10.1590/1980-5373-MR-2016-0898
9. Miranda Jr., E.J.P., dos Santos, J.M.C.: Complete band gaps in nano-piezoelectric phononic crystals. Mater. Res. (2017). https://doi.org/10.1590/1980-5373-MR-2017-0298
10. Miranda Jr., E.J.P., dos Santos, J.M.C.: Phononic band gaps in Al_2O_3/epoxy composite. Mater. Sci. Forum **912**, 112–117 (2018). https://doi.org/10.4028/www.scientific.net/MSF.912.112
11. Olsson III, R.H., El-Kady, I.: Microfabricated phononic crystal devices and applications. Measur. Sci. Technol. **20**(012002), 1–13 (2009)

12. Sigalas, M.M., Economou, E.N.: Elastic waves in plates with periodically placed inclusions. J. Appl. Phys. **75**, 2845–2850 (1994)
13. Timoshenko, S.P.: On the correction for shear of the differential equation for transverse vibrations of prismatic bars. Philos. Mag. **41**(245), 744–746 (1921)
14. Wu, L.-Y., Wu, M.-L., Chen, L.-W.: The narrow pass band filter of tunable 1D phononic crystal with a dielectric elastomer layer. Smart Mater. Struct. **18**(015011), 1–8 (2009)
15. Xiao, Y., Wen, J., Wen, X.: Sound transmission loss of metamaterial-based thin plates with multiple subwavelength arrays of attached resonators. J. Sound Vib. **331**, 5408–5423 (2012)
16. Yang, Z., Dai, H.M., Chan, N.H., Ma, G.C.: Acoustic metamaterial panels for sound attenuation in the 50–1000 Hz regime. Appl. Phys. Lett. **96**(041906), 1–3 (2010)
17. Zhang, Y., Ni, Z.-Q., Han, L., Zhang, Z.-M., Jiang, L.H.: Flexural vibrations band gaps in phononic crystal Timoshenko beam by plane wave expansion method. Optoelectron. Adv. Mater. Rapid Commun. **6**(11-12), 1049–1053 (2012)

Investigating Interface Modes on Periodic Acoustic Waveguides and Elastic Rods Using Spectral Elements

Matheus Inguaggiato Nora Rosa, José Roberto de França Arruda and Massimo Ruzzene

Abstract Due to their particular wave propagation characteristics, phononic crystals (PC's) and acoustic metamaterials have numerous potential applications in passive vibration and noise control. In this context, some geometric phase concepts originally developed in electronics have inspired research in phononics. As a result of the non-trivial topology of the band structure, these concepts allow exploring particular behaviors such as edge and interface modes. For acoustic systems, it has been recently shown that interface modes appear at the boundary separating two PC's having different Zak phases. In this paper, one-dimensional spectral elements are used to investigate interface modes in one-dimensional acoustic (tube) and elastic (rod) systems. The band structure and the forced response are computed using the spectral element method. It is shown that the interface mode appears within the second band gap when a geometrical parameter of one of the connected PC's is varied so that this bandgap closes and reopens, characterizing a change in the Zak phase. The forced response at the interface mode frequency shows that the sound (acoustic) or vibration (elastic) is spatially concentrated (localisation phenomenon) at the interface. Different PC combinations and different excitation locations are investigated. This behavior may have useful engineering applications, such as in sound and vibration energy harvesting.

Keywords Phononic crystals · Interface modes · Acoustic waveguides · Elastic rods · Spectral elements

---

M. I. N. Rosa (✉)
School of Mechanical Engineering, Georgia Institute of Technology, Atlanta, GA 30332-0405, USA
e-mail: mrosa8@gatech.edu

J. R. de França Arruda
Faculty of Mechanical Engineering, University of Campinas, Rua Mendeleyev, 200. Cidade Universitria Zeferino Vaz. Campinas, Campinas, SP 13083-860, Brazil
e-mail: arruda@fem.unicamp.br

M. Ruzzene
School of Aerospace Engineering, Georgia Institute of Technology, Atlanta, GA 30332-0150, USA
e-mail: ruzzene@gatech.edu

A. de T. Fleury et al. (eds.), *Proceedings of DINAME 2017*, Lecture Notes in Mechanical Engineering, https://doi.org/10.1007/978-3-319-91217-2_35

1 Introduction

The study of phononic crystals and acoustic metamaterials for wave control applications is an emerging topic in engineering and physics. A phononic crystal (PC) is a structure with geometric and/or material periodicity, while an acoustic metamaterial is usually characterized by the presence of substructures that work as resonators tuned to a narrow frequency band. In both cases their wave propagation characteristics can be derived from the analysis of a single unit cell. This may be done through the dispersion diagram, which can be computed by different methods and essentially relates the angular frequency to the wavenumber. Wave control can therefore be achieved by manipulating the pass and stop bands on these diagrams, i.e. frequency ranges in which the waves are propagating or evanescent, respectively. This allows a number of applications, such as vibration isolation, vibroacoustic barriers, wave tunneling, and cloaking. A comprehensive review on the subject is presented in [1].

Many applications of photonic and phononic crystals, as well as mechanical metamaterials, were inspired by concepts that were originally developed in electronic band theory. In particular, the geometric-phase concept has been explored in order to produce topological behavior that emerges from the non-trivial topology of some material band structures. One example is the creation of one-dimensional elastic edge waves, which are envisioned for the design of loss-free, one-way acoustic waveguides. In this case, non-trivial band gaps are caused by breaking the time-reversal symmetry, in analogy to the quantum anomalous Hall effect counterpart in electrocnics [2]. The time-reversal symmetry breaking has been achieved on solid lattice structures by inducing Coriolis forces via rotation [3] or by gyroscopic intertial effects [4, 5].

Another recent development of non-trivial topological behavior is the formation of interface modes on periodic acoustic systems. In [6], the authors have shown that its possible to realize the band structure inversion on acoustic PC's, which also has an analogy in electronic systems [7]. In this case, the geometric phase involved is called Zak phase, which is a special type of Berry phase for one-dimensional (1D) periodic systems [8]. The Zak phase of a particular band is related to the symmetry properties of the band-edge states, and the topological characteristics of a band gap are determined by the summation of the Zak phases of all the bands below the gap. Broadly speaking, interface modes are formed on the boundary separating two phononic crystals having different bandgap topological characteristics.

In this paper the interface modes present on these periodic acoustic systems are explored. It is shown that the band inversion concept can be observed in the dispersion diagrams of these PC's by varying one of the system parameters, resulting in the gap closing and reopening process. Therefore, it is not strictly necessary to compute the Zak phase in order to design simple periodic systems with interface modes. It is also shown that the same concepts can be applied to periodic elastic rods, where interface modes can also be created.

To this end, the Spectral Element Method (SEM) [9] is used to compute the dispersion relations and the forced response of elastic rods and acoustic tube PC's. The dynamic stiffness matrices produced by SEM are exact within the scope of the theory used to obtain the equations of motion, and thus the method is a suitable tool for fast and accurate computations of the forced response and dispersion relations of structures with simple geometry and kinematic behavior, such as rods and beams. A one-dimensional spectral element for cylindrical ducts can be derived in analogy to the 1D spectral element for elementary rods, assuming the presence of plane waves only, as their dynamics are both governed by the same 1D wave-equation.

2 Interface Modes on Periodic Acoustic Systems

To demonstrate the presence of interface modes on acoustic systems, simple one-dimensional PC's are used here. Figure 1a shows the geometry of the unit cell that is used for both acoustic and elastic rod systems. Each unit cell is composed of a narrower cylinder (Tube B) of length d_b and radius r_b located between two wider cylinders (Tube A) of length $\frac{1}{2}d_a$ and radius r_a. For acoustic systems, these tubes are hollow cylindrical rigid ducts and are filled with air (mass density ρ_A=1.3 kgm^{-3} and speed of sound $c_A = 343$ ms^{-1}). For the rod systems the cylinders are solid linear elastic isotropic materials (see next section). For both acoustic and elastic rod systems a loss factor $\eta = 0.002$ is used on the wavenumber primarily to avoid numerical instabilities (through a complex wavenumber $k = k(1 + \frac{i\eta}{2})$).

For the examples in this section, the tube radii are set to $r_a = 2.4$ cm and $r_b = 1.5$ cm, and the total length of the unit cell is fixed at $L = 8.5$ cm. Figure 1b shows the band structure, calculated via SEM, for an acoustic PC with $d_a = 2.25$ cm and $d_b = 6.25$ cm. In all dispersion diagrams shown in this article the real part of the wavenumber is plotted on the positive axis while the imaginary part is plotted on the negative axis.

The band inversion for this PC can be observed on the second band gap of the band structure as the system parameter $\Delta d = \frac{d_a - d_b}{2}$ is varied. The topological transition point for this system occurs when $d_a = d_b$ and therefore $\Delta d = 0$. Figure 2 shows the gap closing and reopening process, indicating that the band inversion occurs when Δd shifts from negative to positive values. It should be noted that the band structure for two PC's with the same absolute value for Δd is exactly the same, the only difference being the symmetry properties of the band-edge states. In other words, the topological characteristics of the band gap for two PC's having the same $|\Delta d|$ with opposite signs are different, and their Zak phases, if computed, would also be different.

With the topological transition point for the system defined, interface modes can be observed by calculating the forced response of different PC couplings. In this work, four specific PC's are used to investigate these interface modes: PC$_1$ ($d_a = 2.25$ cm, $d_b = 6.25$ cm and $\Delta d = -2$ cm), PC$_2$ ($d_a = 6.25$ cm, $d_b = 2.25$ cm and $\Delta d = 2$ cm), PC$_3$ ($d_a = 3.75$ cm, $d_b = 4.75$ cm and $\Delta d = -0.5$ cm) and PC$_4$ ($d_a = 4.75$ cm, $d_b = 3.75$ cm and $\Delta d = 0.5$ cm). Each acoustic system analysed consists of ten unit cells of one PC on the left side and ten unit cells of another PC on the right side,

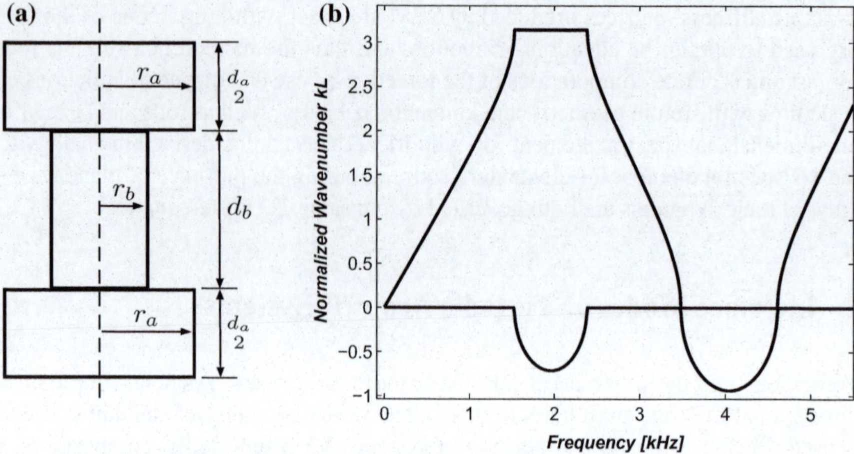

Fig. 1 1D phononic crystal system. **a** Phononic crystal's geometry, **b** band structure for PC with $d_a = 6.25$ cm and $d_b = 2.25$ cm

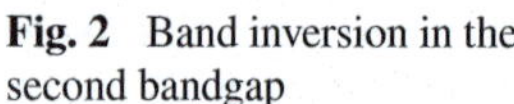

Fig. 2 Band inversion in the second bandgap

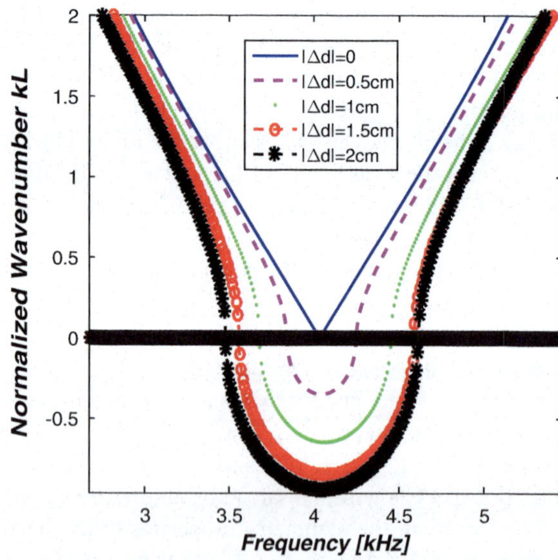

with the interface separating them at the center of the acoustic system. All forced responses are computed via SEM, with free-free boundary conditions.

The interface mode appears when the coupled PC's have different band gap topological characteristics. Figure 3a shows the forced response at the interface for a PC_1-PC_2 coupling excited directly at the interface, where the presence of the interface mode is observed at frequency $f_i = 4035$ Hz. For this frequency, the spacial distribution of the pressure field is concentrated at the interface, which can be observed on

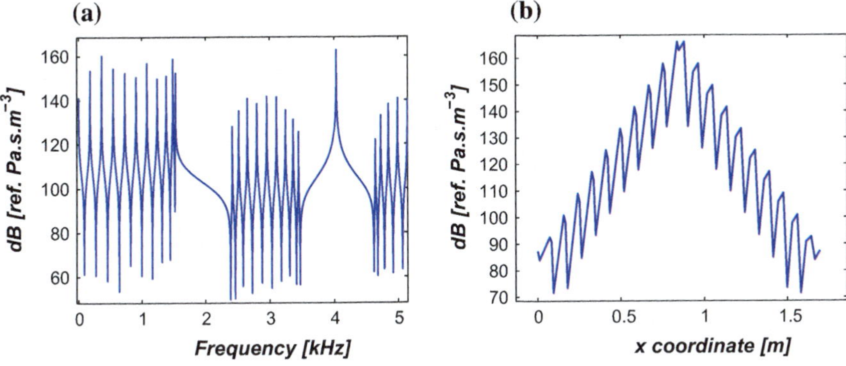

Fig. 3 PC_1-PC_2 excited at interface. **a** FRF at the interface, **b** spatial distribution of pressure at $f_i = 4035$ Hz

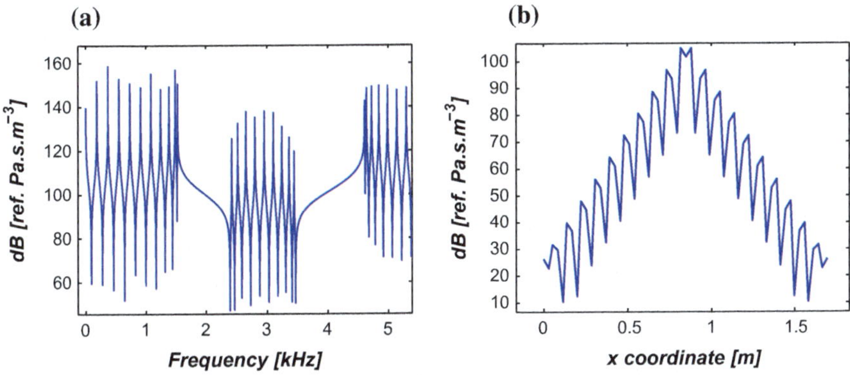

Fig. 4 PC_1-PC_1 excited at interface. **a** FRF at the interface, **b** spatial distribution of pressure at $f_i = 4035$ Hz

Fig. 3b. If a PC_1-PC_1 coupling is used instead (same band gap topological characteristics), the interface mode is not present, which is shown on Fig. 4. In such case, the spatial distribution of the pressure field for frequency f_i is concentrated at the interface but only because the frequency is inside the band gap, and therefore the waves are evanescent. The amplitudes of the pressure field, however, are much lower when compared with the PC_1-PC_2 coupling.

If the same PC_1-PC_2 coupling is excited at the left end instead of directly at the interface, the interface mode is still present, which is shown on Fig. 5. However, the magnitude of the interface peak is much smaller because the frequency f_i lies inside a band gap and, therefore, the waves will undergo significant attenuation before reaching the interface.

For a higher interface peak when the structure is excited away from the interface, a PC_3-PC_4 coupling can be used instead. In this case, the band gap is smaller and,

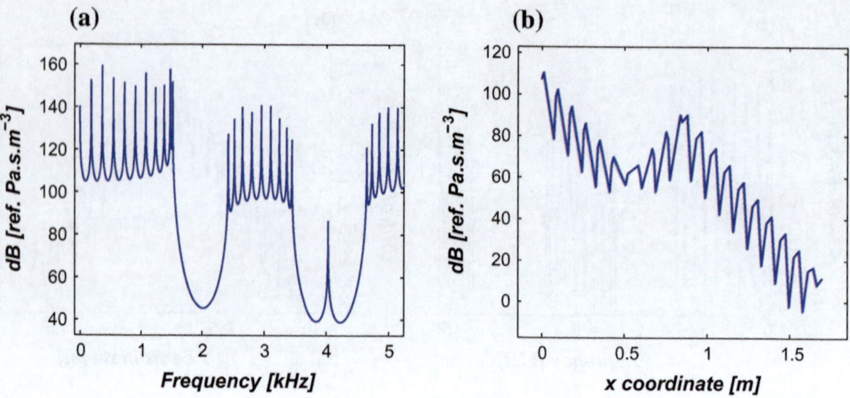

Fig. 5 PC_1-PC_2 excited at left end. **a** FRF at the interface, **b** spatial distribution of pressure at $f_i = 4035$ Hz

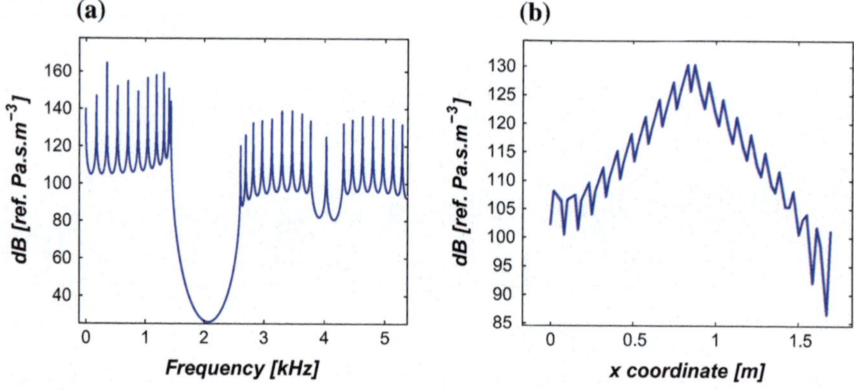

Fig. 6 PC_3-PC_4 excited at left end. **a** FRF at the interface, **b** spatial distribution of pressure at $f_i = 4035$ Hz

therefore, the waves will suffer less attenuation before reaching the interface. The forced response for this coupling when excited at the left end is shown on Fig. 6, which clearly presents a higher interface peak when compared to the PC_1-PC_2 coupling. However, for the same reasons, this coupling has a lower peak when excited directly at the interface, which is shown on Fig. 7.

3 Interface Modes on Periodic Elastic Rods

The same concepts shown for acoustic systems can be extended to similar systems of elastic rods, where interface modes can also be created. To that end, PC's with the same geometry described in the previous section (shown on Fig. 1a) are used, but the tubes are now replaced by elastic rods.

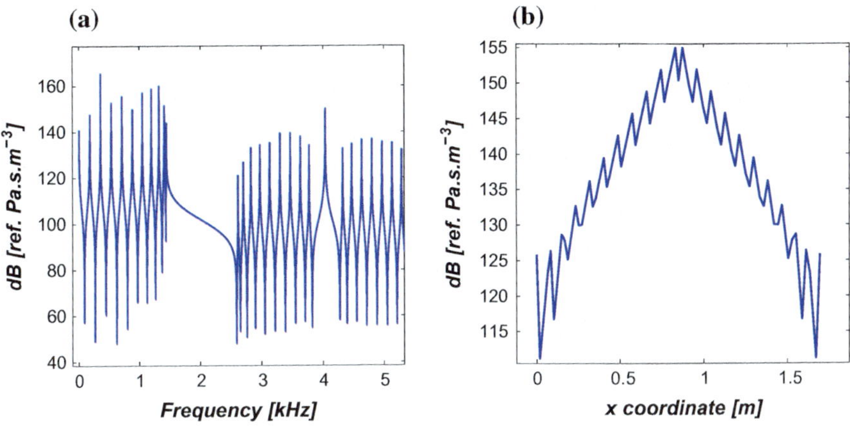

Fig. 7 PC_3-PC_4 excited at interface. **a** FRF at the interface, **b** spatial distribution of pressure at $f_i = 4035$ Hz

The speed of sound in solids is in general much greater than the speed of sound in the air. Therefore, if the same frequency for the interface mode is sought ($f_i = 4035$ Hz), the dimensions of the unit cells need to be much larger. To demonstrate that, two materials are used: steel ($E_s = 210$ Gpa, $\rho_s = 7800$ kgm^{-3} and $c_s = \sqrt{\frac{E_s}{\rho_s}} = 5188.7$ ms^{-1}) and polyacetal ($E_p = 3.3$ Gpa, $\rho_p = 1418$ kgm^{-3} and $c_p = \sqrt{\frac{E_p}{\rho_p}} = 1525.5$ ms^{-1}).

The speed of sound in steel is 15.13 times greater than the speed of sound in the air. Therefore, all the PC's dimensions are multiplied by this factor in order to maintain the bandgaps and the interface modes at the same frequencies as in the acoustic systems. The same is done for polyacetal, but by a factor of 4.45. The tube radii are multiplied by this factor only to keep the same geometric proportions, as they don't influence the frequency of the interface mode when the ratio $\frac{r_a}{r_b}$ is kept constant.

By doing that, the PC dimensions for steel become $L = 1.29$ m, $r_a = 36.31$ cm, $r_b = 22.69$ cm while for polyacetal they are set to $L = 37.8$ cm, $r_a = 10.67$ cm and $r_b = 6.67$ cm. The band gap closing and reopening process of the second band gap for both materials is very similar to what occurs for acoustic systems, and is shown on Fig. 8.

As was the case for acoustic systems, interface modes will appear on systems of elastic rods when the PC's coupled have different band gap topological characteristics. Four different PC's are used in this section: PC_5 (steel, $d_a = 94.55$ cm, $d_b = 34.04$ cm and $\Delta d = -30.25$ cm), PC_6 (steel, $d_a = 34.04$ cm, $d_b = 94.55$ cm and $\Delta d = 30.25$ cm), PC_7 (polyacetal, $d_a = 27.8$ cm, $d_b = 10.01$ cm and $\Delta d = -8.9$ cm) and PC_8 (polyacetal, $d_a = 10.01$ cm, $d_b = 27.8$ cm and $\Delta d = 8.89$ cm). All forced responses are again computed via SEM, with free-free or forced-free boundary

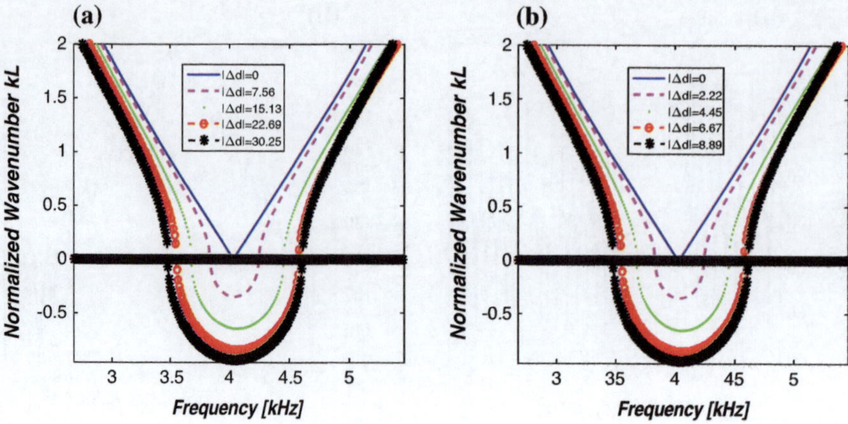

Fig. 8 Band inversion for elastic rods. **a** PC's made of steel, **b** PC's made of polyacetal, Δd dimensions in cm

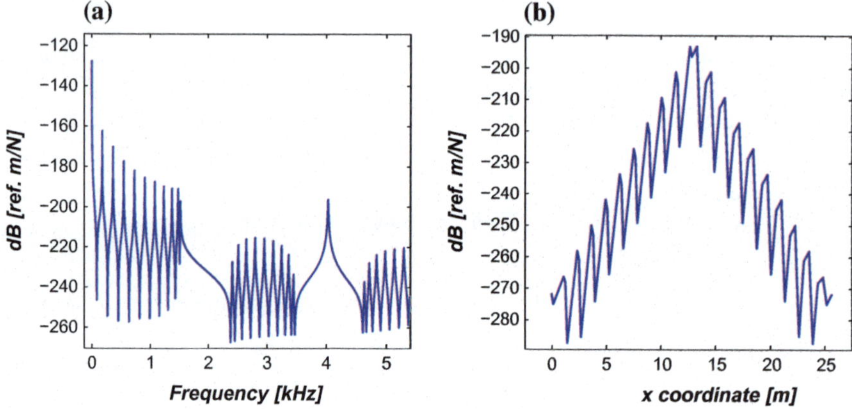

Fig. 9 PC_5-PC_6 excited at interface. **a** FRF at the interface, **b** spatial distribution of displacement amplitude at $f_i = 4035$ Hz

conditions when the applied force is at the interface or at the left end, respectively. Figures 9 and 10 show the forced responses for the PC_5-PC_6 and PC_7-PC_8 couplings, respectively. As expected, the interface modes for both materials occur at the same frequency as in the acoustic systems ($f_i = 4035$ Hz) (Fig. 11).

These interface modes can also be visualized in a surface containing the spatial distribution of displacement amplitudes for a wide frequency range (100 Hz–6 kHz). This is shown for both PC_5-PC_6 and PC_7-PC_8 couplings on Fig. 10, highlighting that interface modes may be seen as structural resonance peaks that are not only concentrated in frequency but also in space.

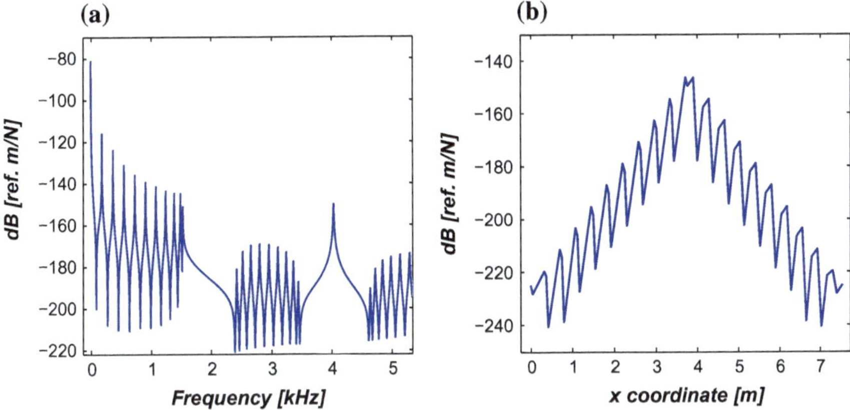

Fig. 10 PC_7-PC_8 excited at interface. **a** FRF at the interface, **b** spatial distribution of displacement amplitudes at $f_i = 4035$ Hz

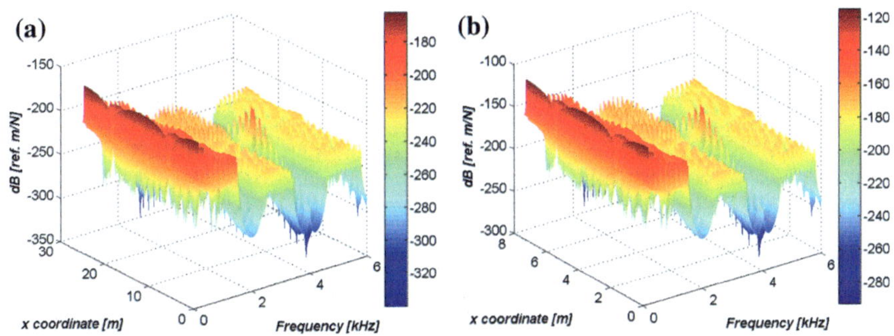

Fig. 11 Surfaces of spatial distribution of displacement amplitudes **a** PC_5-PC_6 coupling, **b** PC_7-PC_8 coupling

4 Conclusions

Interface modes that appear in the interface of two phononic crystals are characterized by a large local dynamic response and fast spatial attenuation. This feature may be useful in applications such as energy harvesting. The topic involves geometrical phase concepts that are new to the structural dynamics community. Using finite elements, Xiao et al. [6] showed that interface modes can be generated in periodic acoustic waveguides. In this paper, one-dimensional spectral elements were used to investigate the interface modes present on acoustic systems and to show they can also be created on elastic structures such as periodic systems of elastic rods. However, as expected, for elastic rods larger dimensions are required for generating interface modes at lower frequencies. This motivates our current research efforts in analyzing more complex geometries and material distributions aiming at producing bandgaps at lower frequencies with smaller dimensions and, therefore, obtaining interface modes at lower frequencies. Due to its low computational cost, the spectral element method was shown to be a suitable tool for a preliminary analysis and optimization of the geometry and materials of the periodic cells, after which a more detailed dynamic analysis (e.g. via the finite element method) and an experimental analysis may be performed.

Acknowledgements The authors would like to acknowledge the financial support of the Brazilian Agencies FAPESP (São Paulo State Research Foundation) through project number 2015/13246-3 and CNPq (National Council for Scientific and Technological Development) through projects number 482351/2013-6 and 304482/2011-0.

References

1. Hussein, M.I., Leamy, M.J., Ruzzene, M.: Dynamics of phononic materials and structures: historical origins. Recent Prog. Fut. Overlook Appl. Mech. Rev. **66**, 040802-1–38 (2014)
2. Haldane, F.D.M.: Model for a quantum hall effect without landau levels: condensed-matter realization of the parity anomaly. Phys. Rev. Lett. **61**, 2015 (1988)
3. Wang, Y.T., Luan, P.G., Zhang, S.: Coriolis force induced topological order for classical mechanical vibrations. New J. Phys. **17**, 073031 (2015)
4. Nash, M.N., Kleckner, D., Read, A., Vitelli, V., Turner, A.M., Irvine, W.T.M.: Topological mechanics of gyroscopic metamaterials. Proc. Natl. Acad. Sci. USA **115**, 14495–500 (2015)
5. Wang, P., Lu, L., Bertoldi, K.: Topological phononic crystals with one-way elastic edge waves. Phys. Rev. Lett. **115**, 104302 (2015)
6. Xiao, M., Ma, G., Yang, Z., Sheng, P., Zhang, Z.Q., Chan, C.T.: Geometric Phase and band inversion in periodic acoustic systems. Nat. Phys. **11**, 240–244 (2015)
7. Pankratov, O., Pakhomov, S., Volkov, B.: Supersymmetry in heterojunctions: band-inverting contact on the basis of $Pb_{1-x}Sn_xTe$ and $Hg_{1-x}Cd_xTe$. Solid State Commun. **61**, 93–96 (1987)
8. Zak, J.: Berry's phase for energy bands in solids. Phys. Rev. Lett. **62**, 2747–2750 (1989)
9. Doyle, J.F.: Wave Propagation in Structures, p. 320. Springer, New York, USA (1997)

Measurement of the Speed of Leak Noise Propagation in Buried Water Pipes: Challenges and Difficulties

Michael John Brennan, Fabrício Lobato César de Almeida, Fábio Kroll de Lima, Pedro Christian Ayala Castillo and Amarildo Tabone Paschoalini

Abstract To accurately determine the position of a leak in a buried plastic water pipe using acoustic correlation, a good estimate of the speed of noise propagation (wave speed) is required. The factors that affect this wave speed, and attenuation of the wave as it propagates along the pipe, include the pipe flexibility and the soil properties. These effects are discussed in this paper, and are illustrated by way of simulations for two different pipe sizes and two different soil types. It is shown that the soil type in Brazil can have a profound effect on the wave speed and hence the accuracy of leak location. Some practical problems in estimating the wave speed from in-situ measurements are also outlined. Although this is relatively simple to measure in principle, in practice it is extremely difficult to do, for a variety of reasons. Some of these are discussed and the reason why this measurement is particularly problematic with plastic water distribution pipes is illustrated.

Keywords Water leak detection · Acoustic correlation · Wave propagation

M. J. Brennan (✉) · F. K. de Lima · P. C. A. Castillo · A. T. Paschoalini
Department of Mechanical Engineering, UNESP, Ilha Solteira, SP 15385-000, Brazil
e-mail: mjbrennan0@btinternet.com

F. K. de Lima
e-mail: famil88@gmail.com

P. C. A. Castillo
e-mail: payala@dimm.com.pe

A. T. Paschoalini
e-mail: tabone@dem.feis.unesp.br

F. L. C. de Almeida
Department of Mechanical Engineering, UNESP, Tupã, SP 17602-496, Brazil
e-mail: fabricio@tupa.unesp.br

A. de T. Fleury et al. (eds.), *Proceedings of DINAME 2017*, Lecture Notes in Mechanical Engineering, https://doi.org/10.1007/978-3-319-91217-2_36

1 Introduction

Leakage occurs in buried water distribution systems through defective joints, or split pipes because of ground movement. It is estimated that 40–50% of drinking water is wasted through leakage in developing countries, and less than 10% is wasted in countries where the utilities are well-maintained, such as Japan [1, 2]. It is costly to locate and repair these leaks.

To determine if a leak is present in a specific part of the network, pressure measurements together with flow measurements are used [3]. To determine a more precise location of a leak, noise correlators are often used [4]. The cross-correlation function between two leak noise signals (acceleration, pressure or velocity) acquired at two different positions (generally hydrants) on the pipe is calculated. The peak in the correlation function is used to determine the difference in propagation times between the leak and the sensors. By combining this with knowledge of the speed at which the leak noise propagates, the location of the leak can be determined. Although correlators work well for metallic pipes, their performance on plastic pipes is more limited [5, 6]. The two main factors that affect correlator performance in this case, are the relatively high rates of attenuation experienced by waves propagating along the pipes and the variability in the speed at which they propagate along the pipe. The wave-speed is heavily influenced by the pipe properties and the surrounding soil [7–11]. The accuracy with which the leak can be located is therefore directly linked to the accuracy with which the wave speed is known. For maximum accuracy, the wave speed should be measured in-situ on the section of pipe in which there is a leak, at the same time as the correlation measurement is made. In nearly all cases, however, the wave speed is estimated from a historical database determined from calculations made using assumed material properties and pipe geometry.

This paper shows why a good estimate of the speed of leak noise propagation is of paramount importance in obtaining an accurate estimate of the location of a leak. The factors affecting the speed of the wave responsible for leak noise propagation, as well as the attenuation of this wave are discussed. A simple expression to predict the wave speed, which is dependent upon both fluid loading and soil loading factors is presented. This shows why the wave speed is found to be very different from location to location, and motivates the need for measurement of the wave speed in-situ [12]. However, there are many practical problems, which make an accurate estimate of wave speed measurements difficult. Some of these issues are also outlined in this paper.

2 An Overview of Leak Detection Using Acoustic Correlation

Figure 1 shows a typical situation in which leak noise is used to detect and locate its position. Acoustic or vibration sensors are attached to convenient access points either side of the suspected leak position. The actuators shown in the figure are not normally used in the field for leak detection, but can be used to measure the speed of the wave responsible for leak noise propagation.

In Fig. 1 the leak position d_2 from the right-hand sensor is given by, [13],

$$d_2 = \frac{d - cT_0}{2} \tag{1}$$

where c is the speed of propagation of the leak noise, $d = d_1 + d_2$ is the total distance between the sensors, and $T_0 = (d_1 - d_2)/c$ is the difference in arrival times of the leak noise at the sensor positions (time delay).

The wave that carries the leak noise in plastic pipes is predominantly a fluid-wave that is strongly coupled to the radial motion of the pipe-wall [7]. The most widely used technique to determine the time delay between sensor signals uses the cross-correlation function (CCF), $R_{12}(\tau)$, between the two measured signals $x_1(t)$ and $x_2(t)$, as shown in Fig. 1. The presence of a leak appears as a distinct peak in the CCF between the measured signals, which is given by, [13],

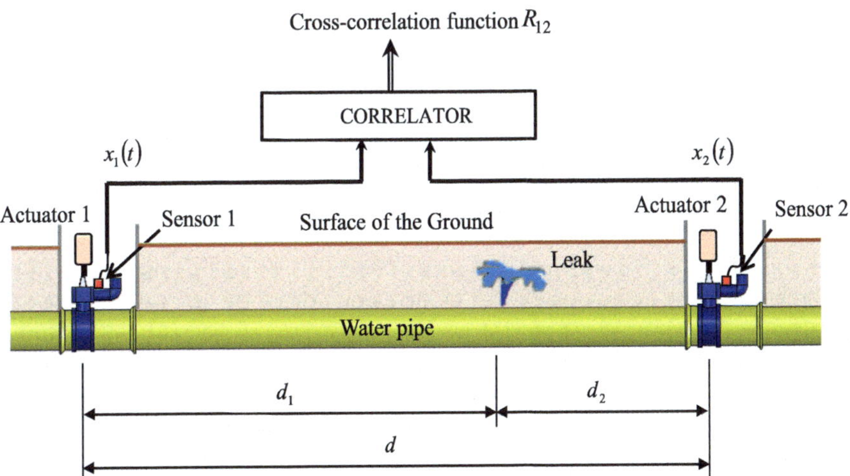

Fig. 1 Schematic of leak detection in a buried plastic water pipe using acoustic/vibration signals with a leak in between the two sensors. The actuators are used for the wave-speed measurement

$$R_{12}(\tau) = F^{-1}\{S_{12}(\omega)\} = \frac{1}{2\pi}\int_{-\infty}^{+\infty} S_{12}(\omega)e^{j\omega\tau}d\omega \tag{2}$$

where $F^{-1}\{\}$ is the inverse Fourier transform, $S_{12}(\omega)$ is the cross-spectral density function (CSD) between the measured signals, ω is circular frequency, and $j = \sqrt{-1}$. This peak in the correlation function gives the time delay estimate between the measured signals $x_1(t)$ and $x_2(t)$. Sometimes, it is preferable to express the cross-correlation function in a normalized form, which has a scale of −1 to +1. This is called the cross-correlation coefficient (CCC) and is given by $\rho(\tau) = R_{x_1x_2}(\tau)/\sqrt{R_{x_1x_1}(0)R_{x_2x_2}(0)}$, where $R_{x_1x_1}(0)$ and $R_{x_2x_2}(0)$ are respectively the autocorrelation functions at positions 1 and 2, when $\tau = 0$.

3 Effect of the Wave-Speed Estimate on Leak Location

It can be seen from Eq. (1) that accurate estimates of c and T_0 are required for an accurate estimate of the leak location. Much research into leak detection has concentrated on the estimation of the time delay, see Gao et al. [14] and the references therein, for example. The greatest error, however, is likely to be in the estimate of the wave-speed as this varies dramatically depending on the geometry and material properties of the pipe and the surrounding soil. In most cases the wave-speed is estimated from tables, which are compiled from simple calculations or from a historical database. An error in the wave-speed estimate used in Eq. (1), produces a corresponding error in the estimate of the distance d_2. The error can be determined from [12],

$$\frac{\Delta d_2}{d} = \left(\frac{1}{2} - \frac{d_2}{d}\right)\frac{\Delta c}{c} \tag{3}$$

where Δc and Δd_2 are the differences between the measured and actual wave speed, and the estimated and actual distance respectively. Equation (3) shows that as the position of the leak becomes closer to one of the measurement points then the error in the wave-speed measurement has an increasing effect. As an example, consider an extreme case of a length of pipe between the measurement positions of 100 m and a 10% error in the wave-speed estimate, with the leak being at one of the measurement positions. The resulting error in the location in this case is 5% of the length of the pipe, i.e. 5 m. Finding a way to reduce this error by more accurate wave-speed estimation is desirable, and some of the issues in doing this are discussed in this paper.

4 Effect of Fluid-Structure-Soil Interaction on the Wave-Speed

For a buried plastic water pipe, the surrounding soil and the material properties of the pipe can have a profound effect on the speed of leak noise propagation. This has been investigated in several papers as mentioned previously, and the key points are summarised here. It will be seen that it is desirable to measure the wave speed in-situ, whenever possible rather than rely on estimates as they are not likely to be very accurate. For a plastic pipe of mean radius a and pipe-wall thickness h, with complex Young's modulus $E^* = E(1 + \mathrm{j}\eta)$, where η is the loss factor, and density ρ, containing water with bulk modulus B_w, which is buried in soil with bulk modulus B_s and shear modulus G, the speed of noise propagation is governed by the speed of a coupled fluid-structural wave in the pipe, which is given by [9].

$$c = \frac{\omega}{\mathrm{Re}\{k\}} \tag{4}$$

where k is a wavenumber given approximately by $k = k_f\left(1 + K_{\mathrm{water}}/\left(K_{\mathrm{pipe}} + K_{\mathrm{soil}}\right)\right)^{\frac{1}{2}}$ in which $K_{\mathrm{water}} = 2B_w/a$ is the dynamic stiffness (pressure/displacement) of the water in the pipe, $K_{\mathrm{pipe}} = E^* h/a^2 - \rho h \omega^2$ is the dynamic stiffness of the pipe-wall and $K_{\mathrm{soil}} = K_c + K_s$ is the dynamic stiffness of the surrounding soil, where K_c and K_s are the dynamic stiffnesses of the compressional and shear waves in the soil, and are given by

$$K_c = \left(B_s - \frac{2G}{3}\right)\frac{k_d^2}{k_d^r}\left(1 - 2\frac{k_1^2}{k_r^2}\right)\frac{H_0\left(k_d^r a\right)}{H_0'\left(k_d^r a\right)} - 2Gk_d^r\left(1 - 2\frac{k^2}{k_r^2}\right)\frac{H_1''\left(k_d^r a\right)}{H_0'\left(k_d^r a\right)} \quad \text{and}$$

$$K_s = -4Gk_r^r\frac{k^2}{k_r^2}\frac{H_1'\left(k_r^r a\right)}{H_1\left(k_r^r a\right)}$$

where the soil radial wavenumbers k_d^r and k_r^r are given by $k_d^r = \sqrt{k_d^2 - k^2}$ and $k_r^r = \sqrt{k_r^2 - k^2}$ respectively, and k_d and k_r are the compressional and shear wavenumbers in the soil respectively; $H_0(\bullet)$ and $H_1(\bullet)$ are Hankel functions of zero order and second kind, and $'$ denotes a spatial derivative. Note that in this formulation, the axial stress at the interface between the pipe and the soil is considered to be negligible, as this has been found to have only a small effect on the wave speed and wave attenuation [11].

Two different pipe sizes are considered. One is found in the UK [9], and one is found in Brazil. As well as this, two soil types are considered for each pipe (soil type A is representative of much of the soil found in the UK, and soil type B is representative of the soil found in São Paulo). The properties of the pipes and the soils are given in Tables 1 and 2. The wave speeds for the two pipes and for the two soil types are shown in Fig. 2a and the attenuation in dB/m, which is given by $20\,\mathrm{Im}\{k\}/\ln(10)$, is shown in Fig. 2b for the frequency range 100 Hz–1 kHz. This

Table 1 Pipe properties

Properties	UK	Brazil
Young's modulus, E (N/m^2)	2×10^9	2×10^9
Density ρ (kg/m^3)	900	900
Loss factor	0.06	0.06
Pipe radius (mm)	84.5	35.8
Pipe-wall thickness (mm)	11	3.4

Table 2 Soil and water properties

Properties	Soil type A	Soil type B	Water
Bulk modulus $B_{s,w}$ (N/m^2)	5.3×10^7	4.5×10^9	2.25×10^9
Shear modulus, G (N/m^2)	2.0×10^7	1.8×10^8	
Density ρ (kg/m^3)	2000	2000	

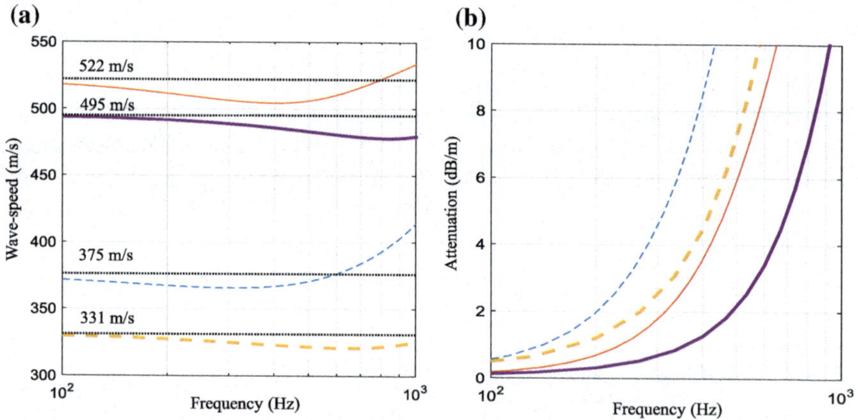

Fig. 2 Noise propagation characteristics **a** wave-speed, **b** wave attenuation. Dashed thin line, UK pipe, soil type A; solid thin line, UK pipe, soil type B; thick dashed line, Brazil pipe, soil type A; thick solid line, Brazil pipe, soil type B

frequency range is chosen as this is the range in which measured leak noise is generally found. The following observations can be made.

- The flexibility of the pipe significantly reduces the wave-speed in the predominantly fluid wave (for a rigid pipe the fluid wave propagates at 1500 m/s). The stiffness of the soil counteracts this to some extent, increasing the wave speed.
- The attenuation of the wave increases with frequency. Some of the attenuation is due to damping in the pipe-wall, but the greatest part of the attenuation at high frequencies is due to the radiation of leak noise into the soil.

Because of the attenuation, a wave cannot travel for long distances along a buried plastic pipe, and the measured vibration on the pipe tends to be at low frequency [10]. Although the wave-speed is function of frequency, it can be seen that it does not change dramatically, and a low frequency approximation to the wavenumber can be used to determine an approximate value. Unfortunately, an approximation cannot be used to predict the wave attenuation.

The approximation for the real part of the wavenumber is given by

$$\mathrm{Re}\{k\} = k_f\left(1 + \frac{\frac{2B_w}{Eh/a}}{1 + \frac{2G}{Eh/a}}\right)^{\frac{1}{2}} \tag{5}$$

where $2B_w/(Eh/a)$ and $2G/(Eh/a)$ are the fluid loading and soil loading terms respectively. It is clear that an increase in the fluid loading term increases the wavenumber, and hence decreases the wave-speed, and an increase in the soil loading term (due to the shear stiffness of the soil), decreases the wavenumber and hence increases the wave-speed. These effects are illustrated in Fig. 3, which shows the wave-speed normalised by the wave-speed in water (1500 m/s) as a function of the soil loading term, for the UK pipe and the Brazilian pipe whose parameters are given in Table 1. It can be seen that both the fluid and the soil loading terms can have a profound effect on the wave-speed. As mentioned previously, accurate knowledge of the wave-speed is needed for an accurate estimate of the leak location. This motivates the need to measure the wave-speed in-situ.

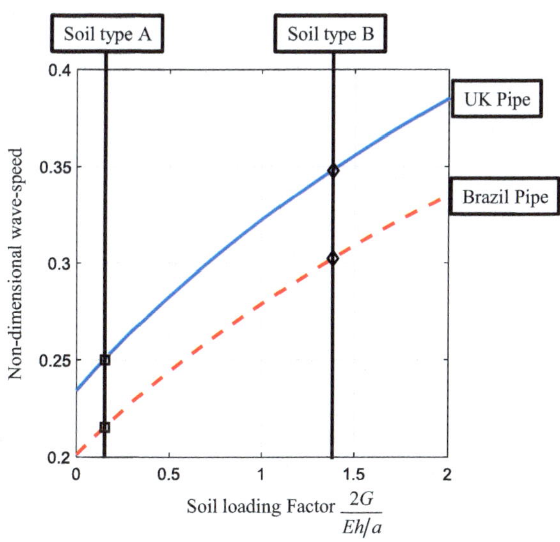

Fig. 3 Non-dimensional wave speed (wave-speed divided by 1500 m/s—wave-speed in water in a rigid pipe) as a function of the soil loading factor for the Brazil pipe (*h*/*a* = 0.095), solid blue line and for the UK pipe (*h*/*a* = 0.132), dashed red line

5 Measurement of Wave-Speed

To measure the wave-speed in a buried pipe, a wave has to be generated in the pipe and two signals measured at access points which are of a known distance apart. The excitation can be done with shakers as in [12] in a configuration as shown in Fig. 1 or by creating a leak at a known position. Typical processed signals are shown for measurements made on a buried plastic pipe rig in the UK using accelerometers. Figure 4a shows the modulus of the CSD normalised by its maximum value, and Fig. 4b shows the phase. Also, shown in Fig. 4b is a straight line corresponding to $\phi = -\omega T_0$. The coherence is shown in Fig. 4c, where it can be seen that the bandwidth over which there is potentially time delay information is about 20–120 Hz corresponding to the frequency range at which the coherence is not close to zero. Finally, Fig. 4d shows the cross-correlation coefficient in which the time delay is indicated.

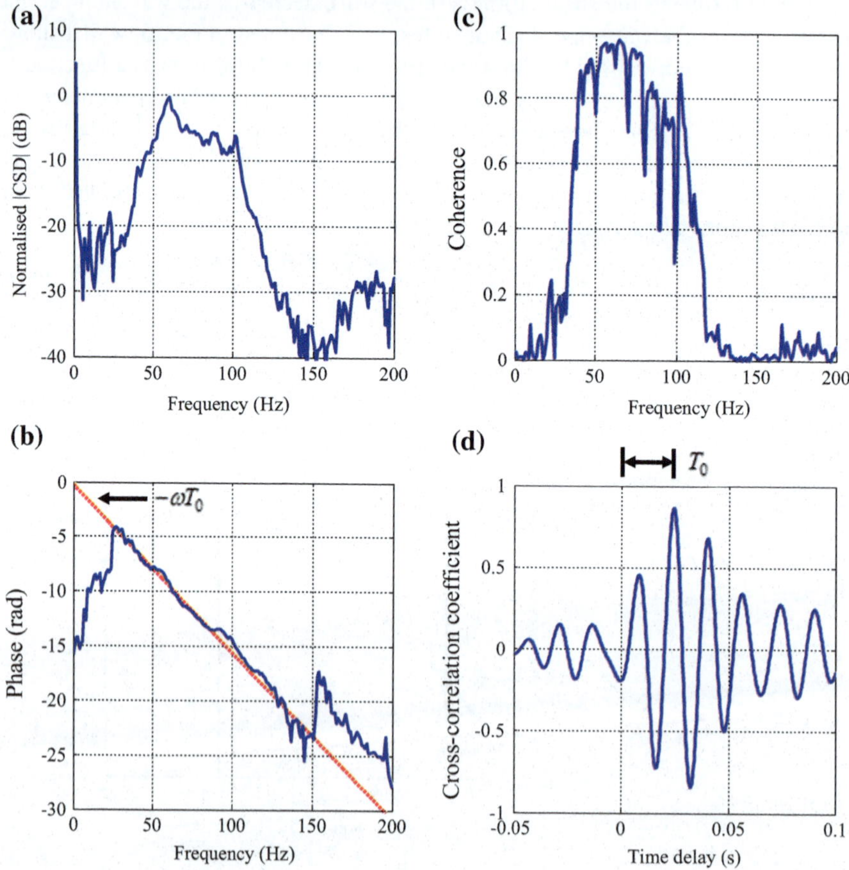

Fig. 4 Processed leak signals from the test-rig shown in Fig. 2. **a** Modulus of CSD normalised by the maximum value, **b** phase, **c** coherence, **d** cross-correlation coefficient

Referring to Fig. 1, $d_1 = 30$ m, $d_2 = 20$ m, and the measured time delay was about 25 ms, which results in a wave speed of about 400 m/s. Of particular note in Fig. 4b is the deviation of the measured phase from the phase that would be measured purely due to a time delay. This deviation can be due to several reasons, including noise, structural dynamics of the pipe system [15], and wave reflections from discontinuities in the pipe system [16]. The effect of a resonance due to structural dynamics is to significantly reduce the bandwidth over with the time delay is estimated by correlation. The effect of reflections is to cause a confusing picture in the cross-correlation function. The effect of bandwidth and the centre frequency of this bandwidth on the ability to determine an accurate estimate of time delay in the presence of reflections is further discussed here.

For simplicity, the attenuation in the pipe is neglected (which is equivalent to using the PHAT estimator [14]), so that the cross-correlation function is given by

$$R_{12}(\tau) = \frac{\Delta\omega}{\pi} \frac{\sin(\Delta\omega(\tau - T_0)/2)}{\Delta\omega(\tau - T_0)/2} \cos(\omega_c(\tau - T_0)) \tag{6}$$

where $\Delta\omega$ is the bandwidth over which there is leak noise and ω_{centre} is the centre frequency of the band. This can be written in non-dimensional form as

$$\hat{R}_{12}(\hat{\tau}) = \frac{R_{12}(\tau)}{\Delta\omega/\pi} = \frac{\sin(\hat{\tau})}{\hat{\tau}} \cos(\alpha\hat{\tau}) \tag{7}$$

where $\hat{\tau} = \Delta\omega(\tau - T_0)/2$ and $\alpha = 2\omega_{\text{centre}}/\Delta\omega$. Equation (7) is plotted in Fig. 5 for $\alpha = 4$.

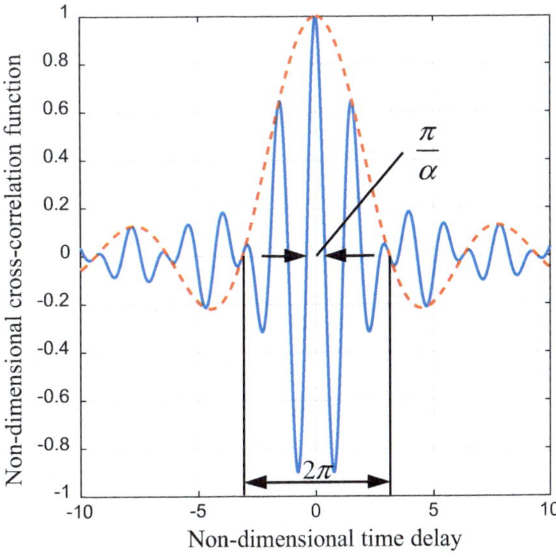

Fig. 5 Normalised cross-correlation function with the time delay set to zero

The envelope $\sin(\hat{\tau})/\hat{\tau}$ is also plotted. It can be seen that the normalized cross-correlation function has a peak at $\hat{\tau}=0$, which corresponds to $\tau = T_0$. It can also be seen that the normalized cross-correlation function oscillates within the envelope. The non-dimensional time of the first crossing point in this function can be determined by setting $\cos(\alpha\hat{\tau})=0$, which occurs when $\alpha\hat{\tau}=\pi/2$. Hence the non-dimensional time between the zero crossing points either side of $\hat{\tau}=0$ is given by π/α. The first zero crossings in the envelope, which governs the shape of the cross-correlation function occur when $\sin(\hat{\tau})=0$, which is when $\hat{\tau}=\pi$ so the non-dimensional time between the first zero crossings in the envelope is 2π. For two time delays to be detected in the cross-correlation function (which correspond to the arrival of two waves, one being the original wave and the second being a reflected wave) requires that the non-dimensional difference in the arrival times $\Delta\hat{T}$ should be such that $\Delta\hat{T} > \pi/\alpha$. Preferably, it should occur after the first zero crossings of the envelopes so that $\Delta\hat{T} > 2\pi$. In dimensional terms this is when $\Delta T > 1/(2f_{\text{centre}})$, where f_{centre} is the centre frequency of the band in Hz, or preferably when $\Delta T > 2/\Delta f$, where Δf is the bandwidth in Hz.

The discrimination of two time delays is illustrated in Fig. 6 for $\alpha=4$. Figure 6a, shows the cross-correlation for a signal where the difference between the two time delays is $\Delta\hat{T}=8$. In this case the criteria given above is fulfilled and hence there are two clear peaks corresponding to $\Delta\hat{T}=0$ and $\Delta\hat{T}=8$. Figure 6b shows a cross-correlation function where the non-dimensional difference between the two time delays is only 0.3. It can be seen that only one peak is apparent and the time delay corresponding to this peak is 0.15. This does not correspond to either time delay. In fact, as the signals corresponding to the two time delays have the same

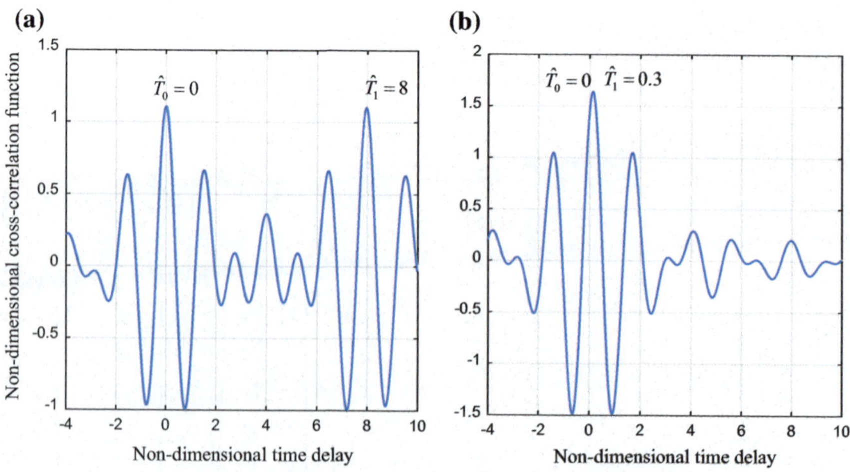

Fig. 6 Illustration of the effect of the signal parameters on the estimation of time delay with system with two time delays (reflections). **a** Two time delays are discriminated, **b** time delays are not discriminated

amplitude then the peak occurs at the mean of the two time delays as discussed in [12]. However, this is unlikely to occur in practice and so a signal that contains a wave reflection in which $\Delta T < 2/\Delta f$ can cause a considerable error in the estimate of time delay and hence the estimate of a wave speed.

6 Conclusions

This paper has discussed the importance of obtaining a good estimate of the speed of leak noise propagation in buried water pipes so as to determine an accurate estimate of the location of a leak. The factors affecting the speed of the wave responsible for leak noise propagation, as well as the attenuation of this wave, as it propagates have been described. A simple expression to predict the wave speed, which is dependent upon both fluid loading and soil loading factors, has been derived (No simple expression is possible for wave attenuation). It has been shown that while the flexibility of the pipe slows down the wave compared to a rigid-wall pipe, the shear stiffness of the soil plays an important role in counteracting this effect, increasing the wave speed. This has been found to be particularly relevant for the type of soil found in Brazil.

Concerning the estimation of wave speed by measurement, some dynamic effects that can cause inaccuracies in the estimate have been highlighted. Among them are resonance effects that can severely limit the bandwidth over which the time delay is estimated using correlation, and the effect of wave reflections in the pipe. The relationship between the bandwidth and the centre frequency of the bandwidth on the ability to differentiate between a direct wave and a reflected wave has been discussed. It has been found that if more than one wave is dominant in the pipe, then a wide bandwidth is necessary in order to obtain an accurate time delay estimate, and hence an accurate estimate of the leak location.

Acknowledgements The authors would like to thank Dr Jen Muggleton from the ISVR, University of Southampton, UK, for helpful discussions on the effects of soil loading on leak noise propagation. The authors would also like to acknowledge the financial support from FAPESP, Process No. 2013/50412-3.

References

1. Kingdom, B., Liemberger, R., Marin, P.: The challenge of reducing non-revenue water (nrw) in developing countries. Water Supply and Sanitation Board Discussion Paper Series, Paper No. 8, The World Bank, Washington (2006)
2. Grant, S.B., Saphores, J.D., Feldman, D.L., Hamilton, A.J., Fletcher, T.D., Cook, P.L.M., Stewardson, M., Sanders, B.F., Levin, L.A., Ambrose, R.F., Deletic, A., Brown, R., Jiang, S. C., Rosso, D., Cooper, W.J., Marusic, I.: Taking the "waste" out of "wastewater" for human water security and ecosystem sustainability. Science **337**, 681–686 (2012). https://doi.org/10.1126/science.1216852

3. Puust, R., Kapelan, Z., Savic, D.A., Koppel, T.: A review of methods for leakage management in pipe networks. Urban Water J. **7**(1), 25–45 (2015). https://doi.org/10.1080/15730621003610878
4. Fuchs, H.V., Riehle, R.: Ten years of experience with leak detection by acoustic signal analysis. Appl. Acoust. **33**, 1–19 (1991). https://doi.org/10.1016/0003-682X(91)90062-J
5. Hunaidi, O., Chu, W.T.: Acoustical characteristics of leak signals in plastic water distribution pipes. Appl. Acoust. **58**(3), 235–254 (1999). https://doi.org/10.1016/S0003-682X(99)00013-4
6. Hunaidi, O., Chu, W., Wang, A., Guan, W.: Detecting leaks in plastic pipes. J. Am. Water Works Assoc. **92**, 82–94 (2000)
7. Muggleton, J.M., Brennan, M.J., Pinnington, R.J.: Wavenumber prediction in buried pipes for water leak detection. J. Sound Vib. **249**(5), 939–954 (2002). https://doi.org/10.1006/jsvi.2001.3881
8. Muggleton, J.M., Brennan, M.J., Linford, P.W.: Axisymmetric wave propagation in fluid-filled pipes: measurements in in-vacuo and buried pipes. J. Sound Vib. **270**, 171–190 (2004). https://doi.org/10.1016/S0022-460X(03)00489-9
9. Muggleton, J.M., Yan, J.: Wavenumber prediction and measurement for buried fluid-filled pipes: inclusion of shear coupling at a lubricated pipe/soil interface. J. Sound Vib. **332**(5), 1216–1230 (2013). https://doi.org/10.1016/j.jsv.2012.10.024
10. Almeida, F.C.L., Brennan, M.J., Joseph, P.F., Whitfield, S., Dray, S., Paschoalini, A.T.: On the acoustic filtering of the pipe and sensor in a buried plastic water pipe and its effect on leak detection: an experimental investigation. Sensors **14**, 5595–5610 (2014). https://doi.org/10.3390/s140305595
11. Gao, Y., Sui, F., Muggleton, J.M., Yang, J.: Simplified dispersion relationships for fluid-dominated axisymmetric wave motion in buried fluid-filled pipes. J. Sound Vib. **375**, 386–402 (2016). https://doi.org/10.1016/j.jsv.2016.04.012
12. Almeida, F.C.L., Brennan, M.J., Joseph, P.F., Dray, S., Whitfield, S., Paschoalini, A.T.: Towards in-situ measurement of wave velocity in buried plastic water distribution pipes for the purposes of leak location. J. Sound Vib. **359**, 40–55 (2015). https://doi.org/10.1016/j.jsv.2015.06.015
13. Gao, Y., Brennan, M.J., Joseph, P.F., Muggleton, J.M., Hunaidi, O.: A model of the correlation of leak noise in buried plastic water pipes. J. Sound Vib. **277**, 133–148 (2004). https://doi.org/10.1016/j.jsv.2003.08.045
14. Gao, Y., Brennan, M.J., Joseph, P.F.: A comparison of time delay estimators for the detection of leak-noise signals in buried plastic water distribution pipes. J. Sound Vib. **292**, 552–570 (2006). https://doi.org/10.1016/j.jsv.2005.08.014
15. Almeida, F.C.L., Joseph, P.F., Brennan, M.J., Whitfield, S., Dray, S.: The dynamic behaviour of a buried water pipe and its effect on leak location using acoustic methods. Key Eng. Mater. **569**, 1194–1201 (2013). https://doi.org/10.4028/www.scientific.net/KEM.569-570.1194
16. Gao, Y., Brennan, M.J., Joseph, P.F.: On the effects of reflections on time delay estimation for leak detection in buried plastic water pipes. J. Sound Vib. **325**(3), 649–663 (2009). https://doi.org/10.1016/j.jsv.2009.03.037

Passive Control of Noise Propagation in Tube Systems Using Bragg Scattering

Vinícius Dias de Lima, José Maria Campos dos Santos and José Roberto F. Arruda

Abstract Noise control in acoustic tube systems is a classical problem. The use of periodic geometries and resonators is also classic in acoustic filter design. The phononic approach to the problem is much more recent. Looking at this classic problem with a novel approach may lead to innovative solutions. This work investigates the band gaps created in acoustic pipe systems using axisymmetric finite element models, wave finite element models and experiments. Periodic geometry variations are investigated. The Floquet-Bloch theorem is used on a transfer matrix of the periodic cell rearranged from a dynamic stiffness matrix to obtain the dispersion diagrams that reveal the band gaps caused by Bragg scattering. Numerical predictions of the forced response obtained with the full finite element axisymmetric model of a duct system with five cells are compared with a wave finite element model and with experimental results.

Keywords Periodic system · Bragg scattering · Passive noise control · Ducts

1 Introduction

The study of periodic structures began with Mead's work in the 70s [5–7]. In the 80s the growing of computational power available allowed the widespread use of numerical methods and the solution of engineering problems without analytical solution. Early this century a new method, called Wave Finite Element (WFE) method was proposed to predict the behavior of a structure by applying the periodicity

V. D. de Lima (✉) · J. M. C. dos Santos · J. R. F. Arruda
University of Campinas UNICAMP-FEM-DMC, Rua Mendeleyev, 200,
Campus Univ. Zeferino Vaz, Campinas, SP CEP13083-860, Brazil
e-mail: vinidiaslima@gmail.com
URL: http://www.fem.unicamp.br/~lva/

J. M. C. dos Santos
e-mail: zema@fem.unicamp.br

J. R. F. Arruda
e-mail: arruda@fem.unicamp.br

A. de T. Fleury et al. (eds.), *Proceedings of DINAME 2017*, Lecture Notes in Mechanical Engineering, https://doi.org/10.1007/978-3-319-91217-2_37

condition of Floquet-Bloch's theorem [8]. This method consists in modeling a periodic cell using conventional Finite Element Method (FEM) and then using propagation models to predict the forced response of a periodic structure.

Acoustic ducts have applications in a large variety of engineering problems. The most ordinary examples are exhaust systems of combustion engines and ventilation systems [10]. The noise propagation in such systems can be controlled via the use of acoustic filters. For this purpose, the design of periodic geometries and Helmholtz resonators is a classical way to reduce noise at a specified frequency bands. Boström [1] studied the wave propagation in ducts with a periodic variation of cross-sectional area. Bradley [2] investigated acoustic wave propagation in periodic waveguides. More recently, Munday et al. [9] addressed the problem of band gaps in periodic waveguides, and Wang and Mak [11] investigated ducts with a periodic array of Helmholtz resonators.

This work investigates band gaps generated by an acoustic tube system, consisting of a five cells constructed with pipes and expansion cavities. A numerical solution by the finite element model is developed using axisymmetric triangular elements. The numerical predictions and experimental results are compared for validating the finite element model.

2 Acoustic Finite Element Formulation

The non-dissipative wave equation can be written in terms of acoustic pressure as [4]:

$$\nabla^2 \boldsymbol{p} = \frac{1}{c^2}\frac{\partial^2 \boldsymbol{p}}{\partial t^2} \tag{1}$$

where c is the velocity of sound, $\boldsymbol{p}$ is acoustic pressure and t is time. The acoustic pressure field in tube systems excited by plane waves can be modeled as axisymmetric. This characteristic allows solving a three-dimensional problem using a two-dimensional model. The problem is formulated using cylindrical coordinates (radial distance r, height z, azimuth θ) with no dependency of θ, and Eq. (1) can be rewritten as

$$\frac{1}{r}\frac{\partial}{\partial r}\left(r\frac{\partial \boldsymbol{p}}{\partial r}\right) + \frac{\partial^2 \boldsymbol{p}}{\partial z^2} = \frac{1}{c^2}\frac{\partial^2 \boldsymbol{p}}{\partial t^2} \tag{2}$$

To solve Eq. (2) in a specified volume, boundary conditions must be applied at its surface boundaries. Applying $\boldsymbol{p} = \boldsymbol{0}$ on a surface implies a free surface of fluid. For a rigid boundary, the boundary condition is

$$\frac{\partial \boldsymbol{p}}{\partial n} = -\rho \ddot{\boldsymbol{u}}_n \tag{3}$$

where ρ is the mass density of the fluid, n is the outward unit surface normal vector and $\ddot{\boldsymbol{u}}_n$ is the boundary acceleration in direction of n.

This boundary value problem is usually solved by finite element analysis, using Galerkin's Method to obtain an approximated solution [3]. After discretization, the system of equations to be solved is

$$\mathbf{M_a}\ddot{\boldsymbol{p}} + \mathbf{K_a}\boldsymbol{p} = \boldsymbol{f} \tag{4}$$

where $\mathbf{K_a}$ is the acoustic stiffness matrix, $\mathbf{M_a}$ is the acoustic mass matrix, $\boldsymbol{f}$ is the acoustic excitation vector and $\boldsymbol{p}$ is the acoustic pressure nodal vector. For an axisymmetric element model, the acoustic element mass matrix, the acoustic element stiffness matrix and the acoustic load vector are given, respectively, by [3]

$$\mathbf{K}_{\mathbf{a}}^{(\mathbf{e})} = 2\pi\bar{r}\int_A \left(s_r^T s_r + s_z^T s_z + s_\theta^T s_\theta\right) dA \tag{5}$$

$$\mathbf{M}_{\mathbf{a}}^{(\mathbf{e})} = \frac{2\pi\bar{r}}{c^2}\int_A \boldsymbol{s}^T \boldsymbol{s} dA \tag{6}$$

$$\boldsymbol{f}^{(e)} = -\rho \boldsymbol{s}^T \ddot{\boldsymbol{u}}_n A \tag{7}$$

where r is the radial distance of element centroid, A is the element revolution area, $\boldsymbol{s}$ is the shape function and $\mathbf{s_r}$, $\mathbf{s_z}$ and $\mathbf{s}_\theta$ are its derivatives with respect to each cylindrical coordinate.

The dynamic stiffness matrix can be obtained as

$$\mathbf{D} = \mathbf{K_a} - \omega^2\mathbf{M_a} \tag{8}$$

which can be partitioned in terms of internal, left-sided and right-sided degrees of freedom by

$$\begin{bmatrix} \mathbf{D_{ii}} & \mathbf{D_{il}} & \mathbf{D_{ir}} \\ \mathbf{D_{li}} & \mathbf{D_{ll}} & \mathbf{D_{lr}} \\ \mathbf{D_{ri}} & \mathbf{D_{rl}} & \mathbf{D_{rr}} \end{bmatrix} \begin{Bmatrix} \boldsymbol{p}_i \\ \boldsymbol{p}_l \\ \boldsymbol{p}_r \end{Bmatrix} = \begin{Bmatrix} \mathbf{0}_i \\ \boldsymbol{f}_l \\ \boldsymbol{f}_r \end{Bmatrix} \tag{9}$$

From Eq. (9), the internal pressures can be obtained as

$$\boldsymbol{p}_i = \mathbf{D_{ii}^{-1}}\left(\mathbf{D_{il}}\boldsymbol{p}_l + \mathbf{D_{ir}}\boldsymbol{p}_r\right) \tag{10}$$

Substituting Eq. (10) into Eq. (9), the condensed acoustic stiffness matrix is obtained as

$$\begin{bmatrix} \mathbf{D_{ll}} & \mathbf{D_{lr}} \\ \mathbf{D_{rl}} & \mathbf{D_{rr}} \end{bmatrix} \begin{Bmatrix} \boldsymbol{p}_l \\ \boldsymbol{p}_r \end{Bmatrix} = \begin{Bmatrix} \boldsymbol{f}_l \\ \boldsymbol{f}_r \end{Bmatrix} \tag{11}$$

where $\mathbf{D}_{ll} = \mathbf{D}_{ll} - \mathbf{D}_{li}\mathbf{D}_{ii}^{-1}\mathbf{D}_{il}$, $\mathbf{D}_{rl} = \mathbf{D}_{rl} - \mathbf{D}_{ri}\mathbf{D}_{ii}^{-1}\mathbf{D}_{il}$, $\mathbf{D}_{lr} = \mathbf{D}_{lr} - \mathbf{D}_{li}\mathbf{D}_{ii}^{-1}\mathbf{D}_{ir}$ and $\mathbf{D}_{rr} = \mathbf{D}_{rr} - \mathbf{D}_{ri}\mathbf{D}_{ii}^{-1}\mathbf{D}_{ir}$.

The periodicity condition allows predicting the behavior under harmonic disturbance of a periodic system modeling a unit-cell only. In this method the dynamic stiffness matrix of a unit-cell modeled by WFE is used to apply the periodicity condition in a harmonic disturbance propagating through the system. Using Floquet-Bloch's theorem [8], the periodicity condition results in an eigenvalue problem. Equation (11) can be rearranged using the Transfer Matrix formulation, resulting in

$$\underbrace{\begin{Bmatrix} \boldsymbol{p}_r \\ -\boldsymbol{f}_r \end{Bmatrix}}_{\boldsymbol{q}_r} = \underbrace{\begin{Bmatrix} -\mathbf{D}_{lr}^{-1}\mathbf{D}_{ll} & -\mathbf{D}_{lr}^{-1} \\ \mathbf{D}_{rl} - \mathbf{D}_{rr}\mathbf{D}_{lr}^{-1}\mathbf{D}_{ll} & -\mathbf{D}_{rr}^{-1}\mathbf{D}_{lr} \end{Bmatrix}}_{\mathbf{T}} \underbrace{\begin{Bmatrix} \boldsymbol{p}_l \\ \boldsymbol{f}_l \end{Bmatrix}}_{\boldsymbol{q}_l} \tag{12}$$

where $\mathbf{T}$ is the transfer matrix that relates the left state vector $\boldsymbol{q}_l$ with the right state vector $\boldsymbol{q}_r$ of the unit-cell.

Considering now two consecutive unit-cells, m and $m+1$, the continuity condition of medium states that $\boldsymbol{p}_r^{(m)} = \boldsymbol{p}_l^{(m+1)}$ and $\boldsymbol{f}_r^{(m)} = -\boldsymbol{f}_l^{(m+1)}$, resulting in

$$\boldsymbol{q}_l^{(m+1)} = \mathbf{T}\boldsymbol{q}_l^{(m)} \tag{13}$$

For wave propagation in an infinite periodic system, Floquet-Blochs theorem produces an eigenvalue problem given by

$$\mathbf{T}\boldsymbol{q}_l = e^{\mu}\boldsymbol{q}_l \tag{14}$$

where e^{μ} is the eigenvalue, $\boldsymbol{q}_l$ is the eigenvector, $\mu = -ikL$ is the attenuation constant, L is the unit-cell length, k is the wavenumber and i is the imaginary unit. This solution provides the behavior in terms of wave propagation.

3 Simulated Model and Experimental Setup

3.1 Simulation Description

The system that was later experimentally verified is numerically modeled with a script implemented in Matlab®. The unit-cell is discretized with 618 triangular elements, and the whole system with five cells was simulated. Dispersion relations and Frequency Response Functions (FRFs) were obtained for each cell. The pipes and cavity walls are assumed rigid. The fluid inside the tube system is air at ambient temperature and atmospheric pressure, with 1.21 kg/m^3 and 20 °C. With these physical proprieties, FRFs and dispersion relations of the system were obtained.

3.2 Experimental Setup Description

A tube system was built with five unit cells made of polyvinyl chloride (PVC), which were constructed with two pipes connected to an expansion chamber. Each pipe has 150 mm length and 37.5 mm internal diameter. The expansion chambers have 165 mm length and 145 mm internal diameter. Table 1 summarizes the geometric properties. A scheme of the experimental setup is shown in Figs. 1 and 2 illustrates the unit-cell and its dimensions.

The system is excited with a volume acceleration at one end. This excitation was applied using a PVC piston with circular cross section of 37 mm diameter, coupled to an electrodynamic shaker. Mounted on the piston, a piezoelectric accelerometer measures the piston acceleration, linearly proportional to the air volume acceleration. The gap between the tube wall and the piston is sealed by a rubber membrane. At the other system termination, a microphone supported on a bar measures the pressure at the system end. Each cavity is simply supported with polypropylene foam. The FRF of pressure caused by volume acceleration was measured with ten averages, a frequency band of 1125 Hz and frequency discretization of 0.625 Hz. The specifications of measurement instruments are summarized in Table 2. Figure 3 shows the experimental setup with all measurement instruments.

Table 1 Tube system geometric parameters

Geometric parameter	Value
Pipe length (m)	0.150
Cavity length (m)	0.165
Total length (m)	2.325
Pipe diameter (mm)	37.5
Cavity diameter (mm)	145
Number of unit-cells	5

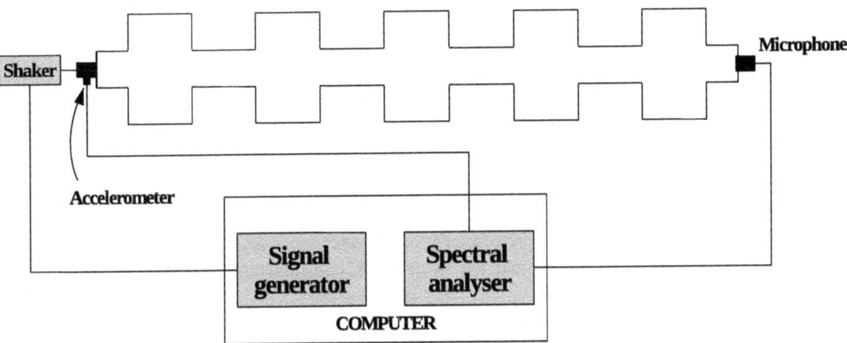

Fig. 1 Schematics of the experimental setup

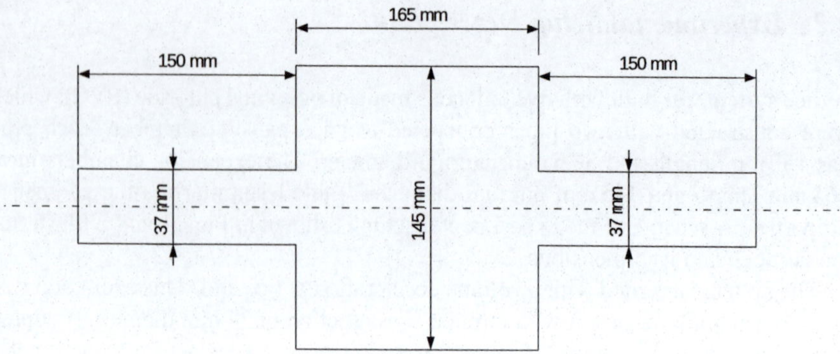

Fig. 2 Unit-cell dimensions

Table 2 Measurement instrument list

Instrument	Manufacturer and model	Sensitivity	Measure range
Accelerometer	Kistler 8614A500M1	3.41 mV/g	± 500 g
Microphone	G.R.A.S 26CA	46.6 mv/Pa	± 0.2 dB
Shaker	TMS K2004E	–	0–11 kHz
Data Acquisition	LDS Dactron Photon II	–	–

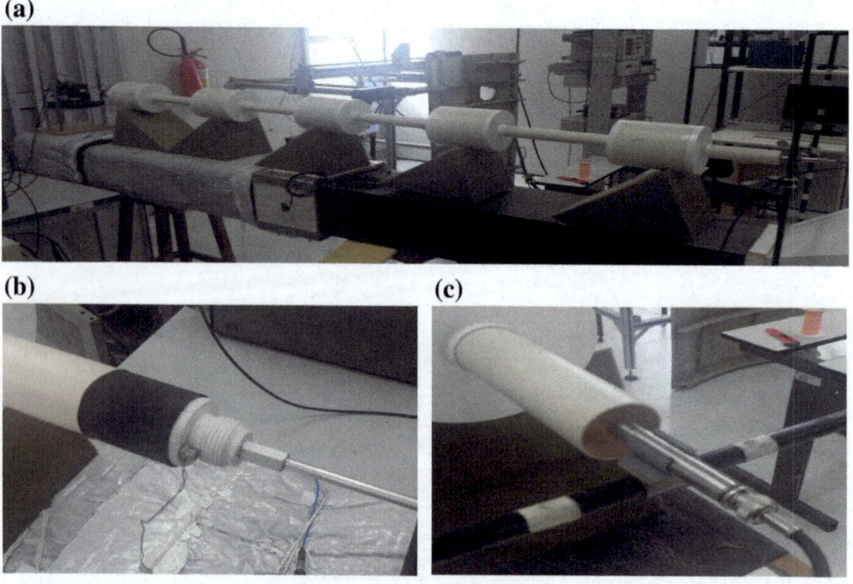

Fig. 3 Experimental setup: **a** system overview; **b** piston with accelerometer; **c** microphone on exit termination

4 Results and Conclusions

Figure 5 shows the dispersion diagram for the acoustic periodic cell. Two band gaps are evident, one in the 100–500 Hz range and one in the 500–900 Hz range. These are typical Bragg scattering band gaps, where the band gap is caused by interference of reflected waves. The FRF in Fig. 4 shows that at the band gaps the response is strongly attenuated. A good agreement is found between the FE and the WFE solutions and a good qualitative agreement is observed between numerical predictions and experiment.

The methodology exposed in this work may be used to design and optimize periodic duct geometries to attenuate duct noise in practical applications. The dispersion analysis of a single periodic cell is sufficient to predict the existence and frequency range of the band gaps. A similar analysis may be conducted with a different strategy consisting of introducing periodic Helmholtz resonators. This may be shown to create band gaps at lower frequencies, but band gaps caused by this local resonance effect are much narrower.

The WFE method may be used to compute the forced response of a finite periodic structure with a lower computational cost compared with a full finite element solution. The experimental results show that numerical methods may be used to predict band gaps that are reasonably robust with respect to small variations of the periodic cell, unavoidable when building the acoustic duct system.

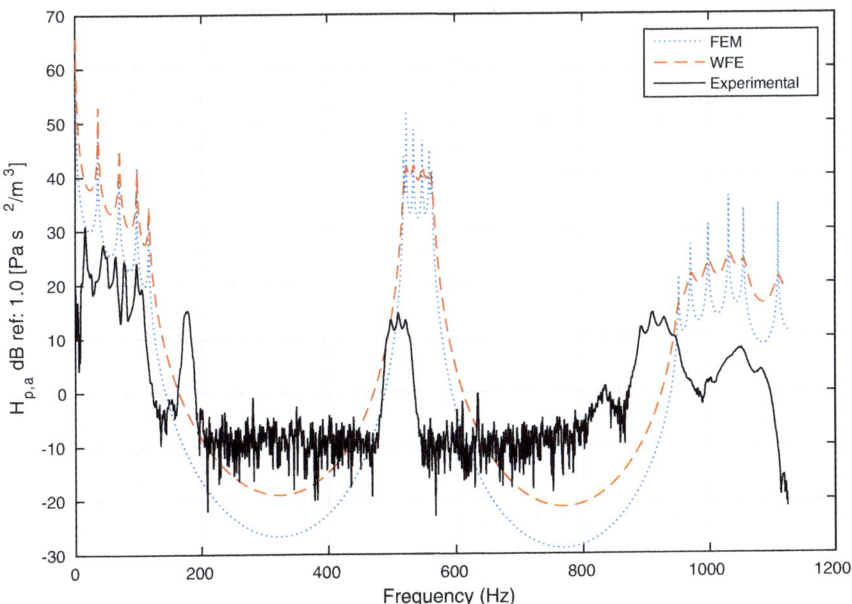

Fig. 4 Simulated and experimental FRF's

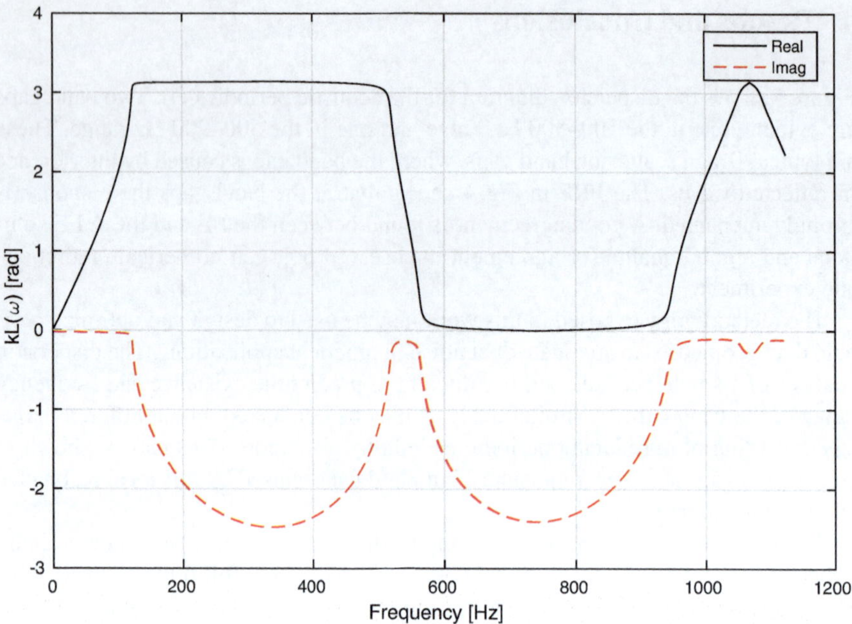

Fig. 5 Dispersion relation

References

1. Boström, A.: Acoustic waves in a cylindrical duct with periodically varying cross section. Wave Motion **5**, 59–67 (1982). https://doi.org/10.1016/0165-2125(83)90007-0
2. Bradley, C.E.: Time harmonic acoustic bloch wave propagation in periodic waveguides. J. Acoust. Soc. Am. **96**(3), 1844–1853 (1994). https://doi.org/10.1121/1.410196
3. Cook, R.D., Malkus, D.S., Plesha, M.E., Witt, R.J.: Concepts and Applications of Finite Element Analysis. Wiley, Chichester, United States (2002)
4. Kinsler, L.E.: Fundamentals of Acoustics. Wiley, New York, United States (1982)
5. Mead, D.: A general theory of harmonic wave propagation in linear periodic systems with multiple coupling. J. Sound Vib. **17**(2), 235–260 (1973). https://doi.org/10.1016/0022-460X(73)90064-3
6. Mead, D.: Free wave propagation in periodically supported, infinite beams. J. Sound Vib. **11**(2), 181–197 (1970). https://doi.org/10.1016/S0022-460X(70)80062-1
7. Mead, D.: Wave propagation and natural modes in periodic systems: I. mono-coupled systems. J. Sound Vib. **40**(1), 1–18 (1975). https://doi.org/10.1016/S0022-460X(75)80227-6
8. Mencik, J.M., Ichchou, M.N.: Multi-mode propagation and diffusion in structures through finite element. Eur. J. Mech.-A/Solids **24**(5), 877–898 (2005). https://doi.org/10.1016/j.euromechsol.2005.05.004
9. Munday, J.N, Bennet, C.B., Robertson, W.M.: Band gaps and defect modes in periodically structured waveguides. J. Acoust. Soc. Am. **112**(4), 1353–1358 (2002). https://doi.org/10.1121/1.1497625
10. Munjal, M.L.: Acoustics of ducts and mufflers with applications to exhaust and ventilation system design. Wiley, New York, United States (1987)
11. Wang, X., Mak, C.M.: Wave propagation in a duct with a periodic helmholtz resonators array. J. Acoust. Soc. Am. **131**(2), 1172–1182 (2011). https://doi.org/10.1121/1.3672692

Dynamic Models for Transmission Lines and Hoses

Petter Krus

Abstract In this paper simplified models for hydraulic transmission lines and hoses, for both time and frequency domain simulation, are presented. Flexible hoses have, in addition to a higher capacitance, also an considerable damping effect, that can reduce noise and vibrations, and in this paper, efficient approximate models for flexible hoses are presented. In hydraulic transmission lines with laminar flow the losses can be divided into two parts. One term that is distributed friction, and one term that is frequency dependent. It is shown that in general, the effect from the hose wall dominate the frequency response characteristics over the frequency dependent friction. A very simple frequency dependent model of the damping term of the hose can then be combined with an equally simple model of the distributed friction to represent a simple but accurate model of a flexible hose for system simulation in the time domain.

Keywords Transmission line · Flexible hose · Dynamics · Simulation model

1 Introduction

Hydraulic transmission lines are omnipresent elements in hydraulic systems. In many cases, when hydraulic systems are analysed and simulated, it is justified to regard pipelines as just lumped restrictors and capacitances. In many cases, however, wave propagation effects in lines may become significant.

The behaviour of fluid lines, with wave propagation, was of interest already for pipes for water distribution where water hammer effects was early identified as a problem. In Stecki et al. [11, 12] an overview of the historic development of models is described, and the different models compared.

Wave propagation is also important in order to predict fluid transients in long lines. One important area is to predict pressure amplitudes in system with periodic

P. Krus (✉)
Division of Fluid and Mechatronic Systems, Department of Management and Engineering, Linköping University, Linköping, Sweden
e-mail: petter.krus@liu.se

A. de T. Fleury et al. (eds.), *Proceedings of DINAME 2017*, Lecture Notes in Mechanical Engineering, https://doi.org/10.1007/978-3-319-91217-2_38

excitations, such as from hydraulic pumps and with appropriate models, efficient attenuators can be designed [10].

Noise in hydraulic systems is one of the major problems with hydraulic systems. For prediction of pump pulsations in hydraulic systems, modelling in the frequency domain is preferable. In this way attenuators can be designed to suppress e.g. frequencies generated by the pump. Here, a transmission line element is modelled with a four-pole equation. This is also described in [14].

In this way complex pipe systems can be analysed at low cost. With appropriate models, pressure amplitudes in system with periodic excitations, such as from hydraulic pumps, can be predicted, and be used to design efficient attenuators [10].

Their impact on hydraulic control systems can also be of importance [14].

2 Basic Equations for a Transmission Line

A general transmission line can be described by the four-pole equation. See e.g. Viersma [14]. Here capitals are used to indicate Laplace transformed variables (Fig. 1).

$$\begin{pmatrix} -Q_2 \\ P_2 \end{pmatrix} = \begin{pmatrix} A_L & B_L \\ C_L & D_L \end{pmatrix} \times \begin{pmatrix} Q_1 \\ P_1 \end{pmatrix} \tag{1}$$

There are two important parameters. These are the time delay, T, due to the limited signal propagation speed, and there is the characteristic impedance Z_c. Here a negative sign has to be put on Q_2 since the definition of flow is always positive entering the line. This convention makes the equations symmetric. According to Viersma [14] frequency dependent friction can be handled by introducing the frequency dependent friction factor N. The elements in the matrix can be written as:

$$A_L = D_L = \cosh Ts\sqrt{N} \tag{2}$$

$$B_L = -\frac{1}{Z_c\sqrt{N}} \sinh Ts\sqrt{N} \tag{3}$$

$$C_L = -Z_c\sqrt{N} \sinh Ts\sqrt{N} \tag{4}$$

The time delay T can be calculated from the oil properties β and oil density ρ and the length l.

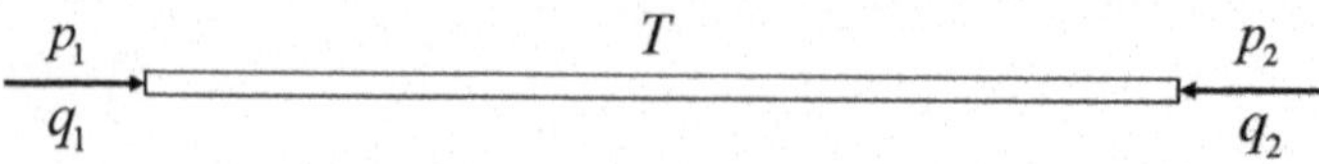

Fig. 1 Transmission line

$$T = l\sqrt{\rho/\beta} \tag{5}$$

The characteristic impedance can be calculated from the oil properties and the cross section area A.

$$Z_c = \frac{1}{A}\sqrt{\rho\beta} \tag{6}$$

Using (1) and multiplying by 2 the following expressions are obtained.

$$\frac{1}{Z_c\sqrt{N}}\left(e^{Ts\sqrt{N}} - e^{-Ts\sqrt{N}}\right)P_1 = 2Q_2 + \left(e^{Ts\sqrt{N}} + e^{-Ts\sqrt{N}}\right)Q_1 \tag{7}$$

$$\left(e^{Ts\sqrt{N}} + e^{-Ts\sqrt{N}}\right)P_1 == 2P_2 + Z_c\sqrt{N}\left(e^{Ts\sqrt{N}} - e^{-Ts\sqrt{N}}\right)Q_1$$

Adding (7) and (8) and dividing with 2 yields

$$P_1 e^{Ts\sqrt{N}} = P_2 + Z_c\sqrt{N}\left(Q_2 + Q_1 e^{Ts\sqrt{N}}\right) \tag{8}$$

Rearranging yields

$$P_1 e^{Ts\sqrt{N}} - Z_c\sqrt{N}Q_1 e^{Ts\sqrt{N}} = P_2 + Z_c\sqrt{N}Q_2 \tag{9}$$

Introducing the wave variables C_1 and C_2 such that

$$P_1 = C_1 + Z_c Q_1 \tag{10}$$
$$P_2 = C_2 + Z_c Q_2 \tag{11}$$

One often used line model is the one proposed by Trikha [13]. This model uses the method of characteristics and is reasonably accurate. The line is divided in sections with the length *ha* where *h* is the time step used in the simulation and *a* is the speed of sound in the line. Pressures and flows are computed for each part of the line. Obviously, this can be time consuming if the line is long.

The model used here have similarities to a model proposed by Karam and Leonard [6]. In this model only the pressures and flows at the ends of the line are computed which greatly reduces the computational effort. A model based on this approach with improved accuracy was also described by Krus and Palmberg [8]. In Krus et al. [9] it was described how this approach can be improved and how it can be mated to the method with characteristics to obtain a very robust, accurate and economical model.

This model gives good results both in the time and frequency domain when compared to more elaborate models. If, for some reason internal state variables are wanted, they can be obtained by representing a line with several line elements.

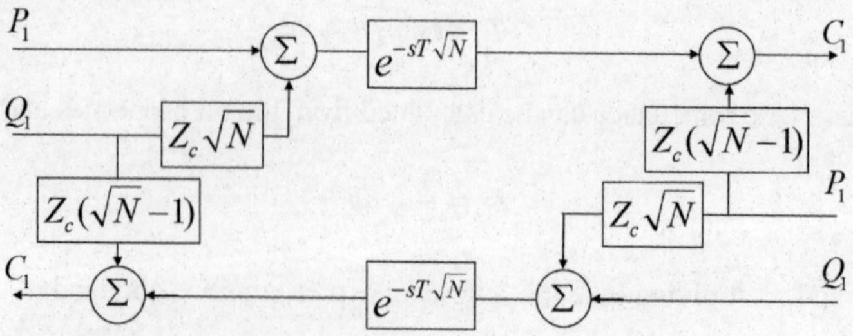

Fig. 2 Block diagram of a transmission line in the frequency domain

The introduction of wave variables is a very effective way to connect different components in simulation of systems. Essentially there are two kinds of components, those who calculate characteristics, such as lines and capacitances, and components that calculate flow and pressure from these characteristics. For a more detailed discussion on this subject see Krus et al. [7].

At each such component, the following system of equations is solved (in the time domain).

$$\begin{aligned} q &= q(p) \\ p &= c + Z_c q \end{aligned} \tag{12}$$

Solving (9) and (11) for C_1 and C_2 yields

$$C_1 = e^{-Ts\sqrt{N}}\left(P_2 + Z_c\sqrt{N}Q_2\right) + Z_c\left(\sqrt{N}-1\right)Q_1 \tag{13}$$

$$C_2 = e^{-Ts\sqrt{N}}\left(P_1 + Z_c\sqrt{N}Q_1\right) + Z_c\left(\sqrt{N}-1\right)Q_2 \tag{14}$$

In Fig. 2 the corresponding block diagram is shown.

2.1 Distributed Resistance

For the case of a uniformly distributed resistance the expression for N is (see e.g. Viersma [14]):

$$N(s) = \frac{\alpha}{s} + 1 \tag{15}$$

where

$$\alpha = \frac{R}{Z_c T} \tag{16}$$

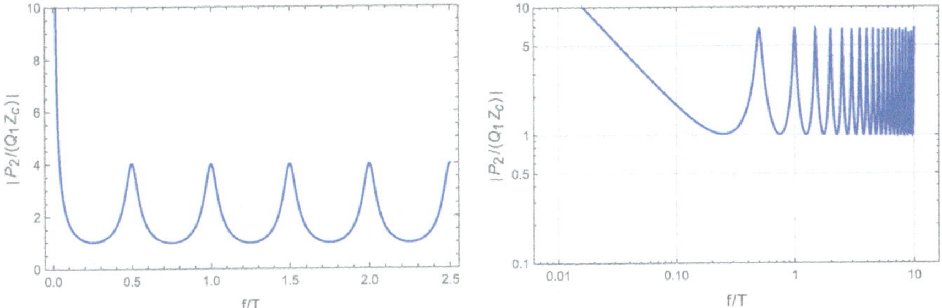

Fig. 3 The frequency spectra of the pressure at the outlet of a transmission line with a blocked outlet in linear and log-log scale

Here, R is the total resistance of the line. See e.g. Viersma [14] for laminar flow.

$$R = \frac{8\pi\eta l}{A^2} \tag{17}$$

The factor N for the distributed friction is hereafter referred to as N_R, in order to distinguish it from the general case with also frequency dependent friction. The frequency response of a transmission line with a blocked outlet can now be evaluated using the expression for N_R for distributed friction. The example in Fig. 3 is shown in dimensionless form. The only quantity that has to be specified is the ratio R/Z_c. In this example, it is set to 0.3. In this way the effect on the reduction of the resonances can be seen. Except for the first peak, corresponding to the DC-level, all the others are having almost exactly the same level.

2.2 *Frequency Dependent Resistance*

One aspect of hydraulic transmission lines that is complicated, is the frequency dependent friction. The exact solution involves expressions with Bessel functions that are relatively costly to evaluate using standard packages. However, very efficient approximations that are valid for the whole frequency range of interest have been developed here. This means that this kind of models can be used also for design optimization. According to Hams [1] and Viersma [14] frequency dependent friction can be handled by calculating N as:

$$N(s/\alpha) = -\frac{J_0\left(i\sqrt{8\frac{s}{\alpha}}\right)}{J_2\left(i\sqrt{8\frac{s}{\alpha}}\right)} \tag{18}$$

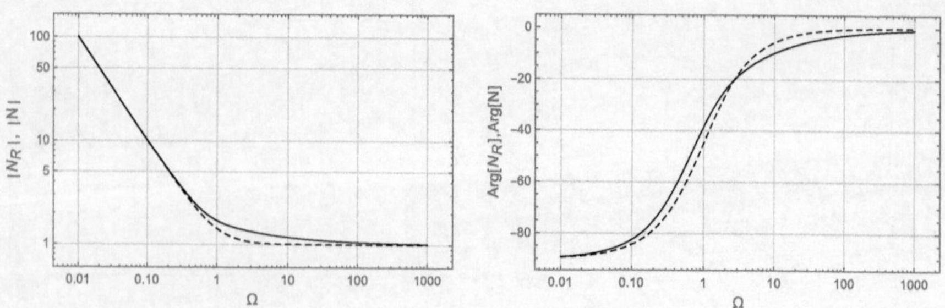

Fig. 4 The absolute values and phase of the factor $N_R(i\Omega)$, dashed line, and for distributed friction and frequency dependent friction $N(i\Omega)$, solid line. For high frequencies the frequency dependent friction has a much higher damping. They are similar except in the midrange where the frequency dependent friction has a higher value

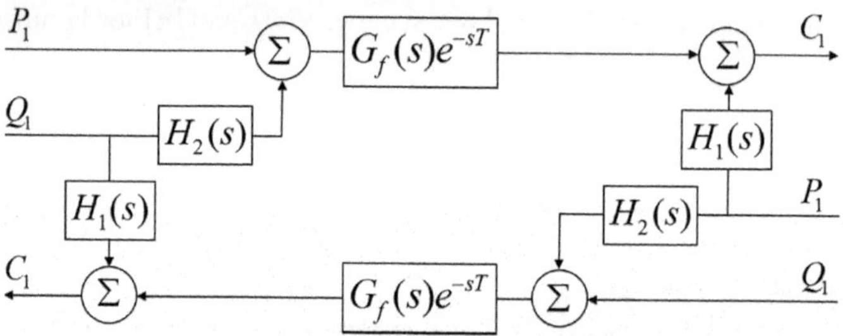

Fig. 5 Block diagram of transmission line with filters introduced for the effects of distributed frequency dependent friction

It can here be convenient to introduce the variable $S = s/\alpha$. To plot the function the Laplace variable s is substituted by $i\omega$, and consequently the non-dimensional frequency Ω is then introduced as (Fig. 4):

$$\Omega = \frac{\omega}{\alpha} \tag{19}$$

In order to be able to deal with the frequency-dependent friction one approach is to introduce the transfer function $G_f(s)$

$$G_f(s) = e^{-Ts\left(\sqrt{N(s/\alpha)}-1\right)} \tag{20}$$

If a pressure pulse is propagating in the line, the $G_f(s)$ represents a filter that is acting on the pressure signal. The block diagram of the transmission line then becomes like in Fig. 5.

Using a frequency domain description of N, the filter $G_f(s)$ can be evaluated. In order to get a general plot of $G_f(i\omega)$. the non-dimensional frequency Ω is used:

Rewriting (20) we obtain if s is substituted for $i\Omega\alpha$

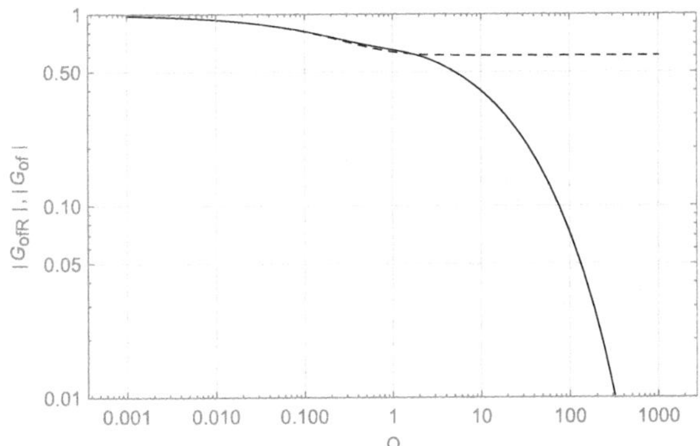

Fig. 6 The transfer function of distributed resistance $G_{ofR}(i\Omega)$, dashed line, and frequency dependent friction $G_{of}(i\Omega)$, solid line. For high frequencies the frequency dependent friction has a much higher damping

$$G_f(i\Omega\alpha) = e^{-Ti\Omega\alpha\left(\sqrt{N(i\Omega)}-1\right)} \tag{21}$$

Introducing the new function $G_{of}(i\Omega)$ and rearranging (21) we obtain

$$G_{of}(i\Omega) = G_f(i\Omega\alpha)^{\frac{1}{\alpha T}} = e^{-i\Omega\left(\sqrt{N(i\Omega)}-1\right)} \tag{22}$$

This is a function that is independent of the properties of the line, since it is only a function of the dimensionless frequency Ω. This function was introduced for the first time in Krus et al. [9].

The plots of this function for N expressed by (15) and (18) is shown in Fig. 6. Clearly there is a dramatic difference, especially compared to the results from just looking at the N functions by themselves, as in Fig. 4.

Another important observation is that $G_f(S)$ is an irrational function of S. Not only is it irrational because it involves the square root of N. Also $N(S)$ for the exact solution of frequency dependent friction for laminar flow, is a function of irrational Bessel functions. Finally, $G_f(S\alpha)$ is an irrational function of $G_{of}(S)$ since the exponent in (22) is a linear function α which is a rational number. This means that it is not possible to find an exact model for time domain simulations.

The expression with the Bessel functions can sometimes be inefficient for computations. Even though they are implemented in standard packages with many software, computations, e.g. simulations in the frequency domain can be much more efficient if good approximations can be found. One approximate expression was developed by Trikha [13]. It is:

$$N_{A1}(i\Omega) = 1 + \frac{1}{i\Omega} + \frac{0.1515}{1 + 0.3030 i\Omega} + \frac{0.1620}{1 + 0.04 i\Omega} + \frac{0.020}{1 + 0.001 i\Omega} \tag{23}$$

Another approximate expression that is introduced here, is:

$$N_{A3} = 1 + \frac{1}{i\Omega} + \frac{1/2}{1 + (k_N i\Omega)^{\xi}} \tag{24}$$

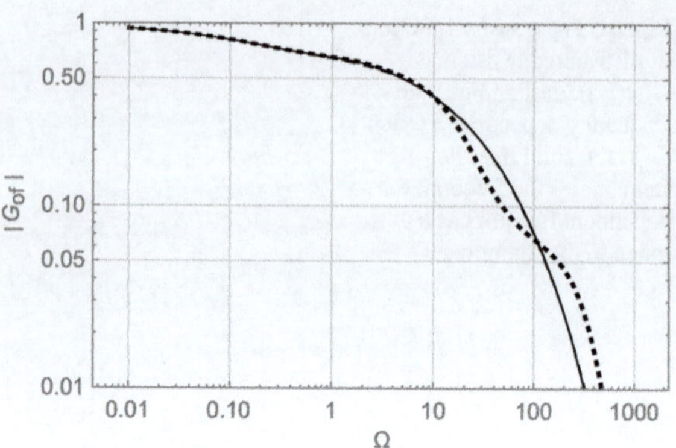

Fig. 7 The transfer function of frequency dependent friction $G_{of}(i\Omega)$ using the exact expression, solid line, and with the two approximate expressions. The Trikha approximation is the dotted line one with visible deviation

where for this case $k_N = 0.37$ and $\xi = 0.54$. This gives a relative error in the $G_{of}(i\Omega)$ function less than 3% for the whole frequency range.

A comparison of the resulting dimensionless transfer functions is shown in Fig. 7. In the figure the approximation based on Eq. (24) cannot be distinguished from the exact solution.

The frequency response of a transmission line with a blocked outlet can now be evaluated using the expression for N. Also in these examples the quotient R/Z_c is set to 0.3.

It can be seen in Fig. 8 that the resonance peaks get progressively more damped at higher frequencies.

The transfer function cannot be inverse transformed exactly into the time domain. It is, however, possible to calculate the time response exactly through convolution of a given input signal with the impulse response that is obtained through inverse Fourier transformation into the time domain. The result is show for a unit input step and an open end, corresponding to the same case as in Fig. 9. In Johnston [5] an approximate model for simulation of frequency dependent friction in the time domain is presented. It is a further development from the model presented in Krus et al. [9].

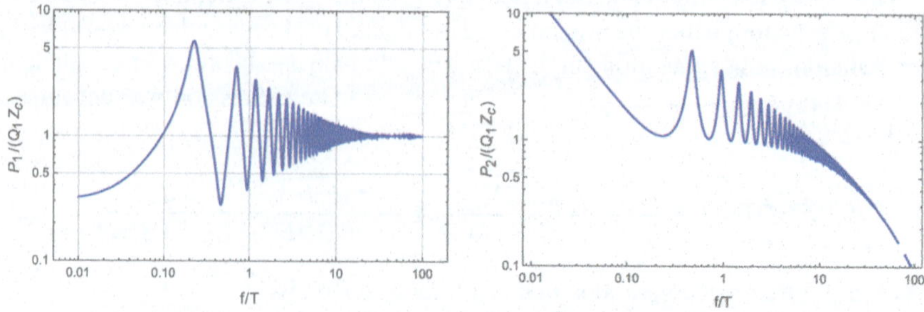

Fig. 8 The frequency spectra of the pressure at the inlet with an open inlet (left) and at the outlet for a transmission line with a blocked outlet (right) in log-log scales

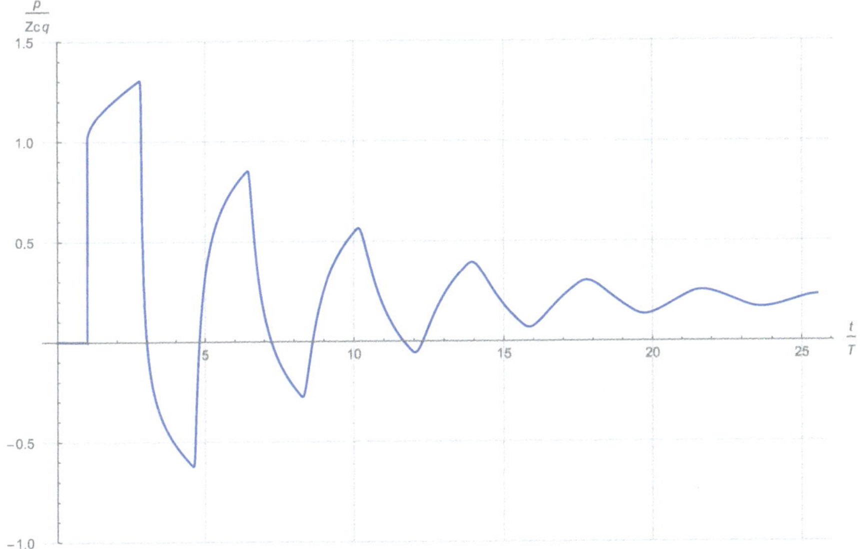

Fig. 9 Step response of a hydraulic transmission line. Inverse transformation from the frequency domain and convolution with a unit input step, at time $t = 1 \times T$

3 Flexible Hoses

Hoses differs from steel pipes in that they have more compliance and also more damping especially at high frequencies. Furthermore, there is also a wave travelling through the wall at a different speed see [4]. Here a model that only look at the dominant effects of compliance and damping which was investigated in Johansson and Nyström [3]. Consider the case where there is some impedance $G'_w(s)$ at each line increment. Figure 10 shows an increment of a line with wall dynamics. The prime "′" indicates that the entity is per line increment. This model is not completely general since no axial propagation in the wall can occur. It is, however, capable of dealing with the capacitance and energy loss in the wall in a very consistent way, and these are the most important effects to deal with in hydraulic systems with hoses.

$G'_w(s)$ is defined as

$$G'_w(s) = P'/Q'_w \tag{25}$$

where Q'_w is the flow communicated through the wall. To establish a model the original Telegrapher's equation developed by Heaviside, [2], where an equivalent model for electrical lines is used.

$$\begin{pmatrix} A_L & B_L \\ C_L & D_L \end{pmatrix} \times \begin{pmatrix} Q_1 \\ P_1 \end{pmatrix} = \begin{pmatrix} -Q_2 \\ P_2 \end{pmatrix} \tag{26}$$

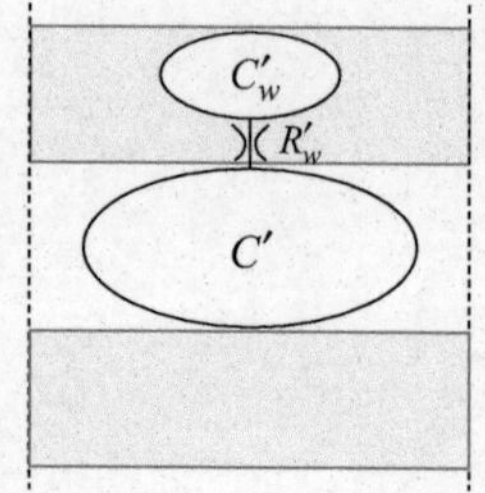

Fig. 10 Increment of a transmission line with wall dynamics

$$A_L = \cosh \gamma l \tag{27}$$
$$B_L = -\frac{1}{Z_{c0}} \sinh \gamma l \tag{28}$$
$$C_L = -Z_{c0} \sinh \gamma l \tag{29}$$
$$D_L = \cosh \gamma l \tag{30}$$

Here

$$\gamma = \sqrt{(sL' + R')(sC' + G'_w(s))} \tag{31}$$

and

$$Z_{c0} = \sqrt{\frac{sL' + R'}{sC' + G'_w(s)}} \tag{32}$$

where $L' = L/l$, $R' = R/l$, $C' = C/l$, $G'_w(s) = G_w(s)/l$. This yelds

$$\gamma = \frac{1}{l}\sqrt{(sL + R)(sC + G_w(s))} \tag{33}$$

and

$$Z_{c0} = \sqrt{\frac{sL + R}{sC + G_w(s)}} \tag{34}$$

An alternative representation is:

$$A_L = = \cosh Ts\sqrt{N_1} \tag{35}$$
$$B_L = -\frac{1}{Z_c\sqrt{N_2}} \sinh Ts\sqrt{N_1} \tag{36}$$
$$C_L = -Z_c\sqrt{N_2} \sinh Ts\sqrt{N_1} \tag{37}$$
$$D_L = = \cosh Ts\sqrt{N_1} \tag{38}$$

This representation has the advantage that the lossless transmission line is obtained by setting $N_1 = N_2 = 1$. Compared to Eq. (4) N has been replaced by N_1

and N_2 in order to handle also the wall dynamics. Equation (33) can be written as:

$$\gamma = \frac{s}{l}\frac{\sqrt{\left(1+\frac{R}{Ls}\right)\left(1+\frac{G_w(s)}{Cs}\right)}}{\sqrt{LC}} \tag{39}$$

Equation (34) can be written as:

$$Z_{c0} = \sqrt{\frac{L}{C}}\sqrt{\frac{1+\frac{R}{Ls}}{1+\frac{G_w(s)}{Cs}}} \tag{40}$$

Identification with Eqs. (35) to (38) yields

$$T = 1/\sqrt{LC} \tag{41}$$

$$Z_c = \sqrt{L/C} \tag{42}$$

$$N_1 = \left(1+\frac{R}{Ls}\right)\left(1+\frac{G_w(s)}{Cs}\right) \tag{43}$$

$$N_2 = \frac{1+\frac{R}{Ls}}{1+\frac{G_w(s)}{Cs}} \tag{44}$$

Introducing

$$N_R = 1+\frac{R}{Ls} \tag{45}$$

$$N_w = \left(1+\frac{G_w(s)}{Cs}\right) \tag{46}$$

This yelds

$$N_1 = N_R N_w \tag{47}$$

and

$$N_2 = N_R/N_w \tag{48}$$

The elements of the four pole equation then becomes:

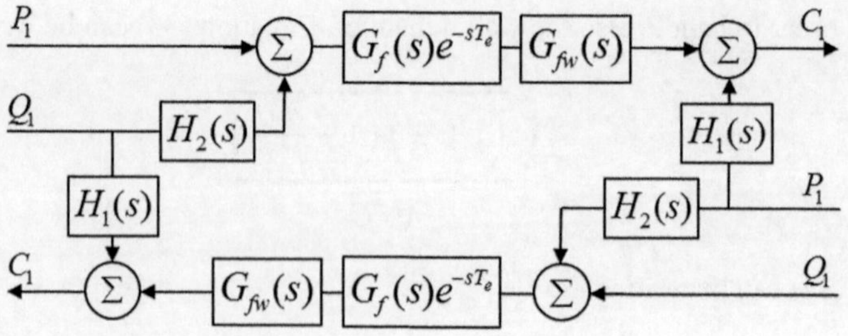

Fig. 11 Block diagram of hose with visco-elastic walls

$$A_L = \cosh Ts\sqrt{N_R N_w} \tag{49}$$

$$B_L = -\frac{1}{Z_c\sqrt{\frac{N_R}{N_w}}} \sinh Ts\sqrt{N_R N_w} \tag{50}$$

$$C_L = -Z_c\sqrt{\frac{N_R}{N_w}} \sinh Ts\sqrt{N_R N_w} \tag{51}$$

$$D_L = \cosh Ts\sqrt{N_R N_w} \tag{52}$$

A first order approximation of the wall behaviour would be to model the capacitance and the energy loss in the wall. This can be done by introducing a wall resistance R_w and a wall capacitance C_w. This is shown in Fig. 10.

For this case

$$G_w(s) = \frac{C_w s}{\tau_w s + 1} \tag{53}$$

where

$$\tau_w = R_w C_w \tag{54}$$

and

$$\kappa = C/(C + C_w) \tag{55}$$

The block diagram of the line can also be represented by Fig. 11. Here T_e is a time delay that is not necessarily the same as T since an arbitrary amount of the delay can be placed in $G_f(s)$.

A comparison of the resulting dimensionless transfer functions is shown in Fig. 7. In the examples the following data is used: $\tau_w = 0.03T$, $\kappa = 0.5$, and $\alpha = 0.2/T$, This means that the quotient $R/Z_c = 0.2$. In Johansson and Nyström [3], the time constant for a typical high-pressure hydraulic hose was reported to be in around 0.05–0.1 ms (Fig. 12).

It can be seen that here the resonance peaks get progressively more damped with the frequency (Fig. 13).

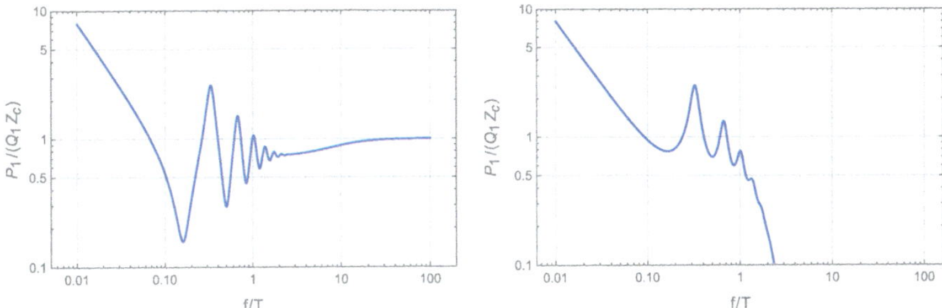

Fig. 12 The frequency spectra of the pressure at the inlet (left) and outlet (right) of a transmission line with a blocked outlet in log-log scales

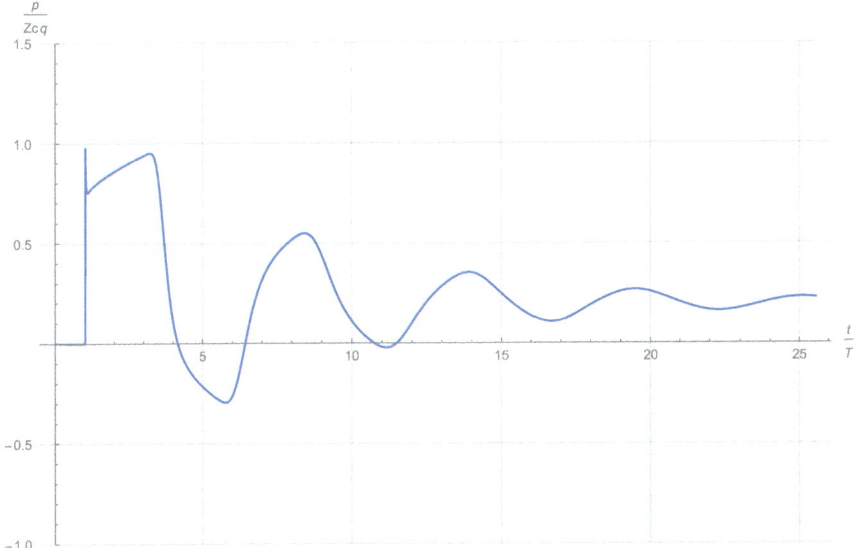

Fig. 13 Step response of hydraulic hose. Inverse transformation from the frequency domain. Unit flow step at time $t = 1 \times T$

4 Approximate Models for the Time Domain

In a transmission line with ideally elastic lossless walls, the capacitance of the walls can simply be added to the fluid capacitance. This gives a longer, total delay time which is referred to her as T_t. This is defined as:

$$T_t = 1/\sqrt{L(C + C_w)} = T/\sqrt{\kappa} \tag{56}$$

The block diagram can be defined using this delay time instead of T. N_w must then be redefined as:

$$N_w = \kappa\left(1 + \frac{G_w(s)}{Cs}\right) \tag{57}$$

Equations (53) in (57) yields

$$N_w = \frac{\kappa\tau_w s + 1}{\kappa(\tau_w s + 1)} \tag{58}$$

At the same time the total effective characteristic impedance Z_{ct} is introduced as:

$$Z_{ct} = \sqrt{L/(C + C_w)} = Z_c\sqrt{\kappa} \tag{59}$$

The part of the $G_f(s)$ filter that is dependent of the wall can be written as:

$$G_{fw}(s)e^{-T_e} = e^{-sT\sqrt{N_R N_w}} = e^{-s\left(T\sqrt{\frac{\kappa\tau_w s+1}{\kappa(\tau_w s+1)}}\right)} = e^{-s\left(T\sqrt{\frac{\kappa\tau_w s+1}{\kappa(\tau_w s+1)}}-T_e\right)}e^{-T_e} \tag{60}$$

Here T_e is an effective time delay in the transmission line. In a transmission line with elastic walls two delay times can be defined. The delay time T correspond to the wave propagation time in a line with stiff walls. Equation (60) can now be written as:

$$G_{fw}(s) = e^{-s\left(T_t\sqrt{\frac{s\kappa\tau_w+1}{s\tau_w+1}}-T_e\right)} \tag{61}$$

Inverting and Taylor expansion around $s = 0$ yields

$$G_w(s)^{-1} = \frac{1}{2}[T_e^2 - 2T_tT_e - (1-\kappa)T_t\tau_w]s^2 + [T_t - T_e]s + 1 + O(s^3) \tag{62}$$

Chose T_e so that

$$T_e^2 - 2T_tT_e - (1-\kappa)T_t\tau_w = 0 \tag{63}$$

The only meaningful root for this is:

$$T_e = T_t - \sqrt{(1-\kappa)\tau_w T_t} \tag{64}$$

This yields

$$G_{fw}(s)^{-1} = \left[\sqrt{(1-\kappa)\tau_w T_t}\right]s + 1 + O(s^3) \tag{65}$$

An approximate expression for $G_{fw}(s)$ is therefore

$$G_{fwa}(s) = \frac{1}{\tau_{we}s + 1} \tag{66}$$

where $\tau_{we} = \sqrt{(1-\kappa)\tau_w T_t}$. This approximation can, however, only be used if. $T_t > (1-\kappa)\tau_w$. In other cases a lumped parameter model is better. The filter $H_1(s)$ can be written as

$$H_1(s) = Z_{ct}\left(\sqrt{\frac{(1+\frac{R}{Ls})(R_w C_w s+1)}{\frac{R_w C_w}{C+C_w}+1}} - 1\right) \tag{67}$$

Equations (54) and (55) in (67) yields the H_2 filter as:

$$H_2(s) = Z_{ct}\left(\sqrt{\frac{(1+\alpha\tau_w+\frac{\alpha}{s}+\tau_w s)}{\tau_w \kappa s+1}} - 1\right) \tag{68}$$

Approximate expressions for both $H_1(s)$ and $H_2(s)$ these filters can be found. Even though the effect in reality is very small from these, and may be of even of lower influence than other effects that have been omitted, they are shown here for completeness.

$$H_1(s) = Z_{ct}\frac{\frac{\tau_w}{\alpha}(1-\kappa)s2+\tau_w s+1}{\left(\frac{\tau_w\kappa}{2}s+1\right)\left(\frac{2}{\alpha}s+1\right)} \tag{69}$$

$$H_2(s) = Z_{ct}\frac{\omega_{21}}{\omega_{22}}\frac{\left(\frac{s}{\omega_{21}}+1\right)\left(\tau_w k_1 s+1\right)}{\left(\frac{s}{\omega_{21}}+1\right)\left(\tau_w k_1\sqrt{\kappa s}+1\right)} \tag{70}$$

4.1 Example

Using the approximate expressions for $H_1(s)$, $H_2(s)$ and $G_f(s)$ from Ref. [7] and $G_{fwa}(s)$ from Eq. (66) the approximate time domain model can be built. The result is shown in Fig. 14. This should be compared to Fig. 13. The spike in the beginning of the step is missing, although this can be included with Eqs. (69) and (70). However, to excite this a very sharp input signal is needed, and it has to the authors knowledge never been observed. Furthermore, there could be other effects that are not considered here, that can have a greater effect. Therefore, it can be omitted for all practical purposes, since the model capture the effect of increased damping at high frequencies that is known.

The main modifications compared to the original transfer functions is that high frequency range is different. There is, however, a great deal of uncertainty concerning the real behaviour of vessels in the high frequency range. Furthermore, excitation is usually limited in frequency range and for system simulation it is usually adequate with to have a good representation of the lower frequency range. The main contribu-

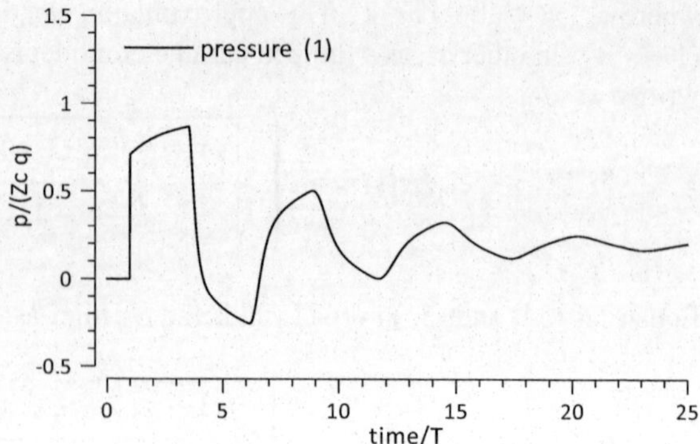

Fig. 14 Pressure response of a transmission line with visco-elastic walls

tion of the visco-elastic walls for such systems is the damping effect (in addition to the added capacitance), and that can be handled very well with just the $G_{fw}(s)$ filter.

5 Discussion

The model for hose dynamics is based on analytical models of assumed mechanisms and effects in a hose. In particular the effects of damping from visco-elastic walls in hoses, can also be modelled. This can have a great effect for damping, especially at high frequencies. Another effect, is the increased compliance which also can be easily implemented as a compliance in series with the oil compliance.

The general approach used to derive models, that is, to make models first in frequency domain and then use them to produce approximate models that can be inverse transformed analytically into the time domain is likely to be valid in any case. The same approach was earlier used to derive the model for a transmission line with distributed resistance. Since the damping of the walls in a hose is so dominant, it is usually sufficient to combine it with the simple model of distributed friction.

6 Conclusions

A simplified frequency dependent friction model was introduced in this paper. This is an analytical function that can be used instead of the expression with Bessel functions that represents the exact solution in the frequency domain. This can substantially speed up simulation in the frequency domain, which can be important for system optimization. Practical models for simulation of flexible hoses has also been derived, both for simulation in the frequency domain and also as an approximate model for simulation in the time domain. This model display higher damping of high frequencies that a pure pipe does, which is consistent with experience of real hoses.

Responsibility Notice

The author(s) is (are) the only responsible for the material included in this paper.

References

1. Ham, A.A.: On the dynamics of hydraulic lines supplying servosystems. Ph.D. thesis, TU Delft (1982)
2. Heaviside, O.: Electrical Papers. Macmillan and Co., (1892)
3. Johansson, A., Nyström, T.: Hydrauliska installationer och yttre läckage, LiTH-IKP-EX-2135. Master thesis, Linköping University (2004)
4. Johnston, D.N.: A time-domain model of axial wave propagation in liquid-filled flexible hoses. Proc. Inst. Mecah. Eng. Part I: J. Syst. Control Eng. **220**(7), 517–530 (2006)
5. Johnston, D.N.: Efficient methods for numerical modelling of laminar friction in fluid lines. J. Dyn. Syst. Meas. Control **128**, 829–834 (2006)
6. Karam, J.T., Leonard, R.G.: A simple yet theoretically based model for simulating fluid transmission line systems. J. Basic Eng. Trans. (1973)
7. Krus, P., Jansson, A., Palmberg, J.-O., Weddfelt, K.: Distributed simulation of hydromechanical systems. In: 3rd Bath International Fluid Power Workshop, Bath, UK (1990)
8. Krus, P., Palmberg, J.-O.: Simulation of fluid power systems in the time and frequency domains, applications on a load sensing system. In: International Conference on Fluid Power, Tampere, Finland (1987)
9. Krus, P., Weddfelt, K., Palmberg, J.-O.: Fast pipeline models for simulation of hudraulic systems. In: 1991 Winter Annual Meeting, ASME, Atlanta, USA (1991)
10. Larsson, P.: Fluid attenuators analysis, measurements and performance. In: Proceedings of the JFPS International Symposium on Fluid Power, pp. 55–62 (1989)
11. Steck, J.S., Davis, D.C.: Fluid transmission lines? Distributed parameter models Part 2: Comparison of models. Proc. Mech. Eng. Part A **200**(4), 229–236 (1986)
12. Stecki, J.S., Davis, D.: Fluid transmission lines-distributed parameter models Part 1: A review of the state of the art. Proc. Mech. Eng. Part A **200**(4), 215–228 (1986)
13. Trikha, A.K.: An efficient method for simulating frequency dependent friction in transient liquid flow. J. Fluids Eng. (1975)
14. Viersma, T.J.: Analysis, Synthesis and Design of Hydraulic Servosystems and Pipelines. Elsevier Scientific Publihsing Company, Amsterdam, The Netherlands (1980)

Part VIII
Uncertainty Quantification and Stochastic Mechanics

Assessment of Uncertainties and Parameter Estimation in a Offshore Gas Pipeline

Elói Rotava, Flavio Celso Trigo and Jorge Luis Baliño

Abstract Natural gas has a great importance in actual economy, and its transport is done usually through pipeline networks. The operation of a gas pipeline uses numerical models for calculation of intermediate properties, prediction of future behavior and estimation of the integrated flow capacity. These models are based on physical assumptions, closure laws and field measurements of boundary conditions such as pressure, flow, temperature and composition of the natural gas. This paper presents a development proposed for state and parameter estimation based on the implementation of an extended Kalman filter, in order to determine appropriate values for the flow parameters and use of complementary measurements in the boundary conditions. These results are compared to the ones obtained by using the Equal Error Fraction Method. It was found reduced pressure and flow systematic errors when the Kalman filter was used to estimate parameters.

Keywords Uncertainties · Estimation · Kalman filter · Gas pipeline · Offshore technology

1 Introduction

To help gas pipeline operation, numerical tools are used to calculate the local flow conditions [1], with a wide variety of commercial software available for this task. These codes can also periodically calculate the flow variables based on the field measurements available for the export and import of the involved actors in a configuration called Pipeline Management System, or PMS.

PMS are based on measurements available for use in predetermined boundary conditions of a flow model, and in adjustable parameters such as pipe roughness

E. Rotava (✉)
Petróleo Brasileiro S. A., UO-BS/ATP-PSMG/EEIS, Santos, SP, Brazil
e-mail: eloi.rotava@usp.br

E. Rotava · F. C. Trigo · J. L. Baliño
Departamento de Engenharia Mecânica, Escola Politécnica, Universidade de São Paulo, São Paulo, SP, Brazil

A. de T. Fleury et al. (eds.), *Proceedings of DINAME 2017*, Lecture Notes in Mechanical Engineering, https://doi.org/10.1007/978-3-319-91217-2_39

and thermal exchange coefficient, in order to minimize errors in dependent variables. Some of the adjustment parameters appears in algebraic relations [2]. These approaches cause difficulties when adjusting simultaneous parameters or when mixing pressure and flow rate measured at the same point in a real time problem.

As redundant measurements are available in the field, it is interesting to develop techniques that allow the use of these measurements not directly used on flow model. This can be done by modifying one or multiple variables for matching a steady state condition based in multiple data acquired in a real pipeline, with manual or automatic operator [2].

Estimation theory allows the use of these redundant measurements to estimate the flow condition in the pipeline. Such a task can be performed by a Kalman filter, a recursive estimator that considers the existence of uncertainties of both model and measurements.

This work describes the implementation of a Kalman filter to provide estimates of flow parameters for a simple pipeline comprising a single fluid inlet and outlet. The results are compared with those provided by a traditional technique, called the Equal Error Fraction Method (EEF method), applied in the pipeline operation, with a specific flow model. This is expected to work also for data reconciliation in a network pipeline that does not achieve steady condition.

2 Flow Model

The first requirement for the implementation of a Kalman filter is building a process model. In this work, the pipeline flow is considered as single-phase, one-dimensional, transient, thermal, compositional in a offshore pipeline with a single inlet and outlet.

2.1 *Conservation Laws*

For the considered flow the model is based on three conservation equations, described below.

Continuity equation. The mass conservation equation can be written as:

$$\frac{\partial \rho}{\partial t} + \frac{\partial}{\partial s}(\rho\, v) = 0 \tag{1}$$

where ρ and v are respectively the gas density and gas speed, t is the time and s is the axial position along the pipeline.

Momentum equation. The linear momentum conservation equation can be written as:

$$\frac{\partial}{\partial t}(\rho\, v) + \frac{\partial}{\partial s}\left(\rho\, v^2\right) = -\frac{\partial P}{\partial s} + \rho\, g_s - \frac{1}{2} f\, \rho \frac{v\, |v|}{D} \tag{2}$$

where D is the pipeline diameter, f is the Darcy friction factor, g_s is the gravity component in the flow direction and P is the gas pressure.

Energy equation. Although the intrinsic thermal effects are negligible, control equipment in the pipeline can cause significant temperature variation, thus justifying the inclusion of energy conservation in the model:

$$\frac{\partial}{\partial t}\left[\rho\left(\hat{h}+\frac{1}{2}v^2+g\,y\right)\right]+\frac{\partial}{\partial s}\left[\rho\,v\left(\hat{h}+\frac{1}{2}v^2+g\,y\right)\right]=\frac{4\,q''}{D} \tag{3}$$

where g is the gravity acceleration, $\hat{h}$ is the specific enthalpy, q'' is the heat flux (positive when added to the fluid) and y is the vertical position. For the purposes of numerical solution this non-linear equation, in terms of temperature and pressure, is linearized around the operational point.

Species conservation equations. Since monitoring of composition is necessary, species conservation equations are also considered:

$$\frac{\partial}{\partial t}\left(\rho X_i\right)+\frac{\partial}{\partial s}\left(\rho\,v\,X_i\right)=0 \tag{4}$$

where X_i is the molar fraction of species i. For N species there are $N-1$ equations, being Eq. (1) the sum over all the species. Due to its low influence on the flow, the compositional field can be solved separately from the previous equations, without major convergence problems.

2.2 Constitutive Equations

For conservation equations solution is necessary to calculate intermediate properties of the fluid, like the compressibility and friction factor.

State equation. As an equation for the calculation of fluid properties, the Peng and Robinson [3] correlation, which is widely used in applications involving natural gas, is adopted.

Friction factor. The pressure drop is quite simple considering only a single phase; the friction factor equation proposed by Swanee and Jain [4] was used for its explicit form.

2.3 Numerical Scheme

The equations are modified for more convenient variables and discretized with the Finite Volume Method (FVM) [5] with proprieties depending on the past flow field evaluated at time t and the actual time step $t + \Delta t$. First the flow field is calculated and then the species field are calculated. The properties of the flow are updated on the actual time step with the new calculated conditions and the iterative process follows to convergence on a full implicit numerical scheme with relaxation when necessary.

The discretized version of the continuous system is represented by:

$$[A] \cdot \{B\} = \{C\} \tag{5}$$

where $[A]$ is the matrix corresponding to the discretized system. The vector $\{B\}$ contains the flow variables evaluated at time $t + \Delta t$, while the vector $\{C\}$ contains elements evaluated at time t.

The assembled matrix $[A]$ is sparse and almost tri-diagonal, which reduces the effort required for inversion. Each row of the matrix corresponds to a conservation equation or boundary condition equation. First, the conservation equations are assembled for each discretization point, and then the boundary equations are assembled for the nodes. This is the motivation for the division of the elements of the flow network into pipelines where the flow occurs, and in nodes where the boundary conditions happen.

3 State Estimation

The approach proposed in this work applies a Kalman filter to a nonlinear flow problem, for the fusion of variables, detection of systematic errors in the measurements and estimation of flow parameters. In order to validate the proposed approach, a comparison with the results provided by the EEF method will be performed. This method adjusts flow rate and pressure at the same point based on errors assumed for each of the measurements.

3.1 Algebraic Method

The use of state estimation for natural gas flows goes back to Van der Hoeven [6], where the proposed method predicted that the differences between measurement and calculation would be managed by an error to be defined. This is the EEF method, that uses associated standard deviations on pressure values σ_i^P and flow rate values σ_i^Q:

$$\frac{P_i^{modeled} - P_i^{measured}}{\sigma_i^P} = \frac{Q_i^{modeled} - Q_i^{measured}}{\sigma_i^Q} \tag{6}$$

It can be noted that the adjustment of one flow variable, such as pressure, depends on the error in another, the flow rate, based in predefined deviations σ_i. This leads to the necessity of systematic adjustments, which can significantly change the calculated inventory for the pipeline.

For estimation of other parameters, such as thermal exchange or pipe roughness, additional variables are compared and small modifications are proposed for better stability [7]. This can be done by using a PID controller with the measured pressure on inlet as set point, the calculated pressure as input value and the output as the modified roughness of pipeline. This approach works also on a real time PMS.

3.2 *Extended Kalman Filter*

The Kalman filter is an optimal stochastic recursive estimator that minimizes the covariance of the estimation error in a least-squares sense. Uncertainties in the deterministic models of the plant and of the measurement are taken into account by the addition of white zero-mean Gaussian noise. In the case of non-linear applications, the so called *Extended Kalman Filter* provides the necessary framework to tackle the problem. In this version, the non-linear model is linearized around the newest estimate of the state. However, due to the linearization, the extended version is a sub-optimal estimator. This drawback is mitigated by the tracking ability of the estimator, that can detect sudden changes in the dynamics of the system.

A thorough discussion on Kalman filtering theory is out of the scope of this work (see, for instance, [8] and references thereon). This way, here we merely state the basic assumptions and present the resulting equations of the discrete-time extended Kalman filter, borrowed directly from [8]. Consider a typical non-linear continuous-time system model given by

$$\dot{x}(t) = f(x(t), t) + \omega(t) \tag{7}$$

in which $f(x(t), t)$ is the non-linear system model and $\omega(t) \sim N(0, \Gamma(t))$ is a white zero-mean Gaussian noise whose covariance matrix is $\Gamma(t)$. The typical discrete-time measurement (or observation) model is described by

$$z_k = h_k(x_k) + v_k \tag{8}$$

where z_k denotes the measurement vector and $v_k \sim N(0, R_k), k = 1, 2, \ldots$ represents a white zero-mean Gaussian noise whose covariance matrix is R_k. Assuming a set of initial conditions $\hat{x}_{k=0}, P_{k=0}$, with P_k representing the estimation error covariance matrix and non-correlated system and measurement errors (a commonly used hypothesis), the model is, then, linearized according to

$$F(\hat{x}(t), t) = \frac{\partial f(x(t), t)}{\partial x(t)}|_{x(t)=\hat{x}(t)} \tag{9}$$

$$H_k(\hat{x}(-)) = \frac{\partial h_k(x_k)}{\partial x_k}|_{x_k=\hat{x}_k(-)} \tag{10}$$

Thus, the discrete model is fully developed. The filter than starts the estimation process by the propagation stage,

$$\hat{x}_k(-) = \boldsymbol{\Phi}_{k-1}\hat{x}_{k-1}(+) \tag{11}$$

$$P_k(-) = \boldsymbol{\Phi}_{k-1}P_{k-1}(+)\boldsymbol{\Phi}_{k-1}^T + \Gamma_{k-1} \tag{12}$$

where $\boldsymbol{\Phi}_{k-1}$ is the state transition matrix. The (−) and (+) signs represent respectively immediately *before* and *after* new measured data is available.

Upon the arrival of a new set of measurements, the state x, the error covariance matrix P, and the Kalman gain K are updated as stated by

$$\hat{x}_k(+) = \hat{x}_k(-) + K_k\left[z_k - h_k(\hat{x}_k(-))\right] \tag{13}$$

$$P_k(+) = \left[I - K_kH_k(\hat{x}_k(-))\right]P_k(-) \tag{14}$$

$$K_k = P_k(-)H_k^T(\hat{x}_k(-))\left[H_k(\hat{x}_k(-))P_k(-)H_k^T(\hat{x}_k(-)) + R_k\right]^{-1} \tag{15}$$

4 Application Case

The pipeline under study is a real one with 175 km long and 16 in. diameter, in operation for more than 25 years with two phase fluid flow and frequent pigging operations. This pipeline has an production unit that injects fluid gas at pipeline inlet and a gas treatment plant receiving the pipeline fluid at outlet. The production unit is offshore and the water depth is 180 m. The volumetric gas flow rate practiced in this pipeline is 1.2×10^6 Sm^3/d, with a condensate flow rate of 400 m^3/d, but this value varies according to the configuration of producing wells, operating difficulties, planned maintenance in the gas treatment unit or in the production plant and other factors.

The gas treatment unit is onshore, and has a simple dew point adjusting device. Both units involved operate with flow rate control where the pressure is the consequence of this operated flow rate. For ease of prediction of future behavior, the numerical model counts with boundary conditions of the flow rate type, instead of the usual flow rate and pressure approach. These conditions do not allow to reach a steady state condition but, because we are dealing with a transient calculation in the current work, this difficulty is overcome by the use of state estimation.

The geometry and flow parameters used with pipe length S, the vertical position variation Δy, the diameter D, the pipeline roughness ϵ, the thermal exchange coefficient U and the ambient temperature T_{amb} are shown in Table 1.

The field measurements available for this pipeline are the pressures, flows rates and temperatures at the pipeline inlet and outlet, while the gas composition is measured periodically for recalibration of the flow rate measurement on production unit. For the purposes of this work, the measurement apparatus at the export of the pro-

Table 1 Pipeline geometry and flow parameters

Pipe	S (m)	Δy (m)	D (in.)	ϵ (mm)	U($\frac{\text{J}}{\text{Km}^2}$)	T_{amb} (°C)
1	200	0	16	0.183	50	18
2	200	−200	16	0.183	50	18
3	175,000	200	16	0.183	50	18
4	200	0	16	0.183	50	18

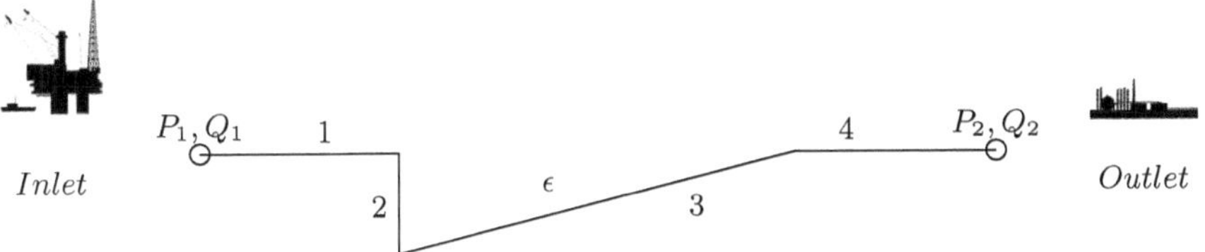

Fig. 1 Pipeline schematic

duction plant is considered the gold-standard; thus, major errors will be credited to the equipment at the import of the processing unit.

In Fig. 1 a schematic drawing of the pipeline in study is shown. It consists of a gas pipeline without large variations in diameter; P_1 and Q_1 are the pressure and flow rate measurements at the inlet, while P_2 and Q_2 are the pressure and flow rate measured at the pipeline outlet.

With the observations described above, the model for the pipeline study and its inlet and outlet conditions was generated, according to Fig. 1. Due to the low influence and simplification of the problem, temperature will not be treated in this case. To evaluate the influence of the pressure drop on the flow in the Kalman filter, state vectors are proposed, including mechanism for change in pressure drop due to friction effects and offset for outlet flow rate.

The observation vector z_k to be used in the Kalman filter is given below, and includes the measured variables that define the flow in the pipeline:

$$z_k = \begin{bmatrix} P_1 & Q_1 & P_2 & Q_2 \end{bmatrix}^T \tag{16}$$

The non-linear observation model is linearized from the discretized continuous model in order to obtain the partial derivatives of Eq. (10).

A random walk model is used to describe the evolution of the discrete-time system (plant model). Thus, matrix F of Eq. (9) is the identity matrix. The state vector is

$$x_k = \begin{bmatrix} \Delta Q_2 & \epsilon^* \end{bmatrix}^T \tag{17}$$

where ΔQ_2 is the flow difference between measured values and those calculated by the evolution model at the pipeline outlet, and ϵ^* is the modified roughness. The first variable is justified since, in order to eliminate data reconciliation problems, an adjustment in output flow can accomplish the task. However, this fit still needs more discussion. The adjustment of this pressure drop through the pipeline equivalent roughness, the second state variable, is also possible.

In relation to the error model, matrix R_k is assumed diagonal and constant. Its elements are the squares of standard-deviations of the added noise, respectively 5% and 1% of the maximum values of the flow rate and pressure. Matrix Γ_k, whose components are the squared standard-deviations of the added noise, 1% of the maximum values assumed by the state variables ΔQ_2 and ϵ^*, is also considered diagonal and constant.

5 Results and Discussion

In this section, results will be presented for the state estimation of the two cases proposed, namely, flow rate and pressure adjustment at the output port through the EEF method plus PID, and estimation of pressure drop at the outlet and modified roughness using the extended Kalman filter, as detailed in Sect. 3. Since these are results obtained from experimental values of a real pipeline, the comparison parameters are based on measured and calculated values of flow rate and pressure for the pipeline inlet and outlet.

5.1 EEF Method and PID

Results of the simulation of the study case using the EEF method are presented in Fig. 2, in which it is possible to check and compare the inlet and outlet pressures and flows. A similar trend can be observed between the estimated state, in segmented lines, and the field measurements, in continuous lines.

The spikes in the curves of measured data can be credited to the periodic pigging operations. When the pig arrives, a large amount of condensate is displaced, thus causing a sudden pressure drop at the duct outlet. This effect was not considered in the flow model due to the complexity involved, and may be the subject of future research.

Figure 3 shows the results for the estimated measurement offset and modified roughness values. Although the solution using the EEF method is stable, during the period of the test, approximately three months, the modified roughness presented a high-amplitude oscillating behavior. The technique allows adjustments, depending on the weighting parameter σ_i of Eq. (6). In the case studied, the error was allocated at the output of the pipeline.

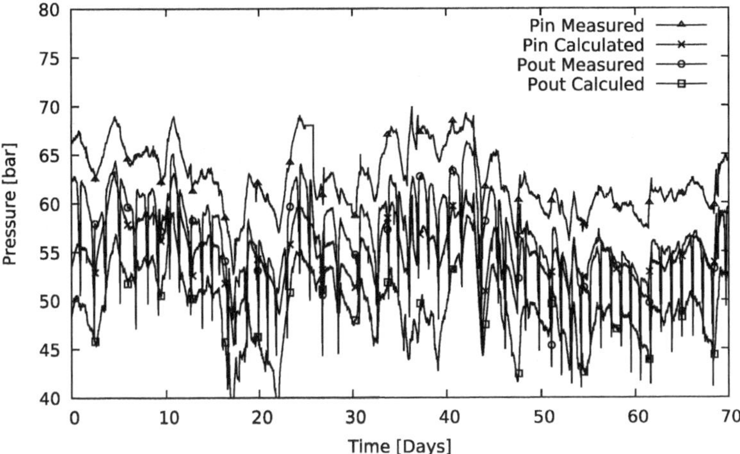

Fig. 2 Output pressure results for EEF method

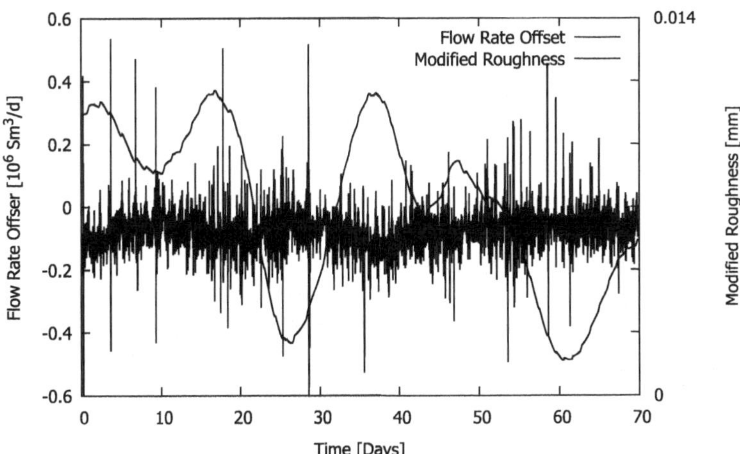

Fig. 3 Estimated parameters by EEF method and PID

According to [7], the method is efficient when the model is used to predict pressure rise or pressure drop depending on unit at outlet position. However, in order to monitor composition in the flow or for inventory calculation, there are drawbacks, since it presents systematic errors in the profile of estimated pressures, with consequences on the calculated inventory and on the transit time of molecules in the pipeline. This is an evidence of the difficulty in applying such approach.

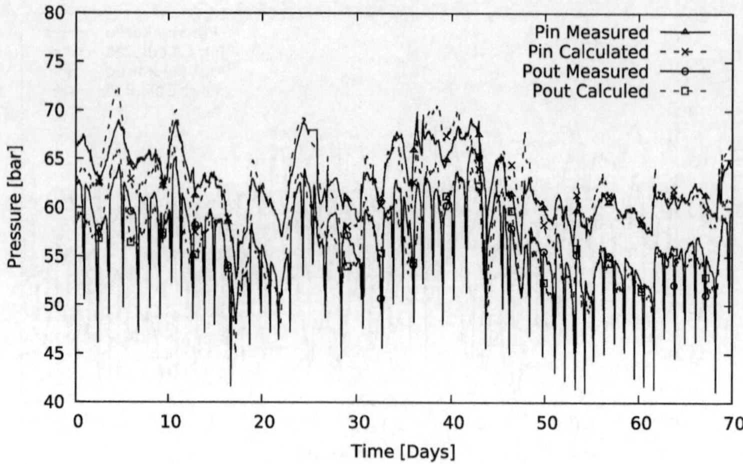

Fig. 4 Pressure and flow rate results for Kalman filter

5.2 *Kalman Filtering*

This study case was simulated for a period of three months and compared to real field data. In Fig. 4 we can check and compare the behavior of the inlet and outlet pressures and flow rates during the test period. It may be observed that the curves representing estimated state variables, in segmented lines, and the field measurements, in continuous lines, are closer than those of the previous case. Again, the pigging operation is responsible for the spikes in the curves that correspond to measured data. In comparison with the algebraic techniques, there was a noticeable change in the difference between calculated and measured pressures, with the Kalman estimates exhibiting good overlap in the inlet pressure and large differences in the output pressure only when the pigging operation takes place.

The time-evolution of the two estimated parameters, the difference in the flow rate ΔQ_2 and the modified roughness ϵ^*, are depicted in Fig. 5. The estimated value for the change in the outflow of the pipeline, -0.05×10^6 Sm^3/d, is compatible with the results of the previous case and with actual operational conditions. The additional estimated parameter, equivalent roughness, showed an initial stabilization, and a change of threshold afterwards. This behavior can be related to unaccounted factors that influence the pressure drop in the flow. Nevertheless, from our field experience, after an initial value adjustment, a relative stabilization was expected, and it effectively occurred from the 55th day.

In Kalman filtering theory, convergence is not guaranteed once estimates presents only small deviations from a certain value during some time; if the statistics of the residuals, $z_k - h_k(\hat{x}_k)(-)$ in Eq. (13), are not consistent with the hypotheses used to develop the filter, divergence effectively occurs. The histogram of the residuals is shown in Fig. 6 as a continuous line, alongside with a Gaussian fit to the simulated

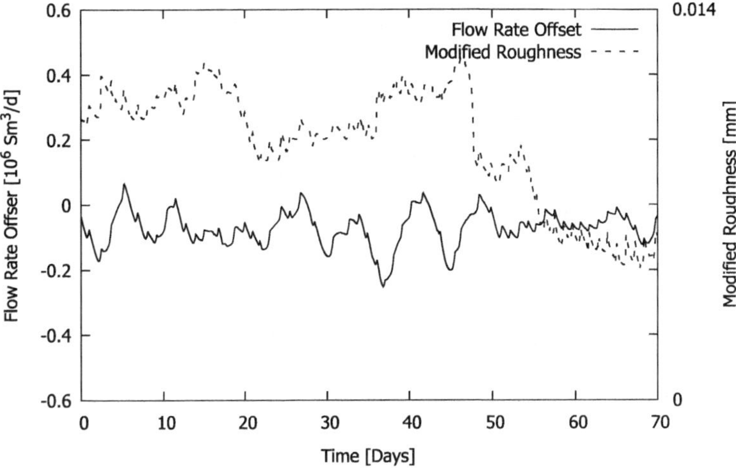

Fig. 5 Estimated parameters by Kalman filter

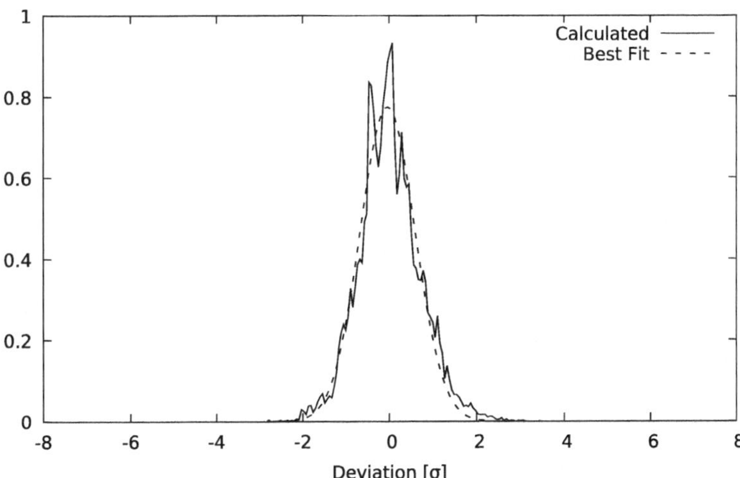

Fig. 6 Residual errors histogram

data, in segmented line. As it is possible to observe in this figure, the simulated data may be regarded as Gaussian; moreover, the mean value is zero, and the standard deviation is 0.64. Since consistency between hypotheses and results was achieved, there is evidence to suggest that the estimates converged.

It is yet important to point out that, during the simulation period, communication failures occurred between the estimator and the field readings. Even under those circunstances, the Kalman filter was able to provide correct estimates of the pressure and flow rate at the pipeline inlet and outlet. Hence, the approach may also be considered robust.

6 Conclusions

The results obtained from state estimation techniques show an improvement when compared to an algebraic approach such as the EEF method, available in the literature, without the necessity for systematic errors for adjustments. In addition, the estimated parameters ΔQ_2 and ϵ^* show a suitable trend, with low variations once convergence is achieved.

On the other hand, the computational cost to calculate the state is increased significantly with the application of extended Kalman filter, by the necessity of determining the local derivatives in the model, which are performed numerically [9], and this effect must be taken into account when using Kalman filtering. This effect can be minimized by using parallel computing for the calculation of partial derivatives.

Pigging operations cause increased differences between estimated and measured conditions that may influence the state estimation. For this reason the development of the technique will be applied to gas networks in periods without pigging until the proposed model is altered in order to detect such behavior.

Based on the results found for a single pipeline, a similar approach is being developed for more complex gas pipelines, which currently operate with two gas treatment units and eight exporters, also from Brazilian Santos Basin, with more than 400 km in length. There is a belief that the proposed approach may lead to the identification of unexpected conditions on production units involved.

Future work may include structuring mechanisms for construction of noise matrices for the plant and for the measurement models. Another possible development is the migration to a multiphase flow model, in order to include pigging operations. Finally, a discussion of the most appropriate approximations for measurement offset is still necessary.

References

1. Al-Rasheed, M., Wallooppillai, R., Wagner, G., Scheerer, H., et al.: Mgs-stat, a pipeline management tool for Saudi aramco's master gas system. In PSIG Annual Meeting, Pipeline Simulation Interest Group (2011)
2. Hanmer, G., Jackson, E., Hansen, D., Velde, B., Losnegard, S.-E., et al.: Tuning of subsea pipeline models to optimize simulation accuracy. In: PSIG Annual Meeting, Pipeline Simulation Interest Group (2012)
3. Peng, D.-Y., Robinson, D.B.: A new two-constant equation of state. Indust. Eng. Chem. Fundament. **15**(1), 59–64 (1976)
4. Swanee, P., Jain, A.K.: Explicit equations for pipeflow problems. J. Hydraul. Div. **102**(5) (1976)
5. Patankar, S.: Numerical Heat Transfer and Fluid Flow. CRC Press (1980)
6. Van der Hoeven, T., et al.: Gas network state estimation with the equal error fraction method. In: PSIG Annual Meeting, Pipeline Simulation Interest Group (1987)
7. Modisette, J., et al.: Automatic tuning of pipeline models. In: PSIG Annual Meeting, Pipeline Simulation Interest Group (2004)
8. Gelb, A.: Applied Optimal Estimation. MIT Press (1974)
9. Modisette, J.P., et al.: State estimation in online models. In: PSIG Annual Meeting, Pipeline Simulation Interest Group (2009)